Alloys and Composites Corrosion and Mechanical Properties

Alloys and Composites Corrosion and Mechanical Properties

Topic Editors

Jingxiang Xu
Zhenhua Chu
Xingwei Zheng

MDPI

Basel • Beijing • Wuhan • Barcelona • Belgrade • Novi Sad • Cluj • Manchester

Topic Editors
Jingxiang Xu
College of Engineering
Science and Technology
Shanghai Ocean University
Shanghai
China

Zhenhua Chu
Department of Mechanical
Engineering
College of Engineering
Shanghai Ocean University
Shanghai
China

Xingwei Zheng
College of Science
Donghua University
Shanghai
China

Editorial Office
MDPI AG
Grosspeteranlage 5
4052 Basel, Switzerland

This is a reprint of the Topic, published open access by the journals *Coatings* (ISSN 2079-6412), *Journal of Manufacturing and Materials Processing* (ISSN 2504-4494), *Materials* (ISSN 1996-1944), *Metals* (ISSN 2075-4701) and *Processes* (ISSN 2227-9717), freely accessible at: https://www.mdpi.com/topics/0HO6EV4F7C.

For citation purposes, cite each article independently as indicated on the article page online and as indicated below:

Lastname, A.A.; Lastname, B.B. Article Title. *Journal Name* **Year**, *Volume Number*, Page Range.

ISBN 978-3-7258-3803-5 (Hbk)
ISBN 978-3-7258-3804-2 (PDF)
https://doi.org/10.3390/books978-3-7258-3804-2

Contents

About the Editors

Jingxiang Xu

Dr. Jingxiang Xu obtained a Bachelor's degree of Engineering in Mechanical Engineering and Automation from Jilin University, China in 2008. He received the Master and Doctor Degrees from Department of Mechanical Engineering, Tohoku University, Japan, in 2012 and 2015, respectively, and worked as a JSPS Postdoctoral Researcher at Institute for Materials Research (IMR), Tohoku University from 2015 to 2016. Then, Dr. Xu was promoted to a specially appointed assistant professor at IMR, Tohoku University. He is currently a Professor at the College of Engineering Science and Technology, Shanghai Ocean University, China. He has published more than 100 papers in international academic journals. His research interests include multiscale materials modeling of complex physical and chemical processes, solid oxide cells, and surface engineering.

Zhenhua Chu

Professor Zhenhua Chu is affiliated with Shanghai Ocean University. She is a Senior Visiting Scholar at the Institute for Materials Research, Tohoku University, Japan. She obtained a Ph.D. in Materials and Metallurgical Engineering from Shanghai Jiao Tong University, a Master's degree in Materials Science and Engineering from Tianjin University, and a Bachelor's degree in Materials Science and Engineering from Hebei University of Technology. As a principal investigator, Prof. Chu specializes in materials used in marine environments, with a particular focus on corrosion and protection. She has led multiple research projects funded by the National Natural Science Foundation of China (General and Youth Programs), the Hebei Outstanding Youth Foundation, and the Tianjin Natural Science Foundation. Prof. Chu has published more than 60 SCI-indexed articles and holds 13 granted patents. She is a member of the Early Career Advisory Board for *Surface and Coatings Technology* and has served as A Youth Editorial Board Member for the *Journal of Chinese Society for Corrosion and Protection* (first and second terms) and *Corrosion Communications*.

Xingwei Zheng

Professor Xingwei Zheng, Ph.D., is a Senior Visiting Scholar at General Motors R&D in the United States, Shanghai Science and Technology Pioneer. After graduating with a PhD in Materials Processing from Shanghai Jiao Tong University in 2012, they worked at COMAC from May 2012 to August 2016, primarily developing ARJ21 and C919. From August 2016 to June 2020, they worked at Shanghai Ocean University, primarily conducting research on lightweight material forming technology, corrosion, and protection. As of July 2020, they have been employed at the School of Physics, Donghua University, primarily developing high-performance lightweight materials and researching their mechanical behavior. They have hosted or participated in 10 provincial- and ministerial-level scientific research projects, including the National Natural Science Foundation General Project. They have published over 60 SCI papers, and have received four provincial, ministerial, or industry research achievement awards.

coatings

Article

The Effect of Nb on the Microstructure and High-Temperature Properties of Co-Ti-V Superalloys

Pengjie Zhou [1], Shenfa Ni [1], Kunjie Luo [2,*], Wanxiang Zhao [2], Enze Liu [3], Yanxin Qiao [1,*] and Shujin Chen [1]

1 School of Materials Science and Engineering, Jiangsu University of Science and Technology, Zhenjiang 212000, China; zhoupengjie@just.edu.cn (P.Z.)
2 Suzhou Nuclear Power Research Institute, Suzhou 215004, China
3 Institute Metal Research, Chinese Academy of Sciences, Shenyang 110016, China
* Correspondence: luokunjie007@163.com (K.L.); yxqiao@just.edu.cn (Y.Q.)

Abstract: The effect of Nb on the microstructure evolution of the γ' phase in Co-Ti-V alloys has been studied. The yield strength and ultimate strength of the alloys are measured via compression at 900 °C. This indicates that the alloys with 1% at. Nb present a dual γ and γ' microstructure. A Co_3V phase exists in 2Nb and 3Nb alloys. The mismatch increases with the increment in Nb content. The yield strength of the alloys rises as the Nb addition increases. After compression at 900 °C, the <101>-type dislocation causes shearing of the γ' phase in the plane of the base alloy. In the 1Nb alloy, the < $1\bar{1}0$>-type dislocation reacts at the γ/γ' interface to generate 1/3<$1\bar{2}1$> super-partial dislocation, and then the γ' phase is cut by this dislocation and stacking faults are formed.

Keywords: Co-Ti-V alloys; lattice misfit; high-temperature deformation; dislocations

1. Introduction

Superalloys are alloys based on nickel, iron–nickel, and cobalt, and are widely employed at temperatures of 540 °C and above. They are widely applied in parts working at elevated temperatures, for instance, aircraft blade, petrochemical equipment, and chemical plant equipment. Recently, they have been used in the automotive industry as materials for turbines. Up to now, nickel-based superalloys have been the most popular in those fields [1].

In 2006, a γ'-Co_3 (Al,W) precipitate which is capable of being stabilized at 900 °C with an $L1_2$ structure was reported by Ishida et al. [2]. Compared with Ni-based superalloys, the melting point of Co is higher than that of Ni by about 40 °C, and conventional Co-based alloys have excellent corrosion resistance and weldability [3], promoting their superiority to the Ni-based superalloy in some applications. Because of the lack of a stable γ' precipitate-strengthening effect, the traditional Co-based superalloy shows inferior strength to that of the Ni-based γ' hardening alloys when operating at elevated temperatures [4,5]. The new discovery of Co_3 (Al,W)- γ' opened up new pathways to developing Co-based alloys with superior high-temperature strength. However, it has been demonstrated that the γ'-Co_3 (Al,W) precipitate is not stable when the temperature is over 1000 °C, so the existing temperature of the γ' is far lower than that of the Ni-based superalloy, indicating its poor high-temperature stability [6]. Increasing the W content can stabilize the strengthening phase, but the higher content of W will increase the mass density and limit the application of such alloys [7–10]. It is necessary to search for a light element to replace W and to ensure the phase stability of the hardening precipitate.

Academic Editor: Vincent Ji

Received: 26 November 2024
Revised: 27 December 2024
Accepted: 2 January 2025
Published: 6 January 2025

Citation: Zhou, P.; Ni, S.; Luo, K.; Zhao, W.; Liu, E.; Qiao, Y.; Chen, S. The Effect of Nb on the Microstructure and High-Temperature Properties of Co-Ti-V Superalloys. *Coatings* **2025**, *15*, 53. https://doi.org/10.3390/coatings15010053

Among those Co-based alloys that have been reported so far, only the γ'-Co_3Ti phase with an $L1_2$ structure has been found to be stable. However, it presents large lattice mismatch, which can affect the properties of the alloy [11]. Liu et al. point out that the alloying of V to the Co-Ti binary system would reduce the mismatch between the dual γ/γ' phases and improve the high-temperature stability of the alloys [12]. It is reported that V, Sc, Y, Cr, Zr, Mo, Ta, Nb, Hf, and W tend to occupy the Ti site in the lattice and degrade the structural stability of $L1_2$-Co_3Ti. In 2014, Liu and Ruan discovered the existence of a stable γ-γ' dual-phase section in a Co-Ti-V ternary alloy at 900 °C [13]. After that, a novel Co-Ti-V ternary alloy was developed which possesses the following advantages compared to Co-Al-W alloys: low density, high γ' solvus temperature, and superior strength at elevated temperatures [14]. The phase equilibria of ternary Co-Ti-V at 873 K was investigated by Zhou [15], and the interdiffusion coefficients of the elements were studied by Zhang, showing that Ti diffuses faster than V [16]. Verma et. al. illustrated that V decreases the solvus temperature of γ', and it modifies the mismatch between γ and γ' from positive to negative [17]. Ni and Al additions were reported to increase the volume fraction and stability of the γ' precipitate [18,19]. Ni is demonstrated to reduce the solvus temperature of γ' in Co-8Ti-11V alloys [19]. Ru seems to have no remarkable influence on the solvus temperature of γ', but modifies the portioning behaviors of Ti and V in the alloy, consequently enhancing the strength of the alloy [20]. Previous research indicates that Nb is a strong γ' former and increases the solvus temperature and volume fraction of the γ' phase; W has similar effect as Nb [21]. Nb is reported to slightly improve the resistance of alloys to hot corrosion, but is less effective than Mo and Ti [22], while Migas and Klein illustrated that Nb degrades the oxidation resistance of Co-Al-W alloys [23,24]. However, their research only covered the composition of Co-10V-4Ti-2Nb (at. %); more studies are necessary to explore the influence of Nb content on the microstructure and properties of Co-Ti-V alloys. This article aims to explore the influence of Nb on microstructure evolution and mechanical improvement and to gain the reliable addition range of Nb in the Co-Ti-V tertiary alloy.

In this work, five Co-6Ti-11V-xNb (x = 0, 0.5, 1, 2, 3) quaternary superalloys are designed, prepared by vacuum arc melting, and investigated. The effect of Nb content on the structure and mechanical properties of Co-Ti-V superalloys is investigated by using the below methods.

2. Materials and Methods

The nominal compositions of the target alloys used in this study are shown in Table 1, and for the convenience of presentation, these materials are abbreviated according to the Nb content in each alloy. High-purity cobalt (99.95%), V (99.95%), and Ti (99.9%) were adopted to prepare the target material. The alloy was firstly melted into ingots using a WK-II vacuum arc-melting furnace. In the melting process, the chamber was vacuumed to 0.1 bar first, then gassed with argon (99.999%); after repeatedly cleaning the atmosphere, the materials were arc-melted under argon atmosphere. The button-shaped sample was turned over and remelted more than 6 times to homogenize the composition of the entire ingot. Each ingot was approximately 70 g. Using this melting method, the cooling rate of the ingot was estimated to be 300 °C/s during the solidification process, and the secondary dendrite arms' spacing was 450 μm on average. The alloys were sealed in quartz capsules for homogenization treatment at 1100 °C for 48 h, then quenched in cold water. After that, the alloys were quartz-sealed again and aged at 870 °C for 72 h, and then cooled in air to room temperature. The specimen was electrolytically etched using 10 mL of HNO_3 plus 20 mL of CH_3COOH and 170 mL of distilled water with 6 V. The morphology of the γ' phase was observed by using a Zeiss–Merlin Compact Field Emission Scanning Electron

Microscope (FE-SEM) (Zeiss: Oberkochen, Germany) to calculate the size distribution and volume fraction of the γ' precipitates. The phase constitutes were observed by using a JEM-2100F Transmission Electron Microscope (TEM) (JEOL Ltd.: Tokyo, Japan), and the distribution of elements in the γ phase and γ' phase was investigated by Oxford Energy Disperse Spectroscopy (EDS) (Oxford Instruments: Abington, UK) with INCA T80 equipped with the TEM. The composition was averaged from ten points of 3 different regions. The evaluation of dislocation after deformation was performed. The film for the TEM analyses was obtained by using twin-jet electropolishing in a solution of 6% perchloric acid and methanol conducted at −30 °C with 25 V. Shimadzu XRD-6000 (Shimadzu: Kyoto, Japan) was employed to investigate the phase constitutes and analyze the lattice mismatch of the γ and γ' phases. Powders used in the XRD test were machined from bulk materials, then sealed in quartz, heated to 900 °C for 3 min, and furnace-cooled to room temperature. A scanning rate of 4°/min was used in the range of 20–90° to analyze the phases. A scanning speed of 1°/min was adopted to measure the mismatch in the interval between 47° and 55°. Origin Pro 9.1 was employed to fit the peak profiles and analyze the mismatch. The compression experiments were performed at a strain rate of $3 \times 10^{-4}s^{-1}$ and a ramping rate of 10 °C/s to 900 °C, using a Gleeble-3800 (Gleeble: Poestenkill, NY, USA) thermal simulator. When the temperature rose to 900 °C, it was maintained for 3 mins before loading. The strain rate used in the study was 3×10^{-4}/s, which was in the range of $(1 \times 10^{-4}/s \sim 1 \times 10^{-3}/s)$ that is widely adopted by previous research [25,26]. The stress corresponding to a 0.2% offset strain was taken as the yield strength. After high-temperature compression, the specimens were water-quenched.

Table 1. Compositions of investigated alloys (at. %).

Alloy	Nominal Composition (at.%)			
	Co	Ti	V	Nb
Base	Bal	6	11	-
0.5 Nb	Bal	6	11	0.5
1 Nb	Bal	6	11	1
2 Nb	Bal	6	11	2
3 Nb	Bal	6	11	3

3. Results and Discussion

3.1. Microstructure of Alloys with Varied Nb Content

Figure 1 shows the XRD patterns of the alloys with varied Nb content. The patterns of the 0Nb, 0.5Nb, and 1Nb alloys are similar, with three relatively obvious peaks which are actually composed of the matrix (γ) and the strengthening precipitate (γ'). The other peaks in the 2Nb and 3Nb alloys indicate the precipitation of Co_3V.

Figure 2 shows the microstructure of the alloys. The 0Nb, 0.5Nb, and 1Nb alloys have no other precipitates except the γ and γ' phases (Figure 2a–c). Meanwhile, the 2Nb and 3Nb alloys exhibit a needle-like precipitate which corresponds to the Co_3V phase in the XRD pattern. The selected area electron diffraction (SAED) of this phase was conducted (Figure 3). The phase was identified as Co_3V by XRD as well as EDS analysis of the TEM and electron diffraction patterns. Therefore, the Nb addition leads to the formation of an x phase (D019) if its content exceeds 1% (at. %). The volume fractions of the x-Co_3V phase in the 2Nb and 3Nb alloys are 14.04% and 37.54, respectively.

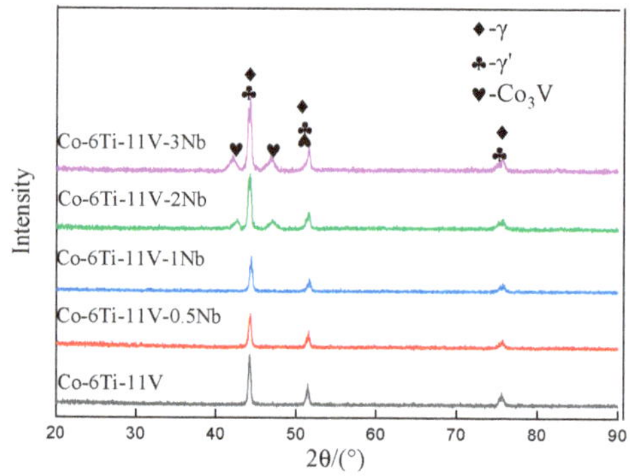

Figure 1. XRD analysis of alloys with various Nb contents.

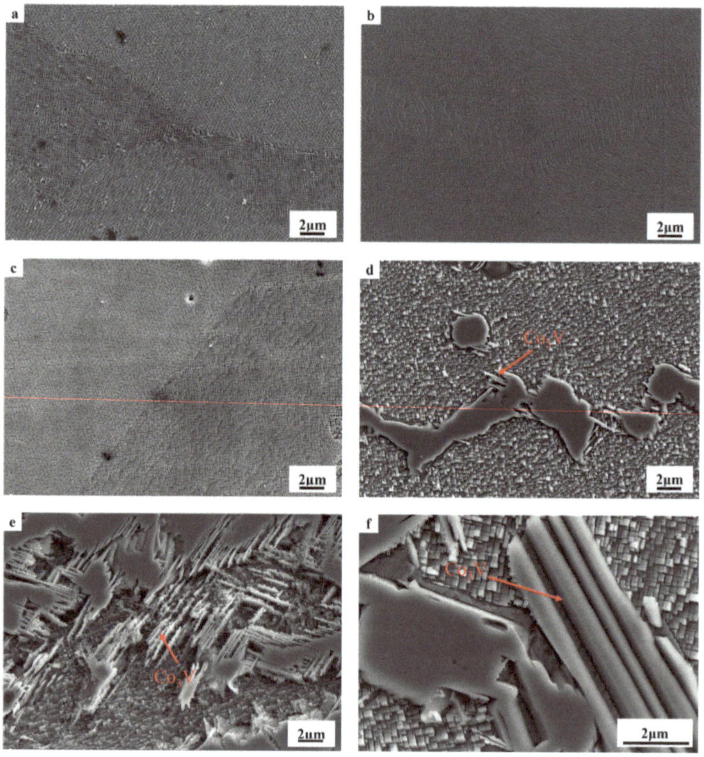

Figure 2. Microstructure of alloys with varied Nb content: (**a**) Nb-free; (**b**) 0.5Nb; (**c**) 1Nb; (**d**) 2Nb; (**e**) 3Nb; (**f**) details of Co_3V phase in 2Nb alloys.

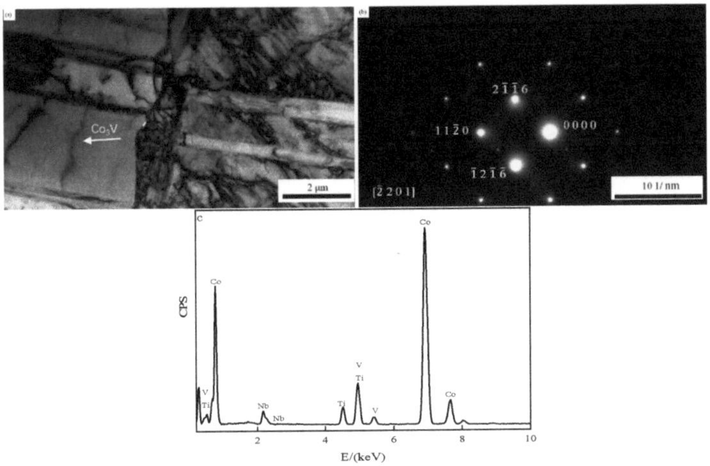

Figure 3. (**a**) TEM image of Co_3V phase; (**b**) selected area electron diffraction pattern of Co_3V phase; (**c**) EDS pattern of Co_3V phase.

3.2. *Lattice Mismatch Between γ/γ' Phases*

Since the γ/γ' phases have similar lattice constants, the observed (002) peaks are doublets due to the overlap from both peaks of the dual phases. Figure 4 shows the separation peaks of the γ phase and the γ' phase and the fitted curves obtained corresponding to the (200) crystal plane. The equations for calculating the lattice constant and mismatch are as follows:

$$\alpha = \frac{\lambda}{2\sin\theta}\sqrt{H^2+K^2+L^2} \quad (1)$$

$$\delta = \frac{2(\alpha_{\gamma'}-\alpha_{\gamma})}{\alpha_{\gamma'}+\alpha_{\gamma}} \times 100\% \quad (2)$$

where α is the lattice parameter of the corresponding phase; θ represents the diffraction angle (°); λ is the wavelength of incident X-rays (Å); H, K, and L are the specific plane indexes; α_γ is the lattice constant of the γ phase (Å); $\alpha_{\gamma'}$ stands for the lattice constant of the γ' phase (Å); δ is defined as the γ/γ' two-phase lattice mismatch (%). The lattice constants of the γ phase and the γ' phase of alloys with various Nb contents are shown in Table 2, and the calculated lattice mismatches were given as well. According to these data, the lattice mismatch gradually rises with the increase in Nb content.

Table 2. Lattice constants of γ and γ' phases and lattice mismatch between dual phases in alloys with various Nb contents.

Alloy (at. %)	γ'-Phase Lattice Constant α_{γ}/Å	γ-Phase Lattice Constant α_{γ}/Å	Lattice Mismatch δ/%
0Nb	0.3594	0.3567	0.76
0.5Nb	0.360	0.357	0.81
1Nb	0.359	0.356	0.82
2Nb	0.3597	0.357	0.85
3Nb	0.3595	0.3564	0.86

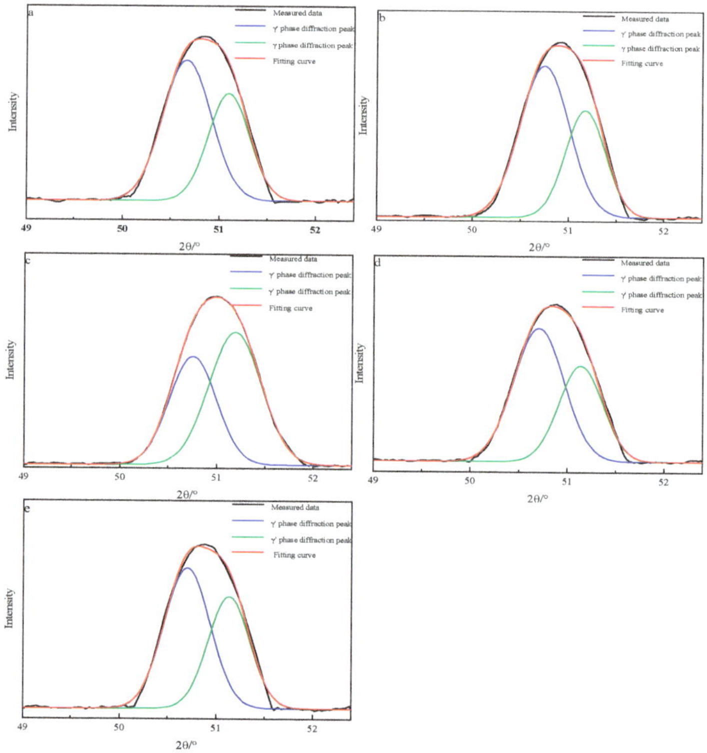

Figure 4. XRD peak-differentiating and -imitating profiles of alloys with varied Nb content: (**a**) Nb-free; (**b**) 0.5Nb; (**c**) 1Nb; (**d**) 2Nb; (**e**) 3Nb.

3.3. *γ′-Phase Morphologies of Alloys with Varied Nb Content*

Figure 5 shows the γ′ morphologies in the alloys with varied Nb content. The γ′ phase is relatively uniformly arranged, and all of them are cuboidal in shape (Figure 5a–c). As illustrated in Figure 5d, when the Nb content increases to 2 at. %, the γ-matrix channel is thin, and the size of the γ′ precipitate varies and drops significantly, but it remains cuboidal in shape. Figure 5e shows the extremely regular shape of the γ′ phase, with a wide range of sizes.

The average size and volume ratio of the γ′ phase in the five materials are shown in Table 3. The volume ratios of the γ′ phase in the 2Nb and 3Nb alloys were not investigated because of the presence of a large amount the Co_3V phase in the two alloys. Since mass precipitation of x–Co_3V will greatly degrade the ductility of the alloys, it is meaningless to analyze the volume ratio of the γ′ phase in the 2Nb and 3Nb alloys. The average size of the γ′ phase increases and then decreases with the rise in Nb content, which peaks at 0.5Nb. With the increased Nb content, the volume ratio of the γ′ phase increases, and the volume fraction of the γ′ phase in the 1Nb alloy reaches the maximum, at 84.7%.

Figure 5. γ'-phase morphology of alloys with different Nb contents: (**a**) Nb-free; (**b**) 0.5Nb; (**c**) 1Nb; (**d**) 2Nb; (**e**) 3Nb.

Table 3. Average size and volume ratios of γ' phase of each alloy.

Alloy	Average Size (nm)	Volume Fraction (%)
0Nb	222.2	84.3
0.5Nb	303.9	84.4
1Nb	252.4	84.7
2Nb	191.0	-
3Nb	189.0	-

3.4. Distribution of Alloying Elements in γ Phase and γ' Phase

Analysis of the elemental distribution behavior in the γ and γ' phases of the alloys was conducted by EDS equipped with a TEM, which was conducted in a STEM model. The expression of the elemental distribution coefficient (K_X) is as follows:

$$K_x = C_{\gamma'-x}/C_{\gamma-x} \tag{3}$$

where $C_{\gamma'-x}$ represents the atomic percent of element X in the γ' precipitate (at. %), and $C_{\gamma-x}$ is the atomic percent of element X in the γ channel (at. %). Table 4 lists the compositions of the γ phase and the γ' phase in the alloys, and the partitioning coefficients of each elements in the alloys are shown in Figure 6. The partitioning coefficients of Ti and Nb are always greater than 1, and they partition preferentially to the γ' precipitate. Furthermore, the partitioning coefficient of Ti increases with the rise in Nb additions, as Nb promotes the enrichment of Ti into the γ' phase. The partition coefficients of V in the 0Nb, 0.5Nb, and 1Nb alloys fluctuate near 1; as the Nb content continues to increase, the partitioning coefficient of V rises gradually. These results show that the addition of Nb leads to Ti and V enrichment in the γ' matrix rather than in the γ matrix, and consequently, Nb addition results in the rise in the γ' volume fraction in the alloys.

Table 4. The elemental components of the γ phase and the γ' phase in the alloys with varied Nb content.

Alloy		**Composition (at. %)**			
		Co	**Ti**	**V**	**Nb**
Co6Ti11V	γ	81	6	13	-
	γ'	79.9	7.1	13	-
Co6Ti11V0.5Nb	γ	78.4	6.9	14.1	0.6
	γ'	78.4	7.4	13.5	0.7
Co6Ti11V1Nb	γ	81.2	4.7	13.4	0.8
	γ'	78.6	6.7	13.4	1.4
Co6Ti11V2Nb	γ	85.5	3.3	10.2	1.1
	γ'	79.3	5.8	12.8	2.1
Co6Ti11V3Nb	γ	86.4	2.5	10.1	1.1
	γ'	76.6	6.1	14.1	3.4

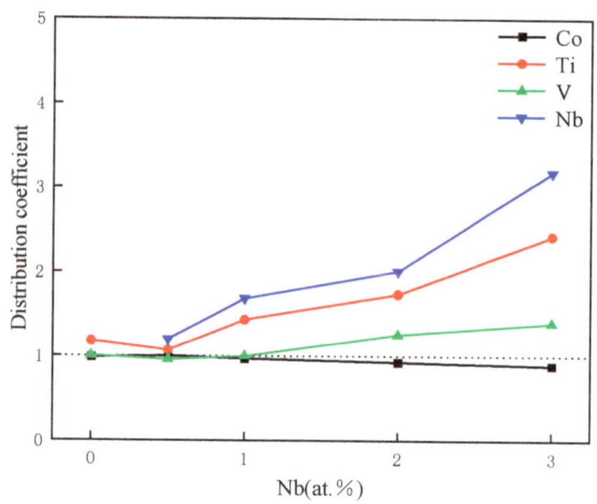

Figure 6. Partitioning coefficients of elements in γ and γ' phases in alloys with various Nb contents.

3.5. Strength at Elevated Temperature

The 0.2-percent offset yield strength of alloys subjected to compression tests at 900 °C is shown in Table 5. The yield strength rises gradually with the increasing Nb content. The 3Nb alloy possesses the highest yield strength at 900 °C, which is 496 MPa, while the 0Nb alloy has the lowest yield strength, which can be related to the lowest γ' volume fraction in the 0Nb alloy. The alloying of Nb can enhance the tensile properties of alloys with proper additions. As listed in Table 5, the yield strength of the 1Nb alloy is higher than that of the conventional Co-based superalloy (Haynes188) and the newly developed Co-9Al-W alloys. However, it is lower than that of the precipitate-hardening Ni-based superalloy (In939).

Table 5. Yield strength at 900 °C of alloys with different Nb contents.

Alloy	Yield Strength (MPa)
Co-6Ti-11V	323
Co-6Ti-11V-0.5Nb	351
Co-6Ti-11V-1Nb	391
Co-6Ti-11V-2Nb	404
Co-6Ti-11V-3Nb	496
Hayness	280 [6]
Mar-M247	392 [6]
In939	520 [18]
Mar-M302	361 [18]
Co9Al9W	263 [27]

3.6. The γ'-Phase Morphologies of Alloys After Compression at 900 °C

The morphologies of the γ' precipitate of the 0Nb, 0.5Nb, and 1Nb alloys after compression at 900 °C are shown in Figure 7, and the average size as well as the volume ratios of the γ' particles in each alloy are listed in Table 6. The γ' precipitate in the 0Nb alloy is not uniformly arranged, and the mean size of the γ' particle increases from 222.2 nm to 271.7 nm compared with before compression. The γ' precipitate in the 0.5Nb alloy is much the same as that before deformation. The γ-matrix channel in the 1Nb alloy is extremely narrow, and the mean size of the γ' particle is only slightly increased in comparison with before compression. The volume fraction of the γ' particle in the 0Nb, 0.5Nb, and 1Nb alloys is substantially smaller than that before the test. This may be due to the fact that parts of the γ' phases were dissolved with load during the hot compression process. During deformation at elevated temperatures, γ' is repeatedly cut by dislocations, especially in the interface of γ' and γ, and some mini γ' particles may be generated by cutting the large γ' particles. And the mini γ' particles dissolve during the compression test. In contrast, the 1Nb alloy achieves the minimal particle size increase and the lowest volume fraction reduction. These results show that the correct addition of Nb is beneficial for improving the phase stability of alloys. Previous research has reported that Nb improves the phase stability of γ' phases in Co-based superalloys [6,21,25].

Table 6. Average dimensions and volume fractions of particles of each alloy after compression at 900 °C.

Alloy	Average Size (nm)	Volume Fraction (%)
Co-6Ti-11V	271.7	58.5
Co-6Ti-11V-0.5Nb	302.1	48.4
Co-6Ti-11V-1Nb	264.4	65.9

Figure 7. The γ' morphologies after compression at 900 °C: (**a**) 0Nb; (**b**) 0.5Nb; (**c**) 1Nb.

3.7. *Deformation Behavior of Alloys*

Figure 8 shows the dislocation configurations of the Co6Ti11V alloy after the compression test at 900 °C. In Figure 8a, a dense dislocation distribution in the γ' phase is observed, indicating that dislocation sliding is the primary deformation mechanism in the γ' precipitate. And shearing of the γ' precipitate by pair dislocations is evident. When imaged with g = $0\bar{2}\bar{2}$, the dislocation A is observed in Figure 8a, while imaging with g = $2\bar{2}\bar{2}$ reveals that the dislocation A is invisible, as shown in Figure 8b; thus, it can be roughly judged that the Berger vector of the dislocation A is b = [101], and this type of dislocation causes shearing of the γ' phase in the {111} plane.

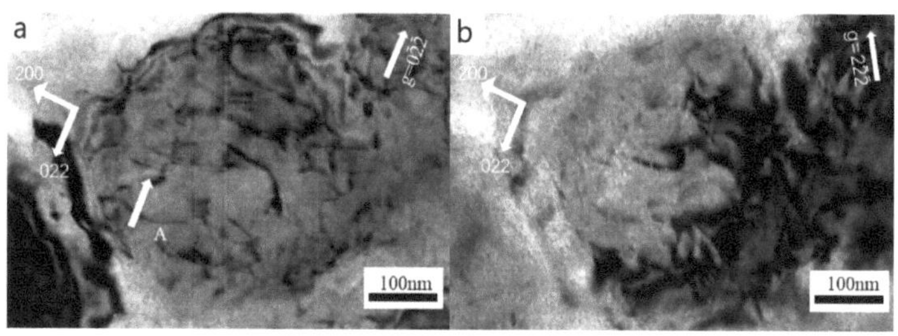

Figure 8. Dislocation configuration of 0Nb alloy after compression: (**a**) g = $0\bar{2}\bar{2}$; (**b**) g = $2\bar{2}\bar{2}$.

A series of typical TEM images of the Co6Ti11V1Nb alloy after compression at 900 °C is shown in Figure 9. As illustrated in Figure 9a, numerous dislocations are visible in the γ' phase when imaged with g = $\bar{2}\bar{2}0$. In contrast, when imaged with g = 220, as shown in Figure 9b, there is a high density of dislocations visible in the γ channel, and a few dislocations in the γ' phase are observed. As illustrated in Figure 9c, dense dislocations are observed in the γ' phases when imaged with g = $\bar{1}11$, whereas the dislocations in the

γ channel are invisable. When imaged with g = $\overline{131}$, the dislocations in the γ channel are visible, while the dislocations in the γ' phases are invisible, which is revealed in Figure 9d. Figure 9c shows that a stacking fault A and dislocation B are observed; based on Figure 9a,d, the type of stacking fault A can be determined to be R (displacement vector) = [$1\bar{1}2$]. According to Figure 9a,b, the Berger vector b = [$1\bar{1}0$] of the dislocation B can be determined. According to Equation (4), it can be seen that the superlattice intrinsic fault A is a result of the reaction of the dislocation B.

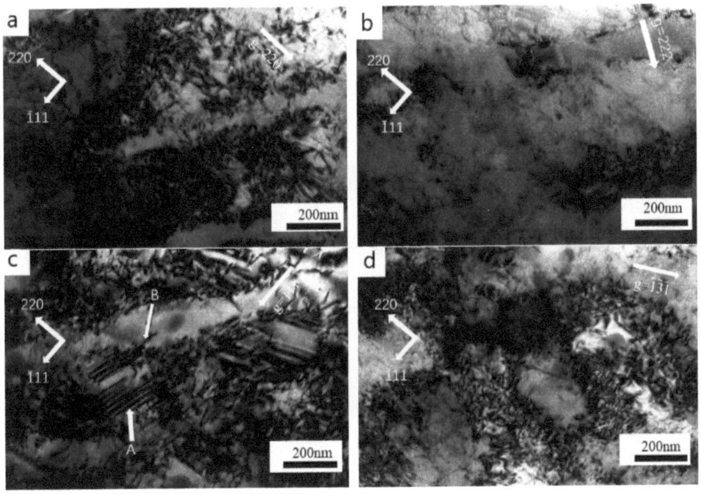

Figure 9. Dislocation configuration of 1Nb alloy after compression: (**a**) two-beam condition, with g = $\overline{22}0$; (**b**) g = $\overline{31}1$; (**c**) g = $\bar{1}11$; (**d**) g = $\overline{131}$.

Suzuki A. et al. have shown that the addition of different elements to alloys will change the elemental distributions between the γ/γ' phases and their lattice mismatch, further affecting the morphology of the γ' particle and thereby modifying the mechanical properties of alloys [6,27]. Previous studies have reported that the strength of an alloy is related to the degree of mismatch, the volume fraction, and the size of the γ' phase [28–30]. Within a certain range, the larger the mismatch, the greater the yield strength of the alloy, similar to nickel-based alloys. As seen in Table 2, as the Nb content increases, the mismatch also rises; so, the mismatch of the 3Nb alloy is the highest, which is 0.86%. However, the dimension of the γ' precipitate of the 3Nb alloy is only 189 nm, which is the minimum among the five alloys. But it can be seen from Figure 1 that there are more Co_3V phases in the 3Nb alloy, which is a hard and brittle precipitation with an ordered close-packed hexagonal structure, so the presence of the Co_3V phase may also contribute to the strength of the 3Nb alloy, but the phase usually precipitates in the form of flakes, which facilitates the formation and propagation of cracks; the close-packed hexagonal structure has a reduced slip system, which is detrimental to the plasticity of the material, so it is not desirable to enhance the strength of the material by introducing large volumes of Co_3V. The yield strength of the 3Nb alloy is the largest among the five alloys, reaching 496 MPa. It has been proven that when the lattice mismatch between γ and γ' is near 0, the γ' precipitate is spherical in shape. If the mismatch of the dual phases is 0.5% ~ 1%, the γ' particle is cuboidal; the mismatch of the five alloys is about 0.8%. Figure 5 shows that the morphology of the γ' phase is typically cubical. Figure 6 shows that Ti, V, and Nb are γ' formers. The addition of γ'-former elements in the alloys enhances the volume ratio of the γ' particle, so it can be concluded that the volume fraction of the γ' phase increases slightly with the rise in Nb content.

After the alloys are compressed at 900 °C, the dimension of the γ' phase in the alloys rises, while the volume ratio of the γ' phase drops substantially. This may be due to the dissolution of some γ' precipitates during cutting by the dislocations at high temperatures. Suzuki studied the compressive deformation mechanism of Co-Al-W alloys and found the following: (i) shearing of the γ' phases by pairs of $\frac{1}{2}$<110>-type perfect dislocations on {111} planes at low temperatures (293–978 K), (ii) at intermediate temperatures (around 1033 K), deformation by shearing of γ' particles by the <112> {111} slide system, and (iii) cutting of the γ' phases by <112> dislocations with the formation of stacking faults (above 1088K) [27]. In this study, after the Co-6Ti-11V alloy was compressed at 900 °C, the [101]-type compact dislocations cut the γ' precipitate along the {111} plane, and two kinds of deformation mechanisms appeared in the Co-6Ti-11V-1Nb alloy, including $[1\bar{1}0]$ dislocations at the interface of γ/γ' reacting to produce $[1\bar{2}1]$ partial dislocations, with the latter cutting the γ' particle and forming SISFs. The reaction [20] is as follows:

$$1/2\left[0\bar{1}\ 1\right] + \rightarrow 1/3\left[1\bar{2}\ 1\right] + 1/6\left[\bar{1}\bar{1}2\right] + \mathrm{SISF} \tag{4}$$

In summary, the dislocation motion of the 1Nb alloy deviates from that of the base alloy because there is no formation of stacking faults in the base alloy, while in the 1Nb alloy, the $[1\bar{1}0]$ dislocation reacts at the frontier of the γ/γ' phases to form a 1/3 <$1\bar{2}1$> partial dislocation, which then shears the γ' phase and forms stacking faults. These can be related to Nb reducing the energy stacking fault of Co-based alloys [31].

The addition of Nb in the Co-Ti-V ternary alloy is desirable, as it improves the phase stability of the γ' phases, as well as enhancing the volume fraction of this precipitate. Consequently, this improves the mechanical properties of the alloys at elevated temperatures. However, the addition should be within 1 at. %.

4. Conclusions

1. After solution treatment at 1100 °C/48 h and aging treatment at 870 °C/72 h, for the alloys containing less than 1Nb, they show a γ and γ' dual-phase microstructure and are free of deleterious phases. The Co_3V phase with a platelet shape is present in the 2Nb and 3Nb alloys.
2. With the increase in Nb content, the lattice mismatch of the γ/γ' two-phase microstructure increases gradually. After compression at 900 °C, the size of the γ' phase in the 0Nb alloy increases from 222.2 nm to 271.7 nm; the volume ratio of the γ' phase in the three investigated alloys decrease gradually. The 1Nb alloy shows the minimum size increase and the smallest volume drop of the γ' precipitate. These results show that Nb improves the phase stability of the γ' phase.
3. The yield strength of the Co-Ti-V alloy increases with the increase in Nb content, and the yield strength of the 3Nb alloy is the highest when compressed at 900 °C, at 496 Pa.
4. When compressed at 900 °C, the <101>-type dislocation causing shearing of the γ' phase in the {111} plane in the Nb-free alloy was observed. For the 1Nb alloy, some <$1\bar{1}0$> dislocations reacted at the γ/γ' interface to generate 1/3 $[1\bar{2}1]$, and then the γ' precipitate was cut by this dislocation, leaving R = <$\bar{1}21$> stacking faults.

Author Contributions: Data curation, S.N., E.L. and K.L.; writing—original draft, P.Z. and W.Z.; writing—review and editing, S.C. and Y.Q. All authors have read and agreed to the published version of the manuscript.

Funding: This research was financially supported by the National Key R&D Program of China [2021YFC2202402] and the National Natural Science Foundation of China (52275339, 51471079). The authors would also like to acknowledge funding support from the Undergraduate Innovation Program of JUST.

Institutional Review Board Statement: Not applicable.

Informed Consent Statement: Not applicable.

Data Availability Statement: The data are contained within the article.

Conflicts of Interest: Authors Wanxiang Zhao and Kunjie Luo are employed by the Suzhou Nuclear Power Research Institute. The remaining authors declare that the research was conducted in the absence of any commercial or financial relationships that could be construed as potential conflicts of interest.

References

1. Donachie, M.; Donachie, S. *Superalloys: A Technical Guide*, 2nd ed.; ASM International: Novelty, OH, USA, 2002; p. 439.
2. Sato, J.; Omori, T.; Oikawa, K.; Ohnuma, I.; Kainuma, R.; Ishida, K. Cobalt-base high-temperature alloys. *Science* **2006**, *312*, 90–91. [CrossRef] [PubMed]
3. Sims, C.T. A contemporary view of cobalt-base alloys. *J. Met.* **1969**, *21*, 27–42. [CrossRef]
4. Pollock, T.M. Alloy design for aircraft engines. *Nat. Mater.* **2016**, *15*, 809–815. [CrossRef]
5. Reed, R.C. *The Superalloys: Fundamentals and Applications*; Cambridge University Press: Cambridge, UK, 2006.
6. Suzuki, A.; Inui, H.; Pollock, T.M. L12-Strengthened Cobalt-Base Superalloys. *Annu. Rev. Mater. Res.* **2015**, *45*, 345–368. [CrossRef]
7. Bocchini, P.J.; Sudbrack, C.K.; Noebe, R.D.; Dunand, D.C.; Seidman, D.N.; Northwestern Univ, E.I.L. Effects of titanium substitutions for aluminum and tungsten in Co-10Ni-9Al-9W (at%) superalloys. *Mater. Sci. Eng. A* **2017**, *705*, 122–132. [CrossRef]
8. Bocchini, P.J.; Sudbrack, C.K.; Sauza, D.J.; Noebe, R.D.; Seidman, D.N.; Dunand, D.C.; Northwestern Univ, E.I.L. Effect of tungsten concentration on microstructures of Co-10Ni-6Al-(0,2,4,6)W-6Ti (at%) cobalt-based superalloys. *Mater. Sci. Eng. A* **2017**, *700*, 481–486. [CrossRef]
9. Ooshima, M.; Tanaka, K.; Okamoto, N.L.; Kishida, K.; Inui, H. Effects of quaternary alloying elements on the γ' solvus temperature of Co-Al-W based alloys with fcc/L12 two-phase microstructures. *J. Alloys Compd.* **2010**, *508*, 71–78. [CrossRef]
10. Pyczak, F.; Bauer, A.; GöKen, M.; Lorenz, U.; Neumeier, S.; Oehring, M.; Paul, J.; Schell, N.; Schreyer, A.; Stark, A.; et al. The effect of tungsten content on the properties of L12-hardened Co–Al–W alloys. *J. Alloys Compd.* **2015**, *632*, 110–115. [CrossRef]
11. Davydov, A.V.; Kattner, U.R.; Josell, D.; Waterstrat, R.M.; Boettinger, W.J.; Blendell, J.E.; Shapiro, A.J. Determination of the CoTi congruent melting point and thermodynamic reassessment of the Co-Ti system. *Metall. Mater. Trans. A* **2001**, *32*, 2175–2186. [CrossRef]
12. Liu, Y.; Takasugi, T.; Izumi, O. Alloying behavior of Co_3Ti. *Metall. Trans. A* **1986**, *17*, 1433–1439. [CrossRef]
13. Ruan, J.J.; Wang, C.P.; Zhao, C.C.; Yang, S.Y.; Yang, T.; Liu, X.J. Experimental investigation of phase equilibria and microstructure in the Co–Ti–V ternary system. *Intermetallics* **2014**, *49*, 121–131. [CrossRef]
14. Ruan, J.J.; Wang, C.P.; Yang, S.Y.; Omori, T.; Yang, T.; Kimura, Y.; Liu, X.J.; Kainuma, R.; Ishida, K. Experimental investigations of microstructures and phase equilibria in the Co–V–Ta ternary system. *J. Alloys Compd.* **2016**, *664*, 141–148. [CrossRef]
15. Zhou, G.J.; Tang, J.G.; Zhou, Y.; An, W.K.; Cai, A.H. Phase equilibria in the Co-Ti-V system at 873 K. *J. Alloys Compd.* **2014**, *602*, 49–52. [CrossRef]
16. Zhang, J.; Du, Y.; Liu, Y.; Zhang, W.; Liu, D.; Chen, C. Assessment of atomic mobilities for fcc Co-Ti-V alloys. *Calphad* **2018**, *61*, 179–188. [CrossRef]
17. Verma, S.K.; Pramanik, A.; Jyothsna, K.; Vamsi, K.V.; Karthikeyan, S. Phase transformation temperatures, γ–γ' lattice parameter misfit, and γ' precipitate morphology in Co–Ti–V alloys. *Metall. Mater. Trans. A* **2022**, *53*, 4011–4022. [CrossRef]
18. Ruan, J.J.; Liu, X.J.; Yang, S.Y.; Xu, W.W.; Omori, T.; Yang, T.; Deng, B.; Jiang, H.X.; Wang, C.P.; Kainuma, R.; et al. Novel Co-Ti-V-base superalloys reinforced by L12-ordered γ' phase. *Intermetallics* **2018**, *92*, 126–132. [CrossRef]
19. Zhou, P.; Gao, X.; Song, D.; Liu, Y.; Cheng, J. Effect of Ni on Microstructure and Mechanical Property of a Co-Ti-V-Based Superalloy. *Scanning* **2021**, *2021*, 678085. [CrossRef]
20. Zhou, P.; Gao, X.; Song, D.; Liu, Q.; Liu, Y.; Cheng, J. Role of Ru on the microstructure and property of novel Co-Ti-V Superalloy. *J. Mater. Res.* **2020**, *35*, 2737–2745. [CrossRef]

21. Liu, X.; Pan, Y.; Chen, Y.; Han, J.; Yang, S.; Ruan, J.; Wang, C.; Yang, Y.; Li, Y. Effects of Nb and W Additions on the Microstructures and Mechanical Properties of Novel gamma/gamma′ Co-V-Ti-Based Superalloys. *Metals* **2018**, *8*, 563. [CrossRef]
22. Xu, Y.; Xia, T.; Yan, J. Effect of alloying elements to hot corrosion behavior of novel Co-Al-W superalloy. *J. Chin. Soc. Corros. Prot.* **2010**, *30*, 457–464.
23. Migas, D.; Moskal, G.; Niemiec, D. Surface Condition of New γ–γ′ Co-Al-Mo-Nb and Co-Al-W Cobalt-Based Superalloys After Oxidation at 800 °C. *J. Mater. Eng. Perform.* **2018**, *27*, 447–456. [CrossRef]
24. Klein, L.; Bauer, A.; Neumeier, S.; Göken, M.; Virtanen, S. High temperature oxidation of γ/γ′-strengthened Co-base superalloys. *Corros. Sci.* **2011**, *53*, 2027–2034. [CrossRef]
25. Makineni, S.K.; Nithin, B.; Chattopadhyay, K. Synthesis of a new tungsten-free γ–γ′ cobalt-based superalloy by tuning alloying additions. *Acta Mater.* **2015**, *85*, 85–94. [CrossRef]
26. Takasugi, T.; Honjo, H.; Kaneno, Y.; Inoue, H. Plastic flow instabilities of $L1_2Co_3Ti$ alloys at intermediate temperatures. *Acta Mater.* **2002**, *50*, 847–855. [CrossRef]
27. Suzuki, A.; DeNolf, G.C.; Pollock, T.M. Flow stress anomalies in γ/γ′ two-phase Co–Al–W-base alloys. *Scr. Mater.* **2007**, *56*, 385–388. [CrossRef]
28. Carroll, L.J.; Feng, Q.; Pollock, T.M. Interfacial Dislocation Networks and Creep in Directional Coarsened Ru-Containing Nickel-Base Single-Crystal Superalloys. *Metall. Mater. Trans. A* **2008**, *39*, 1290–1307. [CrossRef]
29. Zhang, J.X.; Murakumo, T.; Harada, H.; Koizumi, Y. Dependence of creep strength on the interfacial dislocations in a fourth generation SC superalloy TMS-138. *Scr. Mater.* **2003**, *48*, 287–293. [CrossRef]
30. Zhang, J.X.; Koizumi, Y.; Kobayashi, T.; Murakumo, T.; Harada, H. Strengthening by γ/γ′ interfacial dislocation networks in TMS-162—Toward a fifth-generation single-crystal superalloy. *Metall. Mater. Trans. A* **2004**, *35*, 1911–1914. [CrossRef]
31. Lilensten, L.; Antonov, S.; Gault, B.; Tin, S.; Kontis, P. Enhanced creep performance in a polycrystalline superalloy driven by atomic-scale phase transformation along planar faults. *Acta Mater.* **2021**, *202*, 232–242. [CrossRef]

metals

MDPI

Article

Improved Surface Properties of Low-Carbon Steel by Chromizing–Titanizing Coating Using Pack Cementation Process

Ayman Yousef [1], A. M. Bastaweesy [2], Ibrahim M. Maafa [1,*] and Ahmed Abutaleb [1]

[1] Department of Chemical Engineering, College of Engineering and Computer Sciences, Jazan University, Jazan 45142, Saudi Arabia; aymanhassan@jazanu.edu.sa (A.Y.); azabutaleb@jazanu.edu.sa (A.A.)

[2] Department of Chemical Engineering, Faculty of Engineering, Minia University, Minia 60519, Egypt; ali.bastawisy@mu.edu.eg

* Correspondence: imoaafa@jazanu.edu.sa

Abstract: This study investigates the application of chromizing and titanizing coatings on low-carbon steel (LCS) via the pack cementation process, utilizing various compositions, temperatures, and durations. The coating was analyzed using standard techniques, such as X-ray diffraction (XRD), scanning electron microscopy (SEM), energy dispersive X-ray spectroscopy (EDX), and Vickers hardness testing, to determine their characteristics. The kinetics of the pack chromizing, titanizing, and chromotitanizing of low-carbon steel exhibited parabolic behavior, with the rate constant with increasing temperature. The formed diffusion layers primarily consisted of Cr, Ti, $Cr_{1.9}Ti$, FeTi, Al_2O_3, Cr_2O_3, TiO_2, and $Cr_{1.36}Fe_{0.52}$, in addition to Fe. The microhardness reached its highest value of 900 $HV_{0.01}$ Kg_f with 48% FeTi, followed by 790 $HV_{0.01}$ Kg_f with 12% FeCr–36% FeTi, 730 $HV_{0.01}$ Kg_f with 24% FeCr–24% FeTi, 680 $HV_{0.01}$ Kg_f with 36% FeCr–12% FeTi, and 560 $HV_{0.01}$ Kg_f with 48% FeCr. The results indicate a significant enhancement in the mechanical properties of low-carbon steel through the coating process. This study confirms that the pack cementation coatings of chromizing, titanizing, and chromotitanizing significantly enhance the surface hardness and mechanical integrity of low-carbon steel. The controlled diffusion process leads to the formation of robust intermetallic layers, and the variation in FeCr and FeTi composition allows for tailored mechanical properties. Additionally, the results suggest that the interplay between Cr and Ti promotes the development of a complex, multilayered microstructure that balances hardness with potential toughness, providing a broad spectrum of industrial applications. This research underscores the versatility of pack cementation as an effective method to engineer advanced coatings, offering a cost-efficient pathway to enhance the performance of low-carbon steel in demanding environments.

Keywords: pack cementation; low-carbon steel; diffusion coatings; chromizing; titanizing; microhardness

Citation: Yousef, A.; Bastaweesy, A.M.; Maafa, I.M.; Abutaleb, A. Improved Surface Properties of Low-Carbon Steel by Chromizing–Titanizing Coating Using Pack Cementation Process. *Metals* **2024**, *14*, 1456. https://doi.org/10.3390/met14121456

Academic Editors: Koh-ichi Sugimoto, Jingxiang Xu, Xingwei Zheng and Zhenhua Chu

Received: 23 October 2024
Revised: 11 December 2024
Accepted: 17 December 2024
Published: 19 December 2024

1. Introduction

Low-carbon steel (LCS) is widely used in various manufacturing industry sectors, including building components, containers, furnaces, and agricultural machinery [1]. Its popularity is primarily due to its affordability and impressive strength. However, in applications requiring exceptional mechanical properties, bulk hardness alone may not suffice to ensure optimal performance [1]. It is well known that the deterioration and failure of structural materials in engineering are often caused by wear or corrosion, which are mainly influenced by the surface properties of the material rather than its overall performance [2]. Hence, improving surface performance through various techniques while preserving the original characteristics of the material is a promising approach to address these challenges [3,4]. Consequently, numerous efforts have been made to enhance the support performance of LCS through surface treatments, including increasing surface hardness and improving tribological properties [5–9]. Surface coatings provide several benefits in enhancing component performance, particularly in minimizing wear and corrosion [10,11].

Thermochemical treatment is an efficient method for creating protective diffusion coatings on various metallic materials [12,13].

The coating process is conducted at elevated temperatures (often between 800 and 1000 °C) for varying durations (usually between 2 and 10 h), contingent upon the type of steel. The technique is straightforward and cost-effective as it does not require expensive equipment. It can be utilized on substrates with intricate geometries, exhibiting minimal environmental impact as no hazardous vapors are released during the film forming process [14]. It could be utilized to create coatings that enhance their mechanical performance, oxidation resistance, and tribological characteristics [15]. Insufficient attention is devoted to the regulation of the precipitation of reinforcement compounds and their actions, which therefore impacts the even distribution of precipitates and adhesion characteristics. One of the primary limitations of this coating method is the length of time it takes. This may result in inappropriate grain growth, leading to a decline in the mechanical characteristics. These coatings include aluminizing, chromizing, titanizing, carburizing, boriding, and co-diffusion [16–22]. Furthermore, diffusion coatings not only offer exceptional surface effectiveness but also ensure a strong metallurgical bond between the coating and the substrate material, enhancing the protection of the treated component throughout its service life [23]. Fernandes et al. [24] utilized a coating technique to treat AISI 1060 steel samples through pack cementation, subjecting the samples to high-temperature treatment for different durations and temperatures. At 1050 °C, a layer of $(Cr,Fe)_2N_{1-x}$ was formed, while at 1000 °C, a carbide layer $(Cr,Fe)_2N_{1-x}$ was produced. The sample coated at 1050 °C for 9 h exhibited superior characteristics, including enhanced wear resistance and corrosion properties. Lin et al. [2] employed the pack cementation process to apply a titanizing coating onto 316 stainless steel at 900 °C for 6 h, resulting in a titanizing coating approximately 5 μm thick, composed of various intermetallic compound phases, including Ni_2Ti, NiTi, FeNi, Fe_2Ti, and Cr_2Ti. As a result, the titanizing treatment significantly improved the surface hardness of the 316 stainless steel.

Zeng et al. [25] produced two types of Si-Cr coatings, $SiCr_{-1}$ and $SiCr_{-2}$, on AISI 5140 steel using pack cementation with pure silicon and ferrosilicon, respectively. When treated at 950 °C for one hour, $SiCr_{-2}$ demonstrated enhanced characteristics, with a thicker coating (>300 μm), increased Si and Cr content, and the absence of a carbon-deficient layer, in contrast to $SiCr_{-1}$ (150 μm). Microhardness assessments revealed that $SiCr_{-2}$ exhibited superior average hardness, while wear resistance analyses exhibited a reduced friction coefficient of 0.22 compared to $SiCr_{-1}$ at 0.40 and the as-tempered sample at 0.30. These results underscore the influence of feedstock composition, with $SiCr_{-2}$ exhibiting superior surface characteristics attributed to reaction deposition growth and decreased micropore development. Meng et al. [26] performed vanadiumizing treatments on GCr15 steel using a ferrovanadium agent combined with NH_4Cl, La_2O_3, and Al_2O_3 at temperatures between 899.85 °C and 959.85 °C for 1 to 7 h. The resultant vanadiumizing layer, consisting of VC_x and α-Fe, displayed thicknesses ranging from 4.65 μm to 12.65 μm, which increased with elevated temperatures and prolonged durations. Microhardness varied from 1892.3 $HV_{0.02}$ to 2698.6 $HV_{0.02}$, with peak values recorded under the most extreme conditions. Growth adhered to a parabolic curve, exhibiting an activation energy of 164.85 kJ/mol, indicating improved diffusion attributed to rare earth elements. Liu et al. [27] executed chromizing on T9 and SPCC steels using the pack cementation technique with chromium powder and NH_4Cl, applying a force of 8 N for 30 min and a rotational speed of 240 rpm. The chromized outer layers included Cr_7C_3 and $Cr_{23}C_6$ for T9 (approximately 4.19 μm thick) and mostly $Cr_{23}C_6$ for SPCC (approximately 2.89 μm thick), with XRD validating the phase compositions. Microhardness assessments indicated 1737.72 HV in the T9 outer layer, declining to 1378.31 HV in the secondary layer, while SPCC exhibited 1771.91 HV in the outer layer, dropping to 448.52 HV. Friction and wear studies indicated that T9 steel saw a decrease in mass wear from 2 mg to 0.6 mg, while SPCC's wear increased from 2 mg to 2.9 mg, associated with surface fatigue wear. The findings underline the enhanced wear and corrosion resistance of T9, attributed to its durable chromium carbide layer, while SPCC exhibits greater corrosion resistance than wear resistance due to its thinner layer.

Yang et al. [28] explored the wear behavior of a chromizing–titanizing coating on carbon steel through the pack cementation process. Compared to the chromizing coating (1300–1650 Kg_f), the chromizing–titanizing coating (1800–3000 Kg_f) exhibited greater hardness. The enhancement of the wear resistance of carbon steel can be achieved through processes such as chromizing and titanizing. In the current study, one of our goals is to reduce surface damage by producing chromizing, titanizing, and chromotitanizing coatings on LCS using pack cementation. To do this, we systematically investigated the microstructural characteristics of the diffusion layer on LCS and evaluated the microhardness behavior of the coatings to better understand the candidate coatings for LCS, which may provide superior mechanical resistance.

2. Materials and Methods

2.1. Materials

A commercial LCS substrate was selected as the starting material. The Metallurgical Research Institute in Cairo, Egypt, provided ferrochrome and ferrotitanium powder. Alumina (Al_2O_3) was obtained from the Egyptian Aluminum Company, Egyptalum, (Nag Hammadi, Egypt). Ammonium chloride (NH_4Cl) was procured from Aldrich Co., (St. Louis, MO, USA). The chemical compositions of LCS, ferrochrome, and ferrotitanium powder are enumerated in Table 1.

Table 1. The chemical composition of the used materials (wt%).

Elements	LCS (wt%)	Ferrochrome (wt%)	Ferrotitanium (wt%)
C	0.0507	8.9	--
Si	0.016	1.2	3
Mn	0.295	--	--
Al	--	--	6
P	0.024	0.021	--
S	0.018	0.05	--
O	--	--	8.3
Ni	0.001		
Cr	--	61	--
Ti	--	--	27
Fe	balance	balance	balance

2.2. Specimen Preparation

The steel specimens were cut into discs measuring 20 mm in diameter and 2 mm in thickness. The specimens were ground using waterproof SiC abrasive papers of varying grit sizes and then polished with diamond pastes of different particle sizes. The samples were then rinsed in distilled water, and any grease was removed using ethanol. The samples were rinsed again in distilled water and given a final wash with acetone in an ultrasonic bath for 5 min.

2.3. Pack Diffusion Process

The specimens and powders with various compositions were placed inside stainless-steel vessels. The composition of the cementation mixtures utilized is listed in Table 2. Each mixture contained 20 g of ferrochrome powder and/or ferrotitanium. Hard-burned Al_2O_3 was used as an inert filler material to effectively prevent undesired caking, melting, and collapse of the ferrochrome and ferrotitanium. Ammonium chloride (NH_4Cl) served as an activator. The specimens were completely submerged in the powder mixtures, maintaining a distance of approximately 15 mm between the specimens and the walls of the cementation vessels. The boxes were tightly sealed with airtight lids. The wet mixture was composed of cement, refractory clay, asbestos, and sodium silicate, applied along the edges of the lid to prevent oxidation of the contents during diffusion. The sealed vessels were subjected to different temperatures (900 °C, 1000 °C, and 1100 °C) for specific durations (i.e., 3 h, 5 h, and 7 h) in a muffle furnace and then cooled in the furnace after each run. The specimens were carefully removed from the powder, delicately cleaned to eliminate any visible dirt on their surface, brushed, rinsed with distilled water, and weighed to calculate the coat pickup.

Table 2. Chemical composition (wt%) of the cementation mixture for treatment with chromium and titanium. Reprinted from Ref. [29].

Type of Mixture	Coating Materials	Al_2O_3, %	NH_4Cl, %
Mixture A	48% FeCr	50	2
Mixture B	36% FeCr–12% FeTi	50	2
Mixture C	24% FeCr–24% FeTi	50	2
Mixture D	12% FeCr–36% FeTi	50	2
Mixture E	48% FeTi	50	2

2.4. Characterization

The microhardness of the coated and uncoated specimens was determined using a micro-Vickers tester (Shimadzu (HMV) microhardness tester with a 0.01 Kg load, Shimadzu, Kyoto, Japan). The coated layer thickness in µm was measured using a Positector 6000 (model F1 gauge, Cole-Parmer GmbH, Wertheim am Main, Germany). Surface and cross-sectional observations of the coated specimens were conducted using scanning electron microscopy (SEM, JEOL JSM 5410, JEOL, Tokyo, Japan) with energy dispersive spectroscopy (EDS). The composition of the surface layer of as-coated specimens was detected by X-ray diffraction (XRD) (Rigaku Co., Tokyo, Japan) with Cu Kα (λ = 1.54056 Å) in small steps of 2θ = 0.02°. The average crystal size of the phases in the coated layer was calculated using the Debye–Scherrer equation [30]. X-ray fluorescence (XRF) spectroscopy (X900 WDXRF, ThermoFisher Scientific, Waltham, MA, USA) is used for elemental analysis.

3. Results and Discussion

Pack cementation is a complex chemical vapor deposition process that involves a series of thermochemical reactions. These reactions generate diffusion coatings through the production of active diffusion atoms [2]. In this process, some iron atoms are replaced by Cr and/or Ti atoms that diffuse into the lattice of LCS grains. Following the chromizing and/or titanizing treatment, a thermal diffusion layer rich in Cr and/or Ti develops on the surface of LCS. The thermochemical reactions involved in creating the active Cr and/or Ti atoms during chromizing, titanizing, and chromotitanizing are as follows [2,31]:

$$NH_4Cl_{(s)} \rightarrow NH_{3(g)} + HCl_{(g)}, \quad (1)$$

$$2NH_3 \rightarrow N_{2(g)} + 3H_{2(g)}, \quad (2)$$

$$4HCl_{(g)} + M_{(s)} \rightarrow MCl_{4(g)} + 2H_{2(g)}, \quad (3)$$

$$MCl_{4(g)} + 2Fe_{(s)} \rightarrow 2FeCl_{2(g)} + [M], \quad (4)$$

$$MCl4_{(g)} + 2H_{2(g)} \rightarrow 4HCl_{(g)} + [M], \quad (5)$$

where [M] represents the active Cr and/or Ti atom.

The composition at the surface is influenced by the interplay of the supply rates of Cr, Ti, and Cr-Ti from different reactions, as well as their diffusion into the underlying metal. The thickness of the formed layers was measured for each treatment condition. The differences in layer thickness for various process times and temperatures are illustrated in Figure 1a–e. It is noted that the increase in time and temperature substantially increased the thickness of the coating. Data also indicate that as the percentage of Cr in the pack increases, there is a corresponding increase in the thickness of the coating. The coating thickness shows a greater increase at 1100 °C compared to 1000 °C and 900 °C. All curves exhibited a parabolic pattern. Therefore, the thickness of the coated layer (X) in relation to process time follows a parabolic law:

$$X = K_s\sqrt{t}, \tag{6}$$

where X is coating thickness in μm, K_s is the rate constant, $\mu m/\sqrt{h}$, and t is process time in hours.

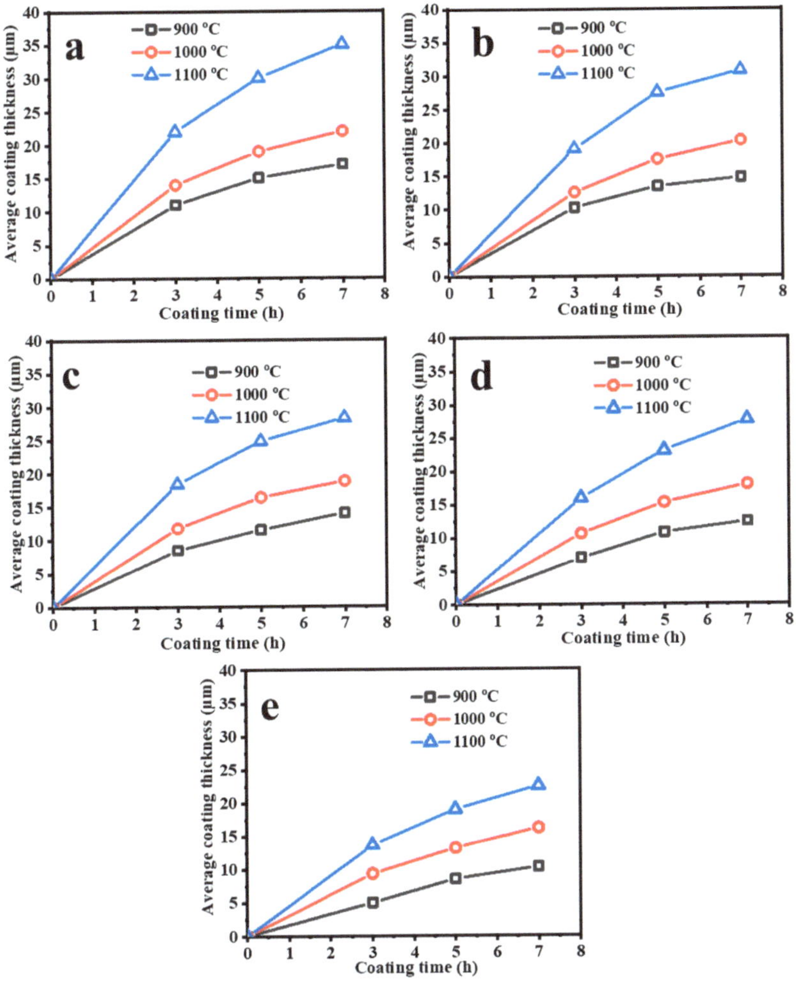

Figure 1. Effect of process time on average coating thickness of coated specimens at different temperatures used: (**a**) 48% FeCr; (**b**) 36% FeCr–12% FeTi; (**c**) 24% FeCr–24% FeTi; (**d**) 12% FeCr–36% FeTi; and (**e**) 48% FeTi.

The results obtained are consistent with previous research [32], which indicated that as the temperature increases, the reaction kinetics between the coating elements (i.e., Cr, Ti, and Cr-Ti) and steel follow a parabolic law. Figure 2a–e demonstrate the linear relationship between X and $\sqrt{t}$ at various temperatures (900 °C, 1000 °C, and 1100 °C). This suggests that the diffusion step is the rate-controlling step. Saturating the surface of steel machine parts with chromium enhances their resistance to corrosion by gases, even at high temperatures of up to 800 °C. High-carbon steels are also treated with chromizing to enhance their hardness and wear resistance. K_s is determined by analyzing the slope of the linear relationship, as depicted in Figure 2a–e. The results presented in Table 3 illustrate the relationship between temperature, chromium percentage, and the rate constant. Generally, it is observed that as the temperature and chromium percentage in the mixtures of chromium and titanium increase, the rate constant also increases.

Table 3. Variation of *E* for different compositions, temperatures, and times.

Type of Mixture (wt%)	T (°C)	K_s ($\mu m/\sqrt{h}$)	*D* (m^2/s) $\times 10^{15}$	*E* (J/mol)
48% FeCr	1100	13.37	12.4	9.434
	1000	8.43	4.93	
	900	6.56	2.99	
36% FeCr–12% FeTi	1100	11.48	9.16	8.909
	1000	7.49	3.9	
	900	5.87	2.39	
24% FeCr–24% FeTi	1100	10.89	8.24	9.664
	1000	7.22	3.62	
	900	5.27	1.93	
12% FeCr–36% FeTi	1100	10.47	7.61	10.544
	1000	6.85	3.26	
	900	4.75	1.57	
48% FeTi	1100	8.55	5.07	10.598
	1000	6.08	2.57	
	900	3.88	1.04	

The reaction and diffusion rate accelerated with increasing temperature, as noted in Figures 1 and 2. This indicates that the thickness of the diffusion layer also increased. Furthermore, raising the temperature increases the presence of highly reactive molecules, which speeds up the diffusion rate. The high level of kinetic energy and activity within the reaction system is a result of the temperature increment. The findings demonstrate that higher temperatures and longer process times lead to increased diffusion. The diffusion coefficient (D) can be determined by measuring the coating thickness. According to Fick's diffusion laws, the depth of penetration (X) into a semi-finite solid is determined by time according to the following expression [33]:

$$X^2 = 4kDt. \tag{7}$$

The constant (k) has a convoluted relationship with the concentration at the surface and the concentration at a depth (X). By analyzing the concentration of Cr and Ti at the surface and interior of coated specimens, it is evident that the value of k is approximately equal to one [14]. Therefore, Equation (7) can be rewritten as follows:

$$X^2 = 4Dt. \tag{8}$$

Equation (8) can be rewritten as follows:

$$D = \frac{X^2}{4t}. \tag{9}$$

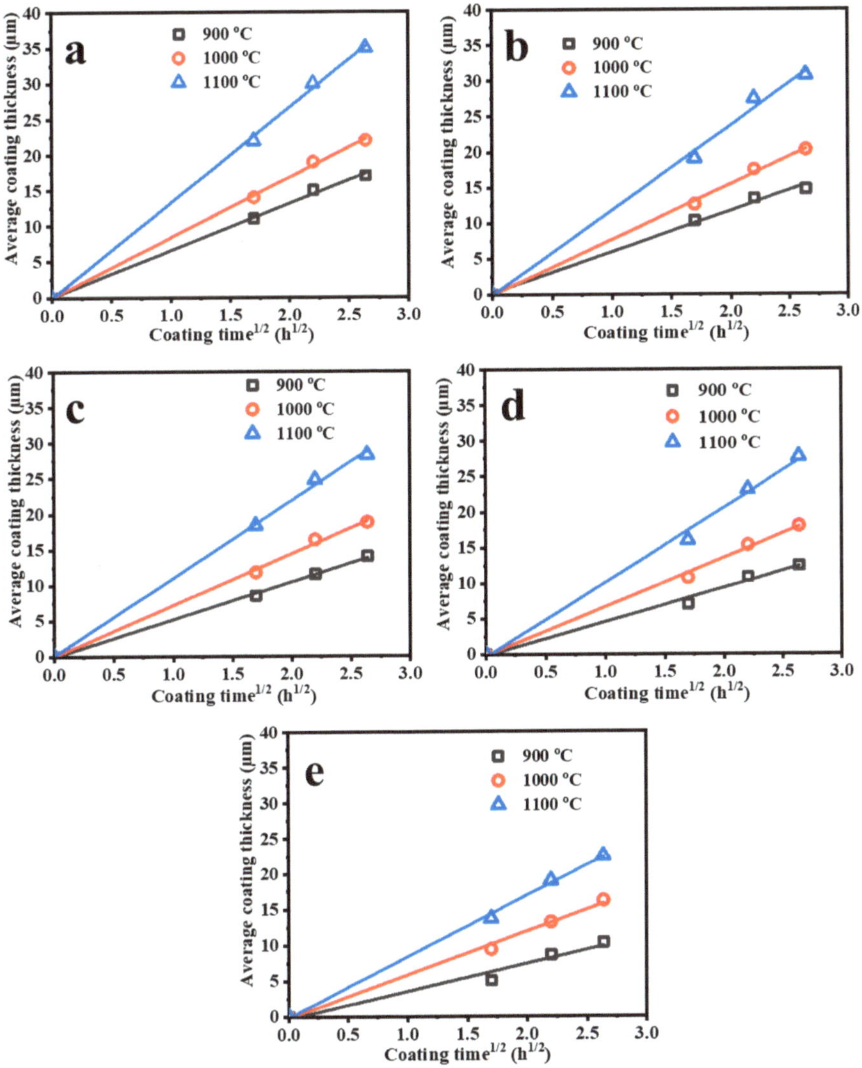

Figure 2. Relationship between the square root of process time and average coating thickness of coated specimens at different temperatures used: (**a**) 48% FeCr; (**b**) 36% FeCr–12% FeTi; (**c**) 24% FeCr–24% FeTi; (**d**) 12% FeCr–36% FeTi; and (**e**) 48% FeTi.

It is possible to calculate the diffusion coefficient (D) based on the known variations of thickness with time; however, it is reasonable to expect that the calculation will only yield approximate values because the derivation ignores the variation of the diffusion coefficient with concentration and assumes that the constant (k) is not dependent on temperature. The value of D was determined by taking the slope of each line in Figure 2a–e and using that information. Table 3 summarizes the values achieved at various temperatures and compositions.

The Arrhenius equation (Equation (10)) was used to calculate the activation energy of the reaction. It shows that D varies with temperature according to the following Arrhenius equation:

$$D = D_0 e^{-\left[\frac{E}{RT}\right]}, \tag{10}$$

where T is the absolute temperature, R is the gas constant, E is the activation energy, and D_0 is the frequency factor.

This equation demonstrates that as the temperature increases, the diffusion coefficient also increases, enhancing diffusion and subsequently increasing weight. Figure 3 illustrates the correlation between ln D and $1/T$. The activation energy (E) was determined using the Arrhenius plot. The slope of the straight line resulted in E values (Table 3) for the diffusion.

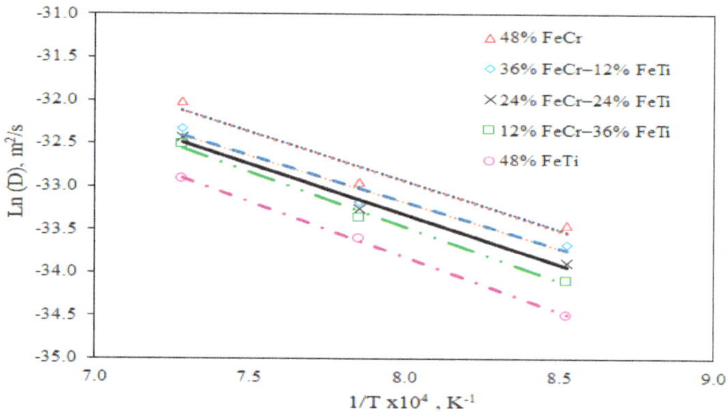

Figure 3. Arrhenius plot of parabolic rate constant versus inverse absolute temperature for average coating thickness of chromized, titanized, and chromotitanized LCS.

3.1. Coating Morphology

3.1.1. Optical Microscope

The majority of coated specimens exhibited changes in microstructure when observed under an optical microscope. It is straightforward to differentiate between the two layers: the base metal and the coated layer. The composition of the coated layer was examined using XRD and SEM. The optical micrographs of coated specimens are displayed in Figure 4. The thickness of the coat is influenced by the temperature and duration of the coating process. The coated layer was smooth and free of inclusions or defects. The thickness of the diffusion layer increases significantly as temperature and time increase, forming a noticeable protective coating due to the accelerated diffusion rate. As the temperature and time were raised to 1100 °C and 7 h, respectively, the thickness of the coating increased. However, the diffusion layer became brittle and developed micro-cracks, which can be attributed to prolonged exposure to high temperatures while being held in pack powder.

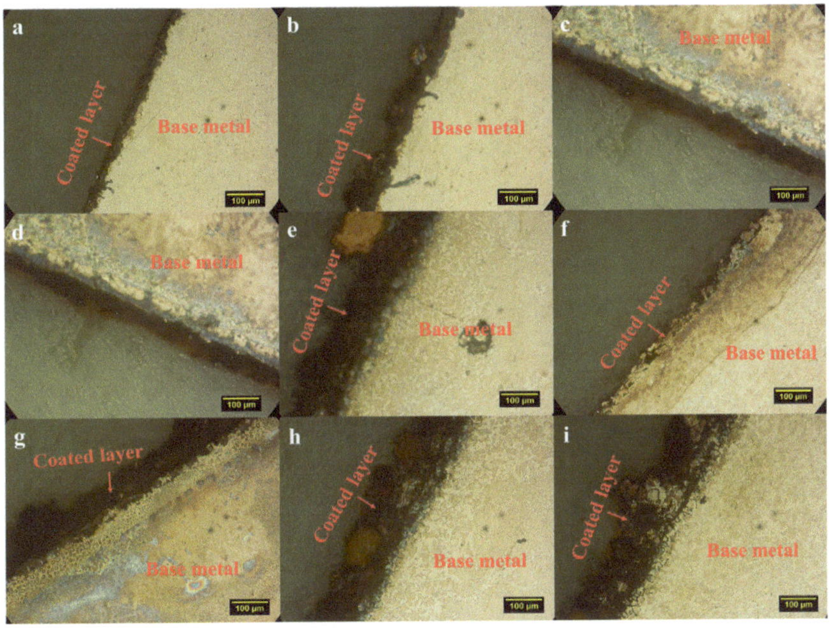

Figure 4. Optical images of pack chromizing specimens at 900 °C for (**a**) 3 h, (**b**) 5 h, and (**c**) 7 h; at 1000 °C for (**d**) 3 h, (**e**) 5 h, and (**f**) 7 h; and at 1100 °C for (**g**) 3 h, (**h**) 5 h, and (**i**) 7 h.

3.1.2. SEM and EDX Characterization

The SEM equipped with EDX was used to examine the cross-sectional morphology and concentration profiles of the elements in the specimens. The analysis was conducted at four different points along the coating thickness, extending from the base metal to the surface. Figure 5a is the SEM cross-section image, pointing out that the coatings possess a uniform layer thickness. Figure 5b–d reveal that the sample does not possess a uniform coating thickness. It can be observed from Figure 5e that the sample does not exhibit a uniform layer thickness but shows the presence of micro-cracks in the coating, which is likely caused by prolonged exposure to the pack powder.

Figure 6 demonstrates the results of EDX analysis conducted at four points across the cross-section. The analysis revealed the presence of Ti and Cr throughout the coating thickness, indicating that these elements are the primary components of the coating. From Figure 6a,b, it is evident that the concentration of titanium increases as the distance from the base metal increases. Chromium follows a similar pattern (Figure 6c), while the concentration of iron decreases with increasing distance from the base metal. EDX analysis also revealed that small amounts of Fe, the primary component of the base metal, diffuse into the layer. The transition between the base and the coating is not abrupt, indicating that some iron diffused into the layer structure. A similar pattern has been observed for the other formulations in Figure 6b–e. The concentration profile results indicate that the top coat layer contains the highest content of Cr and Ti, which gradually decreases towards the substrate. As we move further away from the base metal, the concentration of Ti increases, while Cr follows a similar pattern. Conversely, the concentration of Fe decreases with increasing distance from the base metal.

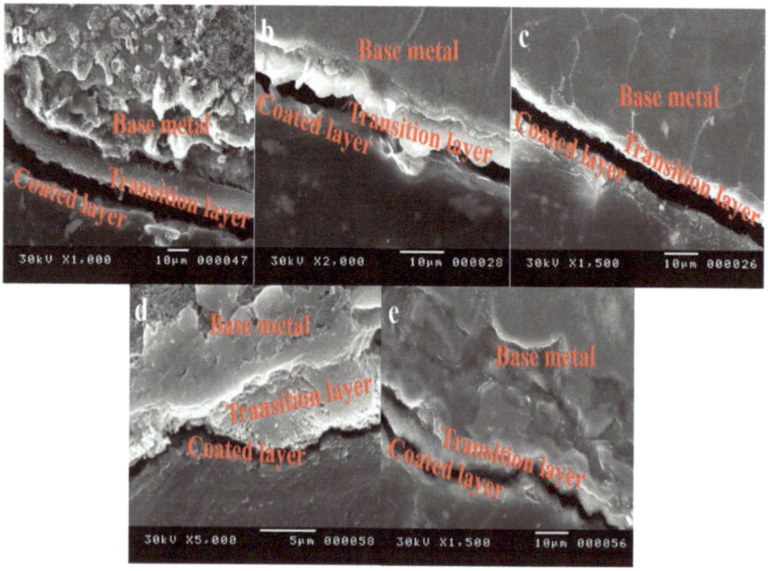

Figure 5. SEM images of coated samples consisting of (**a**) 24% FeTi–24% FeCr at 900 °C for 5 h, Reprinted from ref. [29]; (**b**) 48% FeTi at 1000 °C for 5 h, Reprinted from ref. [29]; (**c**) 48% FeCr at 900 °C for 3 h, Reprinted from ref. [29]; (**d**) 48% FeTi at 1100 °C for 5 h; and (**e**) 24% FeTi–24% FeCr at 1100 °C for 7 h. (Pack composed of 48 wt% (Cr or Ti or CrTi), 50 wt% Al_2O_3 powder, and 2 wt% NH_4Cl).

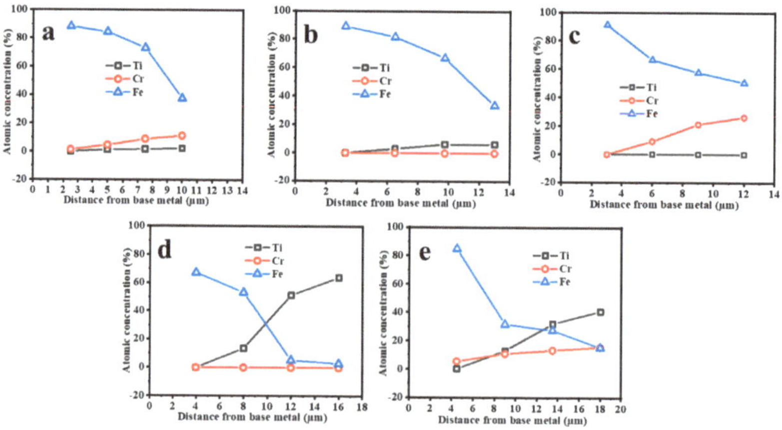

Figure 6. EDX distribution for the corresponding SEM images (**a**) 24% FeTi–24% FeCr at 900 °C for 5 h; (**b**) 48% FeTi at 1000 °C for 5 h; (**c**) 48% FeCr at 900 °C for 3 h; (**d**) 48% FeTi at 1100 °C for 5 h; and (**e**) 24% FeTi–24% FeCr at 1100 °C for 7 h. (Pack composed of 48 wt% (Cr or Ti or CrTi), 50 wt% Al_2O_3 powder, and 2 wt% NH_4Cl).

3.1.3. XRD Analysis

XRD was utilized to examine the phase composition of coated substrates consisting of various compositions at high temperatures for a specific duration. Figure 7 displays XRD graphs of the coated LCS specimens. The XRD analysis of the coated specimen reveals the presence of various compounds, including Cr, Ti, $Cr_{1.9}Ti$, FeTi, Al_2O_3, Cr_2O_3, TiO_2, and $Cr_{1.36}Fe_{0.52}$, in addition to Fe. XRD analyses specify the presence of an oxide layer on the surface under the studied conditions. The formation of this oxide layer is attributed to the selective oxidation of chromium and titanium during the early stages of coating, likely due to their lower vapor pressures. Oxide defects on the chromized surface negatively impact the coating's resistance to corrosion and high-temperature oxidation. Additionally, Al_2O_3 particles may detach from the inert material on the surface. The grain size of the formed phases is determined using the Scherrer equation. The results are summarized in Table 4.

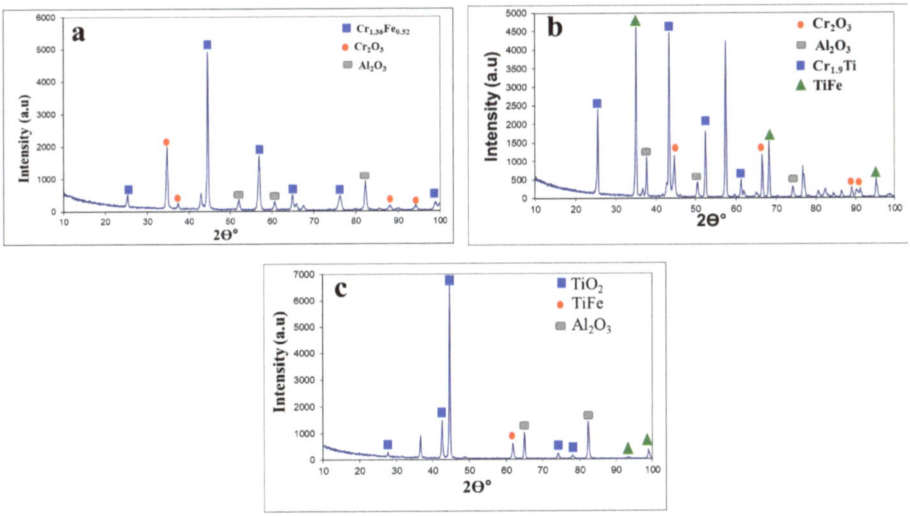

Figure 7. XRD pattern of the coated specimen for (**a**) 48% FeCr, (**b**) 36% FeCr–12% FeTi, and (**c**) 48% FeTi at 1000 °C for 5 h.

Table 4. Grain size of compounds and elements in the thin film of the surface of the coated specimen for 48% FeCr, 36% FeCr–12% FeTi, and 48% FeTi at 1000 °C for 5 h.

Type of Mixture	Compound	Grain Size (μm)	Temperature (°C)	Time (h)
48% FeCr	$Cr_{1.36}Fe_{0.52}$	3.53	1000	5
	Cr_2O_3	3.92		
	Al_2O_3	2.54		
36% FeCr–12% FeTi	$Cr_{1.9Ti}$	17.44	1000	5
	TiFe	6.69		
	Al_2O_3	5.86		
	Cr_2O_3	7.89		
48% FeTi	TiFe	3.55	1000	5
	TiO_2	4.03		
	Al_2O_3	5.92		

3.1.4. Microhardness

The microhardness was measured for five selected specimens to demonstrate the impact of the composition of the coated specimens after being subjected to 1000 °C for 5 h. Figure 8 and Table 5 demonstrate a clear correlation between the titanium percentage in the pack and the resulting increase in hardness value. The hardness value of 900 $HV_{0.01}$ was achieved, significantly higher than the base metal hardness value of 230 $HV_{0.01}$. The results present a substantial increase in hardness for different compositions of FeCr and FeTi alloys. According to the obtained data, the hardness increased by 58% for 48% FeCr, 66% for 36% FeCr–12% FeTi, 68% for 24% FeCr–24% FeTi, 71% for 12% FeCr–36% FeTi, and 74% for 48% FeTi. Based on XRD analysis, at elevated temperatures, Fe can dissolve Cr and Ti, forming intermetallic compounds (specifically TiFe and $Cr_{1.36}Fe_{0.52}$), which enhance the material's strength. The diffusion layer exhibited significantly greater hardness compared to the matrix, enhancing the wear resistance of the samples. The transition layer connected the high-hardness diffusion layer to the matrix, ensuring that the diffusion layer remained intact even under pressure. The highest microhardness achieved was 900 $HV_{0.01}$, which gradually decreased as the chromium content in the pack increased, eventually reaching 560 $HV_{0.01}$. The variations in microhardness values may be attributed to factors such as grain size and the presence or absence of certain elements and phases. This is evident from the analysis conducted using EDX and XRD, which revealed the presence of Cr and Ti in the coating's cross-section and surface.

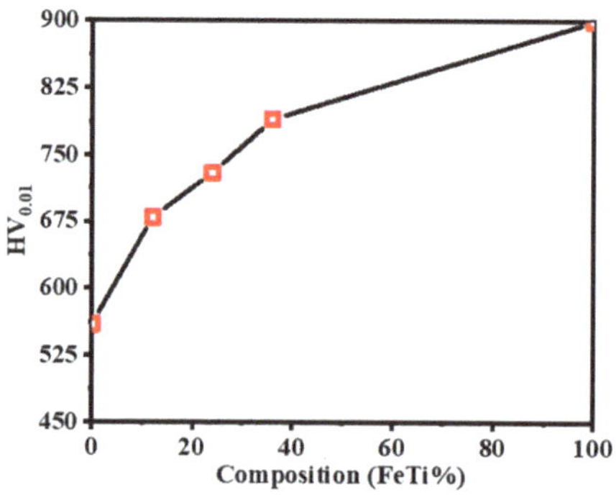

Figure 8. Cross-sectional micro-Vickers hardness profiles.

Table 5. Microhardness values of coated specimens.

Composition	48% FeCr	36% FeCr–12% FeTi	24% FeCr–24% FeTi	12% FeCr–36% FeTi	48% FeTi
$HV_{0.01}$ Kg_f	560	680	730	790	900

4. Conclusions

This research explored the use of the pack cementation technique to improve the surface properties of low-carbon steel through chromizing, titanizing, and chromotitanizing treatments. The findings reveal that both elevated process temperatures and extended durations result in the formation of a progressively thicker diffusion layer, following parabolic growth kinetics. This indicates that the process is diffusion-driven, with the mobility of Cr and Ti atoms increasing at higher temperatures, thereby facilitating the development of a more uniform and durable coating.

Microstructural analysis highlighted the creation of a distinct three-layer diffusion zone, characterized by a gradient distribution of Cr and Ti. The outermost layers were enriched with Cr- and Ti-based compounds such as Cr_2O_3, TiO_2, FeTi, and CrTi, while the

Fe content decreased further away from the low-carbon steel substrate. This compositional gradient promotes a smooth transition in mechanical properties, enhancing the coating's adhesion and minimizing the risk of delamination. The reduction in Fe content in the inner layer supports the formation of intermetallic phases, which contribute to the coating's improved mechanical properties.

The formation of intermetallic compounds like FeTi and CrTi is crucial for enhancing the hardness and wear resistance of the coated steel. The highest microhardness value, 900 $HV_{0.01}$ Kg_f, was achieved with a coating containing 48% FeTi, highlighting the superior hardness contribution of Ti-rich compounds. Conversely, coatings with higher FeCr content exhibited lower hardness, demonstrating the significant role of Ti-based phases in optimizing surface hardness.

These results suggest that the mechanical performance of low-carbon steel can be significantly enhanced through pack cementation coatings, leading to improved wear resistance, corrosion protection, and mechanical durability. The flexibility to adjust the coating composition to achieve a balance between hardness, toughness, and corrosion resistance underscores the broad applicability of this technique. In conclusion, the study establishes pack cementation as a cost-effective and efficient method for upgrading the surface characteristics of low-carbon steel, making it well-suited for use in demanding industries such as automotive, aerospace, and heavy machinery, where enhanced component longevity and performance are essential.

Author Contributions: Conceptualization, A.M.B.; Data curation, A.Y.; Formal analysis, A.M.B.; Funding acquisition, A.A.; Investigation, A.Y.; Methodology, I.M.M.; Project administration, A.Y.; Resources, I.M.M.; Software, I.M.M.; Supervision, A.M.B.; Validation, A.A.; Visualization, A.A.; Writing—original draft, A.Y. and A.M.B.; Writing—review and editing, I.M.M. and A.A. All authors have read and agreed to the published version of the manuscript.

Funding: This research was funded by The Deanship of Graduate Studies and Scientific Research, Jazan University, Saudi Arabia, through project number: GSSRD-24.

Data Availability Statement: The original contributions presented in the study are included in the article, further inquiries can be directed to the corresponding author.

Acknowledgments: The authors gratefully acknowledge the funding of the Deanship of Graduate Studies and Scientific Research, Jazan University, Saudi Arabia, through project number: GSSRD-24.

Conflicts of Interest: The authors declare no conflicts of interest.

References

1. Zhang, S.; Zhang, H.; Zhang, H.; Zhao, X.; Li, Y. Study on diffusion kinetics and law of chromium on the surface of low-carbon steel. *Coatings* **2023**, *13*, 98. [CrossRef]
2. Lin, N.; Zhao, L.; Liu, Q.; Zou, J.; Xie, R.; Yuan, S.; Li, D.; Zhang, L.; Wang, Z.; Tang, B. Preparation of titanizing coating on AISI 316 stainless steel by pack cementation to mitigate surface damage: Estimations of corrosion resistance and tribological behavior. *J. Phys. Chem. Solids* **2019**, *129*, 387–400. [CrossRef]
3. Garcia-Giron, A.; Romano, J.-M.; Liang, Y.; Dashtbozorg, B.; Dong, H.; Penchev, P.; Dimov, S.S. Combined surface hardening and laser patterning approach for functionalising stainless steel surfaces. *Appl. Surf. Sci.* **2018**, *439*, 516–524. [CrossRef]
4. Luo, Q.; Oluwafemi, O.; Kitchen, M.; Yang, S. Tribological properties and wear mechanisms of DC pulse plasma nitrided austenitic stainless steel in dry reciprocating sliding tests. *Wear* **2017**, *376*, 1640–1651. [CrossRef]
5. Kumar, H.; Ramakrishnan, V.; Albert, S.K.; Bhaduri, A.K.; Ray, K.K. Friction and wear behaviour of Ni-Cr-B hardface coating on 316LN stainless steel in liquid sodium at elevated temperature. *J. Nucl. Mater.* **2017**, *495*, 431–437. [CrossRef]
6. Alvarez-Vera, M.; Hdz-García, H.M.; Muñoz-Arroyo, R.; Diaz-Guillen, J.C.; Mtz-Enriquez, A.I.; Acevedo-Dávila, J.L. Tribological study of a thin TiO2 nanolayer coating on 316L steel. *Wear* **2017**, *376*, 1702–1706. [CrossRef]
7. Djellal, R.; Saker, A.; Bouzabata, B.; Mekki, D.E. Thermal stability and phase decomposition of nitrided layers on 316L and 310 austenitic stainless steels. *Surf. Coat. Technol.* **2017**, *325*, 533–538. [CrossRef]
8. Raudoniene, J.; Laurikenas, A.; Kaba, M.M.; Sahin, G.; Morkan, A.U.; Brazinskiene, D.; Asadauskas, S.; Seidu, R.; Kareiva, A.; Garskaite, E. Textured WO_3 and WO_3: Mo films deposited from chemical solution on stainless steel. *Thin Solid Film.* **2018**, *653*, 179–187. [CrossRef]
9. Sun, S.Q.; Ye, Y.W.; Wang, Y.X.; Liu, M.Q.; Liu, X.; Li, J.L.; Wang, L.P. Structure and tribological performances of CrAlSiN coatings with different Si percentages in seawater. *Tribol. Int.* **2017**, *115*, 591–599. [CrossRef]

10. Yang, R.; Yang, H.; Zhang, M.; Shi, X.; Qiao, J. Refractory high-entropy aluminized coating with excellent oxidation resistance and lubricating property prepared by pack cementation method. *Surf. Coat. Technol.* **2023**, *473*, 129967. [CrossRef]
11. Hwang, C.; Shinde, K.P.; Oh, J.; Lee, S.; Chung, C.H.; Park, J.S. Microstructures and high temperature oxidation behaviors of $AlMo_{0.5}NbTa_{0.5}TiZr$ high entropy alloys coated by silicon pack cementation. *Corros. Sci.* **2023**, *219*, 111203. [CrossRef]
12. Liu, S.; Ma, Z.; Liu, Y.; Tian, X.; Zhu, S.; Liu, L.; He, S.; Guangyi, M. Ablation performance of $Ta_{0.8}Hf_{0.2}C$-SiC coating fabricated via pack cementation for carbon/carbon composites. *J. Eur. Ceram. Soc.* **2024**, *45*, 117064. [CrossRef]
13. Li, F.; Zhang, K.; Wang, Z.; Ye, Y.; Li, S.; Tang, Y.; Bai, S. Ir–Zr coating through a pack cementation method for ultra-high temperature applications. *J. Mater. Res. Technol.* **2024**, *28*, 4428–4436. [CrossRef]
14. Stathokostopoulos, D.; Teknetzi, A.; Tarani, E.; Karfaridis, D.; Chrissafis, K.; Hatzikraniotis, E.; Vourlias, G. Synthesis and characterization of nanostructured Mg2Si by pack cementation process. *Results Mater.* **2022**, *13*, 100252. [CrossRef]
15. Najafizadeh, M.; Ghasempour-Mouziraji, M. Multi-objective optimization of process parameters in pack siliconizing on AISI D2 steel. *Silicon* **2021**, *13*, 2233–2242. [CrossRef]
16. Sihai, L.; Weifeng, H.; Liucheng, Z.; Xiangfan, N.; Yinghong, L. Aluminizing mechanism on a nickel-based alloy with surface nanostructure produced by laser shock peening and its effect on fatigue strength. *Surf. Coat. Technol.* **2018**, *342*, 29–36. [CrossRef]
17. Lin, N.; Guo, J.; Xie, F.; Zou, J.; Tian, W.; Yao, X.; Zhang, H.; Tang, B. Comparison of surface fractal dimensions of chromizing coating and P110 steel for corrosion resistance estimation. *Appl. Surf. Sci.* **2014**, *311*, 330–338. [CrossRef]
18. Britchi, M.; Ene, N.; Olteanu, M.; Radovici, C. Titanium diffusion coatings on austenitic steel obtained by the pack cementation method. *J. Serbian Chem. Soc.* **2009**, *74*, 203–212. [CrossRef]
19. Yang, Y.; Yan, M.F.; Zhang, S.D.; Guo, J.H.; Jiang, S.S.; Li, D.Y. Diffusion behavior of carbon and its hardening effect on plasma carburized M50NiL steel: Influences of treatment temperature and duration. *Surf. Coat. Technol.* **2018**, *333*, 96–103. [CrossRef]
20. Mejía-Caballero, I.; Palomar-Pardavé, M.; Trinidad, J.M.; Romero-Romo, M.; Pasten-Borja, R.P.; Lartundo-Rojas, L.; López-García, C.; Campos-Silva, I. Corrosion behavior of AISI 316 L borided and non-borided steels immersed in a simulated body fluid solution. *Surf. Coat. Technol.* **2015**, *280*, 384–395. [CrossRef]
21. Guoxin, H.; Zhengxia, X.; Jianju, L.; Yanhong, L. Microstructure and corrosion resistance of simultaneous Al–Fe coating on copper by pack cementation. *Surf. Coat. Technol.* **2009**, *203*, 3392–3397. [CrossRef]
22. Xiao, T.; Zhang, J.; Zhang, F.; Su, H.; Hu, J.; Guo, N. Deformation and Annealing Behavior of Cr Coating Prepared by Pack-Cementation on the Surface of Austenitic Stainless Steel. *Materials* **2024**, *17*, 3589. [CrossRef]
23. Fan, D.; Liu, X.; Huang, J.; Fu, R.; Chen, S.; Zhao, X. An ultra-hard and thick composite coating metallurgically bonded to Ti–6Al–4V. *Surf. Coat. Technol.* **2015**, *278*, 157–162. [CrossRef]
24. Fernandes, F.A.P.; Heck, S.C.; Picon, C.A.; Totten, G.E.; Casteletti, L.C. Wear and corrosion resistance of pack chromised carbon steel. *Surf. Eng.* **2012**, *28*, 313–317. [CrossRef]
25. Zeng, J.; Hu, J.; Yang, X.; Xu, H.; Li, H.; Guo, N.; Dong, Q. Microstructure and formation mechanism of the Si-Cr dual-alloyed coating prepared by pack-cementation. *Surf. Coat. Technol.* **2020**, *399*, 126142. [CrossRef]
26. Meng, L.; Shang, J.; Zhang, M.; Xie, A.; Zhang, Y. Growth Kinetics and Mechanical Properties of Rare-Earth Vanadiumizing Layer on GCr15 Steel Surface. *Coatings* **2022**, *12*, 1018. [CrossRef]
27. Liu, S.; Yang, J.; Liang, X.; Sun, Y.; Zhao, X.; Cai, Z. Investigation of the preparation, corrosion inhibition, and wear resistance of the chromized layer on the surfaces of t9 and spcc steels. *Materials* **2022**, *15*, 7902. [CrossRef] [PubMed]
28. Lou, B.Y.; Jin, L.C. Wear Behavior Of Chromizing-Titanizing Coating. *Adv. Mater. Res.* **2010**, *97*, 1364–1367. [CrossRef]
29. Matar, S.M.; Yousef, A.; Bastaweesy, A.M.; Qudsieh, I.Y.; Maafa, I.M. Enhancement in Corrosion Resistance of Low-Carbon Steel via Surface Modification. *Coatings* **2024**, *14*, 1518. [CrossRef]
30. Burton, A.W.; Ong, K.; Rea, T.; Chan, I.Y. On the estimation of average crystallite size of zeolites from the Scherrer equation: A critical evaluation of its application to zeolites with one-dimensional pore systems. *Microporous Mesoporous Mater.* **2009**, *117*, 75–90. [CrossRef]
31. Wang, Y.; Liu, D.; Feng, S.; Zhang, Y.; Ouyang, T.; Suo, J. Preparation of tritium permeation barrier consisting of titanium by the pack cementation method. *Surf. Coat. Technol.* **2016**, *307*, 271–277. [CrossRef]
32. Lee, J.-W.; Duh, J.-G.; Tsai, S.-Y. Corrosion resistance and microstructural evaluation of the chromized coating process in a dual phase Fe-Mn-Al-Cr alloy. *Surf. Coat. Technol.* **2002**, *153*, 59–66. [CrossRef]
33. William, D.; Callister, J. *Materials Science and Engineering: An Introduction*, 7th ed.; John Wiley & Sons, Inc.: Hoboken, NJ, USA, 2006.

coatings

MDPI

Article

Enhanced Protective Performance of Carbon Nanotube-Reinforced Waterborne Epoxy Zinc-Rich Coatings for Corrosion Protection of Steel Structures

Xueling Huang [1], Chun Yang [1], Junyu Chen [1], Xinglong Qiao [2], Shuguang Zhang [2,*] and Dan Song [2,*]

1 State Grid Wuxi Power Supply Company, Wuxi 214000, China; hxl_wx@sina.com (X.H.); yangchun@js.sgcc.com.cn (C.Y.); chen_junyu1986@126.com (J.C.)

2 College of Materials Science and Engineering, Hohai University, Nanjing 211100, China; 221608010037@hhu.edu.cn

* Correspondence: shuguang0502@163.com (S.Z.); songdancharls@hhu.edu.cn (D.S.)

Abstract: Carbon nanotube-reinforced waterborne epoxy zinc-rich coatings were developed by modifying waterborne epoxy zinc-rich formulations with varying amounts of carbon nanotubes (CNTs), to improve the coatings' corrosion resistance and their protection for Q355b steel in environments rich in chlorides. A detailed investigation of the microstructural changes in the coatings prior to and following corrosion was conducted through FTIR, SEM, XRD, and XPS analytical techniques. The effectiveness of these innovative coatings in providing corrosion protection for Q355b steel in chloride conditions was assessed via electrochemical corrosion methodologies and neutral salt spray testing. The results indicate that an increase in the CNT concentration led to an initial enhancement in the corrosion resistance of the coatings, followed by a decrease, with optimal performance noted at 0.3 wt.% CNTs. During the electrochemical evaluations, the open circuit potential (OCP) of the coating containing 0.3 wt.% CNTs remained stable below the critical threshold of −0.78 V for an extended period, indicating sustained cathodic protection. In comparison to the coatings with CNT concentrations (wt.%) of 0.1, 0.5, 0.7, and 1.0, the coating with 0.3 wt.% CNTs demonstrated the lowest corrosion current density, measured at 0.0322 $\mu A/cm^2$. Further validation of its exceptional corrosion resistance was provided by the 240 h neutral salt spray tests. This performance can be linked to the capability of the CNTs to improve electrical conductive connectivity between the zinc particles and the Q355b steel substrate beneath them, subsequently enhancing both the cathodic protection of the coating and its physical shielding effectiveness.

Keywords: waterborne epoxy zinc-rich coatings; carbon nanotubes; corrosion resistance; corrosion mechanism; cathodic protection

Citation: Huang, X.; Yang, C.; Chen, J.; Qiao, X.; Zhang, S.; Song, D. Enhanced Protective Performance of Carbon Nanotube-Reinforced Waterborne Epoxy Zinc-Rich Coatings for Corrosion Protection of Steel Structures. *Coatings* **2024**, *14*, 1493. https://doi.org/10.3390/coatings14121493

Academic Editor: Francesco Di Quarto

Received: 6 November 2024
Revised: 23 November 2024
Accepted: 25 November 2024
Published: 27 November 2024

1. Introduction

Zinc-rich coatings are widely applied in various corrosive settings owing to their simple preparation method and remarkable resistance to corrosion. They have become a primary protective measure for engineering structures [1–3]. Waterborne zinc-rich coatings are particularly popular due to their minimal volatile organic compound emissions and eco-friendly characteristics. Specifically, waterborne epoxy zinc-rich variants not only mitigate environmental pollution when compared to traditional organic coatings, but also exhibit enhanced bonding strength and corrosion resistance in practical use [4–6]. Nevertheless, these waterborne epoxy zinc-rich coatings encounter certain performance issues: (1) A significant quantity of zinc powder is added during the formulation stage, leading to settling and agglomeration of the coatings, which can compromise their effectiveness, necessitating rapid application and thereby presenting hurdles to achieving high-quality and efficient coating application. (2) The coatings are microscopically porous, exhibiting inadequate shielding capabilities. The high concentration of zinc powder in waterborne epoxy coatings results in insufficient epoxy resin to adequately fill the spaces between the zinc particles, creating a porous structure. In scenarios requiring prolonged immersion in corrosive environments, harmful agents can infiltrate the coating through these pores and reach the substrate beneath the coatings, resulting in protection failure and corrosion damage.

Traditional epoxy zinc-rich coatings typically contain a zinc concentration exceeding 70%. However, as the zinc powder undergoes oxidation, byproducts like zinc oxide can impede the electrical conductive connections between the zinc particles and the formation of micro-galvanic cells, which compromises the coating's cathodic protection for the substrate [7–15]. When the zinc concentration surpasses 80 wt.%, they provide excellent cathodic protection due to the effective electrical conductive connections between the zinc particles; however, this results in a considerable amount of unused zinc powder within the coatings, leading to resource wastage and potential environmental issues. Additionally, elevated zinc concentrations adversely impact the leveling properties, density, adhesion, and flexibility of the coatings [2,16–19]. Orek et al. have performed theoretical and experimental analyses of the electrical, optical, and structural properties of wurtzite (similar to zinc oxide (ZnO)) nanostructures. The effect of quantum constraints on optical properties and two different ZnO clusters in the nanowire structure were investigated [20].

Furthermore, much of the research surrounding zinc-rich coatings has concentrated on the comparison between spherical and flake zinc powders. A large amount of research on zinc-rich coatings focuses on the selection and application of spherical zinc powder or flake zinc powder. Compared with spherical zinc powder, flake zinc powder has a larger specific surface area and higher surface activity, and the prepared zinc-rich coatings have better corrosion resistance. However, flaky zinc powder has some problems, such as poor dispersion and easy corrosion in water-based coating systems [17,21–24].

Numerous studies have identified the incorporation of fillers into zinc-rich coatings as one of the most economical and effective approaches currently available. Innovative conductive fillers such as CNTs and graphene have been utilized by researchers to partially replace zinc powder in the formulation of nano-composite coatings, showing considerable potential for various applications [21,23–27]. CNTs are known for their remarkable conductivity and nanoscale effects. Their incorporation into coatings can reduce internal porosity, lengthen the route for corrosive agents to access the steel substrate, and facilitate electron transfer during the cathodic protection stage. Moreover, CNTs possess electrochemical characteristics. Their potential surpasses that of the metal substrate, which may result in passivation of the metal, thus improving the substrate's corrosion resistance [28–30]. Park et al. utilized a ball milling dispersion method to first disperse multi-walled carbon nanotubes (MWCNTs) in xylene, to develop MWCNT-enhanced epoxy zinc-rich anti-corrosion coatings and explore the effect of MWCNTs on the coatings' performance [4]. Similarly, Wang et al. noted that incorporating CNTs into epoxy coatings with a low zinc concentration improves their ability to resist corrosion. Additionally, incorporating materials such as graphene and graphene oxide into these zinc-rich epoxy coatings significantly improves their ability to resist corrosion [11,16,21,22,31–34].

In summary, the integration of CNTs significantly enhances the electrical conductive connection among zinc particles and the steel substrate compared to zinc-rich coatings, thereby improving protective characteristics of the coatings. This study primarily focuses on a commercially available waterborne epoxy zinc-rich coating, which is Barrier 80 WF (abbreviated as 80 WF) from Jotun company, and enhances it by incorporating multi-walled carbon nanotubes, to improve its corrosion protection for Q355b steel substrate. A thorough array of evaluations were performed, including tests for pencil hardness, adhesion, impact resistance, electrochemical analysis, SEM, EDS, XPS, XRD, and neutral salt spray, to analyze the structure, composition, and corrosion resistance performance of waterborne zinc-rich coatings containing varying amounts of multi-walled carbon nanotubes. Furthermore, the research examined the anti-corrosion mechanisms associated with these CNT-reinforced waterborne epoxy zinc-rich coatings. The results provide valuable insights for the development of high-performance waterborne epoxy zinc-rich coatings with corrosion protection for Q355b steel substrate.

2. Materials and Methods

2.1. Materials and Preparation of Coatings

The main material utilized was a waterborne coating (80 WF) with a zinc concentration of 70 wt.%, while Q355b steel was chosen as the substrate. Table 1 provides information regarding the chemical composition of Q355b. Following preliminary experiments involving carbon nanotubes and waterborne zinc-rich coatings, multi-walled carbon nanotubes (rolled from multiple layers of graphene) were selected as the filler for modification due to their corrosion resistance and relatively lower cost compared to single-walled carbon nanotubes (rolled from a single layer of graphene). The constituents of the 80 WF utilized in these experiments are outlined in Table 2. To examine how the concentration of multi-walled carbon nanotubes affects the coating's performance, waterborne zinc-rich coatings with different multi-walled carbon nanotube concentration (wt.%) of 0.1, 0.3, 0.5, 0.7, and 1.0 were formulated. The scheme for preparing coatings is depicted in Figure 1. 80 WF is a two-component waterborne epoxy zinc-rich coating cured with polyamine adducts. During the preparation process, components A and B of the commercial 80 WF were firstly combined in a beaker, and then agitated at a speed of 500 r/min to form a thick solution. Afterward, deionized water was added to achieve the desired dilution, and the resulting mixture was stirred at 1000 r/min for a duration of 20 min to guarantee complete dispersion. The uniformly dispersed coating was then applied onto Q355b samples of varying sizes in accordance with different testing standards and cured in an oven at 40 °C for 5 days to create the 80 WF coatings. In a beaker, a mixed solvent with different concentrations of multi-walled carbon nanotubes and dispersants underwent ultrasonic dispersion for 5 min, which was then followed by a low-speed stirring at a rate of 300–500 r/min until homogeneity was reached. Another portion of the previously mentioned uniformly dispersed 80 WF was combined with the multi-walled carbon nanotube solvent and stirred at 500 r/min for 20 min. The coating was subsequently applied to Q355b samples with sizes of 10 mm × 10 mm × 2 mm for electrochemical tests and 50 mm × 50 mm × 3 mm for salt spray tests, and then cured in an oven at 40 °C for a duration of 5 days, leading to the formation of waterborne zinc-rich coatings reinforced with carbon nanotubes (CNTs/80 WF).

Table 1. Chemical composition of Q355b steel.

Element	C	Si	Mn	S	P	Cr	Ni	Cu
wt.%	≤0.24	≤0.55	≤1.60	≤0.035	≤0.035	≤0.24	≤0.30	≤0.30

Table 2. Basic components of 80 WF zinc-rich coating.

Component	Material	wt%
A	Zinc Powder	≥75–≤90
	1-Methoxy-2-Propanol	≤10
	Epoxy Resin (MW 700–1200)	≤5
	Epoxy Resin (MW < 700)	≤4.6
	Zinc Oxide	≤3
	γ-Propyltrimethoxysilane	<3
B	Polyether Polyamine	<25
	1-Methoxy-2-Propanol	≤5
	3-Aminomethyl-3,5,5-Trimethylcyclohexylamine	<2.5
	a,a′-Diaminodimethylbenzene	<2.5
	Sodium Nitrite	≤0.3

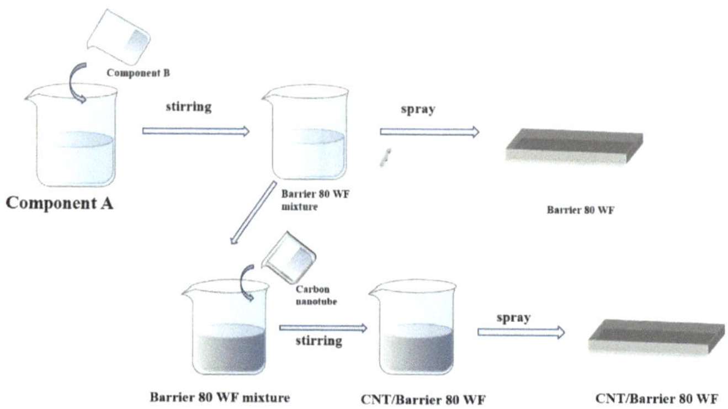

Figure 1. Scheme of preparation of 80 WF and CNTs/80 WF coatings.

2.2. Mechanical Test and Thickness Measurements of Coatings

In accordance with ISO 12944 of "Paints and varnishes-Corrosion protection of steel structures via protective paint systems", various tests for hardness and scratch resistance were carried out on multiple coatings. To assess the thickness of the coatings, an ultrasonic thickness gauge was utilized.

2.3. FTIR Analysis of Coatings

Following the curing of the coatings in an oven at 40 °C for five days, an FTIR test was conducted on the coatings with the Thermo Fisher Scientific Nicolet iS20 instrument (Sydney, NSW, Australia). The samples were prepared utilizing the potassium bromide pellet technique. A strong Globar infrared light source delivered substantial energy. The spectrum's scanning range was set from 4000 cm^{-1} down to 400 cm^{-1}.

2.4. SEM Analysis of Coatings

The morphologies of the sample surface were examined with secondary electron signal imaging by using SEM. This research employed a Sigma 300 SEM (ZEISS, Oberkochen, Germany), operating at a barrel acceleration voltage of 15 kV, achieving a secondary electron image resolution of 0.8 nm. The microscopic structures of the waterborne zinc-rich coatings were analyzed both prior to and following corrosion.

2.5. EDS Analysis of Coatings

In this research, an analysis of different waterborne zinc-rich coatings was performed using the Oxford Ultim Extreme EDS system with SEM capabilities to evaluate the elemental distribution and content both prior to and following corrosion.

2.6. Electrochemical Tests

In this research, various samples of waterborne zinc-rich coatings were immersed in a 3.5 wt.% NaCl solution at approximately 25 °C over a timespan of 28 days. The experimental methods used encompassed open circuit potential (OCP) measurements, electrochemical impedance spectroscopy (EIS), and scanning potentiodynamic polarization (PDP). During the evaluation process, a saturated calomel electrode (SCE) was inserted into the solution and linked to the reference electrode clamp of the electrochemical workstation, while the wire end of the coated sample was connected to the clamp of the working electrode. Once the experiments commenced, EIS and PDP readings were documented once the open circuit potential fluctuations stabilized to below 10 mV. The EIS measurements spanned a frequency range from 10^{-2} to 10^{5} Hz, employing a sine wave signal with an amplitude of

10 mV and collecting data at ten distinct points per decade. Once the tests were conducted, the gathered data were retained, enabling the creation of Nyquist and Bode plots. The impedance data collected were examined with ZSimpWin 3.6 software to understand the corrosion mechanisms involved. To ensure reliable experimental outcomes, three parallel samples of each type of coating were tested.

2.7. Neutral Salt Spray Test

In this study, multiple samples of zinc-rich coatings with size of 50 × 50 × 3 mm underwent a cross-scratching procedure labeled "X". Epoxy resin was used to seal both the edges and the backs of the samples. After the preparation, the samples were placed in a salt spray chamber for the evaluation of neutral salt spray, using a corrosion solution composed of 5 wt.% NaCl and a pH ranging from 6.5 to 7.2. The temperature within the salt spray chamber was maintained at 35 °C, and the air pressure from the pump was controlled to remain between 0.1 and 0.2 MPa. Each set of samples was made up of three parallel specimens, all subjected to uninterrupted spraying for an overall duration of 240 h, during which detailed observations and records of surface changes were carried out.

2.8. XRD Analysis of Coatings

In this study, a Rigaku SmartLab SE XRD instrument (manufactured in Japan, Rigaku Holdings Corporation, Tokyo, Japan) equipped with a copper target was employed, which has a wavelength of 1.5418 Å, operates at a voltage of 40 kV, and a current of 40 mA, while scanning across a range from 5° to 90°. The gathered data were analyzed using Jade 6 software to investigate the changes in the materials located on the surfaces of waterborne zinc-rich coatings, both before and after corrosion.

2.9. XPS Analysis of Coatings

In this study, a device from Thermo Scientific (Sydney, NSW, Australia), namely the K-Alpha XPS (USA), was employed. The X-ray source utilized was a microfocused monochromatic Al Kα source, which allowed for an adjustable spot size between 30 and 400 μm, with a stepping increment of 5 μm. Two scanning methods were executed: a full spectrum scan that spanned an energy range of 100 eV at 1 eV intervals, and a narrow spectrum scan with an energy range of 30 to 50 eV, incrementing by 0.05 to 0.1 eV. To ensure accurate calibration, the reference contaminant C1s (284.8 eV) was used, and the gathered data were analyzed with Avantage 5.9 software to enable peak fitting.

3. Results

3.1. Mechanical Properties of 80 WF Coating and CNTs/80 WF Coatings with Different CNTs Concentration

The test results of the mechanical properties for the coatings are detailed in Table 3. The test of each coating complied with ISO standards regarding flexibility and impact resistance. Initially, as the concentration of CNTs increases, both the hardness and bonding strength of the coatings improve, but they eventually start to decrease. A maximum pencil hardness of 4 H was obtained at a CNTs concentration of 0.3 wt.%, significantly exceeding the national standard of 3 H. This enhancement is largely due to the intrinsic hardness of the CNTs, which elevates the overall hardness of the coating. On the other hand, an overly high concentration of CNTs can lead to agglomeration within the coating, which may decrease hardness, especially in the coating with 1.0 wt.% CNTs. Furthermore, due to the nanoscale properties of CNTs, these materials can efficiently fill the gaps created by steric hindrance between zinc powder and epoxy resin, thus improving the adhesion between layers. However, if the CNT concentration surpasses a specific limit, the agglomeration of these nanoparticles may create localized stress concentrations within the coating, ultimately leading to a decrease in hardness.

Table 3. Mechanical properties of coatings.

Coating Test Items	Coating Type					
	80 WF	0.1 wt.%	0.3 wt.%	0.5 wt.%	0.7 wt.%	1.0 wt.%
Pencil hardness test	2 H	3 H	4 H	3 H	3 H	2 H
Impact testing of coatings (cm)	50	50	50	50	50	50

3.2. FTIR Analysis of 80 WF Coating and CNTs/80 WF Coating

The Fourier transform infrared spectra of the 80 WF and CNTs/80 WF coatings are depicted in Figure 2. The peaks observed at 3426 cm^{-1} are generally linked to the stretching vibration of O-H bonds, potentially originating from hydroxyl groups in water or polyether polyamines. At 2928 cm^{-1}, the spectra reveal a stretching vibration of the C-H bond, particularly linked to both methyl and methylene groups. The peaks observed at 1609 cm^{-1} could denote the C=C stretching vibration, which is commonly found in unsaturated compounds, including the principal monomers used in epoxy resins. At 1506 cm^{-1}, the peaks might indicate C=C stretching vibrations occurring in aromatic rings, which may originate from the fundamental structure of the epoxy resin or compounds such as a,a′-diaminodimethylbenzene. Furthermore, the peaks situated at 1240 cm^{-1} are generally attributed to C-N stretching vibrations, likely arising from amines like polyether polyamines or aminomethyl substances such as 3-amino-3,5,5-trimethylcyclohexylamine. The peaks found at 1183 cm^{-1} and 1098 cm^{-1} may be associated with C-O-C bond stretching vibrations presented in epoxy resins as well as in 1-methoxy-2-propanol. The peaks at 827 cm^{-1} probably correspond to C-H bending vibrations from ring structures or distinctive vibrations attributed to certain substituents, particularly within the frameworks of aromatic rings. Lastly, the peaks at 550 cm^{-1} in the lower wavenumber region signifies Zn-O stretching vibrations. The FTIR spectra for both coatings reveal that incorporating CNTs does not lead to considerable alterations in the peak positions, although some peak areas for the CNTs/80 WF coating appear reduced. These findings imply that the modification of the coatings with CNTs is mainly a physical doping process, and the carbon nanotubes do not react chemically with the components of the coating.

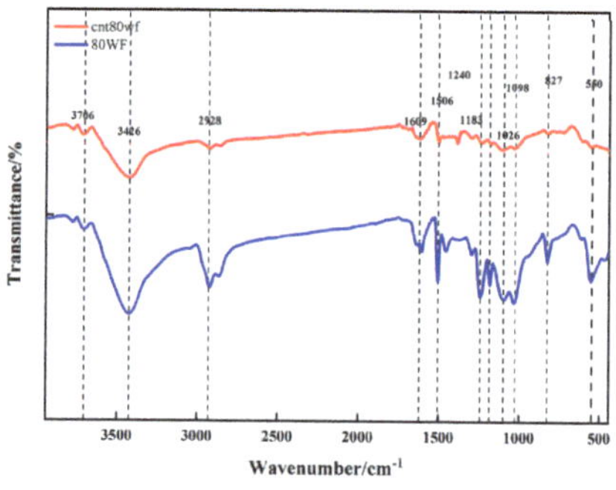

Figure 2. Fourier transform infrared spectra of 80 WF and CNTs/80 WF coatings.

3.3. Surface Morphologies and EDS Analysis of 80 WF Coating and CNTs/80 WF Coatings

An SEM image of the original multi-walled carbon nanotubes is shown in Figure 3. Figures 4–6 depict the surface morphologies and EDS spectra of coatings without CNTs (80 WF), with 0.3 wt.% CNTs and 1.0 wt.% CNTs, respectively. In Figure 4, it is evident that the coating's surface comprises a considerable quantity of spherical zinc particles. The film-forming agents, such as epoxy resin, are capable of adequately encapsulating the zinc particles; however, the coating's surface exhibits irregularities, with the spherical zinc particles distributed unevenly. The zinc particle is isolated and cannot be well connected, and the overall coating is not too dense. Conversely, Figure 5 demonstrates that the incorporation of CNTs leads to their stable adhesion to the coating surface, thereby enhancing the coating's density. This implies that the CNTs occupy the gaps between the zinc particles and epoxy resin, contributing to a shielding effect. The morphology of the surface suggests that the CNTs form connections with the zinc particles, thereby creating a conductive path between neighboring spherical zinc particles. However, as illustrated in Figure 6, a rise in the CNT content results in a markedly uneven distribution and clustering of CNTs on the coating surface containing 1.0 wt.% CNTs, potentially compromising the coating's properties and its resistance to corrosion. Surface analyses conducted using EDS on all three coatings reveal that the concentrations of Zn, C, and O elements are relatively elevated, with these elements uniformly distributed across the coating's surface.

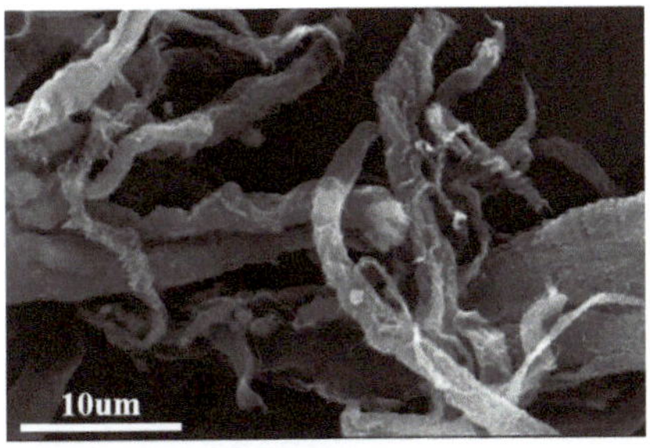

Figure 3. SEM image of the original CNTs.

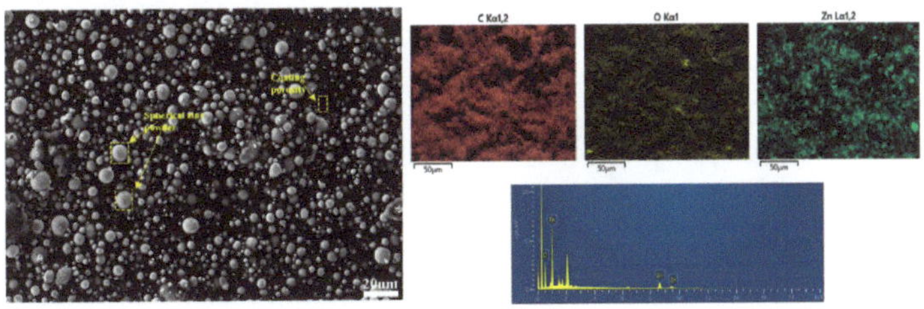

Figure 4. SEM and EDS of 80 WF.

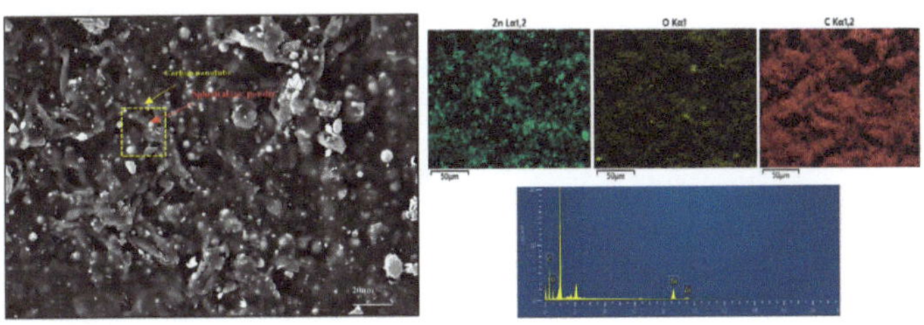

Figure 5. SEM and EDS of 0.3 wt.% CNTs/80 WF.

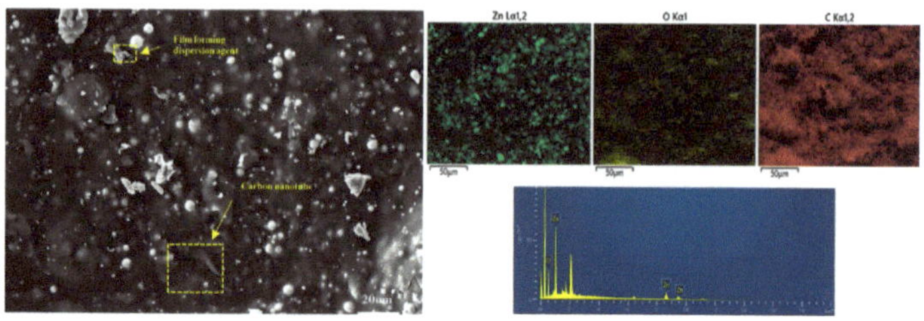

Figure 6. SEM and EDS of 1.0 wt.% CNTs/80 WF.

3.4. OCP, EIS, and PDP Tests of 80 WF Coating and CNTs/80 WF Coatings with Different CNTs Concentration

The 80 WF coating and coatings with five different concentrations of CNTs were subjected to electrochemical testing at different time intervals after immersion. The performance of coatings rich in zinc for cathodic protection is typically assessed by measuring their OCP. When the OCP drops below −0.78 V, it signifies that the coating still possesses a considerable ability to provide cathodic protection to steel substrate. In contrast, an increase in the OCP beyond −0.78 V signifies a decline in the coating's ability to offer cathodic protection, indicating a transition towards a shielding protection phase. Figure 7 demonstrates that with an increase in immersion time, each coating exhibits a gradual rise in OCP. The figure clearly shows that the coating without CNTs (80 WF) and the coating incorporating 0.1 wt.% CNTs offer minimal to no cathodic protection, as indicated by their OCP values staying above −0.78 V. In contrast, the 0.3 wt.% CNTs coating maintains cathodic protection for the maximum duration of 14 days. Nevertheless, a rise in the CNTs concentration results in a shortened period of cathodic protection. A moderate quantity of CNTs improves the conductive connections among zinc particles and the steel substrate, thereby promoting the formation of a strong electrical conductive pathway. Consequently, this improves the efficiency of cathodic protection and increases the corrosion resistance of the coating. Conversely, an overabundance of CNTs leads to a decreased lifespan of cathodic protection, primarily due to the swift removal of spherical zinc particles from the surface of the coating.

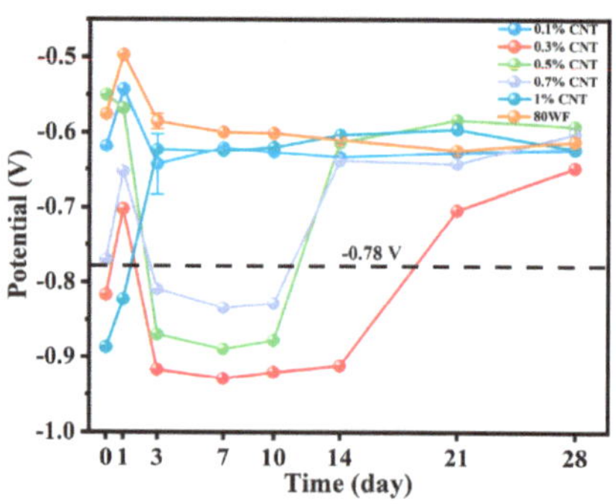

Figure 7. Open circuit potential (OCP) curves of 80 WF and waterborne zinc-rich coatings with 0.1 wt.%–1 wt.% CNTs after 28 days of immersion.

The electrochemical impedance spectra (EIS) of the coatings are presented in Figures 8 and 9. These spectra have been analyzed and fitted using the equivalent circuit shown in Figure 10.

Within this circuit, R_s represents the resistance of the corrosion electrolyte. Furthermore, R_c indicates the resistance of the coating that is applied to the surface. Rather than utilizing a traditional capacitance for the coating, a Constant Phase Element (CPE) is used to more effectively represent the non-ideal behavior in the system, and R_{ct} denotes the charge transfer resistance. Additionally, Warburg impedance is utilized to correct errors associated with diffusion behavior. Examination of the Nyquist plots presented in Figure 8(a1–c1) indicates that the CNTs/80 WF coatings, apart from the 1.0 wt.% CNTs coating, exhibit a larger initial capacitive arc radius in comparison to the unmodified 80 WF coating. Among the six coatings, the coating with 0.3 wt.% CNTs consistently exhibits a larger capacitive arc radius compared to the other five coatings during the same immersion period [34]. Moreover, an increase in the concentration of CNTs from 0.5 wt.% to 1.0 wt.% is associated with a decrease in the capacitive arc radius. As illustrated in Figure 11, the R_{ct} values of the coatings initially rise before experiencing a decrease. The coating incorporated with 0.3 wt.% CNTs exhibits the highest R_{ct} value of 3756, suggesting that its resistance to corrosion is superior to that of the other five coatings. The impedance modulus of the coating with 0.3 wt.% CNTs is observed to be significant at low frequencies. This finding indicates that the coating with 0.3 wt.% CNTs exhibits superior corrosion resistance when compared to the other coatings analyzed in the study. This enhanced performance suggests that incorporating this specific concentration of CNTs into a coating formulation may effectively improve its protective efficiency against corrosion, thereby potentially increasing the longevity and durability of materials subjected to corrosive environments.

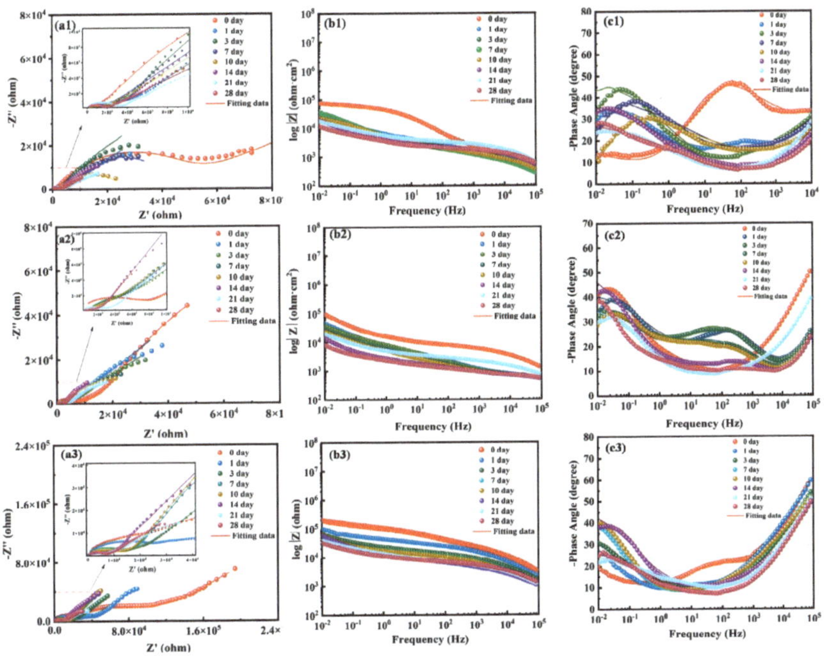

Figure 8. Nyquist plot and Bode plot of electrochemical impedance spectroscopy (EIS) for the 80 WF coating and waterborne zinc-rich coatings with varying CNT content after 28 days of immersion: (**a1–c1**) representing 80 WF, (**a2–c2**) indicating 0.1 wt.% CNTs, and (**a3–c3**) showing 0.3 wt.% CNTs.

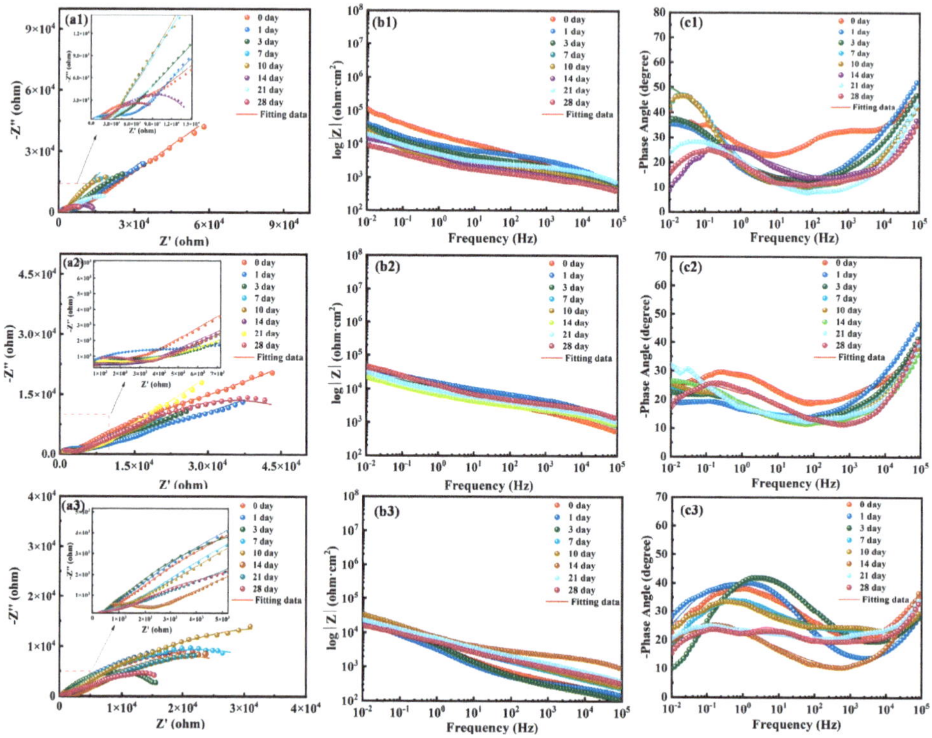

Figure 9. Nyquist plot and Bode plot of EIS for zinc-rich coatings dispersed in water, containing varying amounts of CNTs, examined after 28-day immersion: (**a1–c1**) with 0.5 wt.% CNTs, (**a2–c2**) containing 0.7 wt.% CNTs, and (**a3–c3**) featuring 1.0 wt.% CNTs.

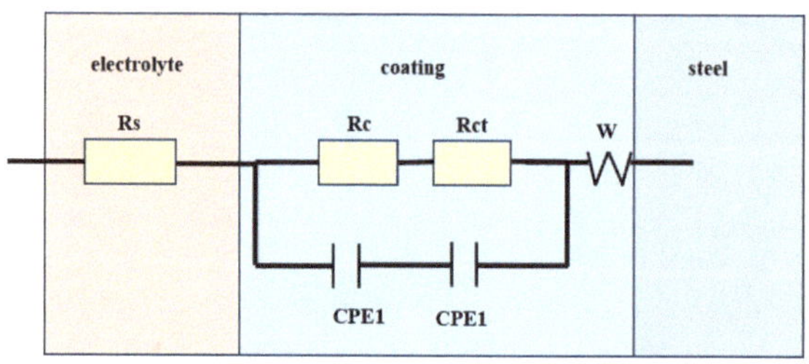

Figure 10. Equivalent circuit diagram for EIS fitting of immersed coatings.

All six coatings demonstrate a notable shift towards higher frequency values with prolonged immersion time, suggesting a decline in their effectiveness of physical protection. This shift represents a transition from sufficient physical protection to potential failure. In addition, trends analogous to those observed in the previous EIS results are present among the six coatings, highlighting that the diffusion of corrosive media in the coating containing 0.3 wt.% CNTs is restricted. This additionally supports the idea that the coating with 0.3 wt.% CNTs provides improved physical shielding capabilities and greater corrosion resistance in comparison to the other formulations [7,35–39].

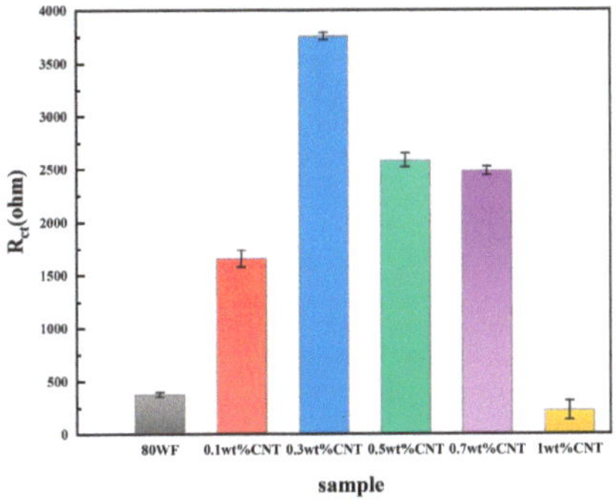

Figure 11. The fitted Rct values for different coatings after 28 days of immersion.

Specifically, a lower self-corrosion current density is indicative of a decreased rate of corrosion, suggesting enhanced corrosion resistance. The findings regarding various coatings after an immersion duration of 28 days are visually represented in Figure 12, which displays the polarization curves corresponding to each coating. The detailed parameters, including corrosion potential (E_{corr}), corrosion current density (i_{corr}), anodic Tafel slope (β_a), and cathodic Tafel slope (β_c) associated with these curves can be found in Table 4. The data presented in the figure reveal that the self-corrosion current density for coatings modified with different concentrations of CNTs initially increases before subsequently decreasing. Table 4 provides the specific parameters associated with these curves. The information depicted in this figure suggests that the self-corrosion current density for coatings modified with different concentrations of CNTs initially increases and then ultimately decreases. Notably, when a concentration of 0.3 wt.% of CNTs is added, the resultant coating demonstrates the lowest self-corrosion current density of 0.0322 $\mu A/cm^2$. This finding indicates that this particular composition offers the most effective corrosion resistance, establishing an optimal balance in the application of CNTs for improving coating performance.

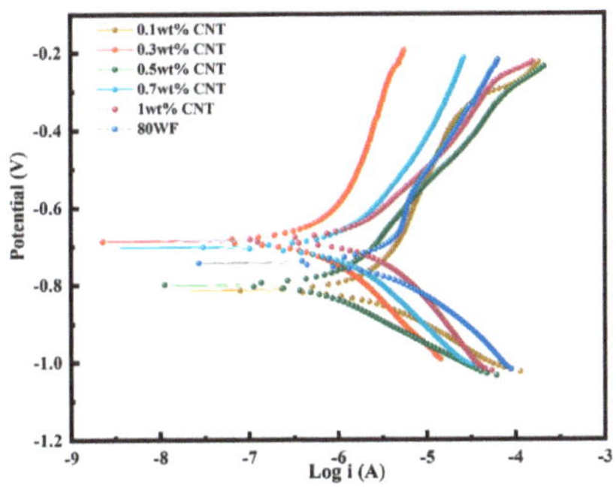

Figure 12. Tafel plots were generated for zinc-rich coatings and 80 WF, incorporating various contents of CNTs (0.1 wt.%–1.0 wt.%) following 28-day immersion.

Table 4. The relevant parameters of the Tafel plots for 80 WF and zinc-rich coatings with different CNT contents.

CNT Concentration (wt.%)	**0.1**	**0.3**	**0.5**	**0.7**	**1.0**	**0**
Corrosion potential E_{corr} (v)	−0.813	−0.685	−0.799	−0.702	−0.684	−0.743
Corrosion current density I_{corr} (A/cm^2)	0.086	0.032	0.054	0.070	0.084	0.155
Anodic Tafel slope β_α	8.090	5.520	5.530	5.170	7.580	7.370
Cathodic Tafel slope β_c	13.570	7.540	7.120	6.520	9.540	10.580

The primary explanation for the previously discussed outcomes could be associated with the ability of the epoxy resin to absorb moisture, leading to its expansion. The expansion, in turn, obstructs the penetration of harmful substances, serving as an effective barrier against the entry of solution. As the immersion period is extended, the zinc-rich coating engages with corrosive substances, allowing zinc particles to provide cathodic protection for the underlying Q355b steel substrate. This engagement generates corrosion byproducts that can physically limit harmful agents from reaching the Q355b steel substrate.

CNTs exhibit remarkable electrical conductivity and possess nanoscale properties. Incorporating CNTs into the 80 WF coating significantly decreases the shrinkage rate during the curing process of the coating, which consequently reduces the formation of micro-pores and effectively seals any pores that were already present. This process slows the infiltration of water and conductive ions within the coating, extending the distance that the corrosive media must travel to reach the Q355b steel substrate. Furthermore, the evenly distributed CNTs can act as electrical conductive pathways in the coating, connecting the spherical zinc particles, which promotes electron transfer during the cathodic protection phase. As the CNTs interconnect the isolated zinc particles, the efficiency of activated zinc powder utilization is improved, resulting in the development of denser or thicker zinc-rich corrosion products within the 0.3 wt.% CNTs coating. This enhances cathodic protection for the Q355b steel and improves the coating's corrosion resistance. During a prolonged immersion duration, the Bode plots indicate a steady decrease in the impedance values, implying either an increase in the size or the quantity of pores that enhance the movement of the corrosive electrolyte. This trend accelerates the corrosion rate at the interface and may prompt the formation of new diffusion layers. Potential degradation of the coating may ensue, leading to a gradual diminishment of its protective performance.

As the quantity of CNTs incorporated into the coating rises, issues such as uneven dispersion and agglomeration of CNTs may arise, which can reduce the coating's performance. Additionally, increased local concentrations of CNTs can elevate the conductivity in these agglomerated regions, leading to a more rapid reaction of the zinc powder at these sites. As a result, the duration of physical shielding offered by zinc corrosion products might be reduced, which could adversely affect the coating's resistance to corrosion. For example, coatings incorporating 0.7 wt.% and 1.0 wt.% CNTs show that a high concentration of CNTs on their surfaces initiates a swift reaction with zinc powder, resulting in a reduced overall protective lifespan of the coating. Furthermore, a considerable amount of CNTs could result in notable interactions between the steel substrate and the CNTs. A significant corrosion current may develop on the coating's surface, which can lead to severe galvanic corrosion that potentially accelerates the degradation of the steel substrate and compromises the cathodic protection provided by the coating. This effect is supported by the observation of surface protrusions on the 1.0 wt.% CNTs coating after undergoing electrochemical testing, demonstrating that an excess of CNT concentration promotes galvanic corrosion with the Q355b steel substrate, contributing to internal rust formation within the substrate, ultimately diminishing its protective qualities.

3.5. Neutral Salt Spray Test of 80 WF Coating and CNTs/80 WF Coatings with Different Carbon Nanotube Concentrations

After conducting a 240 h evaluation with salt spray test, the results regarding the conventional 80 WF coating and the CNTs/80 WF coatings with different CNTs concentration are illustrated in Figure 13. The data indicate that following the 240 h salt spray test, the 80 WF-coated specimen exhibits some blistering, although there is no significant color change. However, substantial rust is observed in the scratched areas, with corrosion products extending laterally and the corrosion depth at the scratch surpassing 1 mm. For the CNTs/80 WF coatings, a higher CNTs concentration correlates with a gradual increase in the coatings' resistance to salt spray. In particular, for the coating with 0.3 wt.% CNTs, the development of white zinc corrosion byproducts progresses slowly on the surface, while minimal rust appears in the scratched regions. This finding implies that the coating maintains a certain level of protective capability and exhibits commendable corrosion resistance. Conversely, when the CNT concentration is raised, a rapid emergence of white zinc on the surface of the 1.0 wt.% CNTs/80 WF coating is noted, leading to considerable corrosion of the underlying substrate. An excess of CNTs suggests a rapid reduction in zinc levels, thereby diminishing the efficacy of cathodic protection. Furthermore, the aggregation of numerous CNTs may adversely affect the crucial properties of 80 WF, potentially accelerating the corrosion of the steel substrate. To summarize, an optimal concentration of CNTs in the coating not only serves as a protective filler but also creates a bridging pathway between zinc particles, thereby enhancing the reactivity and efficient use of zinc.

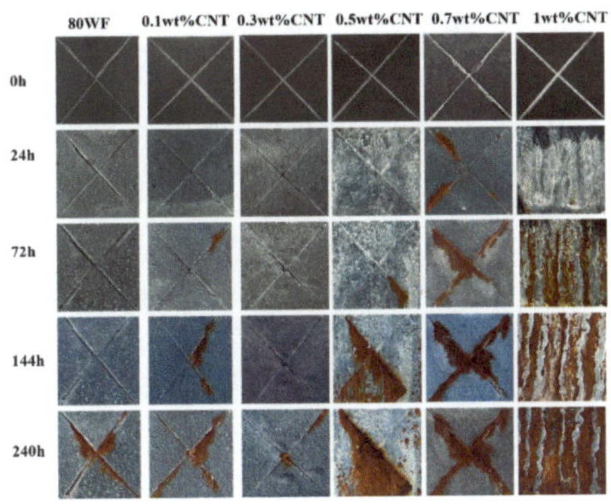

Figure 13. Samples of 80 WF coating and CNTs/80 WF zinc-rich coatings with different CNT contents after 240 h of salt spray test.

3.6. Surface Morphologies and Elemental Changes of 80 WF Coating and CNTs/80 WF Coatings After Corrosion

Figures 14–16 display SEM images and EDS spectra that demonstrate the surfaces of traditional 80 WF coatings as well as those incorporated with 0.3 wt.% and 1.0 wt.% CNTs after immersion. The visuals indicate that after immersion, the surface of the 80 WF coating displayed irregular features. The clustering of zinc powder alongside corrosion byproducts is distinctly observable, alongside a rise in pore dimensions and a reduction in coating density. In Figure 15, the surface morphology of the coating with 0.3 wt.% CNTs post-immersion is shown, highlighting a layered structure of flaky corrosion products on the coating. This layer provides a relatively dense layer across the surface. The EDS analysis conducted in this region indicates that the flaky corrosion byproducts primarily comprise zinc corrosion compounds, including ZnO, $Zn(OH)_2$, and $Zn_5(OH)_8Cl_2$, along with unreacted zinc powder. Furthermore, almost no Fe can be found on the coating surface, suggesting that the substrate stays uncorroded and the coating provides significant protection. The findings indicate that adding CNTs enhanced the conductive connection between the spherical zinc particles and the Q355b steel substrate, thereby creating a continuous conductive pathway among the zinc powder. This enhancement results in the more effective use of zinc and supports the formation of zinc corrosion byproducts. The flaky products produced through corrosion fill the pores in the coating, hindering the penetration of corrosive agents to the substrate. As a result, this improves the effectiveness of cathodic protection along with the coating's barrier properties, leading to a significant enhancement in its resistance to corrosion.

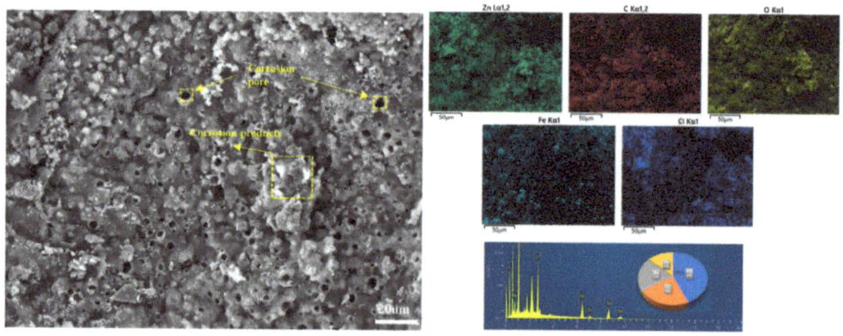

Figure 14. Images from SEM and EDS of the 80 WF coating following a 28-day immersion.

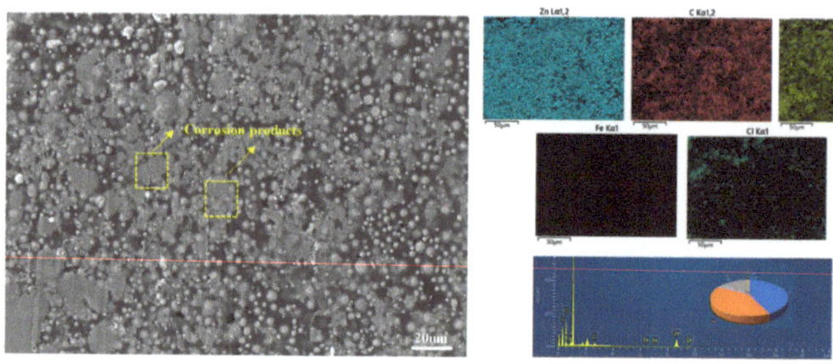

Figure 15. Images from SEM and EDS of the 0.3 wt.% CNTs coating following a 28-day immersion.

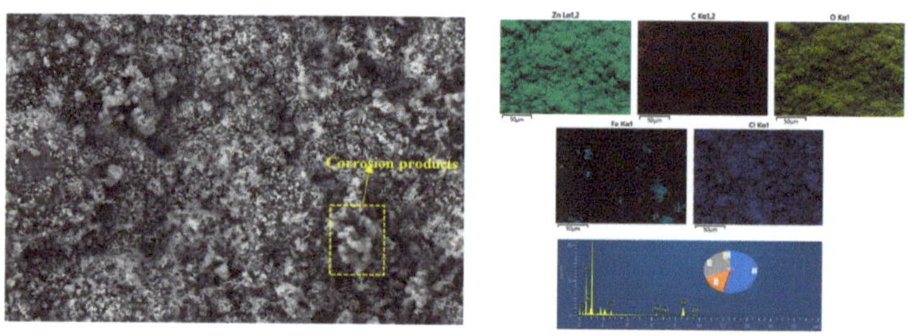

Figure 16. Images from SEM and EDS of the 1 wt.% CNTs coating following a 28-day immersion.

Concerning the coating with 1.0 wt.% CNTs, Figure 16 depicts a significant amount of zinc corrosion products that appear to be unevenly spread across the coating's surface. This phenomenon's emergence is linked to increased concentration of CNTs, which may result in galvanic corrosion between the CNTs and the Q355b steel substrate beneath, consequently accelerating the substrate's corrosion rate. Moreover, the clustering of CNTs causes a swift exhaustion of the nearby zinc powder, and results in corrosion products forming into loose, needle-like structures instead of flaky varieties. The diminished density of the coating enables corrosive substances to infiltrate the coating to reach the substrate, leading to the corrosion of the steel. Analysis of the energy spectrum indicates that corrosion products of iron formed on the coating's surface, suggesting that the substrate is starting to corrode. The effectiveness of the coating as a protective barrier is compromised, resulting in reduced corrosion resistance.

3.7. XRD and XPS Analysis of 80 WF Coating and CNTs/80 WF Coatings After Corrosion

Figure 17 displays the XRD patterns of the coating's surface following 240 h of immersion. The findings reveal that the main corrosion product found in the composite coating is $Zn_5(OH)_8Cl_2$. After the 240 h immersion, a significant reduction in zinc content is observed in the 1.0 wt.% CNTs coating, along with a rise in the intensity of the Fe diffraction peaks. At this stage, the coating ceases to deliver cathodic protection for the underlying material. Conversely, the coating containing 0.3 wt.% CNTs maintains a significant amount of unreacted zinc on its exterior after immersion, with the Fe diffraction peaks showing only a modest intensity. The findings indicate that a coating containing 0.3 wt.% CNTs continues to provide cathodic protection to the substrate. This is evident as corrosion byproducts, such as $Zn_5(OH)_8Cl_2$, efficiently cover the coating's surface and fill the pores. This mechanism successfully separates the substrate from the corrosive environment, improving the coating's physical barrier characteristics. It should be noted that the intensity of the corrosion byproduct, $Zn_5(OH)_8Cl_2$, appears to remain relatively constant despite changes in CNT concentration. This suggests that the corrosion process is affected by CNT concentration, but the formation of $Zn_5(OH)_8Cl_2$ is not significantly altered under the studied conditions.

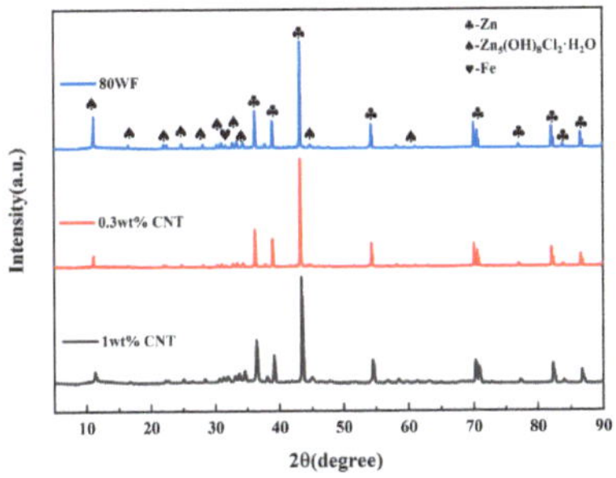

Figure 17. The XRD patterns for zinc-rich coatings containing 80 WF, 0.3 wt.% CNTs, and 1.0 wt.% CNTs following a 240 h immersion.

Figure 18 displays the full XPS spectra of the coatings' surfaces following 240 h of immersion. When compared to the coating with 0.3 wt.% CNTs, the intensity of the Zn2p peak at the binding energy appears to be less prominent for both the 80 WF coating and the 1.0 wt.% CNTs coating. This observation indicates that following corrosion, the zinc on the surfaces of the former two coatings is nearly depleted, while corrosion products originating from the steel substrate accumulate on the coating surface, ultimately leading to its failure. In contrast, the Zn2p peak in the coating with 0.3 wt.% CNTs is distinctly prominent, and no clearly detectable Fe2p peak can be found. This implies that the 0.3 wt.% CNTs coating remains largely preserved on the steel substrate's surface, with a substantial amount of unreacted zinc powder still present there.

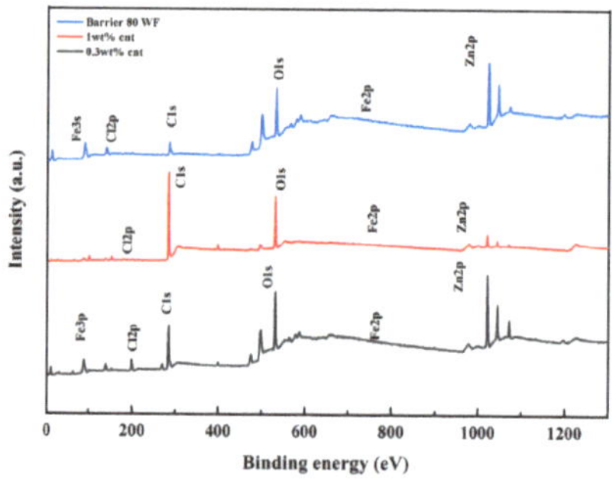

Figure 18. The complete XPS spectra of zinc-rich coatings containing 80 WF, 0.3 wt.% CNTs, and 1.0 wt.% CNTs were obtained following a 240 h immersion.

Figures 19–21 present the XPS fine spectra for the elements Zn, Cl, C, and O in the coatings after 240 h of immersion. The spectral profiles of the elements of all three coatings exhibit significant similarities. In Figure 19a, the presence of the Zn element is predominantly observed as a single compound. A notable peak appears at a binding energy of 1023.1 eV, probably linked to the creation of $Zn(OH)_2$ resulting from the interaction between the zinc powder and OH^- ions in the corrosive environment. Figure 19d reveals three distinct peaks for the O element at varying binding energies. The peak observed at 530.9 eV likely signifies ZnO, which might arise from the decomposition of $Zn(OH)_2$. Furthermore, the peak observed at 532.6 eV is associated with $ZnCO_3$, potentially resulting from the interaction of $Zn(OH)_2$ with CO_2 from the atmosphere. In contrast, the peak at 531.1 eV for $Zn_5(OH)_8Cl_2$ signifies one of the ultimate corrosion products formed from the zinc-rich coating in chloride solutions, generated through the reaction of $Zn(OH)_2$ with $ZnCl_2$.

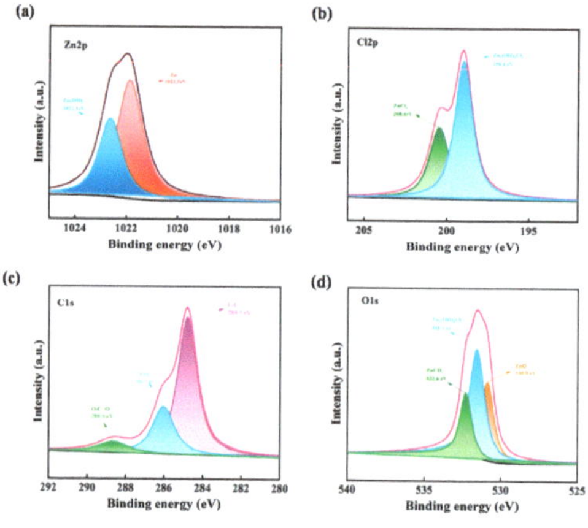

Figure 19. The fine XPS spectra of the 80 WF coating following 240 h of immersion. Fine spectra for the elements (**a**) Zn, (**b**) Cl, (**c**) C, and (**d**) O.

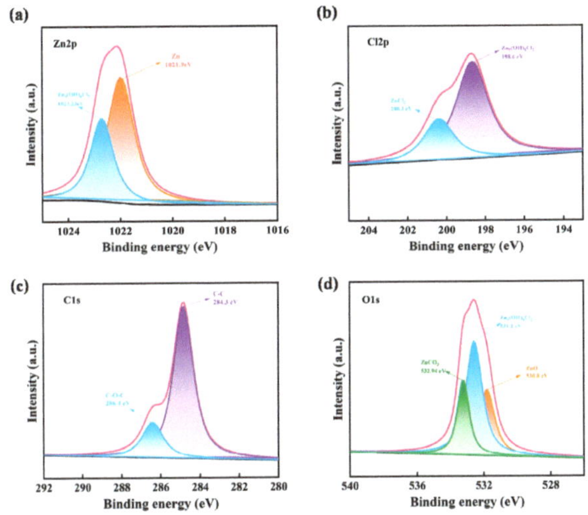

Figure 20. The fine XPS spectra of the 0.3 wt.% CNTs coating following immersion. of immersion. Fine spectra for the elements (**a**) Zn, (**b**) Cl, (**c**) C, and (**d**) O.

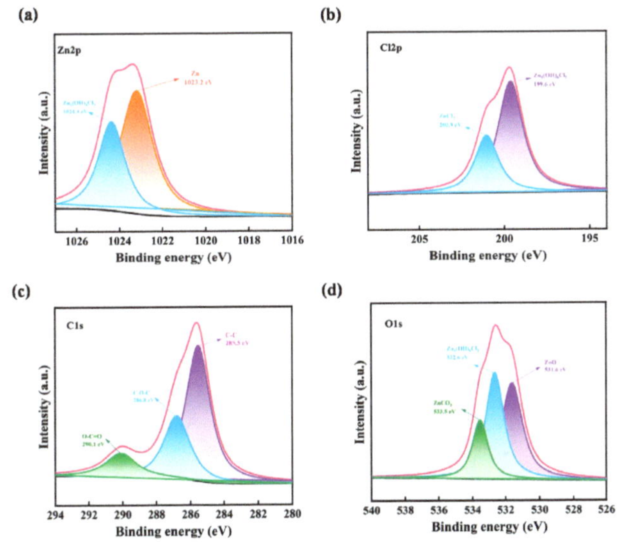

Figure 21. The fine XPS spectra of the 1 wt.% CNTs coating following immersion. Fine spectra for the elements (**a**) Zn, (**b**) Cl, (**c**) C, and (**d**) O.

3.8. The Corrosion Resistance Mechanism of the Coatings

Figure 22 depicts the mechanism of the corrosion reaction, while Figure 22a,b illustrate the processes related to the 80 WF coating without CNTs and the CNTs/80 WF coating after corrosion, respectively. Initially, during the corrosion process, a protective layer is provided by epoxy resin, which shields against corrosive agents such as H_2O, O_2, and Cl^- present in the electrolyte. The epoxy resin protective layer inhibits the infiltration of these substances in the electrolyte into the coating by a physical barrier effect, preventing corrosive agents from reaching the Q355b steel substrate beneath the coating, thereby averting the subsequent corrosion of the Q355b steel substrate. Nonetheless, the formation of micropores on the surface of the coatings is inevitable, and as the immersion duration extends, these corrosive substances begin to permeate through the micropores. In situations where the 80 WF coating does not contain CNTs, the capacity of zinc powder to offer cathodic protection is limited, resulting in the development of zinc corrosion products solely in certain localized areas. The locally generated zinc corrosion products result in an insufficient density within the coating, which cannot effectively prevent corrosive agents from penetrating further, ultimately reaching the Q355b steel substrate and resulting in corrosion. By incorporating CNTs, this issue is alleviated as they fill the pores within the coating and extend the route for corrosive agents to engage with the Q355b steel substrate. This enhancement improves the coating's density and its physical protective characteristics, consequently increasing the longevity of the coating [40,41].

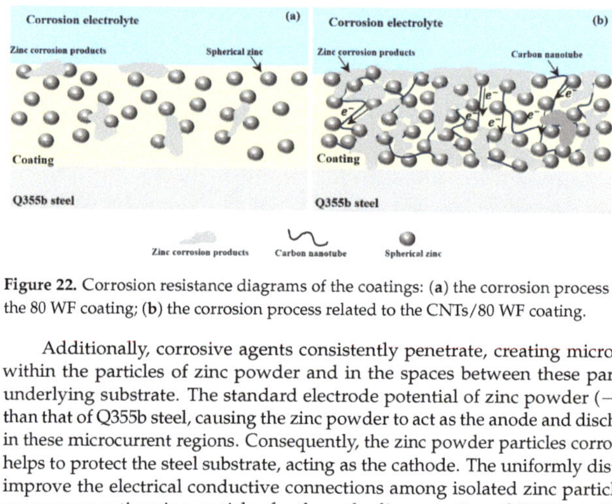

Figure 22. Corrosion resistance diagrams of the coatings: (**a**) the corrosion process associated with the 80 WF coating; (**b**) the corrosion process related to the CNTs/80 WF coating.

Additionally, corrosive agents consistently penetrate, creating microcurrent zones within the particles of zinc powder and in the spaces between these particles and the underlying substrate. The standard electrode potential of zinc powder (−0.78 V) is less than that of Q355b steel, causing the zinc powder to act as the anode and discharge electrons in these microcurrent regions. Consequently, the zinc powder particles corrode first, which helps to protect the steel substrate, acting as the cathode. The uniformly distributed CNTs improve the electrical conductive connections among isolated zinc particles, leading to many more active zinc particles for the cathodic protection of Q355b steel substrate. An increased number of zinc particles further enlarges the specific surface area for both the anode and cathode, subsequently enhancing zinc dissolution and facilitating sustained cathodic protection for the coating [16,20,22,25,35].

When the coating and the substrate are corroded directly, as the anode, the zinc powder corrodes preferentially and plays a role of cathodic protection as the sacrificial anode. This procedure entails a reaction between O_2 in the ambient air and H_2O within the corrosion medium, leading to the formation of $Zn(OH)_2$. Following this, $Zn(OH)_2$ interacts with CO_2 present in the atmosphere to yield $ZnCO_3$. Furthermore, the existence of C and OH^- aids in the development of $Zn_5(OH)_8Cl_2{\cdot}H_2O$.

The findings from the XRD and XPS analyses of the fully immersed samples provide additional insight into the makeup of the corrosion products. Because these corrosion products are insoluble, they occupy the micropores within the coatings, hindering further interaction between the corrosive environment and the substrate. Consequently, the coatings transition from a state of cathodic protection to one of shielding protection.

4. Conclusions

Based on the commercial 80 WF zinc-rich coating system, the main task of this study was to explore the effect of different concentrations of CNTs on the 80 WF coating, and to evaluate the corrosion protection of the coatings with different concentrations of CNTs for Q355b substrate. Electrochemical testing, salt spray evaluation, SEM, EDS, XRD, XPS, and other evaluation methods were used to study the corrosion resistance of coatings with different concentrations of CNTs. This study provides a valuable reference for the corrosion resistance modification of commercial coatings such as 80 WF, which are mature in the market, and subsequent engineering applications. The following is a summary of the findings of this article:

(1) The 80 WF coating without CNTs demonstrates poor resistance to corrosion, with its overall density being insufficient. In the coating before corrosion, the spherical zinc powder on the surface of the 80 WF coating is not uniformly dispersed, and there is a large gap between adjacent zinc powder. The zinc powder is isolated and cannot be well connected, and the overall coating is not too dense. The zinc powder and corrosion products on the 80 WF coating surface agglomerate after corrosion. Corrosion defects and Fe corrosion products appear, and the overall density of the coating is reduced. In the electrochemical testing, the overall OCP of 80 WF is higher than −0.78 V, indicating that the coating has almost no cathodic protection effect, but mainly plays a protective effect of physical shielding. Moreover, after immersion, the impedance arc in the EIS impedance spectrum is small, the R_{ct} value is smaller, and the i_{corr} of the polarization curve is larger than that of the coatings with 0.3 wt% CNTs. After the salt spray corrosion for 240 h in the salt spray test, obvious corrosion occurred on the surface of the coating, and the matrix corrosion was more serious at the scratch at X, indicating that the cathodic protection ability of the coating was weak.

(2) Compared with 80 WF, the CNTs of the 0.3 wt% CNTs coating adhere stably to the surface of the coating, and the coating surface becomes densified, indicating that CNTs fill the pores between zinc powder particles and epoxy resin to a certain extent, and play a certain shielding role. In addition, the surface morphology shows that CNTs and zinc powder are combined in the coating, which also indicates that CNTs form conductive channels between adjacent spherical zinc powder, which increases the amount of effective activated zinc powder, improves its utilization rate, and enhances the cathodic protection ability of the sacrificial anode of the zinc-rich coating. In the electrochemical testing, the overall OCP of the 0.3 wt% CNTs coating was lower than −0.78 V in the first 14 days, indicating that the coating offered efficient cathodic protection. Moreover, the impedance arc in the EIS spectrum was large, and the R_{ct} value after 28 days of immersion reached 3756 ohms, and the i_{corr} of the polarization curve reached 0.032 A/cm^2, showing its electrochemical properties are better than those of the other five coatings. After salt spray corrosion for 240 h, no obvious corrosion occurred on the surface and the substrate at the scratch of X, indicating that the coating played a cathodic protection role.

(3) With the gradual increase in CNT concentration, a large number of CNTs can be obviously observed on the surface of the 1.0 wt.% CNTs coating. The coating's components are unevenly dispersed, which may adversely affect the basic properties and corrosion resistance of the coating. In the electrochemical testing, the overall OCP of the 1.0 wt.% CNTs coating is higher than −0.78 V, indicating that the coating has almost no cathodic protection effect, but mainly has a protective effect of physical shielding. Moreover, the impedance arc in the EIS spectrum is small, the R_{ct} value is smaller, and the i_{corr} of the polarization curve is larger than that of the 0.3 wt% CNTs coating. After salt spray corrosion for 240 h, a large number of new white corrosion

products were generated on the coating surface in a short period of time, and then serious corrosion occurred in the substrate. Excessive CNTs would adversely affect the corrosion resistance of the coating.

Author Contributions: Formal analysis, X.H. and C.Y.; investigation, J.C. and C.Y.; resources, X.H. and J.C.; experimental research, X.H. and X.Q.; writing—original draft preparation, C.Y. and X.Q.; writing—survey and editing, J.C. and X.Q.; visualization, J.C. and X.H.; supervision, S.Z. and D.S.; project administration, D.S.; funding acquisition, D.S. All authors have read and agreed to the published version of the manuscript.

Funding: This research was funded by the State Grid Jiangsu Electric Power Co., Ltd. Technology Project, grant number J2023009.

Institutional Review Board Statement: Not applicable.

Informed Consent Statement: Not applicable.

Data Availability Statement: Data are included in the article and the corresponding references.

Conflicts of Interest: Authors Xueling Huang, Chun Yang, and Junyu Chen were employed by the State Grid Wuxi Power Supply Company. The remaining authors declare that the research was conducted in the absence of any commercial or financial relationships that could be construed as potential conflicts of interest.

References

1. Munger, C.G.; Vincent, L.D. Corrosion Prevention by Protective Coatings. NACE International 1984. Available online: https://trid.trb.org/View/1165252 (accessed on 5 November 2024).
2. Marchebois, H.; Touzain, S.; Joiret, S.; Bernard, J.; Savall, C. Zinc-rich powder coatings corrosion in sea water: Influence of conductive pigments. *Prog. Org. Coat.* **2002**, *45*, 415–421. [CrossRef]
3. Akbarinezhad, E.; Ebrahimi, M.; Sharif, F.; Attar, M.; Faridi, H. Synthesis and evaluating corrosion protection effects of emeraldine base PAni/clay nanocomposite as a barrier pigment in zinc-rich ethyl silicate primer. *Prog. Org. Coat.* **2011**, *70*, 39–44. [CrossRef]
4. Park, S.; Shon, M. Effects of multi-walled carbon nano tubes on corrosion protection of zinc rich epoxy resin coating. *J. Ind. Eng. Chem.* **2015**, *21*, 1258–1264. [CrossRef]
5. Schaefer, K.; Miszczyk, A. Improvement of electrochemical action of zinc-rich paints by addition of nanoparticulate zinc. *Corros. Sci.* **2013**, *66*, 380–391. [CrossRef]
6. Sørensen, P.A.; Kiil, S.; Dam-Johansen, K.; Weinell, C.E. Anticorrosive coatings: A review. *J. Coat. Technol. Res.* **2009**, *6*, 135–176. [CrossRef]
7. Cubides, Y.; Castaneda, H. Corrosion protection mechanisms of carbon nanotube and zinc-rich epoxy primers on carbon steel in simulated concrete pore solutions in the presence of chloride ions. *Corros. Sci.* **2016**, *109*, 145–161. [CrossRef]
8. Yun, T.H.; Park, J.H.; Kim, J.-S.; Park, J.M. Effect of the surface modification of zinc powders with organosilanes on the corrosion resistance of a zinc pigmented organic coating. *Prog. Org. Coat.* **2014**, *77*, 1780–1788. [CrossRef]
9. Volovitch, P.; Vu, T.; Allély, C.; Aal, A.A.; Ogle, K. Understanding corrosion via corrosion product characterization: II. Role of alloying elements in improving the corrosion resistance of Zn–Al–Mg coatings on steel. *Corros. Sci.* **2011**, *53*, 2437–2445. [CrossRef]
10. Volovitch, P.; Allely, C.; Ogle, K. Understanding corrosion via corrosion product characterization: I. Case study of the role of Mg alloying in Zn–Mg coating on steel. *Corros. Sci.* **2009**, *51*, 1251–1262. [CrossRef]
11. Teng, S.; Gao, Y.; Cao, F.; Kong, D.; Zheng, X.; Ma, X.; Zhi, L. Zinc-reduced graphene oxide for enhanced corrosion protection of zinc-rich epoxy coatings. *Prog. Org. Coat.* **2018**, *123*, 185–189. [CrossRef]
12. Shreepathi, S.; Bajaj, P.; Mallik, B.P. Electrochemical impedance spectroscopy investigations of epoxy zinc rich coatings: Role of Zn content on corrosion protection mechanism. *Electrochim. Acta* **2010**, *55*, 5129–5134. [CrossRef]
13. Shen, L.; Li, Y.; Zhao, W.; Miao, L.; Xie, W.; Lu, H.; Wang, K. Corrosion protection of graphene-modified zinc-rich epoxy coatings in dilute NaCl solution. *ACS Appl. Nano Mater.* **2018**, *2*, 180–190. [CrossRef]
14. Ramezanzadeh, B.; Moghadam, M.M.; Shohani, N.; Mahdavian, M. Effects of highly crystalline and conductive polyaniline/graphene oxide composites on the corrosion protection performance of a zinc-rich epoxy coating. *Chem. Eng. J.* **2017**, *320*, 363–375. [CrossRef]
15. Praveen, B.; Venkatesha, T.; Naik, Y.A.; Prashantha, K. Corrosion studies of carbon nanotubes–Zn composite coating. *Surf. Coat. Technol.* **2007**, *201*, 5836–5842. [CrossRef]
16. Wang, D.; Sikora, E.; Shaw, B. A comparison of the corrosion response of zinc-rich coatings with and without presence of carbon nanotubes under erosion and corrosion conditions. *Corrosion* **2018**, *74*, 1203–1213. [CrossRef] [PubMed]
17. Hammouda, N.; Chadli, H.; Guillemot, G.; Belmokre, K. The corrosion protection behaviour of zinc rich epoxy paint in 3% NaCl solution. *Adv. Chem. Eng. Sci.* **2011**, *1*, 51. [CrossRef]

18. Sofian, A.H.; Tanaka, A.; Noda, K. Corrosion performance of zinc-rich paints (ZRP) on mild steel in NaCl solution. *ECS Trans.* **2013**, *45*, 17. [CrossRef]
19. Li, H.; Duan, J.; Wei, D. Comparison on corrosion behaviour of arc sprayed and zinc-rich coatings. *Surf. Coat. Technol.* **2013**, *235*, 259–266. [CrossRef]
20. Orek, C.; Keser, S.; Kaygili, O.; Zuchowski, P.; Bulut, N. Structures and Optical Properties of Zinc Oxide Nanoclusters: A Combined Experimental and Theoretical Approach. *J. Mol. Model.* **2023**, *29*, 227. [CrossRef] [PubMed]
21. Hoang, N.; Khoa, T.A.; Phuong, P.M.; Hang, T.T.X.; Van Chi, N.; Nguyen, T.D. Corrosion protection of carbon steel using a combination of Zr conversion coating and subsequent zinc-rich silicate coating with a flake ZnAl alloy. *Arab. J. Chem.* **2022**, *15*, 103815. [CrossRef]
22. Hayatdavoudi, H.; Rahsepar, M. A mechanistic study of the enhanced cathodic protection performance of graphene-reinforced zinc rich nanocomposite coating for corrosion protection of carbon steel substrate. *J. Alloys Compd.* **2017**, *727*, 1148–1156. [CrossRef]
23. Ge, T.; Zhao, W.; Wu, X.; Lan, X.; Zhang, Y.; Qiang, Y.; He, Y. Incorporation of electroconductive carbon fibers to achieve enhanced anti-corrosion performance of zinc rich coatings. *J. Colloid Interface Sci.* **2020**, *567*, 113–125. [CrossRef]
24. Fukuda, H.; Szpunar, J.A.; Kondoh, K.; Chromik, R. The influence of carbon nanotubes on the corrosion behaviour of AZ31B magnesium alloy. *Corros. Sci.* **2010**, *52*, 3917–3923. [CrossRef]
25. Pourhashem, S.; Ghasemy, E.; Rashidi, A.; Vaezi, M.R. A review on application of carbon nanostructures as nanofiller in corrosion-resistant organic coatings. *J. Coat. Technol. Res.* **2020**, *17*, 19–55. [CrossRef]
26. Cao, X.; Huang, F.; Huang, C.; Liu, J.; Cheng, Y.F. Preparation of graphene nanoplate added zinc-rich epoxy coatings for enhanced sacrificial anode-based corrosion protection. *Corros. Sci.* **2019**, *159*, 108120. [CrossRef]
27. Bai, W.; Ma, Y.; Meng, M.; Li, Y. The influence of graphene on the cathodic protection performance of zinc-rich epoxy coatings. *Prog. Org. Coat.* **2021**, *161*, 106456. [CrossRef]
28. Rastogi, R.; Kaushal, R.; Tripathi, S.; Sharma, A.L.; Kaur, I.; Bharadwaj, L.M. Comparative study of carbon nanotube dispersion using surfactants. *J. Colloid Interface Sci.* **2008**, *328*, 421–428. [CrossRef]
29. Pandey, P.; Mohanty, S.; Nayak, S.K. Tailoring dispersion and interaction of MWNT in polymer nanocomposites, using Triton X-100 as nonionic surfactant. *J. Mater. Eng. Perform.* **2014**, *23*, 4385–4393. [CrossRef]
30. Husanu, M.; Baibarac, M.; Baltog, I. Non-covalent functionalization of carbon nanotubes: Experimental evidence for isolated and bundled tubes. *Phys. E Low-Dimens. Syst. Nanostruct.* **2008**, *41*, 66–69. [CrossRef]
31. Pereyra, A.; Giudice, C.; Herrera, L.K.; Echeverría, F.; Castano, J.G. Tripigmented anticorrosive coatings based on lamellar zinc as inhibitor. *Surf. Coat. Int.* **2006**, *89*, 245–249. [CrossRef]
32. Knudsen, O.Ø.; Steinsmo, U.; Bjordal, M. Zinc-rich primers—Test performance and electrochemical properties. *Prog. Org. Coat.* **2005**, *54*, 224–229. [CrossRef]
33. Kalendová, A. Effects of particle sizes and shapes of zinc metal on the properties of anticorrosive coatings. *Prog. Org. Coat.* **2003**, *46*, 324–332. [CrossRef]
34. Jalili, M.; Rostami, M.; Ramezanzadeh, B. An investigation of the electrochemical action of the epoxy zinc-rich coatings containing surface modified aluminum nanoparticle. *Appl. Surf. Sci.* **2015**, *328*, 95–108. [CrossRef]
35. Gergely, A.; Pászti, Z.; Mihály, J.; Drotár, E.; Török, T. Galvanic function of zinc-rich coatings facilitated by percolating structure of the carbon nanotubes. Part I: Characterization of the nano-size particles. *Prog. Org. Coat.* **2015**, *78*, 437–445. [CrossRef]
36. Xu, L.; Liu, F.; Wang, Z.; Ke, W.; Han, E.-H.; Jie, G.; Wang, J.; Huang, H. The effect of surface modification of zinc particles with phosphoric acid on the corrosion resistance of cold galvanizing coatings. *Prog. Org. Coat.* **2018**, *114*, 90–101. [CrossRef]
37. Hu, J.-M.; Zhang, J.-T.; Zhang, J.-Q.; Cao, C.-N. A novel method for determination of diffusion coefficient of corrosive species in organic coatings by EIS. *J. Mater. Sci.* **2004**, *39*, 4475–4479. [CrossRef]
38. Cubides, Y.; Su, S.S.; Castaneda, H. Influence of zinc content and chloride concentration on the corrosion protection performance of zinc-rich epoxy coatings containing carbon nanotubes on carbon steel in simulated concrete pore environments. *Corrosion* **2016**, *72*, 1397–1423. [CrossRef]
39. Amirudin, A.; Thieny, D. Application of electrochemical impedance spectroscopy to study the degradation of polymer-coated metals. *Prog. Org. Coat.* **1995**, *26*, 1–28. [CrossRef]
40. Wang, F.; Feng, L.; Huang, Y.; Shen, W.; Ma, H. Effect of the gradient distribution of multiwalled carbon nanotubes on the bond strength and corrosion resistance of waterborne polyurethane conductive nanocomposites. *Prog. Org. Coat.* **2020**, *140*, 105507. [CrossRef]
41. Li, A.; Sun, M.; Ma, Z.; Chen, S.; Zhu, G.; Zhang, Y.; Wang, W. Anticorrosion performance of polyvinyl butyral composite coatings improved by polyaniline-multiwalled carbon nanotubes/poly (methylhydrosiloxane). *Thin Solid Films* **2020**, *712*, 138347. [CrossRef]

metals

Article

Investigating the Synergistic Corrosion Protection Effect of an Alloy Element and Corrosion Inhibitor on Steel Reinforcement Using Machine Learning and Electrochemical Impedance Spectroscopy

Cheng Wen [1,2], Baitong Chen [1], Gongqi Lou [1], Nanchuan Wang [1], Yuwan Tian [1,2] and Ningxia Yin [1,*]

1 School of Mechanical Engineering, Guangdong Ocean University, Zhanjiang 524088, China; wheng.3jia@163.com (C.W.)
2 Guangdong Provincial Ocean Equipment and Manufacturing Engineering Technology Research Center, Guangdong Ocean University, Zhanjiang 524088, China
* Correspondence: yinnx@gdou.edu.cn

Abstract: Steel reinforcement in marine concrete structures is vulnerable to chloride-induced corrosion, which compromises its structural integrity and durability. This study explores the combined effect of the alloying element Cr and the smart corrosion inhibitor LDH-NO_2 on enhancing the corrosion resistance of steel reinforcement. Employing a machine learning approach with a support vector machine (SVM) algorithm, a predictive model was developed to estimate the polarization resistance of steel, considering Cr content, LDH-NO_2 dosage, environmental pH, and chloride concentration. The model was rigorously trained and validated, demonstrating high accuracy, with a correlation coefficient exceeding 0.85. The findings reveal that the addition of Cr and application of LDH-NO_2 synergistically improve corrosion resistance, with the model providing actionable insights for selecting effective corrosion protection methods in diverse concrete environments.

Keywords: machine learning; steel reinforcement; corrosion; LDH-NO_2; Cr

Citation: Wen, C.; Chen, B.; Lou, G.; Wang, N.; Tian, Y.; Yin, N. Investigating the Synergistic Corrosion Protection Effect of an Alloy Element and Corrosion Inhibitor on Steel Reinforcement Using Machine Learning and Electrochemical Impedance Spectroscopy. *Metals* **2024**, *14*, 865. https://doi.org/10.3390/met14080865

Academic Editor: Chris Aldrich

Received: 21 June 2024
Revised: 21 July 2024
Accepted: 22 July 2024
Published: 27 July 2024

1. Introduction

In marine concrete structures, the corrosion of steel reinforcement presents a significant challenge, posing threats to the durability and safety of infrastructure worldwide [1]. The corrosive agents in seawater, especially chloride ions, accelerate the degradation process, leading to substantial maintenance costs and structural integrity concerns. This issue is of particular concern in coastal and marine environments, where concrete structures are frequently exposed to harsh conditions. To combat these challenges, various corrosion protection techniques have been developed and employed in marine engineering, including alloy steel reinforcement [2], coatings [3], cathodic protection [4], and corrosion inhibitors [5].

Among the various techniques, the incorporation of the alloying element chromium (Cr) into steel has emerged as a crucial strategy to enhance corrosion resistance. There has been a significant amount of research on the influence of Cr in enhancing the corrosion resistance of steel reinforcement, and there are even corrosion-resistant steel reinforcements with different Cr contents available for purchase [6–10]. The improved corrosion resistance provided by Cr is multifactorial. On the one hand, Cr forms a passive oxide layer on the surface of the steel, acting as a barrier against corrosive agents, thereby extending the chloride-induced de-passivation time of the reinforcement [8]. On the other hand, Cr infiltrates the rust layer on the steel surface, altering its morphology, composition, and structure, thus prolonging the time before concrete cracking occurs [11,12]. As a result, the incorporation of Cr as an alloying element can augment the corrosion resistance of steel reinforcement in harsh environments, thereby prolonging the service life of concrete structures and improving their safety. Therefore, it is evident that Cr-containing steel reinforcement offers considerable potential for utilization in marine engineering projects.

However, it has been observed in engineering applications that a significant issue arises when using Cr-containing steel reinforcement in chloride-contaminated concrete: Cr, as an alloying element, can markedly increase the susceptibility of the reinforcement to pitting corrosion. In other words, while the average corrosion rate of Cr-containing steel reinforcement may not be high in marine environments, severe corrosion and thinning of the steel cross-section can occur in localized areas. To address this problem, the use of smart corrosion inhibitors is considered to be a promising approach to mitigate the shortcomings of Cr-containing steel reinforcement [13]. Smart corrosion inhibitors are designed to store corrosion inhibitors in low-corrosion-risk areas, while rapidly releasing the inhibitors in high-corrosion-risk areas. The synergistic use of Cr-containing steel reinforcement and smart corrosion inhibitors not only maintains the excellent corrosion resistance of Cr-containing steel reinforcement but also further suppresses its localized corrosion.

The "LDH-inhibitor" system is currently one of the research hotspots within the field of smart corrosion inhibitors [14]. In this system, LDH refers to layered double hydroxides (chemical formula of $\left[M_{1-x}^{2+}M_x^{3+}(OH^-)_2\right]^{x+}$, where M^{2+} and M^{3+} represent metal cations, and x is the percentage of M^{3+}), and the "inhibitor" component comprises negatively charged corrosion inhibitor ions (such as nitrite (NO_2^-) and phosphate (PO_4^{3-})). The positively charged LDH layers and the negatively charged inhibitor ions are held together though electrostatic interactions. In the presence of anions in the surrounding environment exerting a stronger electrostatic effect on the LDH layers, the LDH will adsorb these environmental anions and simultaneously release the corrosion inhibitor ions through ion-exchange reactions. In other words, in areas with a high risk of corrosion, where chloride ions are present, the LDH-inhibitor can rapidly release inhibitor ions through ion-exchange reactions (LDH-inhibitor + chloride ions = LDH-chloride + inhibitor ions). Consequently, the LDH in the LDH-inhibitor system functions as an intelligent carrier for inhibitor ions, ensuring that the latter are stored in low-corrosion-risk areas devoid of chloride and released in high-corrosion-risk areas containing chloride. Therefore, the LDH-inhibitor smart corrosion inhibitor can guarantee the exceptional corrosion resistance of Cr-containing steel reinforcement while specifically targeting and mitigating pitting corrosion.

However, existing studies predominantly focus on the impact of individual protective technologies on steel corrosion in concrete. Each technique has its own advantages and limitations, and combining two or more methods can enhance their effectiveness through mutual reinforcement, achieving a cumulative effect greater than the sum of the individual measures. Furthermore, practical engineering applications often require the synergistic use of multiple protective techniques. Therefore, this study investigates the synergistic corrosion protection effect of Cr and LDH-NO_2 (in which the inhibitor ions are NO_2^-) on steel reinforcement. It aims to significantly reduce the average corrosion rate of steel reinforcement through the use of the alloying element Cr, while employing the smart inhibitor LDH-NO_2 to specifically target and repair localized corrosion damage in Cr-containing steel, ultimately ensuring effective corrosion protection over the entire service life of the reinforcement.

In the conventional experimental approach to investigating multi-factor interactions, techniques such as orthogonal experiments and factor analysis are frequently employed. However, these methods have inherent limitations in their ability to handle high-dimensional, multi-variable complex interactions, and they can also be quite cumbersome in terms of experimental design and data analysis. In contrast, advanced machine learning methods offer significant advantages, particularly in terms of automating data processing and model training. Such techniques can identify intricate non-linear relationships and multi-factor interactions that may otherwise remain obscured within the data, thereby facilitating more precise predictions.

Specifically, the aim of this study is to assess the corrosion resistance of steel reinforcement in concrete under different levels of chloride ion contamination and carbonation, considering the content of the alloy element Cr and the dosage of the smart corrosion inhibitor LDH-NO_2. The vast range of environmental factors in concrete and the flexibility in adjusting steel's Cr content and concrete's LDH-NO_2 dosage pose significant challenges for the comprehensive quantification of their relationships. Moreover, due to the complex, non-linear relationship between the corrosion rate and multiple variables, traditional single-variable fitting methods are inadequate. Therefore, advanced machine learning techniques are primarily employed in this research to accurately model these intricate relationships and examine the impact of steel reinforcement Cr content, LDH-NO_2 dosage, concrete chloride concentration, and pore solution pH on corrosion resistance. Correlation models between these factors and steel reinforcement polarization resistance are developed to guide the selection of corrosion prevention methods for concrete structures with varying degrees of corrosion risk.

2. Methods

2.1. *Establishment of Quantitative Predictive Model*

First, a large number of data samples on the influence of steel's Cr content and corrosion inhibitor LDH-NO_2 dosage on the polarization resistance of steel reinforcement were collected from the literature. Then, advanced machine learning techniques were employed to achieve accurate fitting of complex non-linear relationships and explore the influence of Cr content and LDH-NO_2 dosage, environmental chloride concentration, and pH value on the corrosion resistance of steel reinforcement. A correlation model between these factors and the polarization resistance of steel reinforcement was obtained to achieve rapid prediction. Through the preliminary evaluation of commonly used machine learning regression algorithms such as linear regression, neural networks, random forest, support vector machine (linear kernel), support vector machine (Gaussian kernel), and K-nearest neighbors, the final selection for the machine learning prediction model was based on the good performance observed on the data samples, leading to the choice of random forest and support vector machine (Gaussian kernel) models.

Specifically, in the process of evaluating the generalization performance of the machine learning models, 80% of the data samples were used as a training set to train and establish the machine learning model, and the model parameters were optimized by ten-fold cross validation. The other 20% of the data samples were used as a test set to assess the generalized predictive capability of the model. The Pearson correlation coefficient r^2 and the mean absolute error (MAE) were chosen based on the prediction results of the machine learning model for test set samples; as the main evaluation index, the expression for r^2 is as follows:

$$r^2 = \frac{\sum (X - \overline{X})(Y - \overline{Y})}{\left(\sqrt{\sum_{i=1}^{n} (X_i - \overline{X})^2}\right)\left(\sqrt{\sum_{i=1}^{n} (Y_i - \overline{Y})^2}\right)}$$

where $\overline{X}$ and $\overline{Y}$ represent the means of variables X and Y, respectively, while X_i and Y_i represent the variable values corresponding to each sample, denoting the experimental and model-predicted values of the data sample i. From this, the value of r^2 is approximately closer to 1, indicating a better correlation between the model prediction result and the experimental value, implying a better generalization ability of the model. As for MAE, its expression is as follows:

$$\text{MAE} = \frac{\sum_{i=1}^{n} |y - y_i|}{n}$$

where n represents the number of samples, y_i represents the actual value of sample i, and y represents the corresponding model-predictive results. A lower MAE indicates a smaller deviation between the predicted results of the model and the experimental value, signifying a better generalization ability of the model. Furthermore, in order to eliminate the accidental influence of the 80–20% data partition on the model prediction results, we took 100 data partitions, averaging the results, in order to compare the predictive generalization performance of each machine learning model.

Using the collected training data, with the Cr content, LDH-NO_2 dosage, chloride concentration, and pH value as the inputs of the model, and with the experimental value of polarization resistance of the steel reinforcement as the output of the model, we evaluated and selected the machine learning model, and the predictive model of polarization resistance of steel reinforcement was established. In addition, in order to eliminate the effect of the input data dimensions on the model fitting, in the process of evaluating the machine learning models, a 0–1 standardization was applied to the input variables (Cr content, LDH-NO_2 dosage, chloride concentration, and pH value) to eliminate the impact of dimensionality on distance-based machine learning models, such as the K-nearest neighbors model. At the same time, in order to avoid overfitting, the ten-fold cross-validation method was used to select certain machine learning models, executing prediction. Specifically, all data samples were divided into ten equal parts, with the model trained using nine parts and the remaining one part used for prediction. The prediction results of each sample were obtained successively. Each sample was not utilized in the model training process, ensuring the reliability of the model predictions. Furthermore, in order to analyze and ensure the stability of the machine learning models' prediction, 100 cross-validations were performed to validate the model predictions.

2.2. Validation of Quantitative Prediction Model

Steel reinforcements with varying Cr contents were prepared using vacuum melting and rolling processes. The chemical composition is detailed in Table 1.

Table 1. Chemical composition of the steel reinforcement.

Label of Steel	C Content (wt%)	Mn Content (wt%)	Si Content (wt%)	S Content (wt%)	P Content (wt%)	Cr Content (wt%)
0 Cr	0.23	1.57	0.57	0.005	0.024	0.05
1 Cr	0.16	0.58	0.26	0.005	0.004	1.02
2 Cr	0.10	0.69	0.47	0.005	0.003	2.46
5 Cr	0.04	1.07	0.56	0.006	0.003	5.21

The smart corrosion inhibitor LDH-NO_2 was also prepared using the hydrothermal reaction method. A 50 mL mixture of 0.5 mol/L $Zn(NO_3)_2$ and 0.25 mol/L $Al(NO_3)_3$ was slowly added to a 100 mL solution of 1 mol/L $NaNO_2$ at a drip rate of 50 mL/h, with a temperature of 65 °C, maintaining the pH at 10 $\pm$ 0.3 using 1 mol/L NaOH solution and vigorous stirring. After titration, the suspension was crystallized in a high-pressure autoclave at 70 °C for 24 h, followed by filtration, washing, freeze-drying, and grinding [15].

Electrochemical impedance spectroscopy (EIS) was used to test the polarization resistance of steel reinforcements with varying Cr contents under different pH levels and chloride concentrations in simulated concrete solutions with different LDH-NO_2 dosages. The polarization resistance was fitted using ZSimpWin software (version 3.6). Specifically, a three-electrode system was used, with the steel samples employed as the working electrode, a platinum foil as the counter electrode, and a saturated calomel electrode (SCE) as the reference electrode. The working electrodes were all pre-passivated in saturated calcium hydroxide for 7 days before EIS testing. The test solution comprised a chloride-carbonated concrete simulated pore solution, with the pH adjusted using sodium carbonate and sodium bicarbonate. The frequency range for electrochemical impedance spectroscopy testing was from 100 kHz to 10 mHz, with an amplitude of 10 mV.

Finally, the prediction model in Section 2.1 was validated using the measured polarization resistance values from Section 2.2.

3. Results and Discussion

3.1. Preliminary Machine Learning

Table A1 [15–22] presents a total of 106 data samples collected from the literature, encompassing information on steel rebar Cr content, LDH-NO_2 dosage, concrete chloride concentration, and pore solution pH value. In the process of machine learning, the input variables include the Cr content, the LDH-NO_2 dosage, chloride concentration, and pH value. The target variable (output variable) is the polarization resistance of the steel reinforcement. Using the above data, the generalized predictive ability of seven commonly used algorithm models was evaluated, including the linear regression model (LM), linear kernel support vector machine model (SVR.lin), artificial neural network model (NN), Gaussian regression model (GPR), K-nearest neighbors model (KNN), Gaussian kernel support vector machine model (SVR.rbf), and random forest model (RF). The correlation and error of each prediction model are illustrated in Figure 1a, with the performance evaluation metrics, including r^2 and MAE, representing the average of 100 predictions. The volatility of each model is indicated by error bars. From Figure 1a, it is evident that the SVR.rbf model exhibits the poorest prediction performance, with a correlation coefficient below 0.5 between the predicted and experimental values, and relatively large prediction errors, as indicated by the wide error bars, highlighting its instability in predictions. The predictive performance of the LM, SVR.lin, NN, GPR, and KNN models shows no significant differences, with correlations consistently around 0.6. Among these, the artificial neural network model (NN) exhibits slightly less effectiveness. The random forest (RF) model demonstrates the highest performance among all models, achieving a predicted correlation coefficient of 0.65, with an average prediction error of 334 kΩ·cm^2 for polarization resistance. Consequently, the RF model was chosen as the preferred predictive model, and its cross-validation results are depicted in Figure 1b. It can be observed that the model predictions vary significantly across the entire numerical range of polarization resistance, particularly in the low- and high-value ranges, resulting in an overall less-than-satisfactory prediction performance.

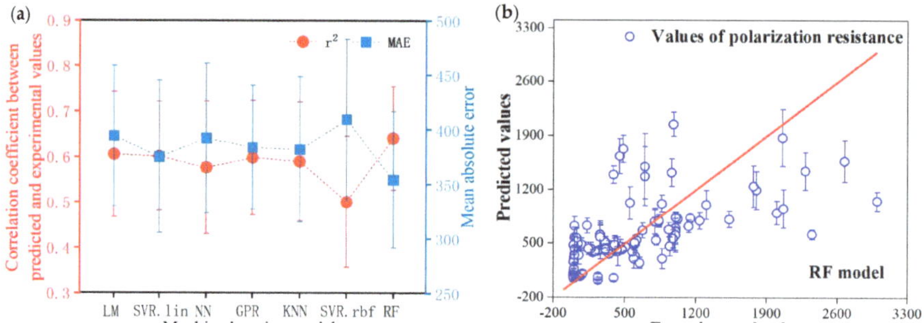

Figure 1. (**a**) Comparison of prediction performance of different machine learning models. (**b**) Prediction results of the random forest model (RF) for polarization resistance.

In-depth analysis of the results from each machine learning model revealed that the differences in their generalization performance are not pronounced. Particularly notable is the Gaussian kernel support vector machine model (SVR, rbf), which is typically esteemed for its performance with small data samples but exhibited the weakest performance in this context. Moreover, the presence of larger error bars suggests potential overfitting in the model predictions. The poor model performance can be attributed to the significant influence of experimental conditions, such as the composition of steel reinforcement, interface status, pre-passivation time, and the test system, on the corrosion resistance of the steel reinforcement. Additionally, variations in the testing conditions for the polarization resistance of steel reinforcement samples were not accounted for in the model fitting process, resulting in negligible differences in model performance. The robust SVR.rbf model, renowned for its fitting capability, paradoxically demonstrates inferior performance. Therefore, to investigate the influence of Cr content and LDH-NO_2 dosage on the polarization resistance of steel reinforcement, it is imperative to re-clean and refine the training data.

3.2. Machine Learning after Data Cleaning

Due to the significant variance in corrosion mechanisms between high-alloy and low-alloy steel, the Cr content range was specified as 0–5 wt%. Considering actual concrete conditions, the LDH-NO_2 and chloride concentrations were defined within the ranges of 0–0.5 wt% and 0–3.5 wt%, respectively, while the pH value ranged from 9.5 to 12.5. Additionally, due to the significant impact of specific experimental procedures on the surface condition and corrosion rates of the steel reinforcement, only the corrosion data from pre-passivated steel reinforcement were selected. After the aforementioned data cleaning process, a total of 52 data samples remained, as detailed in Table A2.

Figure 2a demonstrates the predictive performance of each trained machine learning mode using the cleaned data. A comparison with Figure 1a reveals a notable enhancement in simulation prediction accuracy, stemming from the elimination of test condition influences on the polarization resistance of the steel reinforcement. In Figure 2a, the LM and SVR.lin models demonstrate similar generalization abilities, whereas the random forest (RF) model exhibits the lowest predictive correlation and higher prediction error. The prediction error of the KNN model is comparable to that of the SVR.rbf model, but the SVR.rbf model shows higher correlation in its prediction results. After careful consideration, the SVR.rbf model was chosen as the final predictive model. Figure 2b illustrates the correlation between the polarization resistance predicted by the Gaussian kernel support vector machine model (SVR.rbf) and the experimental data. Despite notable deviations for two samples in the low-value range, the model demonstrates overall excellent predictive performance. The correlation coefficient between the predicted and experimental values exceeds 0.85. For polarization resistances greater than 100 kΩcm^2, the root-mean-square error is only 5.2 kΩcm^2. Furthermore, the prediction error bar, derived from 100 cross-validations, is relatively narrow, indicating the robust predictive stability of the selected support vector machine model.

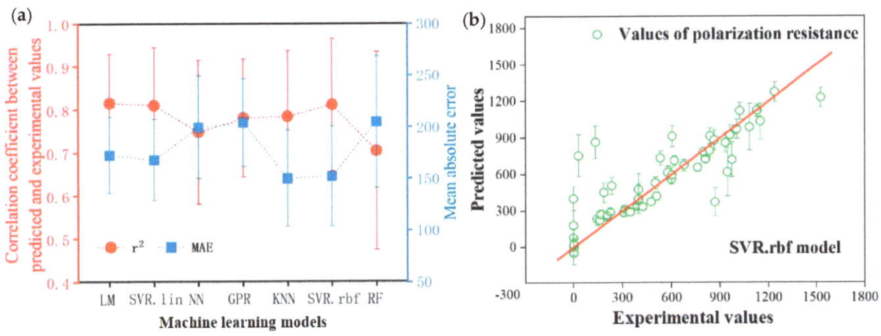

Figure 2. (**a**) Comparison of prediction performance of different machine learning models. (**b**) Prediction results of the Gaussian kernel support vector machine model (SVR.rbf).

3.3. *Establishment of Prediction Model*

Using the predictive capabilities of the Gaussian kernel support vector machine model (SVR.rbf), the distribution of polarization resistance in the steel reinforcement was obtained under varying Cr contents (0–5 wt%), LDH-NO_2 dosages (0–0.5 wt%), concrete chloride concentrations (0–3.5 wt%), and pore solution pH values (9.5–12.5). The original data for this predictive model were derived from electrochemical measurements taken during the direct immersion of steel reinforcement samples in simulated pore solutions. Consequently, this model is only applicable to conditions where the pore solution of the concrete or mortar is water-saturated.

Figure 3 depicts the evolution of the polarization resistance of steel reinforcement with environmental pH value and chloride concentration at a specific steel Cr content and LDH-NO_2 dosage. It is evident that, in the experimental system without the addition of the alloying element Cr and the smart inhibitor LDH-NO_2, the polarization resistance of the steel reinforcement exhibits a gradual decrease with a decrease in pH and an increase in chloride concentration. At a relatively high pH of 12.5, a notable decline in the polarization resistance of the steel reinforcement is only discernible at exceedingly elevated chloride concentrations. Conversely, at a relatively low pH of 10, a notable reduction in polarization resistance is observed at lower chloride concentrations. These findings are in accordance with the empirical observations.

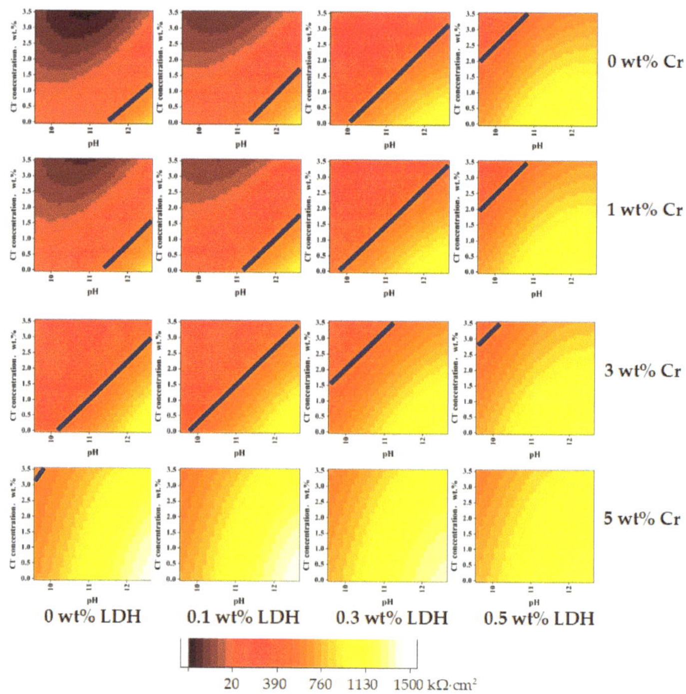

Figure 3. Influence of concrete pH value and chloride concentration on polarization resistance under specific steel reinforcement Cr contents and corrosion inhibitor LDH-NO_2 dosages.

Generally, in steel-reinforced concrete systems, corrosion is considered to be negligible when the corrosion rate is less than 0.1 $\mu A \cdot cm^{-2}$ and the corresponding polarization resistance is 520 $k\Omega \cdot cm^2$ [23]. Therefore, polarization resistances greater than 520 $k\Omega \cdot cm^2$ can be used as an indicator that corrosion has not yet occurred. For carbon steel reinforcement with a Cr content of 0, the polarization resistance starts below 520 $k\Omega \cdot cm^2$ when the environmental pH is below 12 and the chloride concentration exceeds 0.5 wt%. This indicates that carbon steel can only maintain a passive state in concrete with a pH greater than 12 and a chloride concentration less than 0.5 wt%. Therefore, using plain carbon steel reinforcement in marine concrete engineering presents a high corrosion risk.

It is evident that independently increasing the Cr content in the steel reinforcement or the LDH-NO_2 dosage in the concrete can enhance the polarization resistance of the steel reinforcement under identical environmental conditions. When the Cr content in steel increases to 1 wt%, the increase in polarization resistance is not significant, indicating a slight improvement in corrosion resistance compared to carbon steel. However, at a Cr content of 3 wt%, the polarization resistance begins to drop below 520 $k\Omega \cdot cm^2$ in the concrete environment with pH 11.5 and 1 wt% chloride, showing a notable improvement compared to carbon steel. At 5 wt% Cr content, the polarization resistance generally exceeds 520 $k\Omega \cdot cm^2$ within the pH range of 12.5 to 9.5 and chloride concentrations ranging from 0 to 3.5 wt%, indicating excellent corrosion resistance. Using the corrosion inhibitor LDH-NO_2 can also effectively reduce the corrosion rate of steel reinforcement. With an increase in LDH-NO_2 dosage, the polarization resistance of the steel improves relatively linearly. When the LDH-NO_2 dosage is 0.1 wt%, the polarization resistance of the steel reinforcement begins to drop below 520 $k\Omega \cdot cm^2$ in an environment with pH 12 and 0.7 wt% chloride. As the LDH-NO_2 dosage reaches 0.5 wt%, the polarization resistance starts to decrease below 520 $k\Omega \cdot cm^2$ in an environment with pH 10.5 and 2.5 wt% chloride. Moreover, adding 0.2 wt% LDH-NO_2 to the concrete has a similar effect to adding 3 wt% Cr to the steel, while adding 0.5 wt% LDH-NO_2 has a similar effect to adding 4 wt% Cr.

The combined use of Cr and LDH-NO_2 demonstrates greater effectiveness compared to using either alloy elements or corrosion inhibitors alone. In Figure 3, the blue line represents a scenario where the polarization resistance of the steel reinforcement measures 520 $k\Omega \cdot cm^2$. The upper left corner of the blue line indicates corroded steel reinforcement, while the lower right corner signifies non-corroded steel reinforcement. It is evident that the simultaneous application of the alloy element Cr and the corrosion inhibitor LDH-NO_2 expands the area in the lower right corner where the steel reinforcement remains non-corroded. When adding only the alloy element Cr, 5 wt% Cr is required to significantly expand the range of pH and chloride concentrations where the steel reinforcement remains non-corroded. When using only the corrosion inhibitor LDH-NO_2, 0.5 wt% LDH-NO_2 is needed to achieve the same effect. However, when 3 wt% Cr and 0.3 wt% LDH-NO_2 are used synergistically, the range of pH and chloride concentrations where the steel reinforcement remains non-corroded can be significantly expanded.

Figure 4 describes the required Cr content of the steel and LDH-NO_2 dosage to maintain the polarization resistance of steel above 520 $k\Omega \cdot cm^2$ at specific concrete pH values and chloride concentrations. In concrete severely affected by carbonation and chloride intrusion (pH 10 + 3.5 wt% Cl^-), to ensure adequate passivation of the steel reinforcement (polarization resistance > 520 $k\Omega \cdot cm^2$), steel with 5 wt% Cr is required, or alternatively, 3 wt% Cr modified steel combined with 0.4 wt% LDH-NO_2. In environments where carbonation is less severe but chloride intrusion is significant (pH 12 + 3.5 wt% Cl^-), steel with 3 wt% Cr is necessary for ensuring steel's passivation, or alternatively, steel modified with 1 wt% Cr combined with 0.4 wt% LDH-NO_2 can be considered.

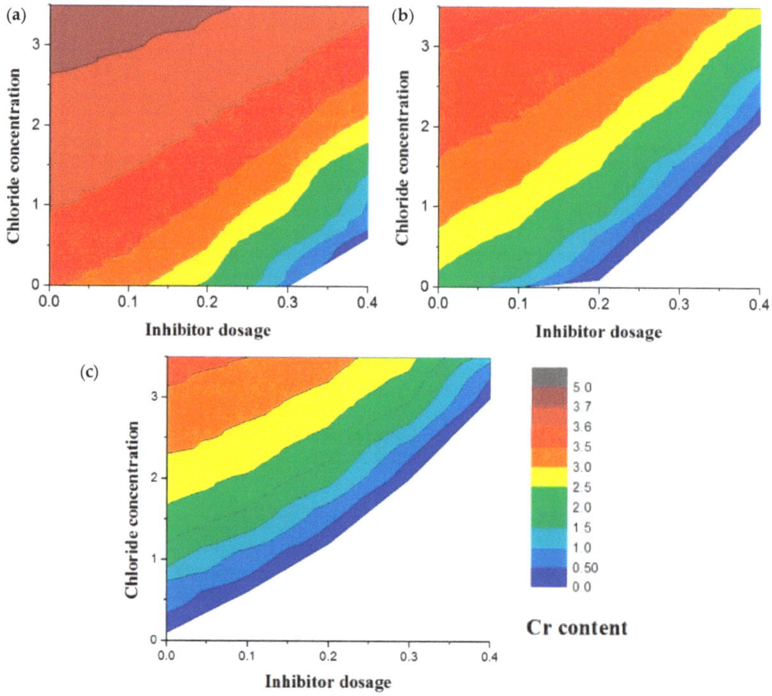

Figure 4. At a specified level of concrete carbonation, the necessary Cr content of the steel reinforcement and the dosage of the corrosion inhibitor LDH-NO_2 required to ensure a polarization resistance exceeding 520 $k\Omega\cdot cm^2$: (**a**) pH 10, (**b**) pH 11, (**c**) pH 12.

3.4. Validation of the Prediction Model

To verify the polarization resistance models obtained from Figure 3, a batch of Cr-modified steel bars and the smart corrosion inhibitor LDH-NO_2 were prepared in this study. Figure 5 shows the EIS of steel bars in the simulated concrete pore solution with different Cr contents and LDH-NO_2 dosages. A fitting was conducted using the equivalent circuit shown in Figure 6, where R_s represents the solution resistance, R_{ct} stands for the charge-transfer resistance, R_f for film resistance, R_p for polarization resistance, Q_{dl} for double-layer capacitance, and Q_f for film capacitance. Due to the rough and uneven electrode interface, a constant-phase element (Q) is employed to replace capacitance, with an impedance of

$$Z_Q = \frac{1}{Q_0(j\omega)^n}$$

where ω is the angular frequency, j is the imaginary unit, and n is the exponent of the constant-phase element. Furthermore, the total impedance (Z) of the system is

$$Z = R_s + \frac{1}{Z_{Q_{dl}} + \frac{1}{R_{ct} + \frac{1}{Z_{Q_f} + \frac{1}{R_f}}}}$$

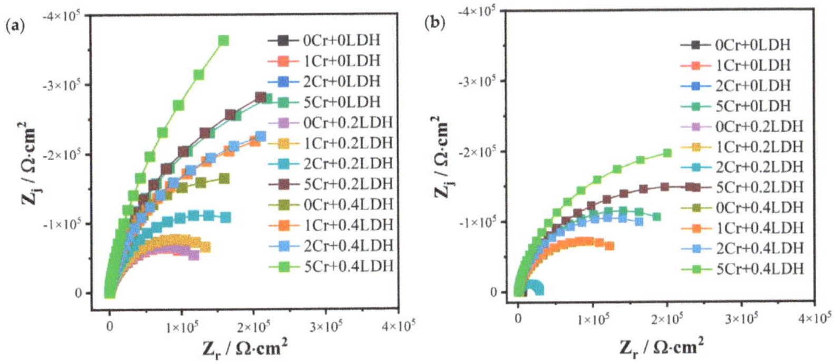

Figure 5. Electrochemical impedance spectra of steel bars in the simulated concrete pore solution with different Cr contents and LDH−NO_2 dosages: (**a**) pH 12 and 3.5 wt% chloride in the solution; (**b**) pH 10 and 3.5 wt% chloride in the solution.

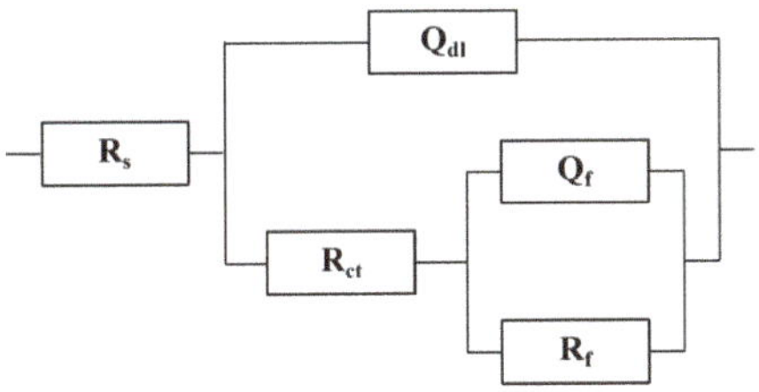

Figure 6. Equivalent circuit.

The fitting results of the EIS in Figure 5 are presented in Tables 2 and 3. Observably, as the Cr content or LDH-NO_2 dosage increases, the polarization resistance of the steel bar increases. The correlation between the measured polarization resistance (R_p) in Tables 2 and 3 and the predicted R_p from Figure 3 is depicted in Figure 7. It is evident that the measured R_p values closely correspond to the predicted ones, further confirming the efficacy of the predictive model.

Table 2. Parameters of electrochemical impedance spectra in Figure 5, where the pH of the solution was 12 and the chloride concentration was 3.5 wt%.

Cr Content	LDH-NO_2 Dosage	Q_f ($\mu\Omega^{-1}\cdot cm^{-2}\cdot s^n$)	n	R_f ($k\Omega\cdot cm^2$)	Q_{dl} ($\mu\Omega^{-1}\cdot cm^{-2}\cdot s^n$)	n	R_{ct} ($k\Omega\cdot cm^2$)	$R_p = R_f + R_{ct}$ ($k\Omega\cdot cm^2$)
0 wt%	0 wt%	59	0.92	91	32	0.89	57	148
1 wt%	0 wt%	2	0.67	0.02	54	0.91	161	161
2 wt%	0 wt%	38	0.91	154	115	1	24	178
5 wt%	0 wt%	35	0.94	140	18	0.81	116	256
0 wt%	0.2 wt%	179	0.96	0.008	204	0.92	278	278
1 wt%	0.2 wt%	15	1	0.1	16	0.84	502	502
2 wt%	0.2 wt%	28	0.93	448	16	0.96	215	663
5 wt%	0.2 wt%	16	1	0.1	14	0.85	694	694
0 wt%	0.4 wt%	1	0.65	0.07	38	0.92	416	416
1 wt%	0.4 wt%	0.02	0.79	0.04	44	0.90	612	612
2 wt%	0.4 wt%	27	0.93	0.001	2	0.44	647	647
5 wt%	0.4 wt%	16	1	0.09	15	0.86	1362	1362

Table 3. Parameters of electrochemical impedance spectra in Figure 5, where the pH of the solution was 10 and the chloride concentration was 3.5 wt%.

Cr Content	LDH-NO_2 Dosage	Q_f ($\mu\Omega^{-1}\cdot cm^{-2}\cdot s^n$)	n	R_f ($k\Omega\cdot cm^2$)	Q_{dl} ($\mu\Omega^{-1}\cdot cm^{-2}\cdot s^n$)	n	R_{ct} ($k\Omega\cdot cm^2$)	$R_p = R_f + R_{ct}$ ($k\Omega\cdot cm^2$)
0 wt%	0 wt%	141	0.9	0.03	194	0.89	1	1
1 wt%	0 wt%	220	0.89	0.02	276	0.88	2	2
2 wt%	0 wt%	126	0.86	0.6	20	0.87	6	7
5 wt%	0 wt%	29	0.92	208	62	0.99	56	264
0 wt%	0.2 wt%	102	0.95	0.09	84	0.92	2	2
1 wt%	0.2 wt%	99	0.93	0.1	58	0.9	2	2
2 wt%	0.2 wt%	4	0.76	0.02	62	0.89	30	30
5 wt%	0.2 wt%	2	0.56	0.04	24	0.94	351	351
0 wt%	0.4 wt%	41	0.92	103	28	0.81	68	171
1 wt%	0.4 wt%	38	0.92	98	34	0.8	75	173
2 wt%	0.4 wt%	35	0.92	232	0.5	0.38	0	232
5 wt%	0.4 wt%	16	0.99	0.1	16	0.83	466	466

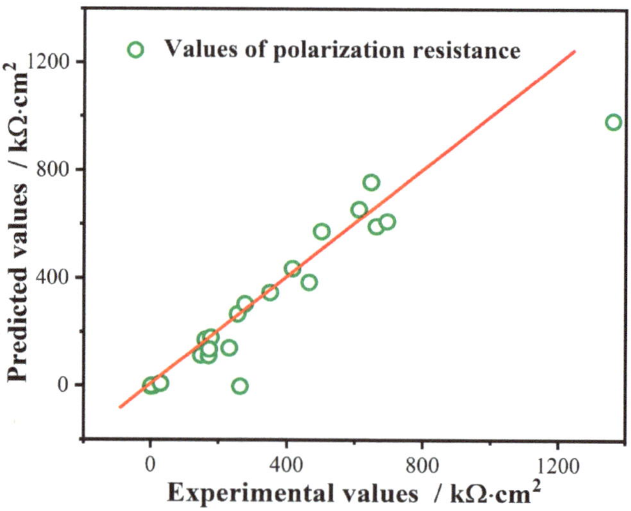

Figure 7. The correlation between the measured polarization resistance (R_p) in Tables 2 and 3 and the predicted R_p from Figure 3.

4. Conclusions

Based on a data-driven research approach, this study employed machine learning algorithms to develop a quantitative predictive model for assessing the correlations between Cr content in steel reinforcement, LDH-NO_2 dosage, concrete chloride concentration, pH value, and the polarization resistance of steel reinforcement. Unlike traditional experimental methods focusing on single control factors, the robust data processing capabilities of machine learning enable rapid analysis of multi-factor interactions affecting steel reinforcement's polarization resistance. It also provides more precise and efficient predictions of the corrosion rate of steel reinforcement under the combined influence of these factors compared to traditional methods. This approach better reflects real-world engineering conditions in steel reinforcement service. The precise and efficient predictive model offers quantitative guidance for selecting alloy elements and corrosion inhibitors in diverse environments, introducing novel methodologies and perspectives to corrosion research in steel reinforcement.

(1) The Gaussian kernel support vector machine model (SVR.rbf), obtained through data processing and the analysis and evaluation of machine learning models, exhibited excellent predictive accuracy and stability. The correlation coefficient between the prediction results and experimental values of polarization resistance of the steel reinforcement exceeded 0.85. This provides an accurate and efficient technical method for the prediction of polarization resistance under multi-factor coupling.
(2) By means of the generalization prediction performance of machine learning models, they provide quantitative reference results for steel reinforcement composition design and corrosion inhibitor selection under actual service conditions. The addition of 1 wt% Cr to steel slightly increases its polarization resistance. However, when the Cr content reaches 3 wt%, the polarization resistance of the steel reinforcement increases significantly; when the Cr content reaches 5 wt%, in the range of pH 12.5~9.5 and containing 0~3.5 wt% chloride, the steel reinforcement's corrosion is essentially negligible.
(3) The addition of the corrosion inhibitor LDH-NO_2 to concrete gradually reduces the corrosion rate of the steel reinforcement. Adding 0.2 wt% LDH-NO_2 in the environment has a similar effect to adding 3 wt% Cr in the steel reinforcement, while adding 0.5 wt% LDH-NO_2 is comparable to adding 4 wt% Cr.
(4) The synergistic use of Cr and LDH-NO_2 is more effective in the inhibition of steel reinforcement corrosion than their individual use. With a Cr content of 3 wt% and an LDH-NO_2 concentration of 0.5 wt%, it can be ensured that the steel reinforcement's corrosion is negligible in the range of pH 12.5~9.5 and containing 0~3.5 wt% chloride.

Author Contributions: Conceptualization, C.W. and Y.T.; methodology, B.C., G.L. and N.W.; investigation, C.W. and Y.T.; writing—original draft preparation, C.W.; writing—review and editing, C.W., B.C., G.L., N.W. and Y.T.; supervision, N.Y.; resources, N.Y.; funding acquisition, C.W. and Y.T. All authors have read and agreed to the published version of the manuscript.

Funding: This research was funded by the Natural Science Foundation of China (No. 52203293), the Natural Science Foundation of Guangdong, China (No. 2021A1515110382), and the Program for Innovation Team of Guangdong Ocean University (CXTD2024008).

Data Availability Statement: The raw data supporting the conclusions of this article will be made available by the authors on request.

Conflicts of Interest: The authors declare that they have no known competing financial interests or personal relationships that could have appeared to influence the work reported in this paper.

Appendix A

Table A1. Training data: correlations between Cr content in steel, LDH-NO_2 dosage, chloride concentration in concrete, environmental pH value in pore solution, and polarization resistance of steel reinforcement in the simulated concrete pore solution.

Cr Content (wt%)	LDH-NO_2 Dosage (wt%)	Cl^- Concentration (wt%)	pH Value	Polarization Resistance (kΩcm^2)	Ref.
0	0	0	13.3	75	[20]
0	0	0	13.3	262	[20]
0	0	0	12.5	1000	[17]
0	0	0	13.5	604	[17]
0	0	0	12.0	457	[20]
0	0	0	10.5	591	[20]
0	0	0	9.0	433	[20]

Table A1. *Cont.*

Cr Content (wt%)	LDH-NO_2 Dosage (wt%)	Cl^- Concentration (wt%)	pH Value	Polarization Resistance (kΩcm^2)	Ref.
0	0	0	13.5	344	[20]
0	0	0	12.0	311	[20]
0	0	0	10.5	42	[20]
0	0	0	9.0	10	[20]
0	0	0.117	12.5	400	[21]
0	0	0.234	12.5	260	[21]
0	0	0.351	12.5	6	[21]
0	0	0.468	12.5	5	[21]
0	0	0.585	12.5	3	[21]
0	0	0.819	12.5	2	[21]
0	0	0.936	12.5	0.8	[21]
0	0	1.000	12.5	60	[21]
0	0	1.000	12.5	70	[21]
0	0	1.170	12.5	0.6	[21]
0	0	2.530	12.5	238	[22]
0	0	2.925	12.5	0.6	[21]
0	0	5.850	12.5	0.8	[21]
0	0	5.850	12.5	1	[21]
0.08	0	0	12.6	867	[7]
0.08	0	0.200	12.6	839	[7]
0.08	0	0.300	12.3	679	[7]
0.08	0	0.400	12.6	813	[7]
0.08	0	0.450	12.2	603	[7]
0.08	0	0.600	11.9	871	[7]
0.08	0	0.750	11.7	2	[7]
0.08	0	0.800	12.6	578	[7]
0.08	0	0.900	10.9	1	[7]
0.08	0	1.000	12.6	400	[7]
0.08	0	1.200	12.6	236	[7]
0.08	0	1.200	10.1	2	[7]
0.08	0	1.350	9.2	2	[7]
0.08	0	1.400	12.6	186	[7]
0.08	0	1.500	8.8	2	[7]
0.08	0	1.650	8.4	2	[7]
0.86	0	0	12.5	1800	[18]
0.86	0	0.117	12.5	2000	[18]
0.86	0	0.234	12.5	1000	[18]
0.86	0	0.351	12.5	8	[18]
0.86	0	0.468	12.5	7	[18]
0.86	0	0.585	12.5	12	[21]
0.86	0	0.819	12.5	4	[18]
0.86	0	0.936	12.5	2	[18]
0.86	0	1.000	12.5	90	[18]
0.86	0	1.000	12.5	240	[18]
0.86	0	1.170	12.5	1	[18]
0.86	0	2.925	12.5	0.7	[21]
0.86	0	5.850	12.5	2	[21]
0.86	0	5.850	12.5	1.5	[21]
1.00	0	3.530	12.5	393	[19]
1.50	0	0	12.6	963	[19]
1.50	0	1.000	12.6	618	[19]
1.50	0	2.000	12.6	508	[19]
1.50	0	2.200	12.6	476	[19]
1.50	0	2.600	12.6	348	[19]
1.50	0	2.800	12.6	211	[19]

Table A1. *Cont.*

Cr Content (wt%)	$LDH-NO_2$ Dosage (wt%)	Cl^- Concentration (wt%)	pH Value	Polarization Resistance ($k\Omega cm^2$)	Ref.
1.50	0	3.000	12.6	161	[19]
3.00	0	4.530	12.5	650	[19]
5.06	0	1.000	12.6	1529	[19]
5.06	0	2.000	12.6	1238	[19]
5.06	0	3.000	12.6	1130	[19]
5.06	0	4.000	12.6	1000	[19]
5.06	0	5.000	12.6	800	[19]
5.06	0	6.000	12.6	600	[19]
5.06	0	7.000	12.6	400	[19]
5.06	0	7.200	12.6	310	[19]
5.06	0	7.400	12.6	170	[19]
5.06	0	7.600	12.6	146	[19]
5.73	0	0	12.5	3000	[18]
5.73	0	0.117	12.5	700	[18]
5.73	0	0.234	12.5	700	[18]
5.73	0	0.351	12.5	800	[18]
5.73	0	0.468	12.5	550	[18]
5.73	0	0.585	12.5	2353	[18]
5.73	0	0.819	12.5	1000	[18]
5.73	0	0.936	12.5	260	[18]
5.73	0	1.170	12.5	260	[18]
5.73	0	2.925	12.5	260	[18]
5.73	0	5.850	12.5	50	[18]
5.73	0	5.850	12.5	9	[18]
6.00	0	0	13.3	447	[9]
6.00	0	0	13.3	2067	[9]
10.36	0	0	9.0	2673	[16]
10.36	0	0	9.0	2055	[20]
10.36	0	0	10.5	2284	[16]
10.36	0	0	10.5	979	[20]
10.36	0	0	12.0	1770	[16]
10.36	0	0	12.0	481	[20]
10.36	0	0	13.3	1305	[16]
10.36	0	0	13.3	388	[20]
0.08	0.5	0.300	12.0	1148	[15]
0.08	0.5	0.450	11.5	1023	[15]
0.08	0.5	0.600	11.0	1004	[15]
0.08	0.5	0.750	10.7	839	[15]
0.08	0.5	0.900	10.7	935	[15]
0.08	0.5	1.200	10.6	974	[15]
0.08	0.5	1.350	10.6	947	[15]
0.08	0.5	1.500	10.6	134	[15]
0.08	0.5	1.650	10.0	31	[15]

Table A2. Training data after re-cleaning: correlations between Cr content in steel, $LDH-NO_2$ dosage, chloride concentration in concrete, environmental pH value in pore solution, and polarization resistance of steel reinforcement in the simulated concrete pore solution.

Cr Content (wt%)	$LDH-NO_2$ Dosage (wt%)	Cl^- Concentration (wt%)	pH Value	Polarization Resistance ($k\Omega cm^2$)
0.08	0	0	12.6	867
0.08	0	0.2	12.6	839
0.08	0	0.4	12.6	813
0.08	0	0.8	12.6	578

Table A2. *Cont.*

Cr Content (wt%)	LDH-NO_2 Dosage (wt%)	Cl^- Concentration (wt%)	pH Value	Polarization Resistance (kΩcm^2)
0.08	0	1.0	12.6	400
0.08	0	1.2	12.6	236
0.08	0	1.4	12.6	186
1.5	0	0	12.6	963
1.5	0	1.0	12.6	618
1.5	0	2.0	12.6	508
1.5	0	2.2	12.6	476
1.5	0	2.4	12.6	424
1.5	0	2.6	12.6	348
1.5	0	2.8	12.6	211
1.5	0	3.0	12.6	161
5.06	0	1.0	12.6	1529
5.06	0	2.0	12.6	1238
5.06	0	3.0	12.6	1130
5.06	0	4.0	12.6	1000
5.06	0	5.0	12.6	800
5.06	0	6.0	12.6	600
5.06	0	7.0	12.6	400
5.06	0	7.2	12.6	310
5.06	0	7.4	12.6	170
5.06	0	7.6	12.6	146
0.08	0.5	0.3	12.0	1148
0.08	0.5	0.45	11.5	1023
0.08	0.5	0.6	11.0	1004
0.08	0.5	0.75	10.7	839
0.08	0.5	0.9	10.7	935
0.08	0.5	1.2	10.6	974
0.08	0.5	1.35	10.6	947
0.08	0.5	1.5	10.6	134
0.08	0.5	1.65	10.0	31
0.08	0	0.3	12.3	679
0.08	0	0.45	12.2	603
0.08	0	0.6	11.9	871
0.08	0	0.75	11.7	1.672
0.08	0	0.9	10.9	1.362
0.08	0	1.2	10.1	1.846
0.08	0	1.35	9.2	1.822
0.08	0	1.5	8.8	2.495
0.08	0	1.65	8.4	1.556

References

1. Ma, D.N.; Zhang, M.H.; Cui, J.Y. A review on the deterioration of mechanical and durability performance of marine-concrete under the scouring action. *J. Build. Eng.* **2023**, *66*, 105924–105953. [CrossRef]
2. Rabi, M.; Shamass, R.; Cashell, K. Structural performance of stainless steel reinforced concrete members: A review. *Constr. Build. Mater.* **2022**, *325*, 126673–126686. [CrossRef]
3. Qureshi, T.; Wang, G.; Mukherjee, S.; Islam, A.; Filleter, T.; Singh, C.V.; Panesar, D.K. Graphene-based anti-corrosive coating on steel for reinforced concrete infrastructure applications: Challenges and potential. *Constr. Build. Mater.* **2022**, *351*, 128947–128965. [CrossRef]
4. Xu, L.; Xin, Y.; Ma, L.; Zhang, H.; Lin, Z.; Li, X. Challenges and solutions of cathodic protection for marine ships. *Corros. Commun.* **2021**, *2*, 33–40. [CrossRef]
5. Wang, X.; Liu, J.; Jin, M.; Yan, Y.; Tang, J.; Jin, Z. A review of organic corrosion inhibitors for resistance under chloride attacks in reinforced concrete: Background, Mechanisms and Evaluation methods. *Constr. Build. Mater.* **2024**, *433*, 136583–136600. [CrossRef]
6. Jin, Z.; Xiong, C.; Zhao, T.; Du, Y.; Zhang, X.; Li, N.; Yu, Y.; Wang, P. Passivation and depassivation properties of Cr-Mo alloyed corrosion-resistant steel in simulated concrete pore solution. *Cem. Concr. Compos.* **2022**, *126*, 104375–104387. [CrossRef]

7. Zhang, L.; Niu, D.; Wen, B.; Fu, Q.; Peng, G.; Su, L.; Blackwood, D.J. Initial-corrosion condition behavior of the Cr and Al alloy steel bars in coral concrete for marine construction. *Cem. Concr. Compos.* **2021**, *120*, 104051–104064. [CrossRef]
8. Ming, J.; Wu, M.; Shi, J.J. Passive film modification by concrete carbonation: Re-visiting a corrosion-resistant steel with Cr and Mo. *Cem. Concr. Compos.* **2021**, *123*, 104178–104193. [CrossRef]
9. Shi, J.; Ming, J.; Wang, D.; Wu, M. Improved corrosion resistance of a new 6% Cr steel in simulated concrete pore solution contaminated by chlorides. *Corros. Sci.* **2020**, *174*, 108851. [CrossRef]
10. Liu, T.; Chen, T.; Zhou, X.; Sun, L.; Yang, W.; Liu, C.; Cheng, X.; Li, X. Investigation of Cr and rare earth (RE) on the corrosion resistance of HRB400 rebar in simulated concrete pore solutions containing chloride and sulfate ions. *Constr. Build. Mater.* **2024**, *423*, 135935–135948. [CrossRef]
11. Pan, D.; Niu, D.T.; Li, Z.J. Corrosion products of low-alloy steel bars and their induction of cracking in seawater sea-sand concrete cover. *Constr. Build. Mater.* **2023**, *389*, 131800–131809. [CrossRef]
12. Wen, C.; Tian, Y.; Mai, Z.; Hu, J.; Wang, G. Effect of macropores at the steel-concrete interface on localized corrosion behaviour of steel reinforcement. *Cem. Concr. Compos.* **2022**, *129*, 104510–104520. [CrossRef]
13. Habib, S.; Shakoor, R.; Kahraman, R. A focused review on smart carriers tailored for corrosion protection: Developments, applications, and challenges. *Prog. Org. Coat.* **2021**, *154*, 106218–106249. [CrossRef]
14. Yang, H.; Xiong, C.; Liu, X.; Liu, A.; Li, T.; Ding, R.; Shah, S.P.; Li, W. Application of layered double hydroxides (LDHs) in corrosion resistance of reinforced concrete-state of the art. *Constr. Build. Mater.* **2021**, *307*, 124991. [CrossRef]
15. Tian, Y.; Wen, C.; Wang, G.; Deng, P.; Mo, W. Inhibiting property of nitrite intercalated layered double hydroxide for steel reinforcement in the contaminated concrete pore solution. *J. Electrochem. Appl.* **2020**, *50*, 835–849. [CrossRef]
16. Ai, Z.; Jiang, J.; Sun, W.; Jiang, X.; Yu, B.; Wang, K.; Zhang, Z.; Song, D.; Ma, H.; Zhang, J. Enhanced passivation of alloy corrosion-resistant steel Cr10Mo1 under carbonation-passive film formation, the kinetics and mechanism analysis. *Cem. Concr. Compos.* **2018**, *92*, 178–187. [CrossRef]
17. Ming, J.; Shi, J.J. Distribution of corrosion products at the steel-concrete interface: Influence of mill scale properties, reinforcing steel type and corrosion inducing method. *Constr. Build. Mater.* **2019**, *229*, 116854–116865. [CrossRef]
18. Shi, J.J.; Ming, J.; Wu, M. Electrochemical behavior and corrosion products of Cr-modified reinforcing steels in saturated $Ca(OH)_2$ solution with chlorides. *Cem. Concr. Compos.* **2020**, *110*, 103587–103598. [CrossRef]
19. Liu, M.; Cheng, X.; Li, X.; Lu, T.J. Corrosion behavior of low-Cr steel rebars in alkaline solutions with different pH in the presence of chlorides. *J. Electroanal. Chem.* **2017**, *803*, 40–50. [CrossRef]
20. Ai, Z.; Jiang, J.; Sun, W.; Song, D.; Ma, H.; Zhang, J.; Wang, D. Passive behaviour of alloy corrosion-resistant steel Cr10Mo1 in simulating concrete pore solutions with different pH. *Appl. Surf. Sci.* **2016**, *389*, 1126–1136. [CrossRef]
21. Shi, J.; Sun, W.; Jiang, J.; Zhang, Y. Influence of chloride concentration and pre-passivation on the pitting corrosion resistance of low-alloy reinforcing steel in simulated concrete pore solution. *Constr. Build. Mater.* **2016**, *111*, 805–813. [CrossRef]
22. Zhang, L.; Niu, D.; Wen, B.; Peng, G.; Sun, Z. Corrosion behavior of low alloy steel bars containing Cr and Al in coral concrete for ocean construction. *Constr. Build. Mater.* **2020**, *258*, 119564–119577. [CrossRef]
23. Stefanoni, M.; Angst, U.; Elsener, B. Corrosion rate of carbon steel in carbonated concrete—A critical review. *Cem. Concr. Res.* **2018**, *103*, 35–48. [CrossRef]

materials MDPI

Article

Fabrication and Characterization of Biomedical Ti-Mg Composites via Spark Plasma Sintering

Taisei Masuda, Minho Oh * and Equo Kobayashi

Department of Materials Science and Engineering, School of Materials and Chemical Technology, Tokyo Institute of Technology, S8-18, 2-12-1 Ookayama, Meguro-ku, Tokyo 152-8552, Japan; equo@mtl.titech.ac.jp (E.K.)
* Correspondence: o.m.aa@m.titech.ac.jp

Abstract: The fabrication of Ti-Mg composite biomaterials was investigated using spark plasma sintering (SPS) with varying Mg contents and sintering pressures. The effects of powder mixing, Mg addition, and sintering pressure on the microstructure and mechanical properties of the composite materials were systematically analyzed. Uniform dispersion of Mg within the Ti matrix was achieved, confirming the efficacy of ethanol-assisted ball milling for consistent mixing. The Young's modulus of the composite materials exhibited a linear decrease with increasing Mg content, with Ti-30vol%Mg and Ti-50vol%Mg demonstrating reduced modulus values compared to pure Ti. Based on density measurements, compression tests, and Young's modulus results, it was determined that the sinterability of Ti-30vol%Mg saturates at a sintering pressure of approximately 50 MPa. Moreover, our immersion tests in physiological saline underscore the profound significance of our findings. Ti-30vol%Mg maintained compressive strength above that of cortical bone for 6-to-10 days, with mechanical integrity improving under higher sintering pressures. These findings mark a significant leap towards the development of Ti-Mg composite biomaterials with tailored mechanical properties, thereby enhancing biocompatibility and osseointegration for a wide range of biomedical applications.

Keywords: spark plasma sintering; Ti-Mg composites; biomaterials; stress shielding; Young's modulus

Citation: Masuda, T.; Oh, M.; Kobayashi, E. Fabrication and Characterization of Biomedical Ti-Mg Composites via Spark Plasma Sintering. *Materials* **2024**, *17*, 3470. https://doi.org/10.3390/ma17143470

Academic Editors: Jingxiang Xu, Xingwei Zheng and Zhenhua Chu

Received: 15 June 2024
Revised: 11 July 2024
Accepted: 11 July 2024
Published: 13 July 2024

1. Introduction

In recent years, with the increase in demand for medical care due to aging and longevity, there is an expectation for the development of high-functioning and high-value-added biomaterials. Metallic biomaterials are often used as load-bearing components, taking advantage of their high fracture toughness and fatigue strength. Alongside existing biomaterials such as stainless steel (SUS316L), Co-Cr alloys, industrial pure titanium, and titanium alloys, functional metallic biomaterials such as shape memory alloys, gradient function materials, and biodegradable Mg materials have been developed [1,2]. Among these, industrial pure titanium and titanium alloys are widely used in dental and surgical implants due to their high specific strength and excellent corrosion resistance. While the Young's modulus of titanium alloys (approximately 110 GPa) is significantly smaller compared to other metallic biomaterials like stainless steel or Co-Cr alloys, it is still much larger than the Young's modulus of cortical bone (7–30 GPa) [2–4]. This difference leads to stress shielding, where a significant portion of stress is preferentially borne by the implant, potentially causing an inhibition of bone growth and a decrease in bone density [5]. Therefore, efforts have been made to develop materials such as β-type titanium alloys and porous materials to decrease Young's modulus [6,7].

It has been found that the Young's modulus of β-type titanium alloys is lower compared to α-type titanium alloys, and the non-toxic Ti-29Nb-13Ta-4.6Zr (TNTZ) alloy, which exhibits the lowest Young's modulus of approximately 60 GPa, has been developed [2,8,9]. Additionally, porous materials have a lower apparent Young's modulus and can enhance osseointegration, where implants bond with living bone at the optical microscope level, allowing for a sufficient fixation of implants [10]. However, there are challenges with β-type titanium alloys, such as the high cost and high melting point of alloying elements like Nb, Ta, Mo, or Zr [11]. On the other hand, porous materials often experience stress concentration in pore regions, leading to inferior mechanical properties, making them more suitable for low-load environments [12].

To address these challenges, research on Ti-Mg composite materials has been conducted [13–15]. Magnesium is an essential element in the body and has a modulus of 41 GPa, which is close to the modulus of cortical bone compared to other metallic biomaterials and is cost-effective [16,17]. Since the modulus of composite materials is roughly proportional to the volume fraction, compounding magnesium with a lower modulus than titanium results in Ti-Mg composite materials having a lower modulus than pure titanium and titanium alloys [18,19]. Moreover, when biodegradable magnesium dissolves in the body, the originally magnesium-containing parts transform into pores, changing Ti-Mg composite materials into porous titanium [20,21]. Therefore, Ti-Mg composite materials exhibit properties that initially have superior strength compared to porous titanium due to the presence of magnesium during implantation. As magnesium dissolves during bone recovery and growth, it transforms into porous titanium with low modulus and excellent osseointegration with living bone [22–24]. The density and melting point of Ti are 4.506 g/cm^3 and 1668 °C, respectively, while the density and melting point of Mg are 1.738 g/cm^3 and 650 °C, respectively. Due to the significant difference in density and melting point, obtaining Ti-Mg composite materials with a uniform structure using conventional casting methods is difficult [25]. Powder metallurgy is practical for manufacturing such metal composite materials with significant differences in properties. Powder metallurgy is a method of producing dense materials by diffusion of metal atoms between powder particles, eliminating the need to melt the metal and allowing for material fabrication at lower temperatures compared to casting methods [26]. In manufacturing Ti-Mg composite materials, the Spark Plasma Sintering (SPS) method has several advantages over other methods [27], such as liquid Mg infiltration, as shown in Table 1, making it suitable for biomedical applications. Specifically, SPS allows for the precise control of sintering temperature and pressure, ensuring uniform microstructural development and enhanced densification. This accurate control minimizes thermal gradients and results in superior mechanical properties. Additionally, SPS significantly reduces processing time compared to liquid infiltration methods and allows for the fine-tuning of microstructural characteristics through adjustable sintering parameters. This flexibility enables tailoring mechanical properties to meet specific biomedical requirements, such as desired Young's modulus and strength [28].

Table 1. Comparison between spark plasma sintering (SPS) and liquid Mg-infiltration methods.

Feature	Spark Plasma Sintering (SPS)	Liquid Mg Infiltration
Temperature Control	Precise, localized heating reduces thermal gradients.	High temperatures required for Mg melting (>650 °C).
Reaction Time	Short sintering times due to efficient heating and pressure.	Longer processing times are needed for complete infiltration.
Microstructure Control	Fine control over microstructure through adjustable parameters.	Dependent on porous Ti structure and Mg flow dynamics.
Porosity Control	Ability to produce dense and controlled porosity composites.	Porosity depends on the initial Ti structure.
Material Homogeneity	Uniform distribution of Mg within Ti matrix.	Potential for uneven Mg distribution.
Mechanical Properties	Tunable properties through precise control of sintering parameters.	Limited by the inherent properties of porous Ti and Mg.
Scalability	Suitable for small-to-medium-scale production.	It can be challenging for large-scale uniform production.
Equipment and Cost	Requires SPS equipment, potentially high initial cost.	Lower equipment cost but higher operational complexity.
Post-processing Needs	Minimal, often no additional machining is required.	It may require additional machining to achieve the final shape.
Applications	Ideal for biomedical implants with tailored properties.	Suitable for large implants where high porosity is needed.

In this study, Ti-Mg composite materials, which are difficult to manufacture using conventional casting methods, were fabricated using SPS, one of the powder metallurgy techniques. SPS involves the intermittent flow of direct current through the sample, causing discharge between powder particles, heating by joule heating, and pressure application, effectively shortening the sintering time due to rapid heating rates [29,30]. Moreover, materials fabricated using SPS typically exhibit high-density uniform structures, and the short sintering time reduces the likelihood of grain coarsening, resulting in sintered bodies with excellent mechanical properties [28,31]. For the Ti-Mg composite materials produced by changing the composition and sintering pressure, their mechanical properties were evaluated by observing the microstructure, compression tests, and Young's modulus measurements. How changes in the amount of Mg added and sintering pressure affect the microstructure and mechanical properties was investigated. Additionally, Mg was dissolved by immersing the Ti-Mg composite materials in a physiological saline solution. Mg toxicity typically results from an excessive intake of medications containing magnesium or impaired kidney excretion [32]. Therefore, it is crucial to regulate and monitor its solubility carefully. The compression test was performed on samples immersed for different durations to investigate the mechanical integrity of the composite materials.

2. Materials and Methods

Using pure Ti powder (purity 99.98%, spherical, maximum particle size 45 μm) and pure Mg powder (purity 99.5%, irregular shape, average particle size 180 μm) as raw materials, three types of powder were prepared: pure Ti, Ti-30vol%Mg, and Ti-50vol%Mg. Pure Ti remained pure Ti powder, while a pulverizing ball mill (Pulverisette 7 classicline, Fritsch, Germany) was used to mix the composite powder. A stainless-steel mixing container was filled with 50 wt% ZrO_2 balls with a diameter of 1 mm and 5 wt% ethanol as the mixing solvent. After mixing for 5 min at 500 rpm, the rotation was stopped for 5 min, then reversed, and mixing was continued for another 5 min at 500 rpm. After mixing, ZrO_2 balls were separated from the mixed powder using a 600 μm sieve. The purpose of the ball mill in this study is to prevent powder aggregation by solvent and aim for a more uniform mixture rather than particle crushing. The powder-mixing operation was conducted in a glove box filled with Ar gas. Before the sintering process, X-ray diffraction (XRD, MiniFlex600 by Rigaku, Tokyo, Japan) analysis was performed with angle conditions set at 20°–90° and a step size of 0.01° on three types of powders (pure Ti powder, pure Mg powder, and Ti-30Mg mixed powder), to verify their composition and crystallinity [33,34]. Carbon paper with a thickness of 0.2 mm was placed on the bottom and sides of the graphite die (NJS-Japan, Tokyo, Japan), and each powder was filled into the graphite die inside a glove box. The powders were then compressed using a hydraulic pump at 20 MPa for 1 min to obtain a compacted body. The sintering container, now containing the compacted body, was carefully installed in the chamber of the SPS device (511S, SPS Syntex, Tokyo, Japan). The vacuum level inside the chamber was meticulously maintained at 50 Pa. The sintering temperature, a critical parameter that directly influences the final properties of the sintered samples, was set to 580 °C. Similarly, the sintering pressure, another key parameter that significantly affects the sintering process, was set to 25, 50, 75, and 100 MPa, respectively. These settings were crucial to achieving the desired sintering results. After sintering, the samples were unloaded without waiting for cooling and cooled while maintaining the vacuum inside the chamber. Table 2 shows the Mg content and sintering conditions of the fabricated samples.

Table 2. Specimen conditions for Mg content and sintering pressure.

Specimen	Mg Content (Vol%)	Sintering Pressure (MPa)
Pure Ti	0	50
Ti-30Mg	30	50
Ti-50Mg	50	50
Ti-30Mg (TM25)	30	25
Ti-30Mg (TM75)	30	75
Ti-30Mg (TM100)	30	100

The sintered specimen was cut using a microcutting system (Accustom-5, Struers, Tokyo, Japan), and the cross-section was polished with #500 to #4000 SiC emery papers, followed by polishing with 3 μm diamond spray as the abrasive. After ultrasonic cleaning with isopropanol for 5 min., observations were made using an optical microscope (OM) (DMI3000M, Leica Microsystems, Wetzlar, Germany) and field emission scanning electron microscopy (FESEM) (JSM-7200F, JEOL, Tokyo, Japan), and an elemental analysis and element distribution map (EDM) were performed using energy dispersive spectroscopy (EDS) (JED-2300 Analysis Station Plus, JEOL, Tokyo, Japan). The powder, fracture surface, and post-infiltration structure were also observed using FESEM without polishing.

The sintered specimens were cut into dimensions of 4 mm × 4 mm × 8 mm and polished using #1000 SiC emery paper. The vertical, horizontal, and height dimensions of the samples were measured three times using a micrometer, and the average values were used to determine the dimensions. The weight of each sample was measured to calculate its density. Compression tests were conducted using an Autograph universal testing machine (AG-1 1000 kN, Shimadzu, Tokyo, Japan). Nominal strain and nominal stress were calculated using the above measurement values in the compression tests. The tests were performed three times for each specimen, and stress–strain curves were plotted based on the results. To measure Young's modulus, an ultrasonic pulse velocity test was conducted. This is a non-destructive technique that involves propagating ultrasonic pulses through the sample using longitudinal and transverse wave transducers. Young's modulus was then determined based on the velocity of the ultrasonic pulses.

Using SiC emery paper ranging from #500 to #4000, the bottom of the sintered specimens was polished to be parallel, then polished with a diamond spray (3 μm). After preparing cylindrical specimens with a diameter of approximately 20 mm and a thickness of about 9 mm, the thickness of the samples was measured at seven points using a micrometer, and the average of five data points, excluding the maximum and minimum values, was taken as the thickness of the sample. For the measurements, an ultrasonic flaw detector (USM35X, GE Measurement and Control, MA, USA), longitudinal wave probe (G5KB, GE Measurement and Control, MA, USA), and transverse wave probe (B2C10SN, ITeS Corporation, Tokyo, Japan) were used. Measurements were conducted seven times for both longitudinal and transverse waves, and the average of five data points, excluding the maximum and minimum values, was taken as the velocity to calculate Young's modulus. Due to the insufficient size of the immersion test samples, a unique approach was taken to calculate Young's modulus [35,36]. The slope in the elastic region of the stress–strain curve obtained from the compression test was used. However, it is important to note that the Young's modulus calculated from this slope in the compression test shows lower values than the usual Young's modulus. Therefore, the evaluation of the Young's modulus change according to the immersion time was conducted by setting 0 days of immersion as 100% and evaluating the decrease rate of Young's modulus.

For the immersion test, two types of sintered materials were prepared for compression testing and observation, respectively, and each sintered material was cut into rectangular shapes of 4 mm × 4 mm × 8 mm. The sides were polished with #1000 SiC emery paper, and then the height, width, and length were each measured three times using a micrometer. Finally, ultrasonic cleaning was conducted for 300 s using isopropanol. A physiological saline solution was prepared by adding 18 g of sodium chloride to 2 L of distilled water. The specimens were immersed in 50 mL of physiological saline solution per 10 mm^2 of sample surface area, and the temperature was maintained at 37 °C using a muffle furnace (FO-60P, Glass Kiki Co., Ltd., Tokyo, Japan) [37]. After the immersion test, the samples were washed with distilled water for 60 s, followed by compression testing. Additionally, the observation samples were immersed in a mixture of chromic acid and nitric acid for 60 s to remove corrosion products, then they were washed with distilled water before observation.

3. Results and Discussion

3.1. Evaluation of Powders

Figure 1a,b, respectively, show the appearance of Ti and Mg powders observed through FESEM. The appearance of the Ti-30Mg mixed powder obtained by mixing this raw material powder is shown in Figure 1c. The portion indicated by the white arrows in this figure is Mg particles, and it was observed that large and irregularly shaped Mg particles are dispersed within the relatively small spherical Ti powder. Mechanical mixing was performed using a ball mill. However, there was no significant difference in the particle size of the Mg particles compared to that of the Mg raw material powder. Due to the short mixing time of 10 min, most of the Mg particles appear to have remained uncrushed. Generally, the effective diameter of pores for bone cells to penetrate and proliferate within pores in porous biomaterials is reported to be between 100 μm and 400 μm [38,39]. Since the average size of the Mg particles remained uncrushed at 180 μm in this mixing condition, good osteoinductivity of the porous body after Mg dissolution can be expected.

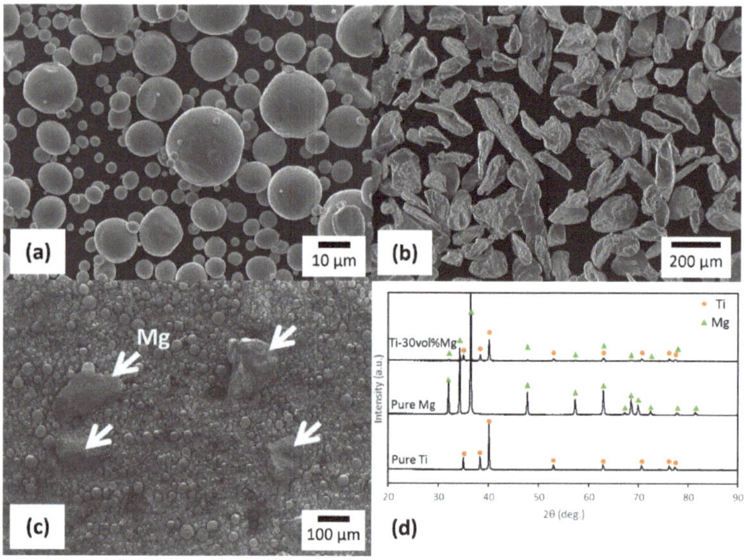

Figure 1. FESEM micrographs of (**a**) Ti powder, (**b**) Mg powder, and (**c**) the Ti-30Mg mixed powder obtained from (**a**,**b**). (**d**) XRD patterns of the powders.

Additionally, the analysis of the Mg particle surfaces in the mixed powder revealed smooth surfaces. However, the Mg particles, after mixing, were found to be covered with fine attachments. The EDS point analysis of these attachments mostly revealed pure Mg, while in some white attachments, an oxygen concentration of 53.41 at% was determined, indicating magnesium oxide. The temperature of the mixing vessel rises due to collisions between the vessel walls, ZrO_2 balls, and the raw material powder in the ball mill, promoting the reaction between ethanol and Mg particles and resulting in the formation of compounds. Figure 1d represents the XRD analysis results for the powder. No peaks other than Ti and Mg were observed in pure Ti and pure Mg. Although peaks of Ti and Mg were observed simultaneously in Ti-30Mg, peaks of oxides were not observed. Although the presence of reaction products was identified through FESEM and EDS analysis, they were only present on the surface of Mg particles and in small amounts overall. Hence, no peaks were observed.

3.2. Microstructure of Sintered Composites

Figure 2a–c depict micrographs of Pure Ti, Ti-30Mg, and Ti-50Mg by optical microscope, respectively. The vertical direction represents the compression direction during sintering. Although 580 °C is a relatively low sintering temperature for Ti, the result of sintering with Pure Ti showed only a small amount of porosity. According to density measurements, the relative density was 91%, meaning the porosity was 9%, indicating sufficient densification of Pure Ti even at 580 °C, as shown in Table 3. With a sintering pressure of 50 MPa, it can be observed that Mg is uniformly dispersed within the Ti matrix in Ti-30Mg and Ti-50Mg. Mg particles appear slightly flattened perpendicular to the compression direction, indicating deformation due to sintering pressure. Based on the calculation of porosity using relative density, the addition of Mg resulted in a decrease in porosity due to its effect of filling the pores between Ti particles. The lowest pressure sample, Ti-30Mg (TM25), exhibited a porosity of 6.4%, while TM50-TM100 showed results close to 0%.

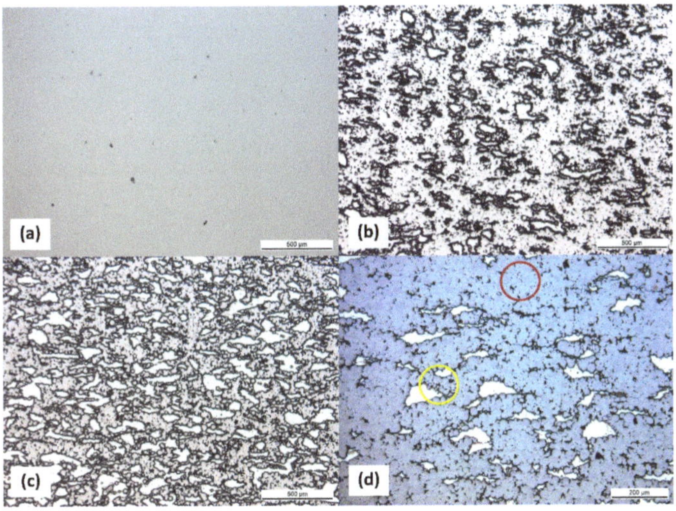

Figure 2. Optical micrographs of (**a**) Pure Ti, (**b**) Ti-30Mg, and (**c**) Ti-50Mg samples. (**d**) Enlarged optical micrographs of Ti-30Mg.

Table 3. Porosity data after sample preparation by SPS.

Specimen	Porosity of Sample (%)
Pure Ti	9.0
Ti-30Mg	0
Ti-50Mg	0
Ti-30Mg (TM25)	6.4
Ti-30Mg (TM75)	0
Ti-30Mg (TM100)	0

Furthermore, an increase in porosity within the Ti matrix compared to Pure Ti can be observed. Figure 2d presents an enlarged image of the microstructure of Ti-30Mg, where, as indicated by the red circle, the regions without Mg show well-progressed sintering with neck growth between Ti particles, whereas, as noted in the yellow circle, Ti particles near the Mg particles do not exhibit neck growth and maintain the shape of the raw powder. This suggests that insufficient sintering stress is applied to Ti particles near Mg particles due to Mg deformation, resulting in delayed sintering progress. The increase in porosity within the Ti matrix is also presumed to be due to the deformation of Mg during sintering. Various parameters affect the structure and mechanical properties in the production of specimens through powder sintering, such as composition, powder particle size, powder mixing method, and sintering conditions. Especially in the case of discharge plasma sintering, the sintering conditions vary, and changes in temperature, holding time, heating rate, pressure, discharge pulse interval, electric current, etc., alter the sintering characteristics. In this study, the sintering temperature was limited to a maximum of 650 °C to conduct sintering at a temperature where Mg does not dissolve. Therefore, to enhance the sintering characteristics of Ti particles, the effect of sintering pressure on the sintering properties of Ti-30vol%Mg composition was investigated in the present study.

Figure 3a–c depict the microstructure of specimens TM25, TM75, and TM100, respectively, which were sintered under different pressures of 25, 75, and 100 MPa, using Ti-30vol%Mg mixed powder raw materials. TM100 sintered at the highest pressure of 100 MPa showed fewer pores within the Ti phase. Through the microstructure, it was possible to confirm that sinterability improves with increasing sintering pressure. To quantify sinterability, the density of each specimen was measured, and density changes with pressure variations were represented, as shown in Figure 3d. The density of TM25 showed a smaller value (3.44 g/cm^3) compared to other specimens, indicating the presence of many pores within the specimen. While the density increased significantly up to 3.70 g/cm^3 in TM50, there was only a slight increasing trend in TM50, TM75, and TM100, with no significant difference in density. When manufacturing Ti-Mg composites using SPS, the porosity of the composite material can be influenced by several factors, in addition to sintering pressure, such as sintering temperature, Mg composition, and the size and shape of the powders [15,40]. Specifically, as the Mg composition increases, the diffusion of Mg, which has a relatively lower melting point, occurs rapidly, leading to a decrease in porosity and an increase in density [15,41].

The sintering process consolidates powder particles into a solid mass by applying heat and pressure. Several mechanisms govern this process, including powder particle interactions, diffusion kinetics, and microstructural evolution [42]. During sintering, powder particles come into contact with each other due to applied pressure. At the contact points, known as necks, atomic bonds form between particles, facilitating the consolidation process. The initial stage involves surface diffusion, where atoms migrate along the particle surfaces to form bonds [43–46]. As sintering progresses, bulk diffusion becomes dominant, with atoms diffusing through the lattice of particles to further densify the material. Diffusion plays a critical role in sintering, as it governs the movement of atoms within the powder compact. The diffusion rate depends on factors such as temperature, pressure, and the chemical composition of the powder [43]. High temperatures increase atomic mobility, promoting faster diffusion and densification. Pressure reduces the activation energy required for diffusion, accelerating the sintering process. Additionally, the chemical composition of the powder influences the diffusion kinetics, as different elements diffuse at varying rates [47]. As sintering progresses, the microstructure of the material undergoes significant changes. Initially, pores between powder particles are eliminated as necks form and grow. As sintering continues, the pores decrease in size and number, increasing material density. The microstructure evolves from a network of interconnected pores to a solid, dense structure. However, excessive sintering can result in grain growth and the formation of large pores, negatively impacting the mechanical properties of the material. The microstructural analysis revealed a uniform dispersion of Mg within the

Ti matrix, which was achieved through mechanical alloying and spark plasma sintering. This uniform distribution is critical for ensuring consistent mechanical properties throughout the composite. The formation of fine Ti-Mg intermetallic phases suggests successful bonding between the titanium and magnesium particles, which is crucial for enhancing the composite's mechanical performance. The presence of these intermetallic phases can be attributed to the high heating rates and localized temperature spikes inherent in the spark plasma sintering process, promoting rapid diffusion and reaction between Ti and Mg.

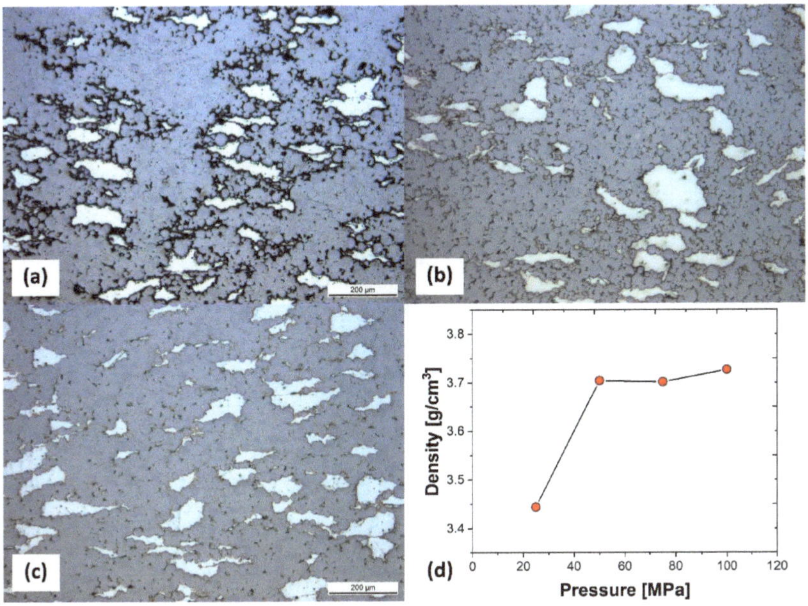

Figure 3. Optical micrographs of Ti-30vol%Mg sintered at (**a**) 25 MPa (TM25), (**b**) 75 MPa (TM75), and (**c**)100 MPa (TM100). (**d**) Relationship between sintering pressure and density.

3.3. *Mechanical Properties Depending on Mg Contents*

Figure 4a shows the stress–strain curves obtained from compression tests of Pure Ti, Ti-30Mg, and Ti-50Mg. In the compression test of Pure Ti, the specimen did not fracture, and the stress continued to increase, so the test was stopped at the point exceeding the fracture strain of Ti-30Mg and Ti-50Mg. The compressive strength of cortical bone is estimated to be approximately 180 MPa, and all specimens exhibited higher strengths than cortical bone. When comparing the stress–strain curves of pure Ti, where Mg compounds are not present in the powder, and Ti-30Mg and Ti-50Mg, where Mg compounds are formed on the surface of Mg, no significant decrease in mechanical properties due to compounds in the powder was observed. The yield stresses were 311 MPa for Pure Ti, 334 MPa for Ti-30Mg, and 239 MPa for Ti-50Mg, with Ti-30Mg showing a higher value than pure Ti. This is believed to be due to the solid solution of oxygen in Ti. According to the Ti-O binary phase diagram, the solubility limit of oxygen in Ti is approximately 33% at both sintering and room temperatures, which is significantly high. It is considered that oxygen solubility occurred due to exposure of the powder to air during transportation until installation in the chamber of the SPS device, as well as oxygen atoms from ethanol, the mixing solvent.

On the other hand, the yield stress of Ti-50Mg was lower than that of Pure Ti and Ti-30Mg. This is attributed to a more significant decrease in strength due to a reduction in the volume fraction of Ti than to solid solution strengthening by oxygen in Ti. Figure 4b illustrates the relationship between Mg content and Young's modulus measured by the ultrasonic pulse method. Meanwhile, the Young's modulus of typical pure Ti is 106 GPa, and that of pure Ti is 92 GPa, which is believed to be due to the presence of 9% porosity in the sintered composite specimen. The Young's modulus of Ti-30Mg and Ti-50Mg were 81 GPa and 75 GPa, respectively, confirming that the Young's modulus decreases with increased Mg content.

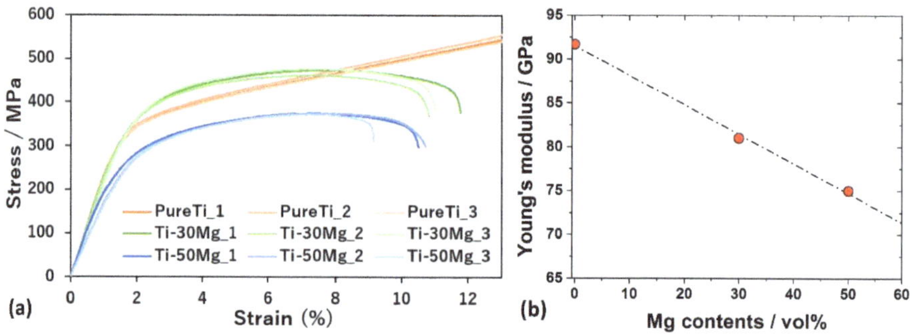

Figure 4. (**a**) Compressive stress–strain curves of pure Ti, Ti-30Mg, and Ti-50Mg. (**b**) Relationship between Mg content and Young's modulus.

The microstructure and phase composition can significantly affect the stress–strain response observed in compression tests of Ti-Mg composites. The microstructure influences the load-bearing capability and deformation mechanisms of the composite [48,49]. A well-sintered microstructure with fewer pores and a uniform distribution of Mg within the Ti matrix tends to exhibit better mechanical properties. Pores act as stress concentrators and can initiate cracks, reducing the material's strength and ductility [50]. The content of Mg in the composite has a direct effect on Young's modulus due to the lower modulus of Mg compared to Ti. Generally, as the Mg content increases, the overall modulus of the composite decreases. This is because the modulus of composite materials is roughly proportional to the volume fraction of the constituent phases [51]. Therefore, with a higher volume fraction of Mg (which has a modulus of 41 GPa) compared to Ti (which has a modulus of about 110 GPa), the composite's modulus will be lower than pure Ti.

Figure 5 depicts SEM images of the fracture surface after compression testing of Ti-30Mg and the results of element distribution mapping by EDS analysis. The fracture surface of the Mg portion exhibits a river line crack pattern, indicating brittle fracture in the Mg portion. In contrast, the Ti portion maintains the shape of the raw powder, suggesting that the fracture in the Ti portion mainly occurred at the particle boundaries of the powder. From these observations, it can be understood that atomic diffusion at the Ti particle boundaries did not sufficiently progress. Optimizing the sintering conditions is expected to enhance the overall mechanical properties of the composite material by improving the sinterability of Ti [52].

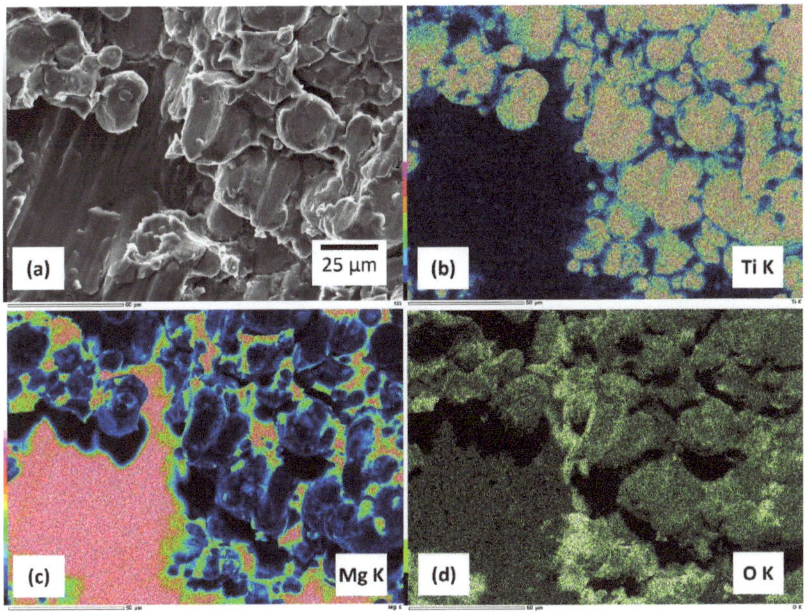

Figure 5. (**a**) Fractography of Ti-30Mg by FESEM. Element distribution maps of (**b**) Ti, (**c**) Mg, (**d**) O by EDS.

Figure 6a plots the changes in uniaxial compressive strength (UCS) and yield strength with variations in sintering pressure, obtained from the stress–strain curves of TM25, TM50, TM75, and TM100 from the compression test results. Only TM25 showed distinctly different compression test results compared to other specimens, with minimum compressive strength and yield strength. TM50, TM75, and TM100 showed no significant differences in compression behavior, compressive strength, or yield strength, similar to density. Additionally, Figure 6b presents the results of Young's modulus measurements at each sintering pressure. Young's modulus increased with increasing sintering pressure, from 57 GPa in TM25 to 91 GPa in TM75 and TM100. Since this study aims to produce Ti-Mg composite materials with low Young's modulus, a lower modulus is desirable. However, considering that Mg dissolves during insertion into the body, leading to porous Ti, and prioritizing the sinterability of the matrix Ti, an increase in Young's modulus due to increased sintering pressure is considered a favorable outcome. Since only TM25 showed significantly lower values in density measurements, and there were no significant differences in TM50, TM75, and TM100, it can be assumed that the sinterability of Ti-30vol%Mg saturates around 50 MPa.

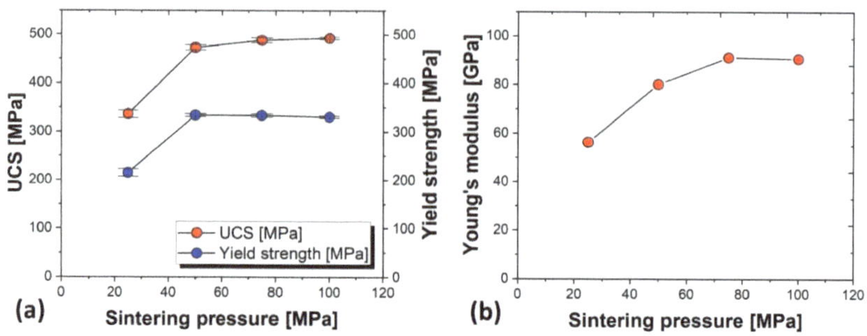

Figure 6. (**a**) Changes in UCS and Yield strength with variations in sintering pressure. (**b**) Relationship between sintering pressure and Young's modulus.

The significant effect of sintering pressure on Young's modulus, compared to UCS or yield strength, can be attributed to the microstructural changes that occur during the sintering process. Higher sintering pressures lead to better densification of the material, reducing porosity and increasing the contact area between grains [28]. This improved grain boundary contact enhances the material's ability to resist deformation, thereby increasing Young's modulus. Sintering pressure helps in achieving a more homogeneous microstructure. A uniform microstructure with fewer defects and voids contributes to a higher Young's modulus because the material can deform elastically more uniformly under stress. Young's modulus is more sensitive to changes in microstructure and density than UCS or yield strength [53]. While UCS and yield strength are influenced by factors like grain size and the presence of flaws, Young's modulus is directly related to the stiffness of the material, which is significantly affected by the degree of densification and the quality of grain boundaries. During sintering, pressure aids in the formation of stronger bonds and larger necks between particles. These stronger inter-particle bonds contribute to a higher elastic modulus, as the material can better resist elastic deformation.

The Mg content and the sintering parameters significantly influence the mechanical properties of the Ti-Mg composites. The addition of Mg reduces the overall density of the composites, making them lighter than pure titanium. This reduction in density is advantageous for biomedical implants, as it can lead to less stress on surrounding bone and tissue. The Young's modulus of the composites can be tailored by adjusting the Mg content. Composites with 5–15 wt% Mg exhibited Young's modulus closer to that of natural bone, which is beneficial in minimizing stress-shielding effects. This characteristic addresses a significant drawback of conventional titanium implants, which have a much higher modulus than bone, leading to stress shielding and bone resorption over time. The compressive strength of the composites decreases with increasing Mg content. However, the values obtained are still within acceptable ranges for load-bearing applications. This trade-off between modulus and strength must be carefully balanced to optimize implant performance.

3.4. Immersion Test in Physiological Saline

Figure 7 shows the microstructure of Ti-30Mg after immersing in physiological saline for 1 day. The observation was conducted after immersion for 1 min in a mixed solution of chromic acid (H_2CrO_4) and silver nitrate ($AgNO_3$) to remove corrosion products. In Figure 7a, the dissolution of Mg and the formation of pores on the sample surface were observed. The micrograph in Figure 7b, magnified inside the pores, shows that most Ti particles maintain the shape of the raw powder. Here, as in the cross-sectional micrograph of OM, it can be observed that the sintering of Ti particles around the Mg powder was not fully completed. Figure 7c shows the element distribution maps of the pore region after the dissolution of Mg. Mg was detected on the pore's surface, and oxygen was distributed along with Mg.

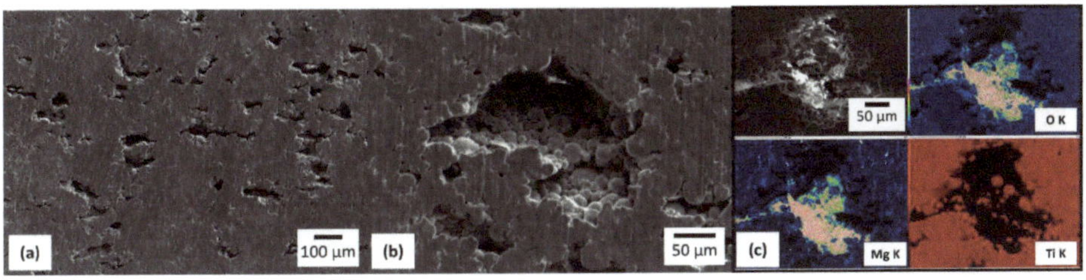

Figure 7. (**a**) A FESEM micrograph of the surface of the Ti-30Mg sample, (**b**) a magnified micrograph for pore observation after 1 day of immersion in saline solution, (**c**) Element distribution maps of O, Mg, and Ti by EDS.

Figure 8a presents the stress–strain curves of compression tests after immersion tests for various durations. At the same time, Figure 8b shows the changes in UCS and fracture strain according to immersion time. After 6 days of immersion, the compression strength was 214 MPa, and Ti-30Mg could maintain a strength higher than that of cortical bone (approximately 180 MPa [54]) until 6 days after immersion. Both compression strength and fracture strain decreased with increasing immersion time. UCS and fracture strain showed nonlinear deceleration with immersion time. In both cases, it was observed that the experimental results closely matched a sigmoidal curve based on the Boltzmann function. Figure 8c shows the stress–strain curves obtained from compression tests of TM100 (Ti-30Mg) fabricated with a sintering pressure of 100 MPa for various immersion times. At the same time, Figure 8d illustrates the changes in compressive strength and fracture strain during immersion. Similar to the compression test results for TM50 (Ti-30Mg) in Figure 8b, the UCS and fracture strain decreased with increasing immersion time due to the dissolution of Mg within the composite material. The UCS on the 9th and 10th days of immersion were 250 MPa and 276 MPa, respectively, maintaining strengths exceeding those of cortical bone until the 10th day. Figure 9a plots the variation in UCS with increasing immersion time for both TM50 (Ti-30Mg) and TM100 (Ti-30Mg), with a green dash-dot horizontal line representing cortical bone strength. It can be inferred that the mechanical integrity improves with increasing sintering pressure, as TM100 maintains strength above that of cortical bone for a longer duration compared to TM50.

Mechanical property changes were relatively significant in the early stages of immersion. However, as time increased, the changes became relatively small, and there was almost no change in fracture strain after the 6th day. This is believed to be because Mg dissolved from the entire sample surface during the early stages of immersion, but subsequently, dissolution of Mg from the center of the sample was required. The dissolution of Mg occurs through the chemical reactions below [55], resulting in the generation of hydrogen gas and corrosion products where Mg is dissolved.

$$Mg + 2H_2O = Mg(OH)_2 + H_2$$

$$Mg(OH)_2 + 2Cl- = MgCl_2 + 2OH^-$$

When hydrogen gas or corrosion products are generated inside the pores, it takes time for the solution to penetrate into the center of the specimen. Therefore, as the immersion time passes, it is expected that mechanical property changes will diminish, especially after a certain period has elapsed.

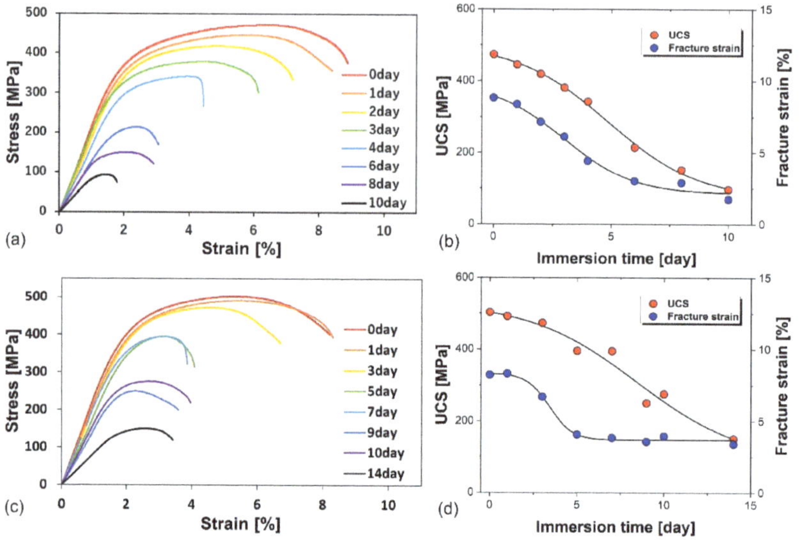

Figure 8. Compressive stress–strain curves of Ti-30Mg samples with (**a**) 50 MPa and (**c**) 100 MPa of sintering pressures immersed in saline for various times up to 10 days. Changes of UCS and fracture strain of Ti-30Mg with (**b**) 50 MPa and (**d**) 100 MPa of sintering pressures increasing immersion times in saline for up to 10 days.

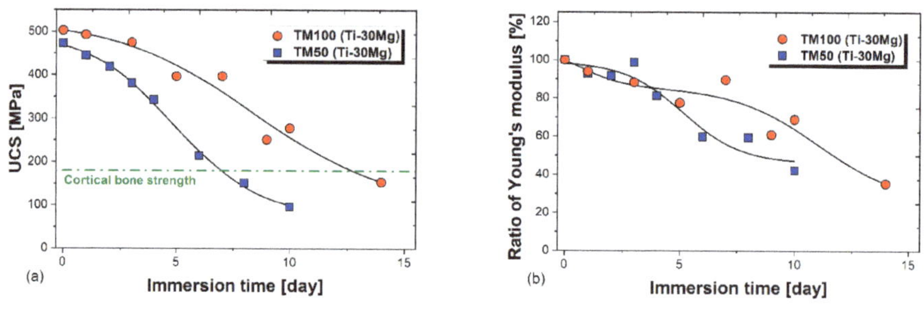

Figure 9. (**a**) Relationship between UCS and immersion time of TM100 (Ti-30Mg) and TM50 (Ti-30Mg). (**b**) Changes in the ratio of Young's modulus to the initial Young's modulus with an increase of the immersion time for TM100 (Ti-30Mg) and TM50 (Ti-30Mg).

At the outset, the immersion phase is crucial as it sets the stage for the corrosion process. Here, the Mg in the Ti-Mg composite swiftly interacts with the saline solution, leading to the creation of $Mg(OH)_2$ and hydrogen gas [56,57]. These reactions initiate the formation of a porous $Mg(OH)_2$ layer on the surface, which initially provides some defense but gradually transforms into more soluble $MgCl_2$ in the presence of chloride ions from the saline solution. As immersion continues, the protective $Mg(OH)_2$ layer gradually breaks down, releasing Mg ions into the solution. The generated hydrogen gas can form bubbles at the interface, further promoting Mg degradation [19,58]. This stage is characterized by significantly reducing mechanical properties as Mg content diminishes. Over time, the corrosion rate slows down as the more easily accessible Mg is depleted, and the remaining Mg is less exposed due to deeper penetration into the composite. The changes in mechanical properties, mainly compressive strength, decrease at this stage [18]. Our immersion tests have yielded significant findings. Ti-30vol%Mg maintained its compressive strength above that of cortical bone for a considerable period of up to 10 days. This behavior suggests that the composite can retain its mechanical integrity for a sufficient duration, potentially supporting the initial bone healing processes. The release of Mg ions plays a critical role in the biodegradation process. Mg ions are known to be biocompatible and can enhance osteogenesis. However, the localized increase in pH due to the formation of OH^- ions needs to be managed to prevent adverse effects on surrounding tissues.

Figure 9b illustrates the variation in Young's modulus during immersion for specimens of TM50 (Ti-30Mg) and TM100 (Ti-30Mg). Here, the reduction in Young's modulus was evaluated with respect to the specimen at 0 days before immersion, which was considered 100%. It can be observed that Young's modulus decreases with increasing immersion time for both specimens. For TM50 (Ti-30Mg), Young's modulus decreased by 40% after 6 days of immersion compared to before immersion, resulting in a modulus of approximately 49 GPa, considering that the modulus of Ti-30Mg was measured to be 81 GPa. In the case of TM100 (Ti-30Mg), the modulus decreased overall by 31% after 10 days of immersion, indicating a modulus of approximately 63 GPa at Day 10 of immersion, considering that the modulus of TM100 measured by ultrasonic pulse method was 91 GPa. It was observed that sintering pressure enhancement led to improved sinterability and mechanical properties. However, in cases where the specimens maintained strengths above that of cortical bone (6 days for TM50 and 10 days for TM100), the modulus after extended immersion was higher for TM100 than TM50. To suppress stress shielding, it is ideal for the modulus after immersion to be as small as possible and closer to that of cortical bone. By adjusting multiple conditions, such as increasing sintering pressure to enhance sinterability and increasing Mg content to reduce the modulus, it is possible to produce Ti-Mg composite materials with higher strength and lower modulus.

4. Conclusions

Uniform Ti-Mg composite materials were fabricated by mixing pure Ti powder with low-modulus and biodegradable Mg powder using spark plasma sintering. The effects of mixing powder, evaluating Ti-Mg composite materials, and the influence of Mg addition and sintering pressure on the properties of Ti-Mg composite materials were investigated systematically.

1. The uniform dispersion of Mg in Ti-30Mg and Ti-50Mg and the confirmation of the effectiveness of powder mixing with ethanol-added ball milling in producing uniform Ti-Mg composite materials further solidify the reliability of our findings.
2. Young's modulus of pure Ti, Ti-30Mg, and Ti-50Mg is 92 GPa, 81 GPa, and 75 GPa, respectively. This indicates a linear decrease in Young's modulus of Ti-Mg composite materials with Mg addition.
3. From density measurement, compression tests, and Young's modulus measurement results, only TM25 (Ti-30Mg) showed significantly smaller values, while TM50 (Ti-30Mg), TM75 (Ti-30Mg), and TM100 (Ti-30Mg) showed no significant differences. Therefore, the sinterability of Ti-30vol%Mg saturates around a sintering pressure of approximately 50 MPa.
4. TM50 (Ti-30Mg) showed a decrease in Young's modulus from 81 GPa to 49 GPa after 6 days of immersion, while TM100 (Ti-30Mg) showed a decline from 91 GPa to 63 GPa after 10 days of immersion, indicating a reduction in stress shielding phenomenon.
5. During immersion in physiological saline, TM50 maintained compressive strength above cortical bone for 6 days and TM100 for 10 days. This confirms that the mechanical integrity of Ti-30vol%Mg improves with increasing sintering pressure.

Author Contributions: Conceptualization, E.K., M.O. and T.M.; methodology, T.M.; software, T.M.; validation, T.M., M.O. and E.K.; formal analysis, T.M.; investigation, T.M. and M.O.; data curation, T.M. and M.O.; writing—original draft preparation, T.M. and M.O.; writing—review and editing, M.O.; visualization, T.M. and M.O.; supervision, E.K.; project administration, E.K. All authors have read and agreed to the published version of the manuscript.

Funding: This research received no external funding.

Institutional Review Board Statement: Nor applicable.

Informed Consent Statement: Not applicable.

Data Availability Statement: The raw data supporting the conclusions of this article will be made available by the authors on request.

Conflicts of Interest: The authors declare no conflicts of interest.

References

1. Otsuka, Y.; Mizutani, M. Introduction of Biomaterials for Beginners: II.Metallic Biomaterials. *Zair. J. Soc. Mater. Sci. Jpn.* **2014**, *63*, 480–486. [CrossRef]
2. Kaur, M.; Singh, K. Review on Titanium and Titanium Based Alloys as Biomaterials for Orthopaedic Applications. *Mater. Sci. Eng. C* **2019**, *102*, 844–862. [CrossRef] [PubMed]
3. Mehta, N.; Hall, D.J.; Pourzal, R.; Garrigues, G.E. The Biomaterials of Total Shoulder Arthroplasty: Their Features, Function, and Effect on Outcomes. *JBJS Rev.* **2020**, *8*, e19. [CrossRef] [PubMed]
4. Lai, Y.S.; Chen, W.C.; Huang, C.H.; Cheng, C.K.; Chan, K.K.; Chang, T.K. The Effect of Graft Strength on Knee Laxity and Graft In-Situ Forces after Posterior Cruciate Ligament Reconstruction. *PLoS ONE* **2015**, *10*, e0127293. [CrossRef] [PubMed]
5. Noyama, Y.; Miura, T.; Ishimoto, T.; Itaya, T.; Niinomi, M.; Nakano, T. Bone Loss and Reduced Bone Quality of the Human Femur after Total Hip Arthroplasty under Stress-Shielding Effects by Titanium-Based Implant. *Mater. Trans.* **2012**, *53*, 565–570. [CrossRef]
6. Sidhu, S.S.; Singh, H.; Gepreel, M.A.H. A Review on Alloy Design, Biological Response, and Strengthening of β-Titanium Alloys as Biomaterials. *Mater. Sci. Eng. C* **2021**, *121*, 111661. [CrossRef] [PubMed]
7. Guo, S.; Meng, Q.; Zhao, X.; Wei, Q.; Xu, H. Design and Fabrication of a Metastable β-Type Titanium Alloy with Ultralow Elastic Modulus and High Strength. *Sci. Rep.* **2015**, *5*, 14688. [CrossRef] [PubMed]
8. Xing, F.; Li, S.; Yin, D.; Xie, J.; Rommens, P.M.; Xiang, Z.; Liu, M.; Ritz, U. Recent Progress in Mg-Based Alloys as a Novel Bioabsorbable Biomaterials for Orthopedic Applications. *J. Magnes. Alloys* **2022**, *10*, 1428–1456. [CrossRef]
9. Lei, S.; Zhang, J.; An, X.; Guo, Y.; Xu, X.; Ma, Z.; Yao, W.; Kong, Q. Mechanical and Corrosion Properties of Ti–29Nb–13Ta-4.6Zr Alloy Prepared by Cryomilling and Spark Plasma Sintering. *Vacuum* **2023**, *215*, 112316. [CrossRef]
10. Putra, N.E.; Mirzaali, M.J.; Apachitei, I.; Zhou, J.; Zadpoor, A.A. Multi-Material Additive Manufacturing Technologies for Ti-, Mg-, and Fe-Based Biomaterials for Bone Substitution. *Acta Biomater.* **2020**, *109*, 1–20. [CrossRef]
11. Niinomi, M.; Nakai, M.; Hieda, J. Development of New Metallic Alloys for Biomedical Applications. *Acta Biomater.* **2012**, *8*, 3888–3903. [CrossRef]
12. Krishna, B.V.; Bose, S.; Bandyopadhyay, A. Low Stiffness Porous Ti Structures for Load-Bearing Implants. *Acta Biomater.* **2007**, *3*, 997–1006. [CrossRef] [PubMed]
13. Liang, L.; Huang, Q.; Wu, H.; Ouyang, Z.; Liu, T.; He, H.; Xiao, J.; Lei, G.; Zhou, K. Stimulation of in Vitro and in Vivo Osteogenesis by Ti-Mg Composite Materials with the Sustained-Release Function of Magnesium Ions. *Colloids Surf. B Biointerfaces* **2021**, *197*, 111360. [CrossRef] [PubMed]
14. Ouyang, S.; Huang, Q.; Liu, Y.; Ouyang, Z.; Liang, L. Powder Metallurgical Ti-Mg Metal-Metal Composites Facilitate Osteoconduction and Osseointegration for Orthopedic Application. *Bioact. Mater.* **2019**, *4*, 37–42. [CrossRef] [PubMed]
15. Xue, J.; Luo, F.; Zhang, Y.; Fang, Y.; Jiang, X. Strengthening Mechanisms of Ti–Mg Composite for Biomaterials: A Review. *Adv. Eng. Mater.* **2023**, *25*, 2300620. [CrossRef]
16. Pham, D.N.; Hiromoto, S.; Yamazaki, T.; O, M.; Kobayashi, E. Enhanced Corrosion Resistance and in Vitro Biocompatibility of Mg-Zn Alloys by Carbonate Apatite Coating. *ACS Appl. Bio Mater.* **2021**, *4*, 6881–6892. [CrossRef] [PubMed]
17. Pham, D.N.; Hiromoto, S.; O, M.; Kobayashi, E. Influence of Substrate Microstructure on Hydroxyapatite Coating and Corrosion Behavior of Coated Mg[Sbnd]Zn Alloys. *Surf. Coat. Technol.* **2021**, *421*, 127414. [CrossRef]
18. Esmaily, M.; Svensson, J.E.; Fajardo, S.; Birbilis, N.; Frankel, G.S.; Virtanen, S.; Arrabal, R.; Thomas, S.; Johansson, L.G. Fundamentals and Advances in Magnesium Alloy Corrosion. *Prog. Mater. Sci.* **2017**, *89*, 92–193. [CrossRef]
19. Pan, S.Q.; Zhang, F.; Wen, C.; Zeng, R.C. Advances in Mg–Al-Layered Double Hydroxide Steam Coatings on Mg Alloys: A Review. *J. Magnes. Alloys* **2023**, *11*, 1505–1518. [CrossRef]
20. Wang, J.; Dou, J.; Wang, Z.; Hu, C.; Yu, H.; Chen, C. Research Progress of Biodegradable Magnesium-Based Biomedical Materials: A Review. *J. Alloys Compd.* **2022**, *923*, 166377. [CrossRef]
21. Sezer, N.; Evis, Z.; Kayhan, S.M.; Tahmasebifar, A.; Koç, M. Review of Magnesium-Based Biomaterials and Their Applications. *J. Magnes. Alloys* **2018**, *6*, 23–43. [CrossRef]

22. Verma, R.P. Titanium Based Biomaterial for Bone Implants: A Mini Review. *Mater. Today: Proc.* **2020**, *26*, 3148–3151. [CrossRef]
23. Meenashisundaram, G.K.; Wang, N.; Maskomani, S.; Lu, S.; Anantharajan, S.K.; Dheen, S.T.; Nai, S.M.L.; Fuh, J.Y.H.; Wei, J. Fabrication of Ti + Mg Composites by Three-Dimensional Printing of Porous Ti and Subsequent Pressureless Infiltration of Biodegradable Mg. *Mater. Sci. Eng. C* **2020**, *108*, 110478. [CrossRef]
24. Miao, L.; Gao, M.; Chen, S.; Wang, J.; Zhang, H.; Tan, L.; Pan, Y. The Mechanical Properties and Bioactivity of Ti–Mg Composites as Orthopedic Implants: A Pilot Study. *J. Mater. Res. Technol.* **2024**, *28*, 1074–1083. [CrossRef]
25. Yang, H.; Chen, X.; Huang, G.; Song, J.; She, J.; Tan, J.; Zheng, K.; Jin, Y.; Jiang, B.; Pan, F. Microstructures and Mechanical Properties of Titanium-Reinforced Magnesium Matrix Composites: Review and Perspective. *J. Magnes. Alloys* **2022**, *10*, 2311–2333. [CrossRef]
26. Eessaa, A.K.; Elkady, O.A.; El-Shamy, A.M. Powder Metallurgy as a Perfect Technique for Preparation of Cu–TiO2 Composite by Identifying Their Microstructure and Optical Properties. *Sci. Rep.* **2023**, *13*, 7034. [CrossRef]
27. Hu, Z.Y.; Zhang, Z.H.; Cheng, X.W.; Wang, F.C.; Zhang, Y.F.; Li, S.L. A Review of Multi-Physical Fields Induced Phenomena and Effects in Spark Plasma Sintering: Fundamentals and Applications. *Mater. Des.* **2020**, *191*, 108662. [CrossRef]
28. Ratzker, B.; Sokol, M. Exploring the Capabilities of High-Pressure Spark Plasma Sintering (HPSPS): A Review of Materials Processing and Properties. *Mater. Des.* **2023**, *233*, 112238. [CrossRef]
29. Liu, W.; An, R.; Wang, C.; Zheng, Z.; Tian, Y.; Xu, R.; Wang, Z. Recent Progress in Rapid Sintering of Nanosilver for Electronics Applications. *Micromachines* **2018**, *9*, 346. [CrossRef] [PubMed]
30. Anselmi-Tamburini, U.; Garay, J.E.; Munir, Z.A. Fundamental Investigations on the Spark Plasma Sintering/Synthesis Process. *Mater. Sci. Eng. A* **2005**, *407*, 24–30. [CrossRef]
31. Liu, S.; Zhang, G.; Shi, M.; Yang, X.; Li, A. Microstructure and Properties of Porous Titanium Prepared by Spark Plasma Sintering. *Metals* **2019**, *9*, 82. [CrossRef]
32. McDonnell, N.J.; Muchatuta, N.A.; Paech, M.J. Acute Magnesium Toxicity in an Obstetric Patient Undergoing General Anaesthesia for Caesarean Delivery. *Int. J. Obs. Anesth.* **2010**, *19*, 226–231. [CrossRef]
33. Van Le, H.; Hiromoto, S.; O, M.; Kobayashi, E.; Nguyen, N.V.; Cao, N.Q. In Vitro Cellular Biocompatibility and in Vivo Degradation Behavior of Calcium Phosphate-Coated ZK60 Magnesium Alloy. *Biomed. Mater.* **2023**, *18*, 035003.
34. Oh, M.; Iwamoto, H.; Kobayashi, E. Influence of Carbon Nanotubes on the Morphology of Cu6Sn5 in Cu/(Sn–Ag–Cu) Solder Joints. *Results Mater.* **2024**, *21*, 100553. [CrossRef]
35. Węglewski, W.; Bochenek, K.; Basista, M.; Schubert, T.; Jehring, U.; Litniewski, J.; Mackiewicz, S. Comparative Assessment of Young's Modulus Measurements of Metal-Ceramic Composites Using Mechanical and Non-Destructive Tests and Micro-CT Based Computational Modeling. *Comput. Mater. Sci.* **2013**, *77*, 19–30. [CrossRef]
36. Rivard, P.; Saint-Pierre, F. Assessing Alkali-Silica Reaction Damage to Concrete with Non-Destructive Methods: From the Lab to the Field. *Constr. Build. Mater.* **2009**, *23*, 902–909. [CrossRef]
37. Trang, L.T.; Quang Cao, N.; Hiromoto, S.; O, M.; Kobayashi, E. Formation and Corrosion Behavior of Calcium Phosphate Coating Layers on ZK60 Alloy Coated at Various PH Conditions by Chemical Conversion Method. *Surf. Coat. Technol.* **2022**, *444*, 128639. [CrossRef]
38. Perez, R.A.; Mestres, G. Role of Pore Size and Morphology in Musculo-Skeletal Tissue Regeneration. *Mater. Sci. Eng. C* **2016**, *61*, 922–939. [CrossRef] [PubMed]
39. Hollister, S.J. Porous Scaffold Design for Tissue Engineering. *Nat. Mater.* **2005**, *4*, 518–524. [CrossRef]
40. Cai, X.; Ding, S.; Li, Z.; Zhang, X.; Wen, K.; Xu, L.; Zhang, Y.; Peng, Y.; Shen, T. Simultaneous Sintering of Low-Melting-Point Mg with High-Melting-Point Ti via a Novel One-Step High-Pressure Solid-Phase Sintering Strategy. *J. Alloys Compd.* **2021**, *858*, 158344. [CrossRef]
41. Oh, M.; Sakaguchi, H.; Kobayashi, E.; Kajihara, M. Understanding Intermetallic Compound Growth at Ag/Zn Interfaces: Kinetics and Mechanisms. *Intermetallics* **2024**, *172*, 108378. [CrossRef]
42. Tavakol, M.; Mahnama, M.; Naghdabadi, R. Mechanisms Governing Microstructural Evolution during Consolidation of Nanoparticles. *Mater. Manuf. Process.* **2015**, *30*, 1397–1402. [CrossRef]
43. Oh, M.; Kim, H.-S.; Kobayashi, E.; Kajihara, M. Understanding Kirkendall Effect in Ni(W) Diffusion-Induced Recrystallization Region. *J. Alloys Compd.* **2024**, *991*, 174556. [CrossRef]
44. Oh, M.; Tokunaga, N.; Kobayashi, E. Reactive Diffusion at the Interface between Cu and Sn–Ag Alloys. *J. Mater. Res. Technol.* **2024**, *30*, 9531–9541. [CrossRef]
45. Mehrer, H. *Diffusion in Solids: Fundamentals, Methods, Materials, Diffusion-Controlled Processes*; Springer Series in Solid-State Sciences; Springer: Berlin/Heidelberg, Germany, 2007.
46. O, M.; Sato, K.; Kobayashi, E. Investigation of the Rate-Controlling Process of Intermetallic Layer Growth at the Interface between Ferrous Metal and Molten Al–Mg–Si Alloy. *Intermetallics* **2023**, *163*, 108069. [CrossRef]
47. Inomata, S.; O, M.; Kajihara, M. Diffusion-Induced Recrystallization in the Cu(Pd) System at Complete Solid-Solution Temperatures. *J. Mater. Sci.* **2011**, *46*, 2410–2421. [CrossRef]
48. Li, X.; Liu, Y.; Si, C.; Xu, Q.; Liu, D.; Le, Q.; Fu, L.; Jia, Z.; Ma, Z.; Yuan, R. The Microstructure Evolution, Mechanical Property and Deformation Mechanism of Rolled Al3La/Mg–Li–Al–Zn Composites Prepared by in-Situ Synthesis. *Mater. Sci. Eng. A* **2024**, *894*, 146182. [CrossRef]

49. Chen, X.; O, M.; Kobayashi, E. Enhanced Mechanical Properties in an Al-Mg-Cu Alloy Processed by the Combination of Cyclic Deformation and Aging Heat Treatment. *J. Alloys Compd.* **2022**, *911*, 165070. [CrossRef]
50. Rezanezhad, M.; Lajevardi, S.A.; Karimpouli, S. Effects of Pore(s)-Crack Locations and Arrangements on Crack Growth Modeling in Porous Media. *Theor. Appl. Fract. Mech.* **2020**, *107*, 102529. [CrossRef]
51. Zhu, H.X.; Fan, T.X.; Zhang, D. Composite Materials with Enhanced Dimensionless Young's Modulus and Desired Poisson's Ratio. *Sci Rep* **2015**, *5*, 14103. [CrossRef]
52. Zeng, F.; Malzbender, J.; Baumann, S.; Nijmeijer, A.; Winnubst, L.; Ziegner, M.; Guillon, O.; Schwaiger, R.; Meulenberg, W.A. Optimization of Sintering Conditions for Improved Microstructural and Mechanical Properties of Dense $Ce_{0.8}Gd_{0.2}O_{2-\delta}$-$FeCo_2O_4$ Oxygen Transport Membranes. *J. Eur. Ceram. Soc.* **2021**, *41*, 509–516. [CrossRef]
53. Choren, J.A.; Heinrich, S.M.; Silver-Thorn, M.B. Young's Modulus and Volume Porosity Relationships for Additive Manufacturing Applications. *J. Mater. Sci.* **2013**, *48*, 5103–5112. [CrossRef]
54. Suchanek, W.; Yoshimura, M. Processing and Properties of Hydroxyapatite-Based Biomaterials for Use as Hard Tissue Replacement Implants. *J. Mater. Res.* **1998**, *13*, 94–117. [CrossRef]
55. Liu, Y.; Li, K.; Luo, T.; Song, M.; Wu, H.; Xiao, J.; Tan, Y.; Cheng, M.; Chen, B.; Niu, X.; et al. Powder Metallurgical Low-Modulus Ti-Mg Alloys for Biomedical Applications. *Mater. Sci. Eng. C* **2015**, *56*, 241–250. [CrossRef] [PubMed]
56. Cai, X.; Ding, S.; Li, Z.; Zhang, X.; Wen, K.; Xu, L.; Zhang, Y.; Shen, T. Mg-Ti Composites Fabricated by a Novel One-Step High-Pressure Sintering: The Correlation between Microstructures and Mechanical Properties. *Compos. B Eng.* **2021**, *215*, 108743. [CrossRef]
57. Zeng, R.C.; Li, X.T.; Li, S.Q.; Zhang, F.; Han, E.H. In Vitro Degradation of Pure Mg in Response to Glucose. *Sci. Rep.* **2015**, *5*, 13026. [CrossRef]
58. Wang, Y.; Liu, B.; Zhao, X.; Zhang, X.; Miao, Y.; Yang, N.; Yang, B.; Zhang, L.; Kuang, W.; Li, J.; et al. Turning a Native or Corroded Mg Alloy Surface into an Anti-Corrosion Coating in Excited CO_2. *Nat. Commun.* **2018**, *9*, 4058. [CrossRef]

coatings MDPI

Article

Preparation and Properties of Environmentally Friendly, Hydrophobic, Corrosion-Resistant, Multifunctional Composite Coatings

Zhenhua Chu, Zhixin Zhang, Wan Tang, Jiahao Lu and Jingxiang Xu *

Department of Mechanical Engineering, College of Engineering, Shanghai Ocean University, Shanghai 201306, China; zhchu@shou.edu.cn (Z.C.); m210811435@st.shou.edu.cn (Z.Z.); wtang@shou.edu.cn (W.T.)
* Correspondence: jxxu@shou.edu.cn; Tel.: +86-22-60204810; Fax: +86-22-26564810

Abstract: With the continuous exploitation of the marine resources, the equipment should meet the marine complex working environment. In this study, a type of environmentally friendly coating was prepared. Based on low surface energy environmental protection and anti-fouling, a film forming material with water-based epoxy-modified silicone resin emulsion was prepared. And industrial fillers were added to give it both inorganic and organic properties. Meanwhile, various contents of graphene oxide (GO) were added in the coating system. The coating properties were comprehensively analyzed, and the optimal GO content was obtained as 0.1 wt. %. The composite coating was studied by seawater immersion experiments, and the failure process of the coating in was proposed. The composite coating prepared in the present study has both environmental protection and hydrophobic anti-fouling characteristics, and its comprehensive performance is excellent through various performance evaluations, i.e., it meets the requirements of long-term coating, environmental friendliness and anti-fouling and corrosion resistance.

Keywords: water-based coating; graphene oxide; antifouling coating; corrosion resistance

Citation: Chu, Z.; Zhang, Z.; Tang, W.; Lu, J.; Xu, J. Preparation and Properties of Environmentally Friendly, Hydrophobic, Corrosion-Resistant, Multifunctional Composite Coatings. *Coatings* **2024**, *14*, 586. https://doi.org/10.3390/coatings14050586

Academic Editor: Luigi Calabrese

Received: 11 April 2024
Revised: 2 May 2024
Accepted: 6 May 2024
Published: 8 May 2024

1. Introduction

During the use of aquaculture network facilities, communities derived from the attachment of contaminated organisms to ship hulls make it easy for bacteria and parasites to breed, which will adversely affect the production of aquatic products [1]. A lack of resistance to biological attachment is one of the serious challenges faced by oceanfront equipment. Biological adhesion not only increases the maintenance cost of equipment, but also leads to equipment failure and corrosion failure. Coatings are among the most widely used methods for marine equipment maintenance. Traditional solvent-based antifouling coatings use organic solvents as dispersing media to prevent microbial attachment through the release of sterilizers in the coating system. However, the use of organic solvents consumes a lot of fossil energy, and organic solvents are harmful to the health of construction workers because of their volatility. In addition, the release of sterilizers causes pollution to the marine environment, and is more likely to accumulate in marine organisms, threatening the safety of humans who eat marine organisms. Sun [2] grafted low-toxic structural compounds as sterilization factors onto the side chain of acrylic resin to control the hydrolysis of sterilization factors to achieve self-polishing of the coating. Compared with heavy metal copper sterilization factors they are green, but they are still toxic and still pose a threat to marine microorganisms. In order to solve this industry dilemma, researchers continue to innovate and devote themselves to the development of new materials and new technologies to realize new coatings that are environmentally friendly and anti-fouling.

The birth of the new water-based coating should meet the requirements of being environmentally friendly, such as the advantages of small pollution and low volatile organic matter emissions. However, most of the protective waterborne coatings cannot possess environmental protection, anti-corrosion and anti-adhesion properties at the same time. Water-based epoxy-modified silicone resin emulsions have the properties of both epoxy resin and silicone resin emulsions, high adhesion and low surface energy, thereby meeting the requirements of environmental protection and anti-fouling. The new water-based coating can realize the performance of traditional sol-based coatings and conform to the concept of environmental protection, and it has promising application prospects in the field of anti-corrosion. Through different modification means and technical optimization, the performance of the new water-based coating can be further improved to meet the needs of more practical applications. Xu [3] synthesized a water-based silicone nanocomposite coating with long-term hydrophobicity and self-cleaning performance by combining a silicone emulsion with a TiO_2 dispersion by using a mechanical blending method. But its long-term self-cleaning performance was lacking in a natural pollution environment. After mixing the cationic silicone emulsion made by Khanjani [4] with an acrylic emulsion the mixed emulsion was cured at room temperature by using the low glass transition temperature of the acrylic emulsion, and the resulting coating had good hydrophobic and mechanical properties. However, the existence of two emulsions in the coating system required consideration of compatibility and the content of the single components, which was a huge workload.

In order to improve the overall performance of the silicone resin emulsion coating, inorganic fillers were added to the coating system to be more suitable for engineering applications. Wang [5] successfully developed a two-component inorganic coating, and systematically studied the influence of film forming additives on the stability of the coating, and then finally determined the best formula and curing time. Shan [6] used a nanosilica sol as the main film forming material and added an appropriate amount of inorganic filler to prepare an inorganic coating with excellent performance. In recent years, graphene showed great application potential in various fields [7]. It is wise to add it as a filler to water-based resin emulsion with low surface energy to improve the corrosion resistance of coatings. However, because the surface of the emulsion particles is rich in hydroxyl groups, it is easy for them to agglomerate and they have a high activity, which will lead to a reduction in the dispersibility and stability. Tian [8] used ammonium-grafted graphene oxide as a dispersant, which effectively solved the problem of microscopic defects in the preparation of water-based epoxy resin, and improved the compatibility and dispersion of graphene nanoparticles in water-based epoxy resin. Huang [9] significantly improved the cathodic protection and barrier properties of the coating by adding different contents of graphene to the water-based epoxy zinc-containing coating, which benefited from the graphene lamellar barrier effect [10] and conductive effect [11]. Chen [12] introduced graphite (g)-C_3N_4 into the aqueous epoxy solution by using an ultrasonic dispersion technology, and the modified water-based composite epoxy coating also showed excellent barrier properties. These research studies focus on the uniform distribution of fillers and the improvement of corrosion resistance of coatings, while there are few research studies on low-surface-energy silicone resins.

Based on the above discussions, a simple mechanical stirring method was used to add the inorganic filler to the water-based epoxy-modified silicone resin emulsion, and the crosslinking agent was formed to cure it at room temperature to prepare a stable coating with environmental protection and corrosion resistance. The microstructure, phase change and corrosion resistance of the coating were analyzed by scanning electron microscopy (SEM), X-ray diffraction (XRD) and electrochemical impedance spectroscopy (EIS). The failure mechanism of organosilicone resin was discussed. The addition of the inorganic filler gives the coating both inorganic and organic properties, i.e., more excellent comprehensive properties. The high silicon content of the epoxy-modified silicone resin emulsion gives the coating better hydrophobic properties. In addition, the flake structure of graphene oxide can effectively alleviate the corrosion of a metal matrix. This work provides a new method against microbial attachment and for protection of metal equipment in a marine environment.

2. Methods

2.1. Materials and Reagents

The epoxy-modified silicone resin emulsion (SH9607) was manufactured by Hubei Longsheng Sihai New Materials Co., Ltd. (Shenzhen, China) The array of active additives encompassed dipropylene glycol methyl ether (DPM), sourced from Shenzhen Jinlong Chemical Technology Co., Ltd. (Shenzhen, China); alkynyl glycol 104E, obtained from Air Chemical Products Co., Ltd. (Tokyo, Japan); the water-based dispersible agent SN5040 and the antifoaming agent J0401, both procured from Shenzhen Jitian Chemical Co., Ltd. (Shenzhen, China); and the leveling agent BYK381, purchased from BYK Chemical (Wesel, Germany). As for fillers, single-layer graphene oxide was acquired from Suzhou Carbonfeng Graphene Technology Co., Ltd. (Suzhou, China), talc was sourced from Guilin Guiguang Talc Development Co., Ltd. (Guilin, China), mica powder was obtained from Chuzhou Wansilk New Material Co., Ltd. (Chuzhou, China), iron red was purchased from Henan Bairun New Material Co., Ltd. (Zhengzhou, China), zinc tetrahydrate phosphate was sourced from Sinopdrug Group Chemical Reagent Co., Ltd. (Shanghai, China), and the curing agent KH550 was procured from Dongguan Kangjin New Material Technology Co., Ltd. (Dongguan, China) Q235 served as the substrate for the coating in the conducted experiments, eliminating the need for any further pretreatment.

2.2. Preparation of Different Coatings

The schematic depiction of the composite coating preparation is presented in Figure 1. The Q235 steel plate, measuring 10 mm × 10 mm × 1 mm, underwent polishing using SiC paper graded from 100 to 800, followed by ultrasonic cleaning with absolute ethanol (99.9%) for 10 min. Subsequently, the plate was dried in an air-drying oven maintained at 60 °C. To ensure the effective dispersion of fillers within the coating system, the active additives were initially introduced into the aqueous solvent, and a pretreatment solution was obtained through uniform stirring. High-speed grinding was then employed to incorporate the fillers, resulting in a uniformly colored paste. This colored paste was subsequently combined with the emulsion in a fixed proportion, completing the preparation of the coating. Prior to spraying, the composite coating and curing agent were mixed in a mass ratio of 20:1 and allowed to age for 20 min. The mixture was then sprayed onto the substrate surface and left for curing at room temperature for 7 days, yielding the desired coating with a thickness of approximately 60 ± 5 μm. To investigate the optimal filler dosage, a series of GO-modified composite coatings containing fillers in varying concentrations (0 wt. %, 0.01 wt. %, 0.05 wt. %, 0.1 wt. % and 0.15 wt. %) were prepared and labeled as G0, G1, G5, G10, and G15, respectively.

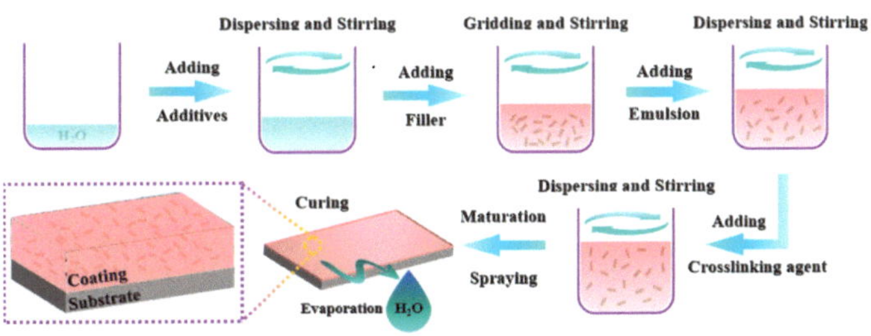

Figure 1. Schematic diagram of composite coating preparation process.

2.3. Characterization

Optical microscopy and scanning electron microscopy (SEM) (condition: EHT = 3.00 KV, WD = 7.4~8.4 mm) were employed to meticulously observe the morphological features of the modified composite coatings with varying concentrations of GO. Precise measurements of the coating's water contact angle were conducted using the droplet angle measurement method on a Shanghai Zhongchen JC2000D3-X (Shanghai, China) contact angle/surface tension measuring instrument. Furthermore, X-ray diffraction (XRD) and SEM (condition: EHT = 3 KV, WD = 8.0~8.4 mm) were utilized to comprehensively characterize the microstructural and phase changes in the G10 coating before and after simulated seawater immersion. Ultrasonic thickness gauges (CT400) were utilized to ensure consistent thickness of coatings with different contents. The corrosion resistance of various coatings was rigorously tested in a 3.5 wt. % NaCl solution using an electrochemical workstation (reference 600+). Additionally, the long-term corrosion behavior of the G10 coating was specifically evaluated in simulated seawater. For the electrochemical measurements, a coated steel sample with an exposed area of 1 cm^2 served as the working electrode, a saturated calomel electrode was used as the reference electrode, and a platinum plate electrode functioned as the counter electrode. The impedance spectrum measurements were conducted within a frequency range of 100 kHz to 0.1 Hz, with an amplitude of 5 mV. To minimize the influence of external factors, the entire experimental process was conducted within a Faraday cage.

3. Discussion and Results

3.1. Status of Filler

The cross section of the composite coating modified with different contents of GO was polished, and the morphology was observed with an optical microscope to judge the compatibility and the distribution of the GO. As shown in Figure 2a–e, the surface morphology of the coating did not show the state of incompatible fillers, indicating that the fillers were evenly distributed in the composite coating and could be well bonded together by the film-forming material. As shown in Figure 2f–j, cross section micro-morphologies of GO composite coatings with different contents can be seen. The lamellar fillers were detected. The film-forming materials on the surface of the G0 coating (due to its poor electrical conductivity and light color) are obvious, but there are more pores between lamellar fillers. The surface density of the G1 coating is improved. Some of the larger fillers are embedded in the surface of the coating, and the smaller fillers are completely wrapped in the film-forming material. The surface of the G5 coating is similar to that of G1. There is a large flake filler on the surface, and the density of the coating is further improved. There is basically no protruding sheet filler on the surface of the G10 coating, which indicates that the compatibility between the filler and the film-forming material is the best, and the coating is the densest. For the G15 coating surface, more pores are in appearance. Meanwhile the density of the coating is poor, due to too much graphene oxide. It results in the agglomeration in the composite, so that its nanoparticle characteristics begin to gradually fail. The distribution of coating filler is also negatively affected. The higher the density of the coating, the stronger the ability of the coating to block the external medium, the stronger the corrosion resistance. It shows that the appropriate addition of graphene oxide filler is beneficial to improve the overall performance of the coating.

3.2. Hydrophobicity

The dynamic water contact angle of the coating can reflect the dynamic anti-fouling performance of the coating. As shown in Figure 3, the dynamic water contact angle changes in composite coatings with different contents of graphene oxide. Because graphene oxide is a hydrophilic nanofiller, the dynamic water contact angle of the coating film shows a decreasing trend with the increase in graphene oxide nanofiller. The reason is that the presence of hydroxyl groups on the edge and base surface of the graphene oxide sheet determines its hydrophilicity. Therefore, with the increase in graphene oxide content in the modified composite coating, the hydrophilicity of the GO was continuously offset by the hydrophobicity of the silicone resin. However, in general, the water contact angle of the composite coating with different graphene oxide contents is greater than 90° at 0 s, and the composite coating has the performance of hydrophobicity at this moment. In the process of a ship traveling, the marine organisms attached to the surface of the ship can be removed by relying on the shear force of the seawater to achieve a better dynamic anti-fouling effect.

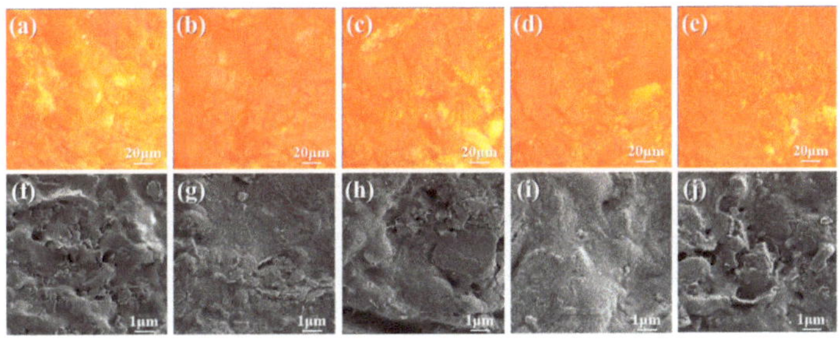

Figure 2. Optical and electron microscope images of cross sections of GO composite coatings with different contents: (**a**,**f**) G0, (**b**,**g**) G1, (**c**,**h**) G5, (**d**,**i**) G10, (**e**,**j**) G15.

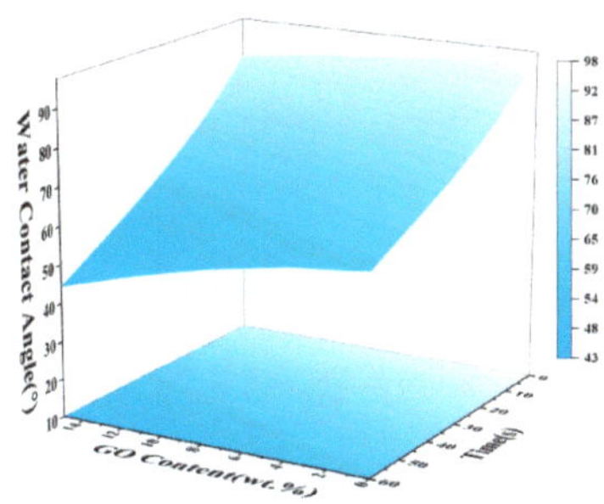

Figure 3. Changes in dynamic water contact Angle of composite coatings with different GO contents.

3.3. *Corrosion Resistance*

As can be seen from Figure 4a,b, the arc resistance of the composite coating is as follows: G10 > G15 > G5 > G1 > G0, indicating that the corrosion resistance of the composite coating increases first and then decreases with the increase in graphene oxide addition. And G10 has the largest arc resistance radius and the best corrosion resistance. As can be seen from the microscopic morphology of the coating, the number of pores on the surface of the coating gradually decreases with the increase in the content of GO. When the content of GO exceeds 0.1 wt. %, the coating is more easily immersed in the external water, and the corrosion resistance of the coating decreases. As shown in Figure 4c,d, with the increase in graphene oxide content, the impedance mode value |Z| in the Bode figure is as high as 10^8 and only one time constant appears in the phase angle figure, indicating that the corrosive medium did not penetrate the coating to have an electrochemical reaction with the metal substrate. Compared with the waterborne inorganic zinc-rich coating modified by rGO studied by Jang [13] et al., the coating developed here has better corrosion resistance. It means that the barrier property of the coating is good. With the increase in graphene oxide content, the impedance mode value |Z| in the low-frequency region of the coating firstly increases and then decreases, and the phase angle diagram shifts first to the left and then to the right, which is in accordance with the change rule of the Nyquist diagram. The impedance mode value of the G10 coating is the largest, and the order of magnitude is 10^8, which is three orders of magnitude larger than that of G0, indicating that the corrosion resistance of the coating can be significantly improved by adding an appropriate amount of graphene oxide to the coating.

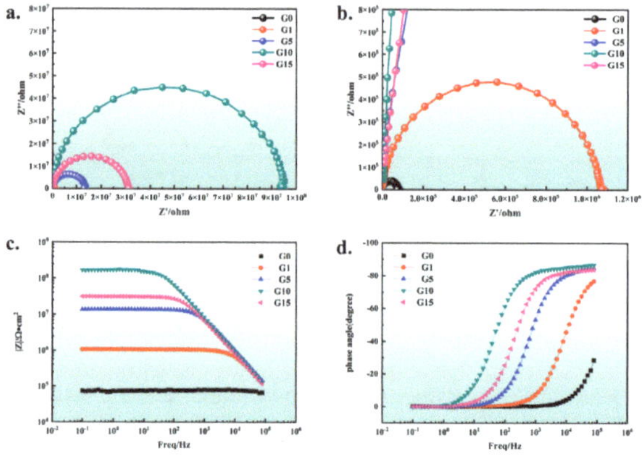

Figure 4. Impedance spectra of modified composite coatings with different GO contents: (**a**) Nyquist diagram; (**b**) Nyquist local magnification diagram; (**c**) impedance modulus |Z| diagram; (**d**) phase angle diagram.

By observing the impedance spectrum of the composite coating, it is found that the coating is a standard charge diffusion. The equivalent circuit diagram of different contents of graphene oxide coating was obtained from the impedance spectrum, as shown in Figure 5. The circuit description code is Rs (CPEdlRct), where Rs is the solution resistance. Rct is the charge transfer resistance, namely the coating resistance. CPEdl is the coating capacitance.

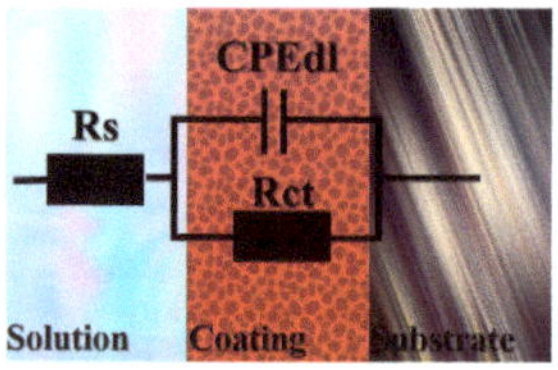

Figure 5. Analog equivalent circuit diagram of modified composite coating with different GO contents.

The impedance spectrum was fitted to obtain the equivalent circuit parameter, Table 1, and the equivalent circuit parameter variation, Figure 6. The increasing capacitance of the coating is caused by the increasing pseudocapacitance provided by the oxygen-containing functional groups on the surface of graphene oxide [14]. It can be seen that the coating resistance first increases and then decreases with the content of GO, reaching the maximum value when the content is 0.1 wt. %. Although the sheet graphene oxide has a barrier effect and can delay the penetration of corrosive media, the presence of hydrophilic groups weakens its anti-corrosion performance. Therefore, when a small amount of graphene oxide is added, it is easy to evenly disperse and can effectively play a blocking role. However, with the increase in the addition, the hydrophilicity of the coating is enhanced, resulting in the agglomeration of the graphene oxide, which leads to the decrease in the coating resistance. Therefore, when the amount of GO was 0.15 wt. %, the corrosion resistance began to decrease. The coating capacitance in electrochemical impedance spectroscopy has been proved to be a key parameter to measure the water permeability of the coating [15]. The coating capacitance CPEdl increases with the increase in the graphene oxide content, because CPEdl represents the dielectric property of the coating, and the coating gradually increases during the process of being gradually permeated by electrolytes [16]. The more graphene oxide in the coating, the easier the electrolytes in the coating will penetrate into the coating due to its hydrophilicity, thus increasing the coating capacitance.

Table 1. Equivalent circuit fitting parameters of different contents of graphene oxide-modified composite coatings.

Content (wt. %)	Rs (Ω·cm^{-2})	Error (%)	CPEdl (F)	Error (%)	Rct (Ω·cm^{-2})	Error (%)	Goodness of Fit
0	0.00126	2.06×10^{7}	1.416×10^{-11}	1.404	75,910	0.5475	0.00133
0.01	846.6	57.38	1.474×10^{-11}	1.647	1.024×10^{6}	1.16	0.00547
0.05	16,860	23.76	1.546×10^{-11}	1.37	1.351×10^{7}	0.9059	0.00404
0.1	11,140	51.88	1.761×10^{-11}	1.823	1.673×10^{8}	1.987	0.011
0.15	17,080	27.35	2×10^{-11}	1.863	3.064×10^{7}	1.621	0.00909

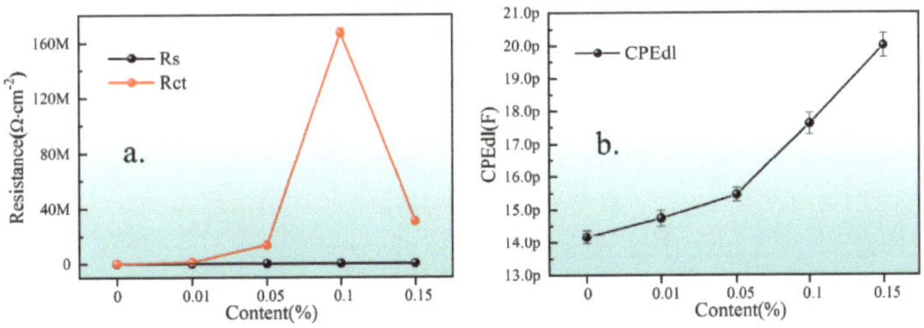

Figure 6. GO equivalent circuit diagram with different content parameter variation diagram: (**a**) coating resistance and solution resistance variation diagram, (**b**) diagram of coating capacitance variation.

3.4. Corrosion Mechanism

The water absorption of the coating has an important impact on the coating/metal interface, which in turn affects the corrosion resistance of the coating [17]. The corrosion resistance of organic coatings is not only achieved by resistance inhibition, but also by the transport of water and corrosive substances. The low permeability of water does not guarantee the best corrosion resistance, because when osmotic pressure is generated due to surface contaminants, it may cause the coating to blister. On the contrary, a certain osmotic pressure is necessary to prevent blistering. The water absorbed in the coating will affect the coating polymer and cause swelling, which is reflected in the change in dielectric properties [18], so the water absorption rate of the coating can be combined with electrochemistry to jointly analyze them. As shown in Figure 7, the water absorption curve of the composite coating with graphene oxide content of 1 wt. % increases with time. It can be seen that the water absorption rate of the coating film is fast in the initial stage of immersion, and then the water absorption rate gradually slows down until it reaches saturation. When immersed in deionized water for 350 h (≈15 d), the water absorption rate no longer increases, and the water absorption rate at this time is 18.6%.

EIS has been widely used in the study of coating protection and aging mechanisms. The resistance of a coating is an important index to evaluate its corrosion resistance, and the capacitance value can reflect the ability of coating to allow electrolytes to penetrate. In addition, the corrosion rate of the metal under the coating can be understood by estimating the reaction resistance [19]. Generally speaking, the arc radius of capacitive reactance in a Nyquist diagram is inversely proportional to the corrosion rate: the smaller the radius, the faster the corrosion rate of the coating. Conversely, the larger the radius, the slower the corrosion rate [20]. As shown in Figure 8a,b, the arc resistance of the composite coating is semi-circular, indicating that the electrode process is mainly controlled by charge transfer. And the effect caused by impedance is negligible. With the extension of soaking time, the arc radius of the composite coating decreases gradually, and the corrosion resistance weakens gradually. In the simulated corrosion experiment of seawater for 28 days, no second arc resistance appeared, indicating that no corrosion products were formed, which confirmed the excellent corrosion resistance of the composite coating. At the initial stage of immersion, seawater quickly permeates the coating, enhancing the ion conductivity. It results in a sharp decrease in the coating resistance and a significant change in the impedance spectrum. With the extension of soaking time, the penetration rate of seawater slows down, and the change in coating resistance is gradually stable. The change in capacitance–reactance spectrum characteristics decreases. When the soaking time exceeds 350 h, the coating resistance shows a downward trend again, which indicates that corrosion occurs at the interface between the coating and the metal, resulting in local peeling of the coating [21]. As shown in Figure 8c,d, the impedance mode value of the composite coating in seawater, |Z|, gradually decreases with the increase in time, and the phase angle curve gradually shifts to the right with the increase in time, which reflects the continuous decrease in the coating resistance and the gradual decline of corrosion resistance, whose changing speed is consistent with that of the Nyquist diagram.

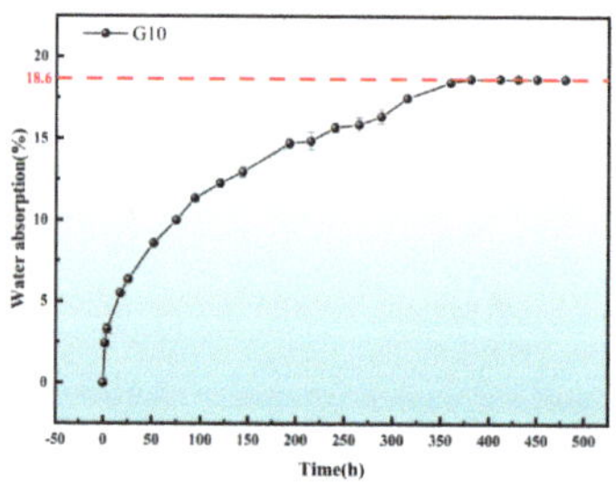

Figure 7. Water absorption of G10 coating—time variation diagram.

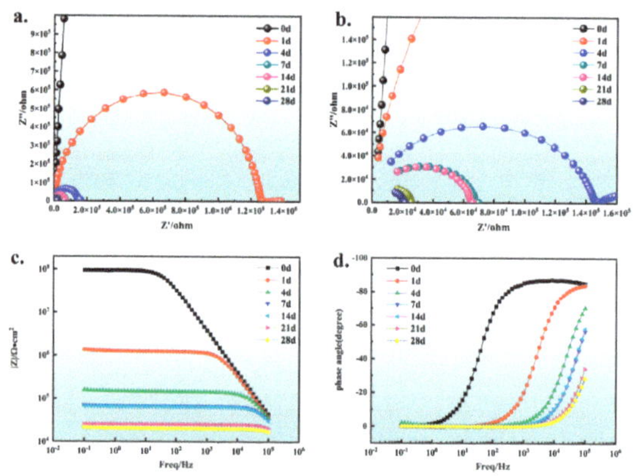

Figure 8. Variation in the impedance spectrum of the composite coating with seawater immersion time: (**a**) Nyquist plot, (**b**) Nyquist local enlarged image, (**c**) Bode modulus |Z| diagram, (**d**) Bode phase angle diagram.

Combined with the characteristics of the impedance spectrum, structural characteristics and existing studies of the composite coating, the equivalent circuit (which is the same as Figure 5) was used to fit the impedance spectrum of the composite coating soaking process. Meanwhile, the circuit description code of the equivalent circuit diagram is Rs (CPEdlRct), where Rs stands for solution resistance, Rct stands for coating resistance (charge transfer resistance), and CPEdl stands for coating capacitance, whose value will change with the infiltration of a corrosive medium [22]. Table 2 lists the fitting parameters of the corresponding electrochemical impedance spectrum equivalent circuit diagram for different immersion periods.

Table 2. The fitting parameters of the equivalent circuit diagram of electrochemical impedance spectra corresponding to different soaking periods.

Time (Day)	Rs ($\Omega\cdot cm^{-2}$)	Error (%)	CPEdl (F)	Error (%)	Rct ($\Omega\cdot cm^{-2}$)	Error (%)	Goodness of Fit
0	5563	25.62	4.104×10^{-11}	1.028	9.353×10^{7}	1.164	3.62×10^{-3}
1	4587	19.99	4.426×10^{-11}	1.054	1.266×10^{6}	0.6227	1.72×10^{-3}
4	5157	17.16	4.566×10^{-11}	1.926	142,000	0.7506	1.34×10^{-3}
7	4070	10.42	4.686×10^{-11}	1.561	63,960	0.6546	2.196×10^{-4}
14	2810	12.09	4.823×10^{-11}	1.283	62,420	0.5377	1.562×10^{-4}
21	2109	12.94	5.298×10^{-11}	2.398	23,670	1.133	4.756×10^{-5}
28	3313	20.03	6.327×10^{-11}	7.795	17,570	3.72	2.71×10^{-4}

Figure 9 shows the variation trend of CPEdl, Rs and Rct in the composite coating equivalent circuit diagram. At the initial stage of immersion, water permeates the coating evenly, and the double electric layer capacitance of the coating increases slowly. Before 350 h, the water absorption in the composite coating gradually reaches saturation. The capacitance gradually increases, and the increase rate gradually slows down. This corresponds to the ion transport process inside the composite coating, which is the incubation stage when the coating substrate begins to corrode. The coating capacitance begins to change greatly, indicating that the interface between the coating and the substrate at this time is locally foamed and stripped due to the immersion of the corrosive medium, which leads to a substantial increase in the coating capacitance, and the coating begins to fail at this time and the substrate corrosion begins [17]. The coating resistance Rct changes greatly in the initial stage. The reason is that at the initial stage of composite coating immersion the penetration of water to the coating can significantly promote the ion conductivity of the coating, resulting in a large change in the coating resistance after one day and four days of immersion. After that, the water molecules in the coating gradually approach saturation, and the ion conductivity changes little, resulting in a small change in the coating resistance.

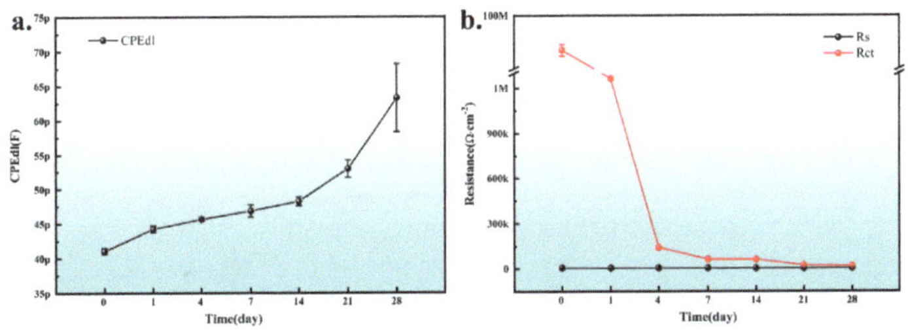

Figure 9. Variation trend of (**a**) parameters CPEdl, (**b**) Rs and Rct in equivalent circuit diagram.

Figure 10 shows the microstructure of composite coating G10 before and after it was soaked in simulated seawater for 28 days. After soaking, the number of pores in the composite coating increased, resulting in increased surface roughness of the composite coating, and the remaining materials were mainly inorganic fillers with large sheets, because organic film-forming materials on the surface of the composite coating decomposed under the action of the seawater. The tiny inorganic fillers were dissolved in the simulated seawater, resulting in a lot of tiny pores. The water absorption of the composite coating reached saturation. Due to the existence of these pores in seawater, the seawater filled these tiny pores, resulting in an increase in the overall ion conductivity of the composite coating, a decrease in the resistance of the composite coating and a decrease in the corrosion resistance of the composite coating.

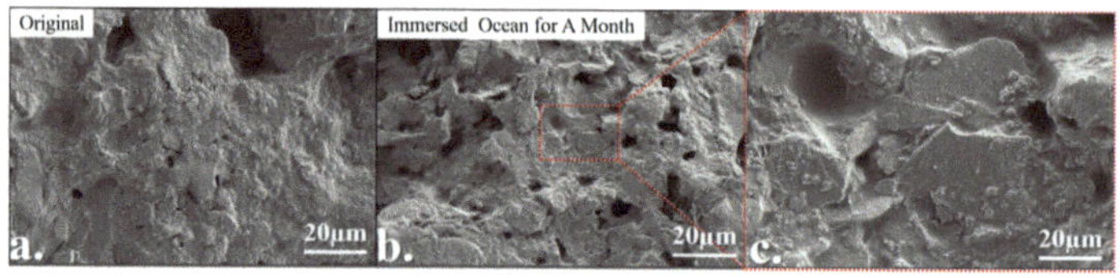

Figure 10. Comparison of the micromorphology of composite coating G10 after 28 days of immersion in simulated seawater: (**a**) The original morphology of the coating, the morphology of the coating after 28 days of immersion in simulated seawater figure (**b**), and the local enlarged figure (**c**).

Phase analysis of modified composite coating G10 before and after seawater immersion is shown in Figure 11. It can be seen that the phase composition of the composite coating did not change before and after seawater immersion, indicating that no corrosion products were produced on the surface of the coating. During seawater immersion, the structural stability of the composite coating was good, and no different phases were produced. The stable structure of the coating can reduce the dissolution of the internal GO as much as possible, thus reducing the damage caused by GO to marine life [23]. According to the above electrochemical impedance spectroscopy analysis, it can also be seen that the interface between the composite coating and the substrate has just bubbled or spalled, the corrosion behavior is in the initial stage, and the corrosion products still do not appear on the surface of the coating.

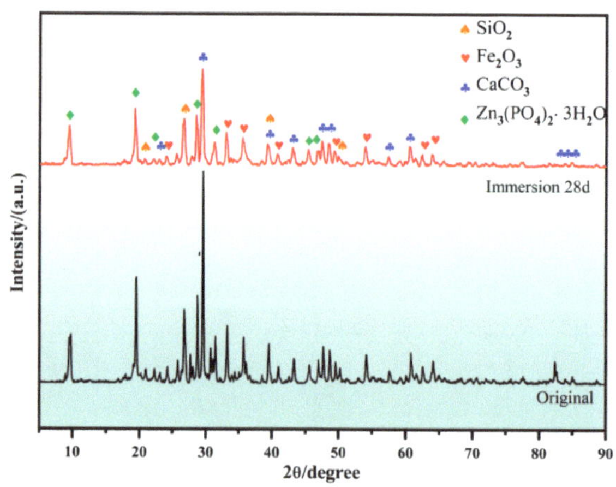

Figure 11. Phase analysis of composite coating G10 before and after 28 d seawater immersion.

3.5. Failure Mechanism Analysis of Coating

Based on the water absorption, electrochemical impedance spectroscopy, phase analysis and microscopic morphology observation and analysis of the composite coating, the failure mechanism diagram of the coating in simulated seawater, as shown in Figure 12, was obtained. At the initial stage of immersion of the composite coating, the corrosive medium stops on the surface of the coating. At this time, the corrosion resistance of the coating is the maximum, and the double electric layer capacitance of the composite coating is the minimum. As the corrosive medium begins to penetrate the composite coating, it gradually dips into the interior of the coating. In the middle stage of corrosion, the penetration rate of the corrosive medium gradually changes from fast to slow and gradually tends to saturation. Water molecules in the corrosive medium will affect the polymer and cause the coating to swell. At this time, the ion conductivity of the composite coating increases significantly, the resistance of the composite coating decreases sharply, and the double electric layer capacitance of the composite coating increases slowly until 350 h later. The corrosive medium in the composite coating reaches saturation, which belongs to the gestation stage of matrix corrosion. At the end of immersion, local bubbling and spalling occur at the interface between the composite coating and the substrate. The double electric layer capacitance increases sharply with the process at the interface between the composite coating and the substrate, from local spalling to whole spalling, and the coating resistance gradually decreases. From then on, the corrosion of the substrate begins, and the protective effect of the coating begins to fail until it completely fails.

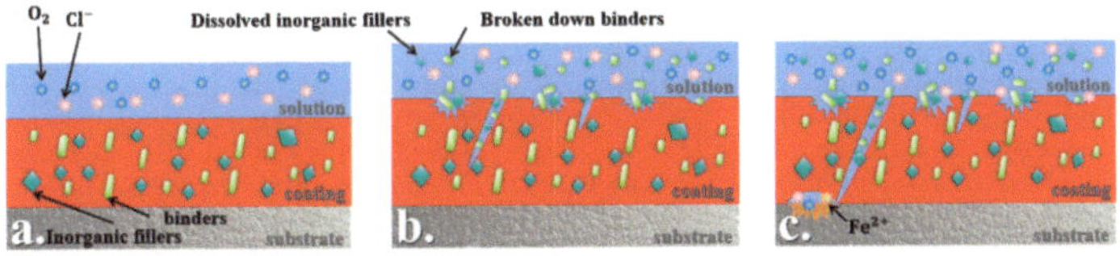

Figure 12. Failure mechanism of composite coating G10 coating in simulated seawater. (**a**) At the initial stage of immersion; (**b**) at the middle stage of immersion; (**c**) at the beginning stage of late immersion.

4. Conclusions

With the enhancement of environmental protection awareness, water-based coatings will gradually replace traditional oil-based coatings. This paper focuses on the study of the structure and properties of water-based epoxy-modified silicone coatings, and the research results provide reference values for the exploration of water-based silicone coatings. In this paper, graphene oxide as a nanofiller was introduced into the composite coating for modification, and the performance of the graphene oxide-modified composite coating was systematically studied. The composite coating with the best comprehensive performance was obtained when the amount of graphene oxide was 0.1 wt. %, and the best modified composite coating was immersed in simulated seawater to study its failure behavior. The filler, different contents of GO, in the modified composite coating is uniformly distributed, all states are compatible, and the impedance modulus of the coating $|Z|$ can reach 10^8 $\Omega \cdot cm^2$. The composite coating shows good hydrophobic properties at 0 s water contact, i.e., an angle greater than 90°, which also results in a good dynamic anti-fouling effect. The structure of the composite coating is relatively stable before and after immersion in seawater. The long-term corrosion rule of the composite coating in simulated seawater has been deeply studied. Through the data analysis of the test results, the failure process of the composite coating is rationally explained, and the mechanism diagram of the failure process is drawn.

Author Contributions: Conceptualization, J.X.; Formal analysis, J.L.; Data curation, W.T.; Writing—original draft, Z.C.; Writing—review & editing, Z.Z. All authors have read and agreed to the published version of the manuscript.

Funding: This work was supported by the National Nature Science Foundation of China (No. 51872072), Special funding for the development of science and technology of Shanghai Ocean University (No. A2-0203-00-100231 and A2-2006-00-200371).

Institutional Review Board Statement: Not applicable.

Informed Consent Statement: Not applicable.

Data Availability Statement: Data are contained within the article.

Conflicts of Interest: The authors declare no conflict of interest.

References

1. Zhang, K.; Cong, W.; Gui, T.; Xing, T.; Wu, L. Effect and Remediation of Biofouling on Marine Aquaculture. *Mater. Rep.* **2020**, *34*, 78–81.
2. Sun, B.-K.; Fan, H.-S.; Pan, X.-L.; Lu, A.-D.; Hu, J.-K. Development of Copper-free Self-polishing Anti-fouling Paints by Using Acrylate Resin. *Surf. Technol.* **2022**, *51*, 280–286.
3. Xu, F.; Wang, T.; Bohling, J.; Maurice, A.M.; Chen, H.; Wu, L.; Zhou, S. Extended hydrophobicity and self-cleaning performance of waterborne PDMS/TiO_2 nanocomposite coatings under accelerated laboratory and outdoor exposure testing. *J. Coat. Technol. Res.* **2018**, *15*, 1025–1034. [CrossRef]
4. Khanjani, J.; Pazokifard, S.; Zohuriaan-Mehr, M.J. Improving, dirt pickup resistance in wat erborne coatings using latex blends of acrylic/PDMS polymers. *Prog. Org. Coat.* **2017**, *102*, 151–166. [CrossRef]
5. Wang, Q.; Li, C.; Li, G. Study on stability of water-based two-component, i norganic coatings based on silica sols. *Electroplat. Finish.* **2021**, *40*, 654–659.
6. Shan, X.; Wang, S.; Zhang, L.; Jiang, L. Preparation of silica sol-based water-based pure in organic coatings. *Appl. Chem. Ind.* **2018**, *47*, 1195–1199.
7. Wang, C.; Zhao, L.; Jia, J.; Wang, D.; Peng, Z. Effect of mixed acid functionalized carbon nanotubes doping on the electrical and thermal conductivity of epoxy resin. *Trans. China Electrotech. Soc.* **2019**, *34*, 457–464.
8. Tian, Y.; Xie, Y.; Dai, F.; Huang, H.; Zhong, L.; Zhang, X. Ammonium-grafted graphene oxide for enhanced corrosion resistance of waterborne epoxy coatings. *Surf. Coat. Technol.* **2020**, *383*, 125227. [CrossRef]
9. Huang, S.; Kong, G.; Yang, B.; Zhang, S.; Che, C. Effects of graphene on the corrosion evolution of zinc particles in waterborne epoxy zinc-containing coatings. *Prog. Org. Coat.* **2020**, *140*, 105531. [CrossRef]
10. Li, S.; Zhang, S.; Yang, B.; Guo, H.; Li, Y.; Wen, F. Research progress of graphene anticorrosive coatings. *Corros. Sci. Prot. Technol.* **2019**, *31*, 455–461.
11. Chen, S.; Brown, L.; Levendorf, M.; Cai, W.; Ju, S.Y.; Edgeworth, J.; Li, X.; Magnuson, C.W.; Velamakanni, A.; Piner, R.D.; et al. Oxidation resistance of graphene-coated Cu and Cu/Ni alloy. *ACS Nano* **2011**, *5*, 1321–1327. [CrossRef]
12. Chen, C.; He, Y.; Xiao, G.; Zhong, F.; Xia, Y.; Wu, Y. Graphic C_3N_4-assisted dispersion of graphene to improve the corrosion resistance of waterborne epoxy coating. *Prog. Org. Coat.* **2020**, *139*, 105448. [CrossRef]
13. Jiang, Y.; Jiang, H.Y.; Xiong, S.P. Effect of Reduced Graphene Oxide on Corrosion Resistance of Waterborne Inorganic Zinc-rich Coatings. *Paint. Coat. Ind.* **2023**, *53*, 68–72.
14. Xiao, P.; Wang, D.-H.; Lang, J.-W. Comparison in Factors Affecting Electrochemical Properties of Thermal-Reduced Graphene Oxide for Supercapacitors. *J. Electrochem.* **2014**, *20*, 553–562.
15. Darowicki, K. The application of impedance measurements for the determination of the probability of the course of corrosion processes. *Corros. Sci.* **1997**, *39*, 1087–1092. [CrossRef]
16. Wang, W.J.; Guan, Z.C.; Han, J.C.; Xu, H.; Jin, X.; Di, Z.G. Application of Electrochemical Impedance Spectroscopy in the Evaluation of Protective Performance of Organic Coating. *Paint. Coat. Ind.* **2023**, *53*, 58–64.
17. Gao, L.X.; Zhang, D.Q.; Zhou, G.D.; Li, H.G. Study on water absorption and corrosion protection properties of modified epoxy coatings. *Chin. J. Corros. Prot.* **2002**, *22*, 41–43.
18. Van Westing, E.P.M.; Ferrari, G.M.; De Wit, J.H.W. The determination of coating performance with impedance measurements—II. Water uptake of coatings. *Corros. Sci.* **1994**, *36*, 957–977. [CrossRef]
19. Cai, G.; Zhang, D.; Zhao, W.; Dong, Z. Research progress on protection performance and failure evaluation of organic coatings. *Corros. Prot.* **2017**, *38*, 657–664.
20. Yuan, S.J.; Chong AM, F.; Pehkonen, S.O. The influence of the marine aerobic Pseudomonas strain on the corrosion of 70/30 Cu-Ni alloy. *Corros. Sci.* **2007**, *49*, 4352–4358. [CrossRef]
21. Yang, L.H.; Liu, F.C.; Zhang, Z.Q.; He, P.; Han, E.H.; Qi, L. Research on the Anti-Diffusion Behavior of ZnO/Polyurethane Coatings. *J. Peking Univ. Nat. Sci.* **2006**, 91–96.

22. Wu, C.; Lin, Z.; Li, X.; Ma, B.; Xu, L. Electrochemical Corrosion Behavior of Zinc Powder Sherardized Carbon Steel. *Corros. Sci. Prot. Technol.* **2014**, *26*, 441–445.
23. De Medeiros, A.M.; Khan, L.U.; da Silva, G.H.; Ospina, C.A.; Alves, O.L.; de Castro, V.L.; Martinez, D.S. Graphene oxide-silver nanoparticle hybrid material: An integrated nanosafety study in zebrafish embryos. *Ecotoxicol. Environ. Saf.* **2021**, *209*, 111776. [CrossRef] [PubMed]

coatings

Article

A Study on the Corrosion Behavior of RGO/Cu/Fe-Based Amorphous Composite Coatings in High-Temperature Seawater

Zhenhua Chu, Yunzheng Zhang, Wan Tang, Yuchen Xu and Jingxiang Xu *

Department of Mechanical Engineering, College of Engineering, Shanghai Ocean University, Shanghai 201306, China; zhchu@shou.edu.cn (Z.C.); yz_zhang7@163.com (Y.Z.); wtang@shou.edu.cn (W.T.); 17749450088@163.com (Y.X.)
* Correspondence: jxxu@shou.edu.cn; Tel.: +86-22-60204810; Fax: +86-22-26564810

Abstract: In this paper, based on an Fe-based amorphous alloy, four kinds of RGO/Cu/Fe-based amorphous composite coatings with mass ratios of 5%, 10%, 15%, and 20% of RGO/Cu were prepared on the surface of 45# steel by using high-velocity oxy-fuel (HVOF) spraying. The coatings were immersed in simulated seawater at room temperature and at 90 °C for different lengths of time, and their corrosion resistance was tested using electrochemical impedance spectroscopy (EIS), scanning electron microscopy (SEM), and X-ray diffraction (XRD), and the surface morphology and phase distribution of the samples were observed. The results showed that with the increase in the introduction ratio of RGO/Cu, when the addition ratio reached 15%, the composite coating had the best corrosion resistance. After soaking in simulated seawater at 90 °C for 18 days, the surface of the coating showed slight peeling and crack propagation, but no obvious pitting phenomenon occurred. The corrosion mechanism of the RGO/Cu/Fe coating in high-temperature seawater is mainly that high temperature causes the cracking of the coating, which opens up a transport channel for corrosion media. However, due to the addition of RGO, the corrosion has a certain self-limitation effect, which is mainly due to the toughening effect of RGO on the coating and its effect on extending the corrosion channel.

Keywords: Fe-based amorphous composite coating; graphene; high-temperature seawater; corrosion

Citation: Chu, Z.; Zhang, Y.; Tang, W.; Xu, Y.; Xu, J. A Study on the Corrosion Behavior of RGO/Cu/Fe-Based Amorphous Composite Coatings in High-Temperature Seawater. *Coatings* **2024**, *14*, 556. https://doi.org/10.3390/coatings14050556

Academic Editor: Tadeusz Hryniewicz

Received: 11 April 2024
Revised: 25 April 2024
Accepted: 28 April 2024
Published: 1 May 2024

1. Introduction

As land resources are increasingly depleted, people have turned their attention to the development of marine resources. Human development of the ocean relies on various marine equipment. However, in the process of service, marine equipment is subject to severe corrosion due to the presence of moist, salt-laden air and water vapor in seawater. Marine engineering equipment such as ships and offshore platforms work under severe corrosive environments such as salt spray corrosion (caused by solid NaCl and water vapor) in high-temperature environments, which poses challenges to marine equipment materials [1–3]. Although traditional metal materials have been studied to improve the corrosion resistance, their protective effect is limited in such harsh marine environments [4–6]. In order to improve the durability of such equipment, spraying high-performance coatings has become a common and effective protective method, aiming to form a strong barrier to isolate the direct erosion of corrosive media and metal substrates [7–10].

Compared with metal alloys, amorphous alloys exhibit remarkable advantages in terms of the physical, chemical, and mechanical properties, such as the high strength and hardness, high elastic strain limit, and excellent wear resistance and corrosion resistance. In 2013, Ye and Shin [11] synthesized Fe–Cr–Mo–W–Mn–C–Si–B metallic glass composite materials containing a large amount of amorphous phase using the laser direct deposition method. They found that the microhardness ($HV_{0.2}$ 1591) of the amorphous phase was significantly higher than that of the crystalline phase ($HV_{0.2}$ 947), and the wear resistance increased significantly with the increase in the amorphous phase ratio. However, due to the limitation of the glass-forming ability (GFA), it is very difficult to fabricate large-scale BMG workpieces or directly use them as structural materials, which limits the application of amorphous alloys. Fortunately, amorphous coatings prepared by thermal spraying have demonstrated their advantages in corrosion resistance [12], showing great potential in industrial applications. Lin et al. [13] prepared $Fe_{40}Cr_{19}Mo_{18}C_{15}B_8$ Fe-based amorphous coating on 316 stainless steel using the high-velocity oxy-fuel (HVOF) spraying method. The experimental results showed that the Fe-based amorphous alloy coating exhibited a transition behavior of activation, passivation, and overpassivation in seawater and 3.5% NaCl solution. After alternating cold and hot salt-spray corrosion and high-speed water erosion tests, although the coating surface showed rusting after 12 weeks, its weight loss was minimal and a stable passive film was formed. This indicates that Fe-based amorphous alloys have excellent stable passivation and corrosion resistance properties. Wang et al. [14] prepared Fe-based amorphous coatings on an AISI 1020 steel tube surface using high-speed laser cladding technology, showing excellent corrosion resistance. The corrosion potential of this coating was as high as -0.471 V, the corrosion current density was as low as 2.7×10^{-6} A/cm^2, and the polarization resistance value was as high as 22,149 $\Omega \cdot cm^2$. These excellent corrosion resistance properties are mainly due to the high content of amorphous phase (up to 95%) in the coating and the protection of the Cr oxide layer formed on the surface.

Adding a second phase to Fe-based amorphous alloys is often an effective means of improving their corrosion resistance. Chu et al. [15,16] prepared TiN/Fe-based amorphous composite coatings and AT13/Fe-based amorphous composite coatings through plasma-spraying technology. From the morphology point of view, the typical layered structure of the thermal spray coatings and the close combination of the two phases in these two composite coatings have no obvious structural defects, and they have good corrosion resistance and wear resistance.

In recent years, graphene, as a new emerging material, has attracted worldwide attention for its unique properties and wide application prospects. In the field of corrosion protection, graphene is a cutting-edge material that can be used as a nano-filler reinforcement material to enhance its anti-corrosion performance [17,18]. Various research results reveal that the effectiveness of graphene in preventing corrosion is mainly attributed to its unique "maze effect" [19–21]. When corrosive media attempt to penetrate the graphene structure, its intricate layout makes the diffusion path of the corrosive media extremely tortuous and difficult. In addition, graphene can significantly fill the tiny pores in composite coatings, thereby reducing the porosity of the coating and enhancing its compactness [22]. Moreover, graphene is also known for its excellent mechanical properties. Introducing graphene as an additive into composite coatings can greatly improve the wear resistance and other mechanical properties of the coating [23,24].

In fact, the corrosion damage of high-temperature seawater to materials is more severe. For marine equipment that must operate in high-temperature environments, they need to be exposed to high-temperature seawater for long periods, such as tens or even hundreds of degrees Celsius. For example, the service environment of the armor layer of submarine oil and gas pipelines is generally between 20 °C and 130 °C [25]. Such extreme conditions can easily lead to equipment failure due to corrosion, so there are extremely stringent requirements for its corrosion resistance.

There have been a lot of studies on the corrosion resistance of Fe-based amorphous coatings in ambient seawater [13–16], but there are few research articles on the high-temperature seawater corrosion resistance of Fe-based amorphous coatings. For the high-temperature corrosion resistance of Fe-based amorphous coatings, some scholars believe that high temperatures affect the formation of passive films [26], while other scholars' research has proved the negative impact of high temperatures on metal passive films [27,28]. The working conditions of marine equipment are complex, and a large number of corrosion studies focused on ambient seawater cannot meet the service needs of certain special equipment, so it is very important to study the corrosion behavior of Fe-based amorphous coatings in high-temperature seawater.

In this study, Fe-based amorphous composite coatings of reduced graphene oxide (RGO)/copper (Cu) were prepared using plasma-spraying technology. The Fe-based amorphous composite coating containing 15% RGO/Cu was immersed in simulated seawater at 90 degrees Celsius for up to 18 days. Its resistance to high-temperature seawater corrosion was comprehensively evaluated and the protective mechanism of the coating was reviewed.

2. Experimental Materials and Methods

2.1. Preparation of the Coating

GO/Cu composite powder was prepared by the gas-atomizing drying method, in which the mass ratio of GO to Cu was 1:9. Then, the GO in the GO/Cu composite powder was reduced by the thermal reduction method to obtain RGO/Cu composite powder. The Cu powder and Fe-based amorphous powder required for this experiment were purchased from Shanghai Naio Nanotechnology Co., Ltd. (Shanghai, China). The purity of the Cu powder was 99.9%, and the particle size was 1 μm. The chemical formula of the Fe-based amorphous composite powder was $Fe_{45}Cr_{16}Mo_{16}C_{18}B_5$, with a particle size ranging from 15 μm to 45 μm. The GO was purchased from Suzhou Carbon Feng Technology Co., Ltd. (Suzhou, China), with a layer count of 1–2 and a purity of over 98%. The sheet diameter ranged from 0.2 μm to 10 μm. The mechanical mixing of the $Fe_{45}Cr_{16}Mo_{16}C_{18}B_5$ amorphous powder with m (RGO/Cu mass accounting for 5%, 10%, 15%, and 20% of the total) resulted in RGO/Cu/Fe-based amorphous composite powder. Finally, an RGO/Cu/Fe-based amorphous composite coating was prepared on the surface of 45# steel (0.45 wt.% C) with a size of 10 mm × 10 mm × 12 mm by plasma spraying. The powder used for spraying is shown in Figure 1, and the plasma spraying parameters are shown in Table 1 [29].

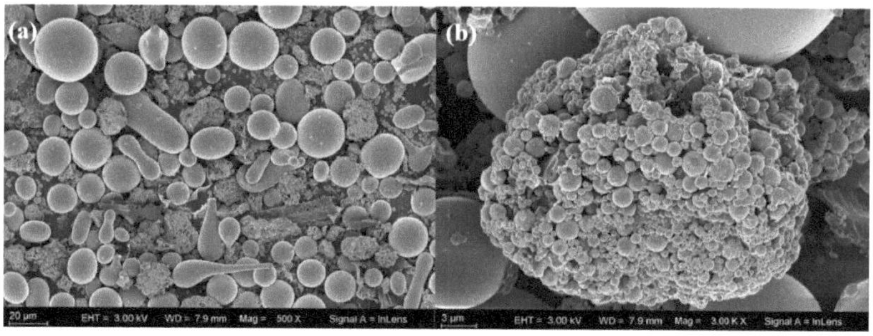

Figure 1. RGO/Cu/Fe-based amorphous composite powder: mixed powder (**a**); and single RGO/Cu particle (**b**).

Table 1. Spraying process parameters.

Process Parameters	RGO/Cu/Fe-Based Amorphous Composite Coating
Arc voltage (V)	70
Arc current (A)	500
Gun distance (mm)	100
Movement speed of spray gun (m/min)	5–7
Argon flow rate (dm^3/min)	30
Nitrogen flow rate (dm^3/min)	120
Coating thickness (μm)	300

In order to confirm the introduction of graphene, Raman spectroscopy (as shown in Figure 2a), a conventional method for characterizing graphene, and infrared spectroscopy (as shown in Figure 2b) were used for the composite powder. It can be calculated from the Raman spectrum that the ID/IG value of RGO/Cu was 1.14, and the ID/IG value of GO/Cu was 1.10. After the thermal reduction of the powder, the ID/IG value increased slightly, indicating a slight increase in disorder. The content of various oxygen-containing functional groups is shown in Figure 2b. It can be seen that at 3440 cm^{-1}, 1630 cm^{-1}, 1400 cm^{-1}, and 1060 cm^{-1}, oxygen-containing functional groups corresponding to -OH, C=O, C–OH, and C–O–C appeared, respectively. Moreover, the intensity of the oxygen-containing functional groups of the reduced RGO/Cu composite powder decreased, indicating the presence of GO, and GO forms RGO through reduction.

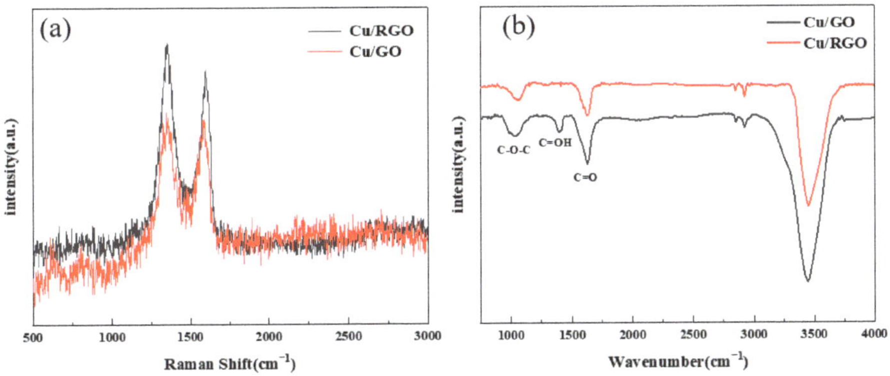

Figure 2. Raman spectrum of the composite powder (**a**); and infrared spectrum (**b**).

2.2. Experimental Route

In this paper, firstly, potentiodynamic polarization curve scanning was conducted on the coatings with different RGO/Cu addition ratios after being immersed in simulated seawater at room temperature for 30 days. The long-term corrosion performance of the four coatings in simulated seawater at room temperature was compared, and the optimal RGO/Cu addition ratio was obtained. Based on this optimal ratio coating, it was immersed in simulated seawater at 90 degrees Celsius for 18 days, and its high-temperature corrosion performance was tested. A series of characterizations were performed on its microstructure, phase, etc. The simulated seawater composition is shown in Table 2 [15].

Table 2. Artificial seawater formula (g/L).

NaCl	$MgCl_2$	Na_2SO_4	$CaCl_2$	KCl	$SrCl_2$	$NaHCO_3$	KBr	H_3BO_3	NaF
24.530	5.200	4.090	1.160	0.695	0.025	0.201	0.101	0.027	0.003

2.3. Coating Performance Testing Method

The electrochemical workstation interface 1010E produced by Gamry was used. During the test, a three-electrode system was used, with the coating sample as the working electrode, the saturated calomel electrode as the reference electrode, and the platinum sheet electrode as the counter electrode. Before performing the potentiodynamic polarization curve scan and electrochemical impedance spectroscopy (EIS) experiments, the coating was polished and buffed until it was in a mirror state, and it was then encapsulated with epoxy resin, leaving only a 1 cm^2 surface to be tested. The measurement of the potentiodynamic polarization curve was performed after the open circuit potential (EOCP) stabilized, with a scan rate of 1 mV/s and a scan range set at EOCP ± 500 mV. The test frequency of the electrochemical impedance spectroscopy ranged from 100,000 Hz to 0.1 Hz, with an additional sine wave AC perturbation frequency of 5 mV. The coating was subjected to polarization curve and EIS tests after being immersed in high-temperature simulated seawater for 0 day, 1 day, 4 days, 7 days, and 18 days. Each set of samples was tested three times and averaged to avoid accidental errors. After the testing was completed, data analysis was performed using Gamry Echem Analyst software (version 7.0.0.7). For each sample, the potentiodynamic polarization curve test was conducted after the EIS test.

For the composite coatings with different soaking times, scanning electron microscopy (SEM, S4800, Hitachi, Tokyo, Japan) was used to observe the surface morphology of the coatings. For the composite coatings soaked for 18 days, energy dispersive spectrometer (EDS) was used to analyze the elemental distribution of the coating surface, and X-ray diffraction (XRD, Bruker D8 Focus, Billerica, MA, USA) and X-ray photoelectron spectroscopy (XPS, Thermo Scientific K-Alpha, Thermo Fisher Scientific, Waltham, MA, USA) were used to analyze the phase composition of the coating surface. In the XRD test, the specific scanning angle was 20°–80°, and the scanning speed was 2°/min. In the XPS test, the fine spectra of the Cu and Fe elements were tested.

3. Results

3.1. *Characterization of RGO/Cu/Fe-Based Amorphous Composite Coating*

The surface morphology of the composite coatings with different RGO/Cu addition ratios was observed, as shown in Figure 3. As the RGO/Cu ratio increases, the dark-colored portion of the coating surface increases and is more evenly distributed. At the same time, it was observed that all four coating surfaces have a certain porosity, which increases with the increase in the RGO introduction ratio. This may be due to the fact that during plasma spraying, due to the high melting point of RGO, some RGO is not completely melted but is deposited on the substrate surface in a solid form, resulting in an increase in surface defects.

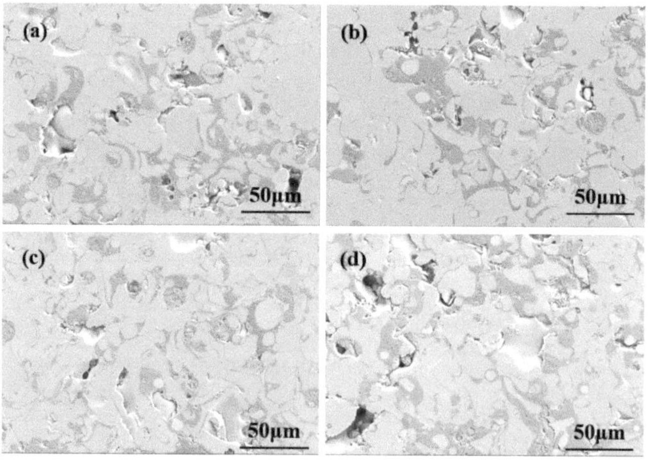

Figure 3. Surface morphology of composite coatings with different RGO/Cu addition ratios: 5% RGO/Cu/Fe-based coating (**a**); 10% RGO/Cu/Fe-based coating (**b**); 15% RGO/Cu/Fe-based coating (**c**); and 20% RGO/Cu/Fe-based coating (**d**).

Figure 4 shows the XRD scan results of the RGO/Cu/Fe-based amorphous composite coating after thermal spraying and polishing. It can be seen that the coating forms a typical amorphous diffraction peak, with a relatively high degree of amorphization. The signal peak of Cu is relatively strong. Figure 5 shows the EDS scanning results of the composite coating. The main elements are evenly distributed, and the Cu element has not been agglomerated seriously in a few areas.

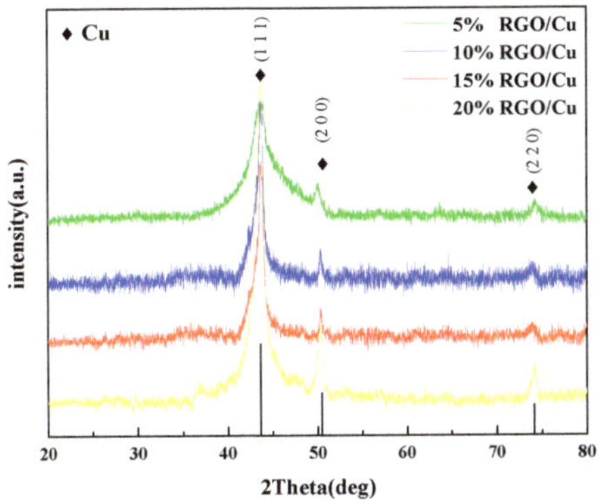

Figure 4. XRD scanning results of RGO/Cu/Fe-based amorphous composite coating.

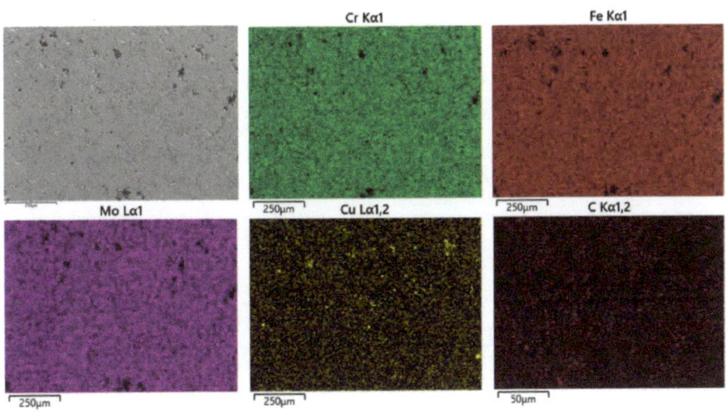

Figure 5. EDS results of composite coating.

3.2. Analysis of Potentiodynamic Polarization Curve

In this experiment, the content of the powder accounted for 5%, 10%, 15%, and 20% of the total mass of the powder and was named G1/Cu1, G2/Cu2, G3/Cu3, and G4/Cu4, respectively. The potentiodynamic polarization curves of the four composite coatings without immersion are shown in Figure 6a, and the fitting results (as shown in Table 3) indicate that the G3/Cu3 group has the highest self-corrosion potential (−371.3 mV) and the lowest self-corrosion current density (2.22 $\mu A \cdot cm^2$). After soaking these four composite coatings in simulated seawater at room temperature for 30 days, their potentiodynamic polarization curves were measured (as shown in Figure 6b). The fitting results of the potentiodynamic polarization curves of the four coatings are shown in Table 4. It can be seen that after 30 days of immersion, the G3/Cu3 composite coating has the highest corrosion potential (−557.8 mV) and the smallest self-corrosion current density (12.04 μA·cm^2). It is also found that as the content of RGO/Cu composite powder increases, the corrosion resistance of the coating shows a trend of increasing firstly and then decreasing. Compared with the G3/Cu3 composite coating, the corrosion resistance of the G4/Cu4 composite coating decreases. Considering the influence of porosity on the performance of the coating, as well as the high melting point of RGO, this may be due to the high content of RGO addition causing the thermal spray powder to not melt well, resulting in an increase in porosity in the coating, which in turn affects its corrosion resistance.

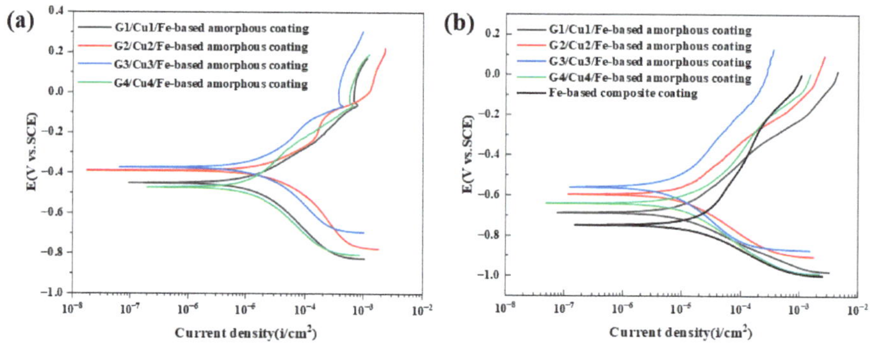

Figure 6. Polarization curves of composite coatings with different proportions of added phases after soaking in simulated seawater for 0 day (**a**) and 30 days (**b**).

Table 3. Fitting results of dynamic potential polarization curves for composite coatings with different components (soaking in artificial seawater for 0 day).

Samples	E_{corr} (mV)	i_{corr} (μA·cm^{-2})	CorrRate (mpy)
G1/Cu1/Fe-based amorphous coating	−449.9	5.22	1.95
G2/Cu2/Fe-based amorphous coating	−387.5	2.23	0.83
G3/Cu3/Fe-based amorphous coating	−371.3	2.22	0.83
G4/Cu4/Fe-based amorphous coating	−469.9	8.36	3.13

Table 4. Fitting results of dynamic potential polarization curves for composite coatings with different components (soaking in artificial seawater for 30 days).

Samples	E_{corr} (mV)	i_{corr} (μA·cm^{-2})	CorrRate (mpy)
G1/Cu1/Fe-based amorphous coating	−686.7	18.95	8.66
G2/Cu2/Fe-based amorphous coating	−593.8	16.00	7.30
G3/Cu3/Fe-based amorphous coating	−557.8	12.04	5.50
G4/Cu4/Fe-based amorphous coating	−638.2	15.33	7.00
Fe-based amorphous coating	−746.5	25.58	11.69

Given that the 15% RGO/Cu/Fe-based amorphous composite coating shows the best corrosion resistance among the four coatings, a high-temperature simulated seawater corrosion experiment was conducted using this coating. The potentiodynamic polarization curves of the 15% RGO/Cu/Fe-based amorphous composite coating immersed in simulated seawater at 90 °C for different days are shown in Figure 7. Meanwhile, the Tafel fitting data of the polarization curves are given in Table 5. It is shown that during the first day of corrosion, the self-corrosion potential of the coating slightly increases, while the self-corrosion current density decreases. At the same time, there is also a significant passivation plateau on both polarization curves, indicating that the coating improves its own corrosion resistance at the beginning stage of the corrosion process. Until the 18th day of immersion, the self-corrosion potential of the coating shows a downward trend, while the magnitude of the self-corrosion current also shows an upward trend, indicating that after the initial stage of corrosion, the corrosion resistance of the coating continues to decrease.

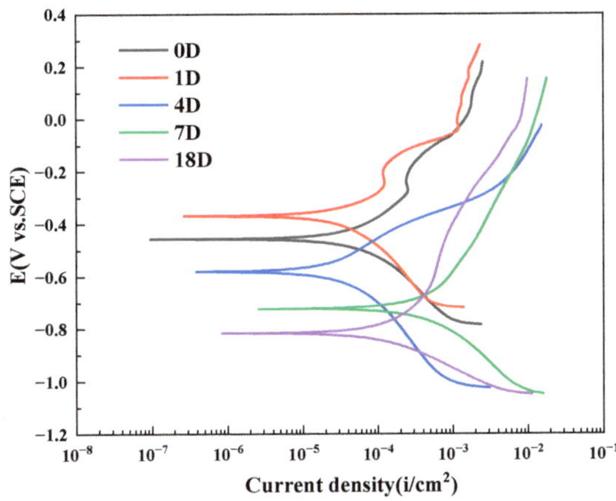

Figure 7. Dynamic potential polarization curve of RGO/Cu/Fe-based amorphous composite coating after soaking in simulated seawater at 90 °C for different days.

Table 5. Tafel fitting results after soaking in simulated seawater at 90 °C for different days.

Samples	E_{corr} (mV)	i_{corr} (μA·cm^{-2})	CorrRate (mpy)
0D	−457.2	7.65	3.50
1D	−368.2	7.41	3.39
4D	−579.8	10.19	4.66
7D	−772.6	21.64	9.89
18D	−816.7	20.12	9.19

The reason for the increase in corrosion resistance at the beginning of corrosion may be that at the beginning of corrosion, the effect of high temperature on the composite coating is not yet obvious. At the same time, the composite coating undergoes simulated seawater corrosion, and corrosion products are generated at the pores of the coating, forming a passive film on the surface of the coating, blocking the pores, reducing the passage of corrosion media into the interior of the coating through pores and other defects, playing a protective role for the substrate, and reducing the corrosion efficiency of the coating. As the immersion time increases, the coating gradually shows pitting corrosion or even cracks under the influence of high temperature. At this time, the protective effect of the passive film on the coating begins to weaken until disappearing, resulting in a decrease in the self-corrosion potential and an increase in the self-corrosion current density.

3.3. Electrochemical Impedance Spectroscopy Analysis

In order to further investigate the corrosion behavior of the composite coatings in high-temperature seawater, electrochemical impedance spectroscopy was performed on RGO/Cu/Fe-based composite coatings immersed in simulated seawater at 90 °C for different days. Figure 8 shows the Nyquist and Bode plots of the composite coating. From the Nyquist plot, it can be seen that the size of the capacitive arc of the coating after immersion for different days is ranked as follows: 1 D > 0 D > 4 D > 7 D > 18 D. The radius of the capacitive arc increases on day 1, and then the radius of the capacitive arc shows a continuous shrinking trend, indicating that the coating has a passivation phenomenon on the first day, and its corrosion resistance increases in the initial stage of corrosion, but then rapidly decreases, which is also consistent with the trend of changes in the self-corrosion potential and self-corrosion current in the polarization curve.

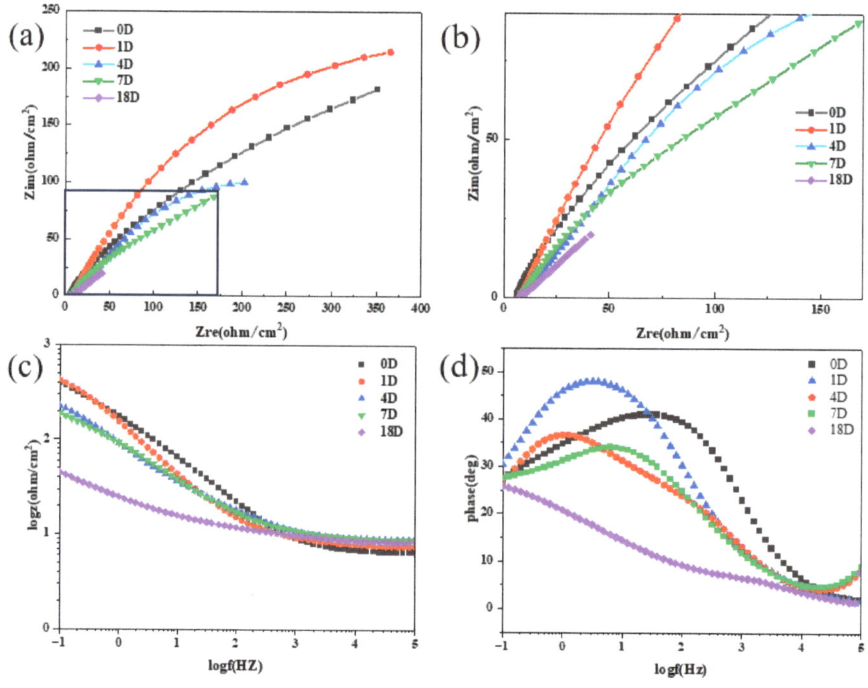

Figure 8. 15% RGO/Cu/Fe-based amorphous composite coating soaked in 90 °C simulated seawater for different days: Nyquist plot (**a**,**b**) and Bode diagram (**c**,**d**).

Figure 8c,d show the relationship between the scanning frequency and the impedance mode value and phase angle. As shown in Figure 8d, within the scanning frequency range, there is a peak on the frequency–phase angle diagram for 0 D and 1 D, which represents a time constant. However, for 4 D and 7 D, there is a trend of a second time constant appearing in the high-frequency range. The number of time constants is closely related to the corrosion behavior. The first time constant represents the nature of the coating, such as passivation, while the emergence of the second time constant often indicates the failure of the coating. Combined with the Nyquist diagram, it can be seen that the coating maintains a passive state during the initial soaking period. After the initial period, the coating begins to exhibit surface defects, which continuously reduce its protective ability. In the frequency–phase angle diagram, it appears that a low-frequency time constant appears on the 18th day for the coating in the mid-frequency range, which may indicate that some significant defects appear on the coating surface at later stages of high-temperature erosion.

The fitting results of the composite coating circuit after soaking in high-temperature seawater for different days are shown in Figure 9, and the fitting data of each component are shown in Table 6. The circuit codes of the three circuits are Rs(CPEf(Rf(CPEdlRct))), Rs(CPEfRf)(CPEdlRct), and Rs(CPEf(Rf(CPEdlRct)))W. It can be seen that Rct shows a trend of first rising and then falling, with its value reaching the highest at 1 D. This may be due to the intense corrosion process, where the corrosion products continue to precipitate, providing a certain degree of protection to the coating, and then start to drop sharply, which may be due to the stress cracking of the coating due to high-temperature heating at this time. The Warburg impedance appearing in the later fitting circuit may also be due to this reason. Until day 7, the Rct value on day 18 is basically the same as that on day 7 due to a balance between corrosion cracking and corrosion products clogging the cracks, at which point the coating still has a certain degree of protection. Rf shows an unstable fluctuation trend, which should be related to the formation and dissolution of surface corrosion products. In this fitting circuit, CPEf represents the coating capacitance, which is related to the coating thickness, surface roughness, and degree of corrosion. Generally speaking, the thicker the coating, the lighter the degree of corrosion and the higher the value. In this fitting result, the value shows a trend of first increasing, then decreasing, and then increasing again. Combined with the fitting results of the potentiodynamic polarization curve and SEM results, this may also reflect the changing trend of the degree of corrosion on the coating surface. CPEdl represents the double-layer capacitance, which is related to the charging and discharging ability of the matrix-coating electric double-layer capacitance and the electrolyte ion concentration. In this fitting result, the value changes in a trend of first increasing and then decreasing, which reflects the process of accumulation–dissolution of corrosion products on the coating surface.

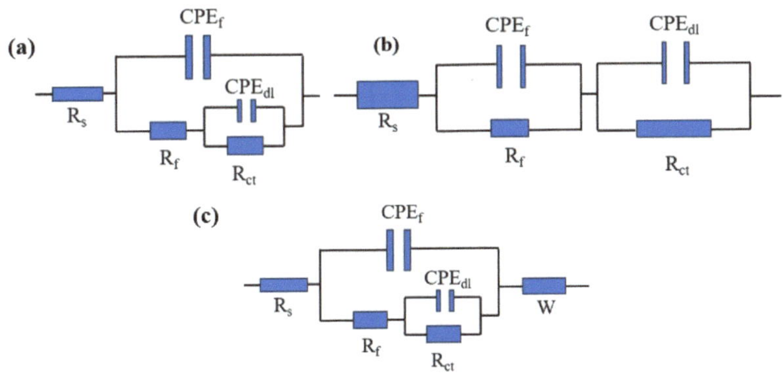

Figure 9. RGO/Cu/Fe-based composite coating circuit fitting results after soaking in simulated seawater at 90 °C for different days: 1 D (**a**); 0 D and 4 D (**b**); 7 D and 18 D (**c**).

Table 6. Fitting results of various components.

Immersion Time	R_s ($\Omega\cdot cm^2$)	CPE_f (s-sec^n)	R_f ($\Omega\cdot cm^2$)	CPE_{dl} (s-sec^n)	R_{ct}($\Omega\cdot cm^2$)	Chi-Squared
0D	6.18×10^0	1.46×10^{-3}	6.41×10^1	2.36×10^{-3}	6.36×10^2	3.35×10^{-4}
1D	7.25×10^0	1.48×10^{-3}	8.48×10^1	3.95×10^{-3}	8.04×10^2	6.10×10^{-5}
4D	8.50×10^0	3.44×10^{-3}	1.81×10^1	3.80×10^{-3}	3.68×10^2	8.18×10^{-4}
7D	1.07×10^0	2.56×10^{-4}	8.53×10^0	4.10×10^{-3}	1.18×10^2	4.19×10^{-4}
18D	7.87×10^0	3.25×10^{-3}	3.50×10^0	2.86×10^{-3}	1.02×10^2	2.87×10^{-4}

3.4. Surface Morphology Analysis

To further investigate the performance of the RGO/Cu/Fe-based amorphous coating in high-temperature corrosive environments, the surface of the coating after soaking for different days was carefully observed. The observation results are shown in Figure 10. At the beginning of the soaking period, on the first day, the surface of the coating is smooth, with only a small amount of flaky corrosion products scattered on it. However, as the soaking time extended, from the fourth day onwards, the number of corrosion products shows a significant increase trend. Under high magnification, it can be clearly observed that small cracks have begun to appear on the surface of the coating, which is caused by the coating cracking phenomenon in high-temperature environments, indicating that the protective ability of the coating has begun to decrease. By the seventh day, the increase in corrosion products and the expansion of cracks are more evident, and the protective performance of the coating is further weakened. When the soaking time reached the 18th day, the corrosion condition on the surface of the coating did not change significantly, except for a slight increase in the number of corrosion products and cracks. This series of observation results indicate that in high-temperature corrosive environments, the protective ability of the RGO/Cu/Fe-based amorphous coating gradually decreases over time, but after prolonged soaking, the corrosion and self-healing abilities of the coating can reach a balance, preventing further corrosion of the coating.

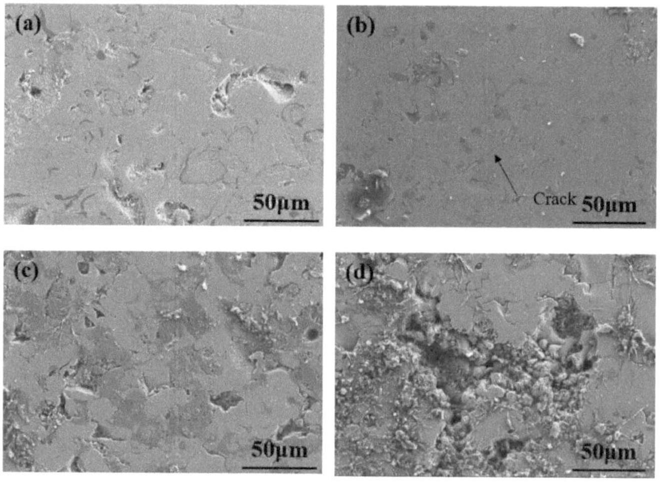

Figure 10. Surface morphology of RGO/Cu/Fe-based amorphous coatings soaked in simulated seawater for different days: 1 D (**a**), 4 D (**b**), 7 D (**c**), and 18 D (**d**).

In order to observe the damage degree of the coating surface more intuitively, the RGO/Cu/Fe-based composite coating immersed in high-temperature simulated seawater for different days was observed, as shown in Figure 11. Compared to the pitting failure mode in normal-temperature seawater, the failure mode of the coating in high-temperature seawater is mainly cracking and surface peeling. Except for the good surface morphology of the coating on the first day, from the fourth day onwards, small cracks and small areas of peeling begin to appear on the coating surface. At this point, the protective effect of the coating begins to decrease gradually. The appearance of cracks creates conditions for corrosion agents to penetrate the interior of the coating, forming micro-battery structures locally and accelerating further corrosion of the coating. As the corrosion process progresses, from the seventh day onwards, the cracks gradually increase and expand, while the degree of peeling increases. At this point, the protective effect of the coating is relatively low, and until the eighteenth day, the degree of crack expansion and surface peeling does not significantly deteriorate, indicating that the failure process of the coating is close to completion. This is consistent with the results of the electrochemical experiments, which show that the coatings will undergo a passive state during the initial stage of high-temperature corrosion, then lose their protective effect and finally tend toward a stable corrosion stage. Meanwhile, Figure 12 shows the surface roughness of the coating before and after high-temperature seawater corrosion. As shown in Figure 12b,d, it can be seen that after 18 days of high-temperature seawater corrosion, the surface roughness of the coating increases, reflecting the damage to the coating surface caused by high-temperature corrosion and the accumulation of corrosion products.

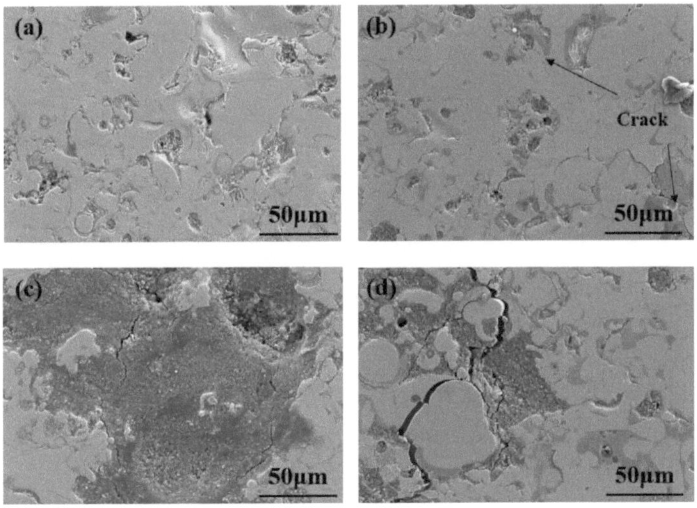

Figure 11. Surface morphology of RGO/Cu/Fe-based amorphous coatings after soaking in simulated seawater for different days and removing corrosion products: 1 D (**a**), 4 D (**b**), 7 D (**c**), and 18 D (**d**).

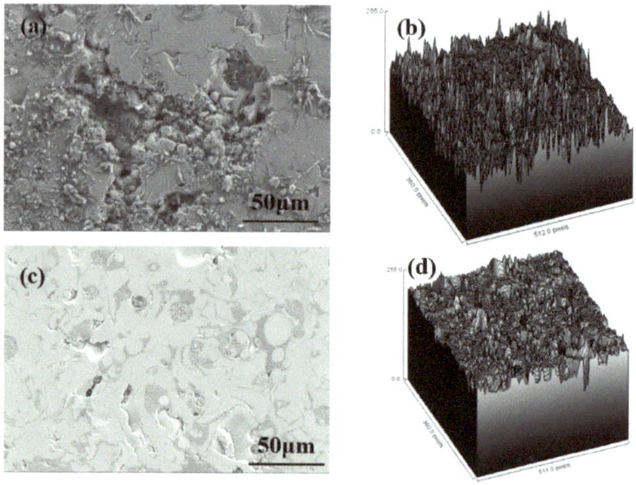

Figure 12. Schematic diagram of the surface roughness of the coating after different days of corrosion: surface morphology (**a**) and roughness (**b**) after 18 days of corrosion; and surface morphology (**c**) and roughness (**d**) after 0 days of corrosion.

To determine the distribution of elements on the surface of the coating after high-temperature corrosion, an EDS scan was performed on its surface, as shown in Figure 13. Through observation, it is seen that the coating surface has peeled off after 18 days, and as the surface coating peels off, the metal elements also escape, which causes microscopic galvanic corrosion. The coating defects also occur in the initial pitting or cracks and gradually expand, slowly reducing the protective ability of the coating.

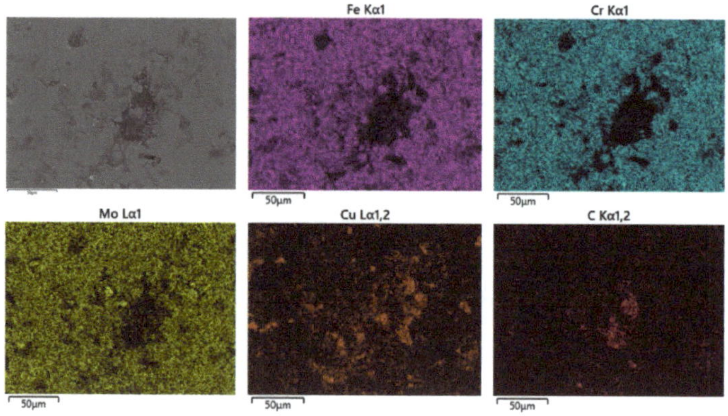

Figure 13. EDS image of RGO/Cu/Fe-based amorphous composite coating after immersion in high-temperature seawater for 18 days.

3.5. *Phase Analysis of Corrosion Products*

To further investigate the high-temperature seawater corrosion behavior of the composite coatings, XRD scanning was performed on the surface of the coating immersed in high-temperature simulated seawater for 18 days to determine the corrosion products generated on its surface. The XRD scanning results are shown in Figure 14. It can be seen that after prolonged immersion, the main corrosion product is Fe_3O_4, accompanied by a small amount of CuO and FeO(OH). Similar to the corrosion products in normal-temperature seawater, some Fe_3O_4 is generated. It is worth noting that the bimetallic interface exhibits special properties in catalytic reactions. Some studies have shown that reactive metal oxides are formed through surface recombination in bimetallic nanoparticles [30–32], demonstrating the interaction between metals, which may include electron transfer, synergistic effects, etc., which may affect the oxidation tendency of the interface, making one of the metals in the bimetallic interface more susceptible to oxidation. Some scholars have studied the role of SiO_2-supported AuCu nanocatalysts in CO oxidation reactions [33]. The study showed that under CO and O_2 oxidation conditions, the surface of oxygen-pretreated AuCu nanoparticles undergoes atomic recombination, in which Cu_2O forms clusters of CuO. The XRD scanning results for the coating also show the formation of metal oxides such as CuO and Fe_3O_4, which may indicate that the formation of corrosion products in the composite coating is related to this bimetallic interface.

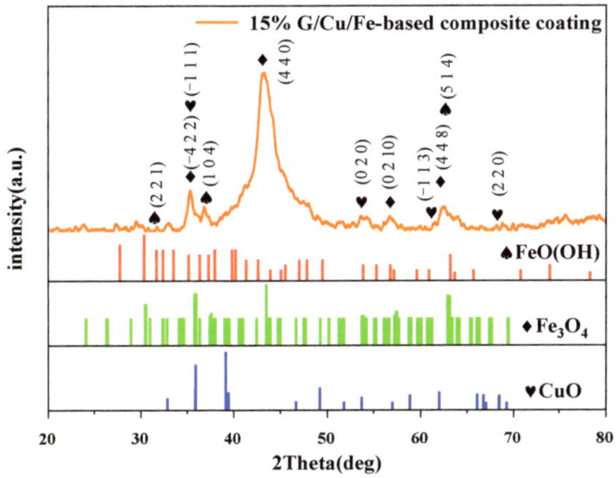

Figure 14. XRD scan results of 15% RGO/Cu/Fe-based amorphous coating soaked in high-temperature simulated seawater for 18 days.

To further confirm the formation of corrosion products on the surface of the coating, XPS scanning was performed on the surface of the composite coating after soaking in high-temperature simulated seawater for 18 days. The scanning results concerning the Cu and Fe elements are shown in Figure 15a,b, respectively. In the scanning result for the Fe-2p orbital, there are three valencies of the Fe element: 0 valency, 2 valency, and 3 valency. It can clearly be seen that the peak area of the 2-valent Fe is the largest, followed by the 3-valent Fe, and the peak area of the 0-valent Fe is very small. The curve only shows a small fluctuation near 714 eV, and satellite peaks of 2-valent Fe and 3-valent Fe appear in the curve, confirming the reliability of the scanning results. Combined with the XRD phase results, there are phases of Fe_3O_4 and Fe on the surface of the coating, and the valencies in the compound can correspond to this scanning result.

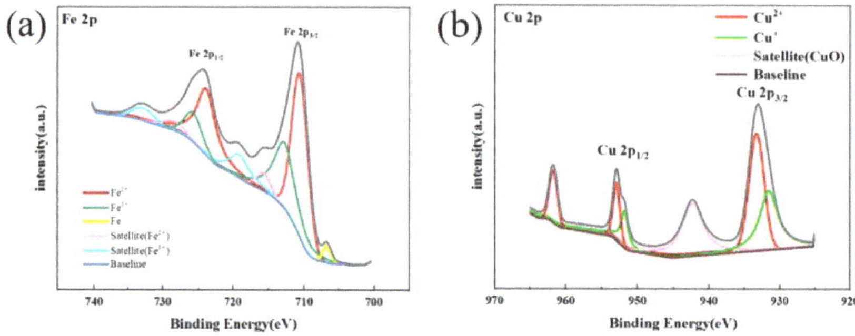

Figure 15. XPS scan results of composite coatings soaked in high-temperature simulated seawater for 18 days. Fe-2p (**a**); Cu-2p (**b**).

A large area of Cu^{2+} peaks and a small area of Cu^{+} peaks appears on the Cu-2p orbital. After a period of high-temperature corrosion, the Cu element is oxidized. Combined with the XRD scanning results, there is a certain amount of CuO in the phase, and the XRD peak intensity also corresponds to the peak area in the XPS results. The satellite peaks of CuO are also shown in the XPS diagram, which confirms the reliability of the test results.

4. Discussion

In order to further reveal the corrosion mechanism of RGO/Cu/Fe-based amorphous composite coatings, a schematic diagram is shown in Figure 16. During the spraying process, it is inevitable that some pores will appear in the single Fe-based amorphous coating, even if the spraying quality of the coating is excellent. Pitting defects are prone to occur and expand at the pores. In high-temperature seawater, the difference in thermal expansion coefficients between different phases can easily lead to the expansion and cracking of the coating, resulting in the corrosion medium reaching the coating–substrate interface through defects [34]. Copper in the composite coating is beneficial for sealing the pores generated by thermal spraying, reducing the stress concentration at the pores, and reducing the occurrence of pitting corrosion. This may be why there are few pitting corrosion defects on the surface of the coating after removing the corrosion products in the SEM images. Graphene can increase the toughness of the coating and hinder the generation of cracks during exposure to high-temperature environments [35]. At the same time, due to its "maze effect", the corrosion medium is difficult to penetrate through graphene to reach the substrate, which is beneficial for forming a physical barrier, extending the corrosion channel, and prolonging the service life of the coating.

In high-temperature seawater, the following chemical reactions mainly occur on the surface of the coating:

$$4Fe + 3O_2 + 2H_2O = 4FeO(OH) \tag{1}$$

$$2Fe + O_2 + 2H_2O = 2Fe(OH)_2 \tag{2}$$

$$4Fe(OH)_2 + O_2 + 2H_2O = 4Fe(OH)_3 \tag{3}$$

$$2Fe(OH)_3 \rightarrow Fe_2O_3 + 3H_2O \tag{4}$$

$$2Fe_2O_3 + 2Fe + O_2 = 2Fe_3O_4 \tag{5}$$

$$2Cu + O_2 + 2H_2O \rightarrow 2Cu(OH)_2 \quad (6)$$

$$Cu(OH)_2 \rightarrow CuO + H_2O \quad (7)$$

In high-temperature seawater, the high temperature accelerates the dissolution of the corrosion product films $Fe(OH)_2$ and $Cu(OH)_2$, resulting in the formation of more Fe_3O_4 and CuO on the coating surface, which is also consistent with the phase analysis results.

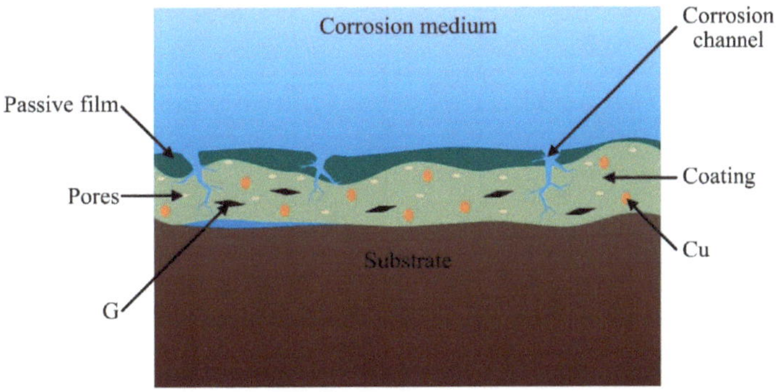

Figure 16. Schematic diagram of the corrosion behavior of RGO/Cu/Fe-based amorphous composite coating in high-temperature seawater.

5. Conclusions

This paper studies the corrosion behavior of RGO/Cu/Fe-based amorphous composite coatings under simulated seawater at 90 °C, and reaches the following conclusions:

(1) With the increase in the proportion of RGO/Cu introduced, the corrosion resistance of the coating increases firstly and then decreases, with the best result occurring at a doping ratio of 15 wt.%.
(2) The main failure modes of Fe-based amorphous composite coatings under high-temperature seawater are coating cracking and peeling, with limited pitting corrosion.
(3) The introduction of RGO can effectively increase the toughness of the coating, improve its mechanical properties, and help to suppress the propagation of cracks at high temperatures.
(4) In high-temperature seawater, the corrosion products on the coating will block cracks and pores, and the corrosion rate will decrease after reaching a maximum value, as a result of the balance between the corrosion and the blocking effect of the corrosion products.

Author Contributions: Formal analysis, Z.C., W.T. and J.X.; Investigation, Z.C.; Writing—original draft, Z.C. and Y.Z.; Writing—review & editing, Z.C. and Y.X. All authors have read and agreed to the published version of the manuscript.

Funding: This research was funded by the National Nature Science Foundation of China (No. 51872072) and by special funding for the development of science and technology of Shanghai Ocean University (No. A2-0203-00-100231 and A2-2006-00-200371).

Institutional Review Board Statement: Not applicable.

Informed Consent Statement: Not applicable.

Data Availability Statement: Data are contained within the article.

Conflicts of Interest: The authors declare no conflict of interest.

References

1. An, Y.; Hou, G.; Chen, J.; Zhao, X.; Liu, G.; Zhou, H.; Chen, J. Microstructure and tribological properties of iron-based metallic glass coatings prepared by atmospheric plasma spraying. *Vacuum* **2014**, *107*, 132–140. [CrossRef]
2. Cui, S.; Zhai, H.; Tong, W.; Li, W.; Li, X.; Fan, X.; Xiong, D. Effect of hydration reaction on the wear behavior of Fe-based amorphous coating in different wear conditions. *J. Non-Cryst. Solids* **2023**, *602*, 122093. [CrossRef]
3. Li, X.; Zhai, H.; Li, W.; Cui, S.; Ning, W.; Qiu, X. Dry sliding wear behaviors of Fe-based amorphous metallic coating synthesized by d-gun spray. *J. Non-Cryst. Solids* **2020**, *537*, 120018. [CrossRef]
4. Al-Abboodi, H.; Fan, H.; Mhmood, I.A.; Al-Bahrani, M. The dry sliding wear rate of a Fe-based amorphous coating prepared on mild steel by HVOF thermal spraying. *J. Mater. Res. Technol.* **2022**, *18*, 1682–1691. [CrossRef]
5. Mahade, S.; Awe, S.A.; Björklund, S.; Lukáč, F.; Mušálek, R.; Joshi, S. Sliding wear behavior of a sustainable Fe-based coating and its damage mechanisms. *Wear* **2022**, *500–501*, 204375. [CrossRef]
6. Wang, Y.; Li, K.; Scenini, F.; Jiao, J.; Qu, S.; Luo, Q.; Shen, J. The effect of residual stress on the electrochemical corrosion behavior of Fe-based amorphous coatings in chloride-containing solutions. *Surf. Coat. Technol.* **2016**, *302*, 27–38. [CrossRef]
7. Wang, Y.; Li, M.; Sun, L.; Zhang, X.; Shen, J. Environmentally assisted fracture behavior of Fe-based amorphous coatings in chloride-containing solutions. *J. Alloys Compd.* **2018**, *738*, 37–48. [CrossRef]
8. Widjajanto, T.; Darmadi, D.B.; Irawan, Y.S.; Gapsari, F. Comparative microstructure characteristics and properties of arc-sprayed Fe-based and HVOF-sprayed Ni-based coatings on ASME SA 210 C steel tube. *Results Eng.* **2023**, *17*, 100985. [CrossRef]
9. Nayak, S.K.; Kumar, A.; Sarkar, K.; Banerjee, A.; Laha, T. Mechanistic insight into the role of amorphicity and porosity on determining the corrosion mitigation behavior of Fe-based amorphous/nanocrystalline coating. *J. Alloys Compd.* **2020**, *849*, 156624. [CrossRef]
10. Zhang, S.; Zhang, W.; Wang, S.; Gu, X.; Wang, J. Characterisation of three-dimensional porosity in an Fe-based amorphous coating and its correlation with corrosion behaviour. *Corros. Sci.* **2015**, *93*, 211–221. [CrossRef]
11. Ye, X.; Shin, Y.C. Synthesis and characterization of Fe-based amorphous composite by laser direct deposition. *Surf. Coat. Technol.* **2014**, *239*, 34–40. [CrossRef]
12. Huang, C.; Wang, J.; Yan, J.; Liao, L. Evaluation of corrosion and wear performance of Fe-based coating and Hastelloy C276 in H2S environment on 304 stainless steel substrate. *Colloids Surf. A Physicochem. Eng. Asp.* **2023**, *674*, 131871. [CrossRef]
13. Lin, T.-J.; Sheu, H.-H.; Lee, C.-Y.; Lee, H.-B. The study of mechanical properties and corrosion behavior of the Fe-based amorphous alloy coatings using high velocity oxygen fuel spraying. *J. Alloys Compd.* **2021**, *867*, 159132. [CrossRef]
14. Wang, H.; Cheng, Y.; Wan, Y.; Jeyaprakash, N.; Wang, Y.; Ma, K.; Yang, J. Influence of scanning speed on microstructure and corrosion resistance of Fe-based amorphous coatings by high-speed laser cladding. *Surf. Coat. Technol.* **2024**, *479*, 13044. [CrossRef]
15. Chu, Z.; Wei, F.; Zheng, X.; Zhang, C.; Yang, Y. Microstructure and properties of TiN/Fe-based amorphous composite coatings fabricated by reactive plasma spraying. *J. Alloys Compd.* **2019**, *785*, 206–213. [CrossRef]
16. Chu, Z.; Zheng, X.; Zhang, C.; Xu, J.; Gao, L. Study the effect of AT13 addition on the properties of AT13/Fe-based amorphous composite coatings. *Surf. Coat. Technol.* **2019**, *379*, 125053. [CrossRef]
17. Yang, Z.; Che, J.; Zhang, Z.; Yu, L.; Hu, M.; Sun, W.; Gao, W.; Fan, J.; Wang, L.; Liu, G. High-efficiency graphene/epoxy composite coatings with outstanding thermal conductive and anti-corrosion performance. *Compos. Part A Appl. Sci. Manuf.* **2024**, *181*, 108152. [CrossRef]
18. Li, L.; Nian, F.; Zhang, S.; Xu, Y.; Ma, S.; Li, Y. Hydrophobic and anti-fouling novel anti-corrosion coatings of graphene quantum dots in situ doped with polyphenylene sulfide. *Surf. Coat. Technol.* **2024**, *479*, 130527. [CrossRef]
19. Liu, C.; Qiu, S.; Du, P.; Zhao, H.; Wang, L. An ionic liquid–graphene oxide hybrid nanomaterial: Synthesis and anticorrosive applications. *Nanoscale* **2018**, *10*, 8115–8124. [CrossRef]
20. Wen, J.-G.; Geng, W.; Geng, H.-Z.; Zhao, H.; Jing, L.-C.; Yuan, X.-T.; Tian, Y.; Wang, T.; Ning, Y.-J.; Wu, L. Improvement of corrosion resistance of waterborne polyurethane coatings by covalent and noncovalent grafted graphene oxide nanosheets. *ACS Omega* **2019**, *4*, 20265–20274. [CrossRef]
21. Bai, T.; Lv, L.; Du, W.; Fang, W.; Wang, Y. Improving the tribological and anticorrosion performance of waterborne polyurethane coating by the synergistic effect between modified graphene oxide and polytetrafluoroethylene. *Nanomaterials* **2020**, *10*, 137. [CrossRef]
22. Mourya, P.; Goswami, R.N.; Saini, R.; Ray, A.; Khatri, O.P. Epoxy coating reinforced with graphene-PANI nanocomposites for enhancement of corrosion-resistance performance of mild steel in saline water. *Colloids Surf. A Physicochem. Eng. Asp.* **2024**, *687*, 133500. [CrossRef]
23. Pandey, U.; Sharma, C. Comparative study of graphene oxide-multifunctional oxide doping on corrosion resistance of electrodeposited nickel coatings in saline environments. *Int. J. Hydrog. Energy* **2024**, *60*, 165–179. [CrossRef]
24. Pang, W.; Jiang, H.; Wang, S.; He, T.; Chen, H.; Yan, T.; Cheng, M.; Sun, S.; Li, C. Graphene oxides enhanced polyurethane based composite coating with long term corrosion resistance and self-healing property. *Eur. Polym. J.* **2024**, *207*, 112825. [CrossRef]
25. Taravel-Condat, C.; Desamais, N. Qualification of High Strength Carbon Steel Wires for Use in Specific Annulus Environment of Flexible Pipes Containing CO_2 and H_2S. In Proceedings of the 25th International Conference on Offshore Mechanics and Arctic Engineering, Hamburg, Germany, 4–9 June 2006; pp. 585–591.
26. Wu, L.; Zhou, Z.; Zhang, K.; Zhang, X.; Wang, G. Electrochemical and passive film evaluation on the corrosion resistance variation of Fe-based amorphous coating affected by high temperature. *J. Non-Cryst. Solids* **2022**, *597*, 121892. [CrossRef]

27. Escrivà-Cerdán, C.; Blasco-Tamarit, E.; García-García, D.; García-Antón, J.; Akid, R.; Walton, J. Effect of temperature on passive film formation of UNS N08031 Cr–Ni alloy in phosphoric acid contaminated with different aggressive anions. *Electrochim. Acta* **2013**, *111*, 552–561. [CrossRef]
28. Carranza, R.; Alvarez, M. The effect of temperature on the passive film properties and pitting behaviour of a FeCrNi alloy. *Corros. Sci.* **1996**, *38*, 909–925. [CrossRef]
29. Zhang, C.; Chu, Z.; Wei, F.; Qin, W.; Yang, Y.; Dong, Y.; Huang, D.; Wang, L. Optimizing process and the properties of the sprayed Fe-based metallic glassy coating by plasma spraying. *Surf. Coat. Technol.* **2017**, *319*, 1–5. [CrossRef]
30. Wu, C.H.; Liu, C.; Su, D.; Xin, H.L.; Fang, H.-T.; Eren, B.; Zhang, S.; Murray, C.B.; Salmeron, M.B. Bimetallic synergy in cobalt–palladium nanocatalysts for CO oxidation. *Nat. Catal.* **2019**, *2*, 78–85. [CrossRef]
31. Song, Y.; Kim, D.; Hong, S.; Kim, T.-S.; Kim, K.-J.; Park, J.Y. Bimetallic synergy from a reaction-driven metal oxide–metal interface of pt–co bimetallic nanoparticles. *ACS Catal.* **2023**, *13*, 13777–13785. [CrossRef]
32. Pan, Y.; Xu, L.; Huang, L.; He, W.; Li, H.; Wang, S.; Long, Z.; Sun, Z. Identification of active sites in pt–co bimetallic catalysts for co oxidation. *ACS Appl. Energy Mater.* **2021**, *4*, 11151–11161. [CrossRef]
33. Kim, T.-S.; Choi, H.; Kim, D.; Song, H.C.; Oh, Y.; Jeong, B.; Lee, J.; Kim, K.-J.; Shin, J.W.; Byon, H.R.; et al. Catalytic boosting on AuCu bimetallic nanoparticles by oxygen-induced atomic restructuring. *Appl. Catal. B Environ.* **2023**, *331*, 122704. [CrossRef]
34. Qian, H.; Xu, Z.; Chen, S.; Liu, Y.; Yan, D. Silicon carbide/enamel composite coatings for steel corrosion protection: Microstructure, thermal expansion behavior, and anti-corrosion performance. *Surf. Coat. Technol.* **2022**, *434*, 128172. [CrossRef]
35. Singh, P.; Satish, C.; Islam, A.; Prasad, S.; Pandit, N.; Keshri, S.; Keshri, A.K. Multifunctional hybrid composite: Plasma sprayed Al and graphene reinforced alumina coating. *J. Alloys Compd.* **2024**, *984*, 173984. [CrossRef]

metals

MDPI

Article

Experimental Study of Performance of Ti-6Al-4V Femoral Implants Using Selective Laser Melting (SLM) Methodology

Wenjie Zhang, Hongxi Liu, Zhiqiang Liu *, Yuyao Liang and Yi Hao

School of Mechanical Engineering, Jiangsu University of Science and Technology, Zhenjiang 212000, China; hongxi_liu@just.edu.cn (H.L.)
* Correspondence: zhiqiangliu@just.edu.cn

Abstract: Selective laser melting (SLM) technology used for the design and production of porous implants can successfully address the issues of stress shielding and aseptic loosening associated with the use of solid implants in the human body. In this paper, orthogonal experiments were used to optimize the process parameters for SLM molding of Ti-6Al-4V (TC4) material to investigate the effects of the process parameters on the densities, microscopic morphology, and roughness, and to determine the optimal process parameters using the roughness as a judging criterion. Based on the optimized process parameters, the mechanical properties of SLM-formed TC4 alloy specimens are investigated experimentally in this paper. The main conclusions are as follows: the optimal combination of roughness is obtained by polar analysis, the microhardness of SLM-molded TC4 alloy molded specimens is more uniform, the microhardness of specimens on the side and the front as well as the abrasion resistance is higher than that of casting specimens, the yield strength and tensile strength of specimens is higher than that of ASTM F136 standard and casting standard but the elongation is not as good as that of the standard, and the elasticity and compressive strength of porous specimens are higher than that of casting specimens at different volume fractions. The modulus of elasticity and compressive strength are within the range of human skeletal requirements. This work makes it possible to fabricate high-performance porous femoral joint implants from TC4 alloy SLM-molded materials.

Keywords: process parameters; SLM; TC4; femoral implant; molding performance; mechanical performance

Citation: Zhang, W.; Liu, H.; Liu, Z.; Liang, Y.; Hao, Y. Experimental Study of Performance of Ti-6Al-4V Femoral Implants Using Selective Laser Melting (SLM) Methodology. *Metals* **2024**, *14*, 492. https://doi.org/10.3390/met14050492

Academic Editor: Antonio Mateo

Received: 19 March 2024
Revised: 17 April 2024
Accepted: 19 April 2024
Published: 23 April 2024

1. Introduction

In today's medical field, most of the solid metal implants used are prone to aseptic loosening, leading to a decrease in the service life of the implant [1–4]. It is possible to design the implant surface or all of it as a porous structure, but conventional manufacturing techniques are not able to achieve complex porous structures [5,6]. Selective laser melting (SLM) technology can achieve the molding of almost any complex structural parts with good mechanical properties and high molding accuracy. In terms of implant molding, the molding performance of SLM technology has met the requirements, but the performance of SLM-molded parts needs further research [7]. Titanium alloy material has been widely used in the field of bone implants such as in artificial scaffolds and artificial joints and dentures because its elastic modulus is close to that of autologous bone [8,9]. Powder material properties play an important role in the SLM forming process and the density of the formed parts and good powder properties are the basic guarantee of SLM forming technology.

Elshaer et al. [10] and Sangali et al. [11] investigated the corrosion properties of Ti-6Al-4V alloys fabricated by casting and SLM methods in 0.9% NaCl, respectively, in which the corrosion rate was 0.000171 mm/yr in case of casting, and corrosion products appeared on the surface of the samples in the case of SLM, and corrosion became more pronounced the faster the scanning speed. A lot of research has been conducted on cast Ti-6Al-4V. Feng et al. [12], Zhu et al. [13], Lee et al. [14], and Wen et al. [15] have determined the mechanical properties of Ti-6Al-4V cast alloys and the mechanical properties of Ti-6Al-4V casting alloys in terms of the machining conditions, the addition of Ti-TiB2 nanoparticles, vacuum centrifugation, Ni sandwich method from a single point of view, and the mechanical properties of Ti-6Al-4V casting alloys and the mechanical properties of Ti-6Al-4V casting alloys from a single point of view, in addition to 4V casting alloy's mechanical properties as well as strength. In addition, some scholars have also investigated the properties of Ti-6Al-4V alloys made by the SLM method: Banu et al. [16] investigated the chemical properties of SLM Ti alloys in a lactic acid environment and observed that the corrosion resistance rate of Ti alloys by the SLM method was lower than that of wrought alloys; Cheng et al. [17] investigated Ti alloys with 0.3 wt.% Zn additions, and concluded they had better impact toughness and stable high-temperature tensile mechanical properties; Ben-Hamu et al. [18] investigated the corrosion resistance of Ti-6Al-4V in 0.9 M NaCl solution under EBM and SLM fabrication methods, and the results show that the corrosion resistance of SLM method Ti-6Al-4V alloys is slightly superior to that of the EBM method Ti-6Al-4V in both the XY and XZ planes. In recent years, the corrosion resistance of the SLM method Ti-6Al-4V alloys for medical articles has also been studied: Tamaddon et al. [19], Mondal et al. [20], and Kutsukake et al. [21] investigated the performance of SLM method Ti alloys in the field of implants, and initially confirmed that SLM method Ti-6Al-4V alloys have a certain development potential in the field of medical implants.

To summarize, the difficulty is in utilizing the important properties of the good biocompatibility and structural or mechanical compatibility of titanium, and the excessive densification of the SLM method may cause problems such as stress shielding or aseptic loosening of the implant, and the current research lacks an analysis of the properties of the SLM method in making multi-hollow femoral implants out of Ti-6Al-4V alloys; in particular, the topography of the material properties of the SLM method Ti-6Al-4V alloys.

The main objective of this study was to investigate the obtaining of high-performance porous femoral implants from TC4 alloy SLM-molded materials. Firstly, the best process parameters for SLM forming of Ti-6Al-4V materials were optimized by orthogonal test, and the best process parameters based on roughness evaluation criteria were determined. Then, based on the optimized process parameters, the microstructure, microhardness, tensile properties, compressive properties, and friction and wear properties of the SLM-molded specimens were systematically analyzed, and it was discovered that the microhardness, yield strength, tensile strength, and wear resistance of the TC4 alloy SLM-molded specimens were better than those of the casting specimens, but the elongation at break was slightly lower than that of the ASTM F136 standard [22]. The feasibility of molding porous femoral implants with SLM TC4 alloys was verified. This study lays the foundation for the application of high-performance porous femoral implants.

2. Materials and Methods

2.1. Materials

The experiments use TC4 powder from EOS company in Krailling, Germany, and its main chemical composition is shown in Table 1, which meets the requirements of China GB/T 13810-2007 [23] and the American ASTM F136 standard for the use of surgical implantation of titanium and titanium alloy materials. Table 2 shows the basic characteristics of TC4 powder. The bulk density of the powder is 2.45 g/cm^3.

Table 1. Comparison of Ti-6Al-4V powder with GB/T 13810-2007 and ASTM F136 standard. Reprinted from Ref. [24].

Ingredients	Ti	Al	V	Fe	C	O	N	H
ASTM F136	Residual	5.5–6.5	3.5–4.5	≤0.25	≤0.08	≤0.13	≤0.05	≤0.012
GB/T 13810	Residual	5.5–6.75	3.5–4.5	0.3	0.08	0.02	0.05	0.015
TC4	Residual	6.1	3.95	0.1	0.004	0.01	0.01	0.001

Table 2. Basic characteristics of TC4 powder.

D3	D10	D50	D70	D90	D95	D97
8.942 μm	18.214 μm	35.106 μm	43.330 μm	58.394 μm	65.242 μm	73.408 μm

2.2. Experimental Apparatus and Experimental Methods

2.2.1. Orthogonal Test of Process Parameters

The essence of SLM technology molding is to use the laser to melt the metal powder and then use a molding process; in this process, the laser power, powder properties, scanning parameters, and other parameters will have different degrees of influence on the molding quality of the specimen. Thus, the introduction of the concept of laser energy density per unit volume and the size of the laser energy density per unit volume directly affects the molding quality. Laser unit volume energy density expression is shown in Equation (1) [25].

$$\varphi = \frac{\mathrm{P}}{\mathrm{vdh}} \tag{1}$$

According to Equation (1), it can be seen that laser power (P), scanning speed (v), scanning pitch (d), and powder spread thickness (h) are the direct factors affecting the laser energy density per unit volume. It is also found by referring to the previous experimental study that laser power, scanning speed, scanning spacing, and powder spreading thickness are the most important process parameters affecting the quality of SLM molding [26]. This chapter focuses on optimization experiments for these four process parameters. The design of the orthogonal experimental program with four levels for four factors is shown in Table 3.

Table 3. Factors and horizontal design of orthogonal experiment. Reprinted from Ref. [24].

Number	1	2	3	4
Laser power P/(W)	300	320	340	360
Scanning speed v/(mm/s)	1050	1250	1450	1650
Scan spacing d/(mm)	0.08	0.09	0.10	0.11
Thickness of powder printing layer h/(mm)	0.04	0.05	0.06	0.07

The purpose of the orthogonal test is to analyze the effects of different factors and different levels on the experimental results, and then determine the best level for each factor. Extreme variance analysis, analysis of variance, and signal-to-noise ratio analysis are commonly used to analyze data from orthogonal tests, and in this chapter, the extreme variance method was used to analyze the densification and surface roughness (using the profile arithmetic mean deviation Ra) of the 16 groups of specimens.

According to the process parameters designed in Table 3, 16 groups of specimens were printed using the EOS M290 molding equipment with a specimen size of 10 mm × 10 mm × 10 mm. The display effect of the specimens in Materialise Magics 24.0 software before printing is shown in Figure 1a, and the effect display after the specimen printing is completed is shown in Figure 1b.

2.2.2. Densification Experiments Based on the Drainage Method

Based on the drainage method, this experiment uses a high-precision electronic balance with an accuracy of up to 0.0001 g to measure the quality of SLM-molded specimens. The specific steps are as follows: (1) clean the surface of the specimen by, firstly, placing the specimen in the ultrasonic cleaner with alcohol solution, cleaning the specimen for ten minutes according to the specification of the manual, and then taking the specimen out and drying it for subsequent measurement of its mass; (2) measure its mass in the air; (3) measure its mass after being completely submerged in water; and (4) calculate its densification according to the formula shown below.

$$\rho_r = \frac{\rho_0}{\rho_{th}} \times 100\% \tag{2}$$

where ρ_r is the densification of the specimen, ρ_0 is the actual density of the specimen, and ρ_{th} is the theoretical density of the specimen.

$$\rho_0 = \frac{m_{air} \times \rho_{H_2O}}{m_{air} - m_{H_2O}} \tag{3}$$

where m_{air} is the mass of the specimen in air, m_{H_2O} is the mass of the specimen completely submerged in water, and ρ_{H_2O} is the density of water taken as 1.0 g/cm^3.

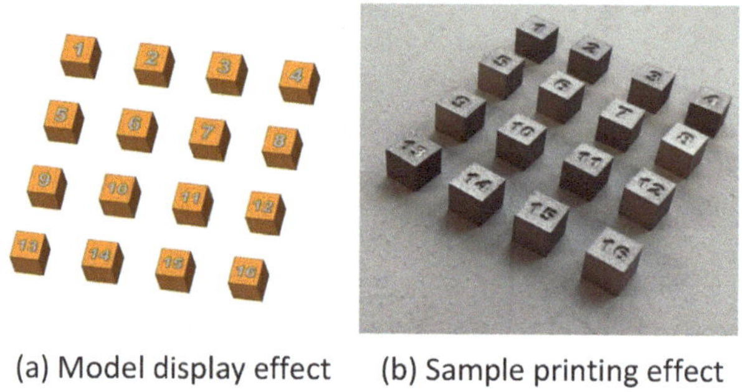

Figure 1. Model display effect. Adapted from Ref. [24].

2.2.3. Micromorphological Analysis of Materials

In this experiment, the shape measurement laser confocal microscope of VK-X1000 which manufactured by Keyence Corporation, Osaka, Japan (the microscope has an integrated magnification of up to 28,800×; a minimum field of view from 11 μm to 7398 μm; laser measurement speeds from 4 to 125 Hz, 7900 Hz; a height measurement resolution of 0.5 nm; a dynamic range of 16 bit; an accuracy of 0.2 + L/100 μm; a width measurement resolution of 1 nm; and the ability to observe images that include ultra-high resolution color CMOS images, 16 bit laser color confocal images, confocal and ND filter optics, and C-laser differential interference images) series was used to scan the morphology of SLM-molded specimens, which mainly included focus variation and laser confocal with a display resolution of 0.001 μm.

A scanning electron microscope (SEM) uses secondary electrons that are very sensitive to the surface state of the material and backscattered electron imaging that can produce specific compositional information. In this experiment, a FEI Quanta FEG 250, Hillsboro, OR, USA (the FEI QUANTA FEG250 field-emission scanning electron microscope has a secondary electron image resolution of 1.04 nm, a magnification of 15 to 300,000 times, an accelerating voltage of 0.2 to 30 kV, and a detection current of 0.3–22 nA) scanning electron microscope was used to observe and analyze the microscopic morphology of SLM-molded TC4 specimens.

The microstructure observation of SLM-molded TC4 specimens was carried out by a German Zeiss Observer D1m, Jena, Germany (micrometer: grid value of 0.001 mm/1 mm; differentiated eyepiece: 10× flat field zoom, grid value of 0.1 mm; and total magnification: 100–1000×; focusing mechanism: coaxial coarse micromanipulation; limit protection, lifting range of 30 mm, fine-tuning 0.002 mm) metallographic microscope. It is worth noting that before the metallographic observation of the TC4 specimen, it is first necessary to prepare a metallographic specimen, which needs to be sanded with different grits (320#, 600#, 1200#, 1500#, 2000#) of sandpaper, so that the surface of the metallographic specimen should be characterized by a basically smooth surface, free of stains and obvious scratches. After the initial formation of metallographic specimens, in order to be able to more clearly observe the microstructure of the specimen, a metallographic specimen corrosion treatment must be used. The corrosion solution chosen is Kroll reagent, the solvent ratio of the reagent is 5 mL HF:10 mL HNO_3:85 mL H_2O, and the corrosion time of the metallographic specimen using Kroll reagent is set to 20 s.

2.2.4. Measurement of Surface Roughness

In this experiment, the surface roughness of the SLM-molded TC4 specimen was measured using the roughness tester Mitutoyo (Kawasaki, Japan). During the measurement, three sampling points were selected on the surface of the TC4 molding specimen for measurement, and the average of the surface roughness of the three points was calculated, and this value was taken as the surface roughness value of the surface of the TC4 molded part.

2.2.5. Micro-Vickers Hardness Determination

This experiment used an HVS-1000AT/EOS100B which manufactured by NANBEI Instruments Ltd., Zhengzhou, China (measuring range of hardness value: 5–5000 HV; optical measuring system objective lens: 10× (observation), 40× (measurement); total magnification of optical measuring system: 100× (observation), 400× (measurement); measuring range: 200 μm; index value: 0.025 μm; light source: 12 V/20 W) automatic microhardness tester to test the surface hardness of SLM-molded specimens. In order to ensure the accuracy of the specimen Vickers hardness test data, before the Vickers hardness test, first of all, sandpaper of the mesh numbers 280 mesh, 400 mesh, 600 mesh, 800 mesh, 1000 mesh, 1500 mesh was used. The specimen was polished, and then the polishing paste was used to polish the specimen on the polishing machine to ensure that the surface of the specimen was clean and flat without large scratches. For the test, three parallel lines were selected in the measurement plane of the specimen, and five consecutive points with the same distance between each point were selected for measurement, with the distance between each point being 1 mm, and the distance between the starting point and the edge of the specimen in the measurement plane being more than 1 mm. The load size selected for the test was 9.8 N, and the loading duration was set to 15 s. After the load was unloaded, the microscopic images were transferred to a data transmission channel Based on the diagonal length of the rhombus measured by the four-line method, the computer automatically displayed the micro-hardness values, and the average of the micro-hardnesses of the five points was finally taken as the micro-hardness of the measurement plane of the specimen.

2.2.6. Tensile and Compression Tests

We used CMT 5105 microcomputer-controlled electronic universal testing equipment, which manufactured by JINSON Optical Instruments Ltd. in Nanjing, China, to measure the tensile and compressive properties of the specimen. The maximum test force of 100 KN was used, and was tested using constant speed loading mode: for tensile test, a loading rate of 0.5 mm/min; for compression test, a loading rate of 1 mm/min. The equipment recorded the entire experimental process of force and displacement ($F - \Delta s$) relationship. According to Equations (2)–(5), the force–displacement ($F - \Delta s$) relationship can be transformed into the actual stress–strain ($\sigma - \varepsilon$) relationship, and the transformation formula is as follows:

$$\sigma_0 = \frac{F}{A} \tag{4}$$

$$\varepsilon_0 = \frac{s}{\Delta s} \tag{5}$$

$$\sigma = \sigma_0(1+\varepsilon_0) \tag{6}$$

$$\varepsilon = \text{In}(1+\varepsilon_0) \tag{7}$$

where σ_0 is the nominal stress, ε_0 is the nominal strain, s is the initial length of the specimen, and A is the initial cross-sectional area of the specimen.

2.2.7. Friction Wear Test

This experiment uses the FTM M30 modular-controlled lubrication multifunctional friction and wear tester developed by Nanjing YuanSheng Shen Intelligent Technology Company (Nanjing, China) to carry out friction and wear tests on SLM-molded titanium alloy specimens. For the friction wear test, the reciprocating motion function module is selected, with a stroke of 10 mm, a frequency of 1 Hz, a speed of 10 mm/s, an experimental time of 60 min, and an experimental loading force of 50 N. The upper fixture model is FUS063, and the specification is a flat column specimen with a diameter of 6 mm and a length of 13 mm. According to the previous research experience, the upper fixture material used in this friction wear test is polyethylene molecule (UHMWPE), and the lubricant is deionized water [27].

3. Results and Discussion

3.1. Analysis of Orthogonal Test Results

According to Table 4, it can be seen that the influence on the side roughness of the specimen is ordered in the following sequence: scanning spacing > scanning speed > laser power > paving powder layer thickness, in which the scanning speed and laser power influence the side roughness in basically the same way, although the former is slightly higher than the latter. The optimal combination of the four process parameters is as follows: laser power is 300 W, scanning speed is 1250 mm/s, scanning distance is 0.09 mm, and the thickness of powder layer is 0.05 mm.

Table 4. Analysis results of the side roughness of the sample.

Factors (Process Parameters)	Laser Power P	Scanning Speed v	Scan Spacing d	Thickness of Powder Layer h
Level index and K1	35.911	38.532	41.6	40.585
Level index and K2	37.331	36.045	36.431	36.142
Level index and K3	40.807	39.328	37.988	37.502
Level index and K4	40.937	41.081	38.967	40.757
Horizontal k1	8.978	9.633	10.400	10.146
Horizontal k2	9.333	9.011	9.108	9.036
Horizontal k3	10.202	9.832	9.497	9.376
Horizontal k4	10.234	10.270	9.742	10.189
Average index R	1.256	1.259	1.292	1.153

According to Table 5, it can be seen that on the front roughness of the specimen, the order of the size of the influence is scanning speed > scanning distance > laser power > paving powder layer thickness; the optimal combination of the four process parameters is laser power of 300 W, scanning speed of 1250 mm/s, scanning distance of 0.09 mm, paving powder layer thickness of 0.05 mm.

Table 5. Analysis results of the front roughness of the sample.

Factors (Process Parameters)	Laser Power P	Scanning Speed v	Scan Spacing d	Thickness of Powder Layer h
Level index and K1	42.422	43.578	44.745	44.677
Level index and K2	44.501	42.426	42.028	43.356
Level index and K3	45.59	44.506	45.433	45.365
Level index and K4	46.867	48.87	47.174	45.982
Horizontal k1	10.606	10.895	11.186	11.169
Horizontal k2	11.125	10.607	10.507	10.839
Horizontal k3	11.398	11.127	11.358	11.341
Horizontal k4	11.717	12.218	11.794	11.496
Average index R	1.111	1.611	1.287	0.657

3.2. Influence of Process Parameters on Molding Quality

3.2.1. Influence of Process Parameters on Densification

Densification is not only an important index for assessing the quality of SLM-molded parts, but also a prerequisite and basis for certain mechanical properties of molded parts [28].

In order to analyze the effects of process parameters (laser power, scanning speed, scanning spacing, and powder layer thickness) on the densities of molded specimens, each level of different influencing factors was set as the independent variable, and the densities of the molded specimens were set as the dependent variables, where the molded specimen densities were the mean values of the densities corresponding to each level of each factor.

The effect of the experimental levels of factors on the densities of the molded specimens is shown in Figure 2, and it can be seen that the densities of the molded specimens are above 95%.

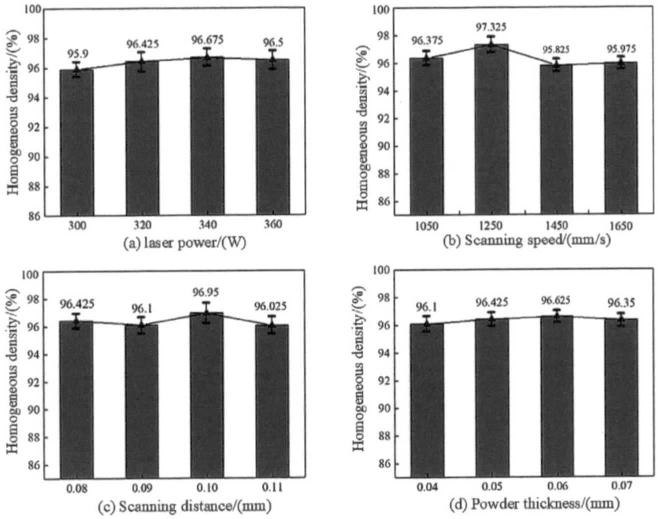

Figure 2. Influence of each factor level value on density in orthogonal experiment.

3.2.2. Influence of Process Parameters on Micromorphology

During the SLM molding process, the process parameters should be set to match the characteristics of the selected powder material, otherwise it will lead to an increase in the internal porosity of the molded specimen, which will reduce its density and affect the molding performance. In order to study the effect of different process parameters on the micromorphology of SLM-molded specimens, this paper adopts the Keens VK-X1000 series shape measurement laser confocal microscope to observe the micromorphology of SLM molding of TC4 alloy specimens; Figure 3 shows the micromorphology of 16 groups of molding specimens on the side and front of the Keens VK-X1000 series shape measurement laser confocal microscope.

According to Figure 3, it can be seen that the overall micromorphology of the 16 groups of samples is relatively good, in which the surface of the No. 5 specimen molding quality is high, the surface is relatively flat, with basically no hole defects. The surface morphology of the remaining 14 groups of specimens mainly shows unfused defects, because in the process of SLM molding, a deep melt pool is produced. The emergence of a deep melt pool can promote the melting of TC4 metal powder and, thus, less molded specimen surface defects, but due to the plateau Raleigh capillary instability, being the reason for the melt pool elongation [29], the melt pool penetration is insufficient, so the melt pool means mobility and wettability is poorer and unable to fill the molded layer overlap of the concave region of the molded layer, thus, leading to the emergence of irregular fusion of the defects. The surface morphology of specimen No. 1 is relatively poor, and more irregular hole defects appear on the surface. The reason for this phenomenon is that the laser and the metal powder in the SLM molding process are unstable, and the deep melting pool generated promotes the volatilization of elements to induce porosity, and the continuous growth of columnar grains is hindered by the plateau Raleigh capillary instability, which ultimately leads to the appearance of irregular holes in the specimen.

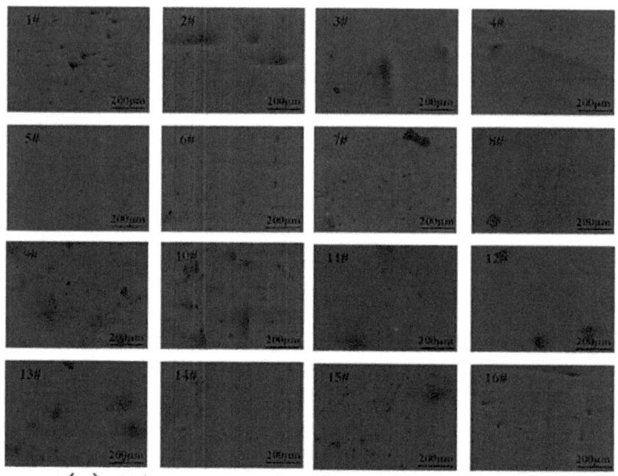

(a) Side micrographs of 16 groups specimens

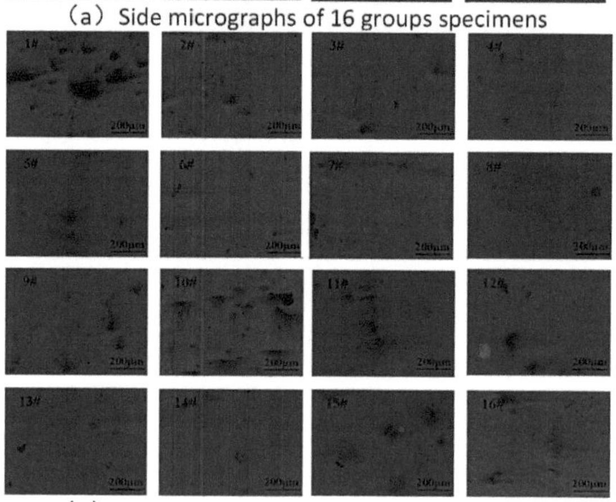

(b) Frontal micrographs of 16 groups specimens

Figure 3. Micromorphology of 16 groups of samples.

3.2.3. Influence of Process Parameters on Roughness

Surface roughness is one of the important parameters for evaluating the performance of implants, and SLM molding technology has a broad development prospect in the biomedical field. For artificial implants, less roughness is conducive to an improvement in the fatigue strength and wear resistance of implants, and greater roughness is a favorable basis for ensuring the attachment and proliferation of osteoblasts of artificial implants. There is currently a great deal of controversy about the optimal roughness of implants, and the differences in roughness due to different additive manufacturing methods is an important parameter to consider; all the values found are biocompatible for osteoblast adhesion and value-addition, and due to the heterogeneity of the studies, it is not possible to determine an optimal roughness [30,31]. Therefore, it is necessary to optimize the process parameters so as to obtain a better surface roughness.

Figure 4 shows the roughness curves of the side (perpendicular stacking direction) and front (parallel stacking direction) of the same specimen molded by SLM, and the roughness of the side of the molded specimen is measured to be 7.796, and the roughness of the front is 12.499. A reason for this result may be that the SLM molding layer-by-layer stacking process means that the melt channel and the real position have a certain deviation, and in the stacking direction of the phenomenon there is sticky powder, making the specimen surface uneven and rougher.

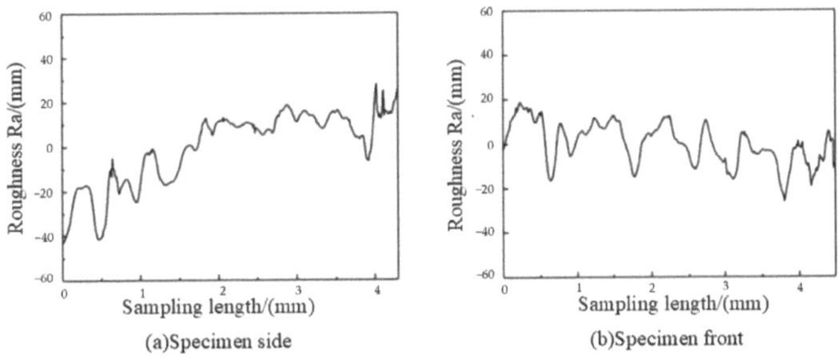

Figure 4. Roughness curve.

3.3. *Analysis of the Results of the Experimental Study on the Evaluation of Properties of SLM-Molded TC4 Alloy*

3.3.1. Microstructure Analysis of SLM-Molded TC4 Alloy

On the basis of optimizing the process parameters, the TC4 specimens were molded by SLM, and microscopic observation of the molded specimens in cross-section and longitudinal section was carried out by metallurgical microscope and scanning electron microscope, respectively. Shown in Figures 5 and 6 are the micrographs of the cross-section and longitudinal section of the molded specimens, respectively, from which it can be seen that the micrographs of the molded specimens on the side and the front side show obvious differences, which is mainly due to the influence of the molding characteristics of the SLM technology (layer-by-layer powder laying, layer-by-layer deposition).

Figure 5a,b show the microstructure morphology of the longitudinal section of the molded specimen under optical microscope, and Figure 5c,d show the microstructure morphology of the longitudinal section of the molded specimen observed by SEM. Figure 5 shows that the cross-section microstructure of the TC4 molded specimen consists of several thick β columnar crystals with epitaxial growth of the fusion cladding layer, the growth direction of the columnar crystals is perpendicular to the scanning direction of the laser beam and along the direction of the stacked layer growth, and the inside of the original thick β columnar crystals consists of $\alpha\prime$ martensite almost completely, and $\alpha\prime$ martensite is widely present inside the grain boundaries. This is due to the fact that in the additive manufacturing process, in order to ensure the continuity between adjacent layers, the laser energy will penetrate the powder layer to melt the top of the last deposited layer [32], resulting in the remelting of the previous layer of powder, the temperature gradient in the direction of powder layer deposition, and the bottom of the melting pool has the largest temperature gradient, and the remelted grains will appear to be epitaxial along the direction of deposition, resulting in the formation of the columnar crystals [33,34].

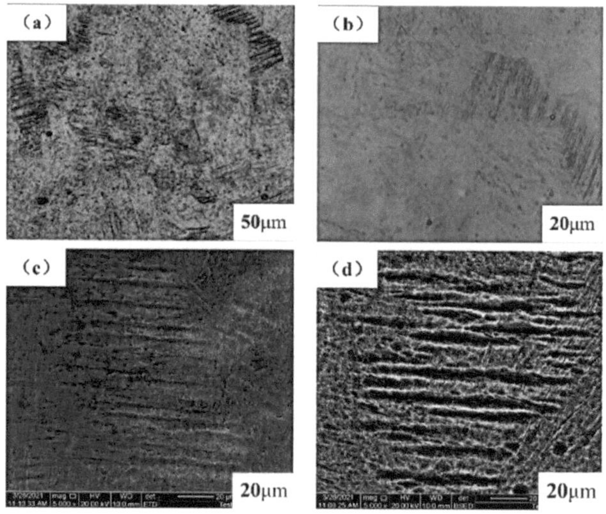

Figure 5. Micromorphology of longitudinal section of forming sample. (**a**,**b**) show the microscopic morphology of the longitudinal section of the specimen under optical microscope, and (**c**,**d**) show the microscopic morphology of the longitudinal section of the specimen observed by SEM.

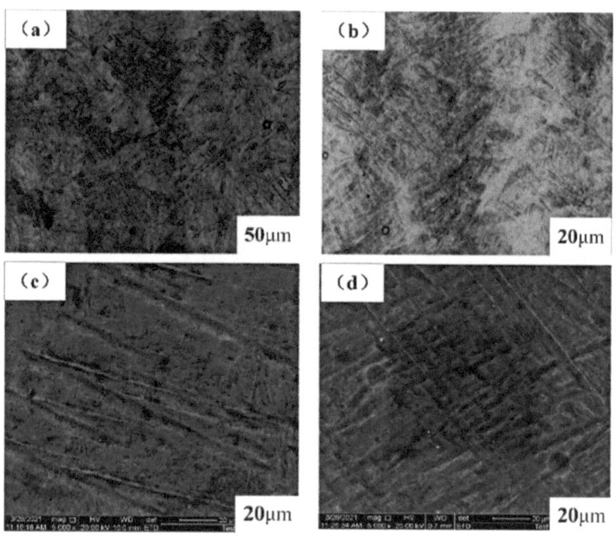

Figure 6. Micromorphology of cross-section of molding sample. (**a**,**b**) are the microscopic morphology of the specimen cross-section under optical microscope, and (**c**,**d**) are the microscopic morphology of the specimen cross-section observed by SEM.

At the same time, it can be clearly seen that along the direction of the stacking layer β columnar crystal grain boundaries, adjacent to the two β columnar crystal grain boundaries within the fine needle-like β martensite, growth direction is different; the growth of β columnar crystal grain boundaries stops growing. This is because the titanium alloy goes through a phase transformation process: the α phase from the parent phase β tethers in line with the Burgers orientation relationship as shown in Equations (1) and (2); in the cooling process, the β phase into the α phase has 12 different orientation relationships, and the martensitic lath bundle in the same β crystal has a specific orientation difference to localize the shape of the strain, which is reduced to a minimum [35].

From Figure 6, it can be seen that the microstructure is not completely uniform, which is because the melting and solidification of the powder material in the SLM molding is faster, and the solidification rate is higher than the rate of martensite phase transformation, which leads to the transformation of the β phase in the organization to the α phase too late, so that the β phase is transformed to the martensite $\alpha\prime$ phase. Due to the local cooling rate and residual heat, there is a certain difference, so the martensite shows a different morphology. Some black particles can also be observed in Figure 6a,b, which may be caused by the oxidized metal powder splashing into the molten pool during the high-speed laser scanning in the SLM molding process.

In order to analyze the physical phase distribution of the specimens after SLM molding, the surface of the molded specimens was subjected to XRD analysis, and the XRD patterns of the molded specimens are shown in Figure 7. According to Figure 7, it can be seen that the peaks of SLM-molded TC4 alloy specimens are mainly Ti peaks, and no other substances were detected. Compared with the data of the standard (PDF#89-2762) card [36], it was found that the position of SLM-molded specimens was positively shifted, which means that the spacing of grain boundaries had become smaller, and, thus, it is speculated that the reason for this phenomenon may be the occurrence of the martensitic phase transition.

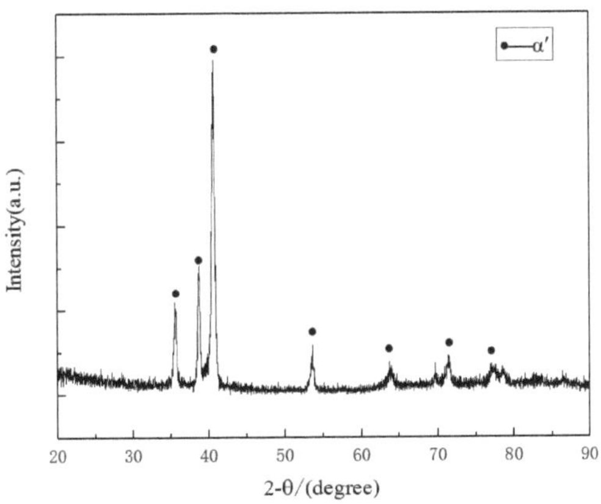

Figure 7. XRD pattern of SLM-formed sample.

3.3.2. Testing and Analysis of Mechanical Properties of SLM-Molded TC4 Alloy

1. Microhardness Analysis

Figure 8a shows the microhardness of the side of the SLM-molded specimen. The microhardness of the side of the specimen is between 363~376 HV, and the average microhardness is 369.13 HV. It can be seen from Figure 8a that the side of the microhardness shows a waveform change of firstly increasing, then decreasing, and then increasing; the microhardness of the measuring point 2 reaches the maximum value, and the microhardness of measuring point 4 is the smallest, and the reason for this change may be that measurement point 2 is in the middle of the melt channel and the powder in the middle of the melt channel is more fully melted, resulting in a more detailed granular organization, so that the microhardness of the specimen is higher. Measurement point 4 is in the overlap of the melt channel, and the powder at the overlap of the melt channel will appear to be melted for the second time, resulting in a larger granular organization, so that the microhardness of the specimen is reduced. However, the overall difference in the hardness of the specimen is not very large, so the overall hardness of the specimen is more uniform.

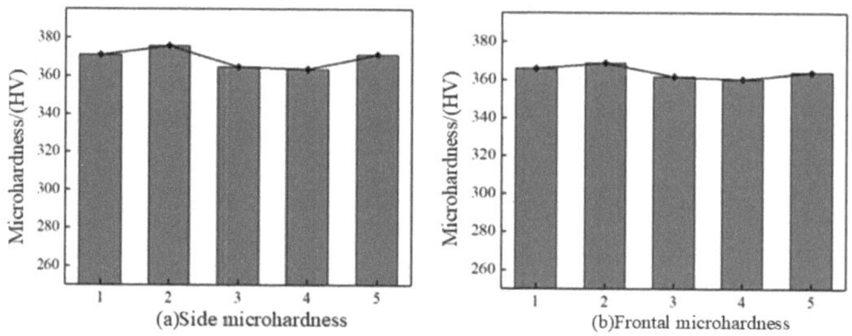

Figure 8. Microhardness values of molding samples.

Figure 8b shows the microhardness of the front side of the SLM-molded specimen. The microhardness of the front side of the specimen is between 360 and 370 HV, and the average microhardness is 364.06 HV. The difference in microhardness at different measurement points may be due to the temperature effect on the microhardness produced during the SLM molding process. If the measurement point occurs between the molding layers (measurement point 2), the laser energy absorbed by the powder here is greater, the melting is more adequate, and the microhardness is higher; if the measurement point occurs at the combination of the molding layers (measurement point 4), the energy absorbed by the powder here is less than that between the molding layers, the melting of powder is not enough, and microhardness is decreased. However, from the Figure, it can be seen that the difference between the maximum and minimum microhardness values is very small, so the microhardness is more uniform.

The microhardness of TC4 alloy SLM-molded specimens is higher than casting (320 HV), which may be due to the fact that SLM molding is a fast melting and fast solidification process, the number of grains is larger, the growth space between adjacent grains is narrow, and the growth time is short, resulting in insufficient grain growth, which results in a large number of fine grains, and, thus, the hardness of the specimen is increased.

2. Tensile Properties Analysis

The mechanical properties of SLM-molded specimens with different densities are also different, and it is generally considered that high densities are the prerequisite for guaranteeing mechanical properties. Based on the optimization of process parameters, tensile specimens were molded and tensile performance tests were conducted to verify whether the optimized process parameters could meet the requirements of mechanical properties. The preparation dimensions of the tensile specimens refer to the ASTME8/E8M-15a [37] "Standard Test Methods for Tensile Testing of Metallic Materials" standard (small-size specimens), in which the marking distance G is taken to be 25.0 mm, and the width of the cross-section T is taken to be 6 mm; the specific dimensions are shown in Figure 9.

The researchers found that the tensile properties of the molded specimens were closely related to the direction of metal powder accumulation during molding, and by placing the specimens horizontally and performing SLM molding along their thickness direction (i.e., the thickness direction is the direction of powder accumulation), the molded specimens with the highest tensile and yield strengths were obtained [38], and the molding effect is shown in Figure 10a. Figure 10b shows the effect of the molded specimen after tensile fracture, according to the recorded load–displacement ($F - \triangle s$) curve, and converted into the actual stress–strain ($\sigma - \varepsilon$) curve.

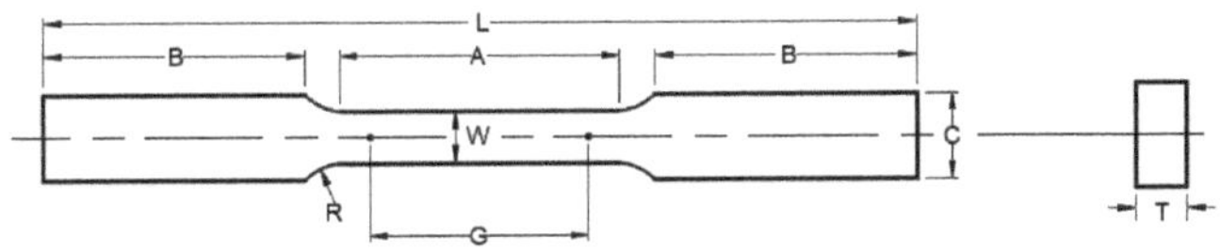

$W = 6.0 \pm 0.1\text{mm}$, $R = 6\text{mm}$, $L = 100\text{mm}$, $A = 32\text{mm}$, $B = 30\text{mm}$, $C = 10\text{mm}$

Figure 9. Dimensions of tensile specimens.

(a) SLM-molded tensile specimens (b) Specimen after tensile fracture

Figure 10. Tensile specimens.

Figure 11 shows the stress–strain curves of three molded specimens obtained from the tensile experiment. The stress–strain curves show that the molded specimens in the tensile process experience a total of three stages: the elastic stage, the strengthening stage, and the deformation and fracture stage. In the elastic stage, the change curves of the three molded specimens basically overlap, and the stress–strain curve is similar to a straight line. After the elastic stage, the molded specimen enters the strengthening stage, and there is no obvious yield stage, so the yield strength of the specimen is characterized by $\sigma_{0.2}$ (0.2% residual deformation of the stress value), the specimen passes through the maximum tensile strength of the stress σ_b, and the phenomenon of local deformation occurs until the specimen is fractured.

The average yield strength of the three specimens was 1217.40 MPa, the tensile strength was 1315.32 MPa, and the elongation at break was 9.6%. Table 6 shows the test results of tensile properties of SLM-molded titanium alloy specimens. From the table, it can be seen that the yield strength and tensile strength of SLM-molded titanium alloy tensile specimens are higher than those of ASTM F136 and casting standards, but the elongation of SLM-molded titanium alloy specimens is slightly lower relative to the ASTM F136 standard. The reason for this phenomenon may be that the SLM molding process is a rapid melting and rapid solidification process, the metal grains cannot grow up in time to be limited by the surrounding grains, and then the formation of a very small and balanced organization, fine grain size, and a greater number of grain boundaries more effectively prevent the occurrence of dislocations, improve the SLM molding specimens yield strength and tensile strength. However, at the same time, it may lead to certain defects at the grain boundaries, which makes the molded specimen plasticity decrease and elongation decrease.

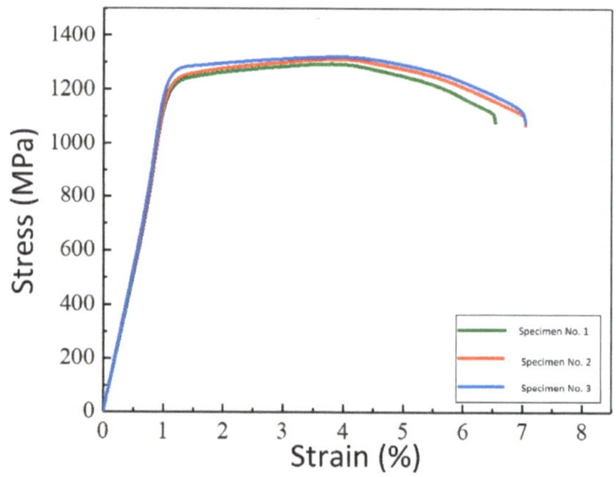

Figure 11. Stress–strain curve.

Table 6. Tensile test results and comparison of SLM-molded samples.

Type	Yield Strength (MPa)	Tensile Strength (MPa)	Elongation (%)
ASTM F136	795	860	10.0
Casting	847	976	5.1
SLM molding	1217.40	1315.32	9.6

3. Compression Performance Analysis

The femur is an important part of the human lower limb, and the study of its stress state is of strategic significance for artificial joint replacement. Some studies have shown that the maximum stress on the femur during standing can reach 27.70 MPa [39], and the compressive strength of implants with a porous structure is lower than that of implants with a solid structure. In implant applications, porous structures also need to meet the requirements of mechanical properties (e.g., compressive strength), and, therefore, compressive properties are an important mechanical property index for evaluating porous implants.

Based on the topology-optimized reconstruction model, the compression specimen model was constructed by arraying the porous unit structures with volume fractions of 10%, 20%, 30%, and 40% through Materialise Magics 24.0 software, as shown in Figure 12.

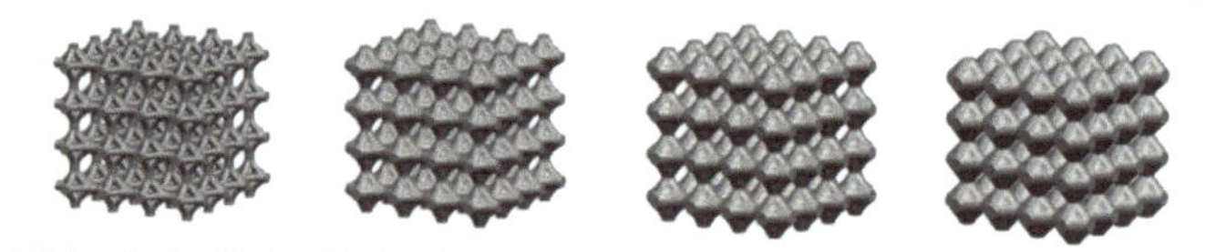

Figure 12. Compressed sample model.

Based on the optimized process parameters, the porous compression specimen is molded by SLM technology. When the compression specimen is molded, the specimen is firstly cut from the substrate by using the wire cutting technology to ensure the bottom surface of the specimen is flat. Then the compressed specimen is subjected to a series of post-treatment processes, such as ultrasonic cleaning, blowing, sandblasting, etc., in order to remove the metal powder that is not firmly bonded inside and on the surface of the porous structure. The final molded specimen was obtained as shown in Figure 13, and from left to right, the volume fractions of the porous compression specimen were 10%, 20%, 30%, and 40%.

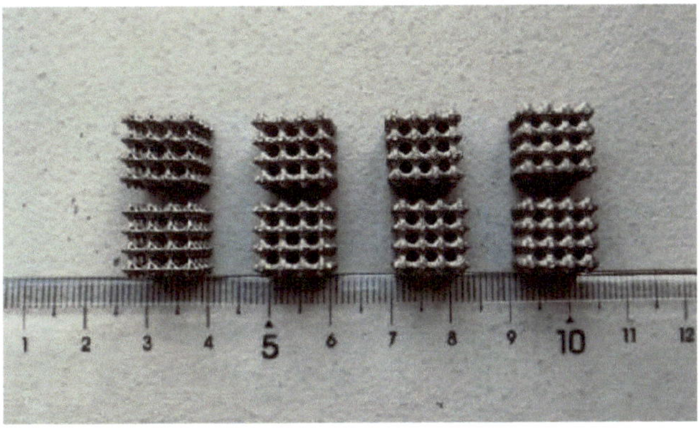

Figure 13. Compression sample molding effect.

According to the ISO 13314 standard [40] for compression testing of porous and honeycomb metals, the compression properties of SLM-molded porous specimens of TC4 alloy were analyzed by compression testing. It was found that specimens with different volume fractions had different failure locations and different resistance to deformation when compressed, and Figure 14 shows the effect diagram of the specimens after being compressed.

Figure 14. Sample effect after compression test.

The corresponding force–displacement ($F - \triangle s$) curves were obtained by compression tests of specimens with different volume fractions, and the stress–strain ($\sigma - \varepsilon$) curves shown in Figure 15 were transformed according to equation. Generally speaking, when the compression test starts, the stress–strain curve will have a linear phase, but according to Figure 15, it is found that although the volume fractions of the compression specimens are different, the compression specimens all have a nonlinear concave tendency at the beginning part, which can be attributed to the local deformation of the bottom of the specimen when the specimen on the substrate is wire-cut or the porous structure of the surface of the compression specimen after molding is not flat enough, leading to the stress at the initial stage not being smooth, which is not enough.

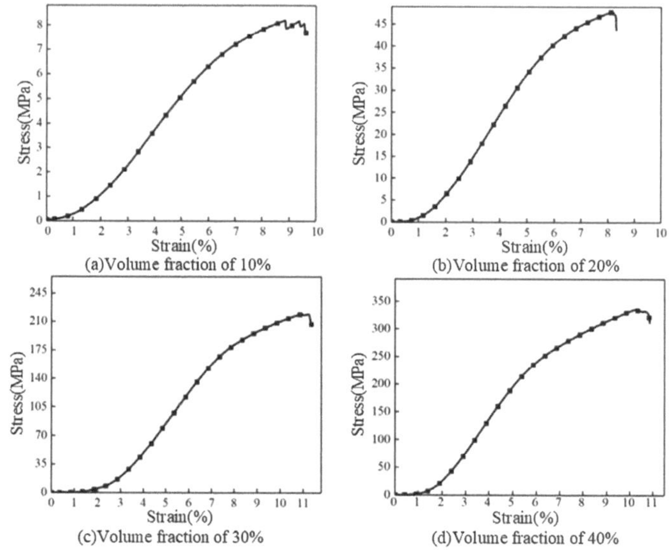

Figure 15. Stress–strain curve.

The surface of the compressed specimen after molding is not flat enough, which leads to the uneven force in the initial stage. After a small phase of nonlinear changes, the structure enters the elastic deformation phase, and the equivalent elastic modulus of the structure can be obtained from the slope of the stress–strain curve in this phase. As the strain increases, the stress appears to change nonlinearly, and then enters the plastic deformation stage of the stress–strain curve. The ability to resist deformation of the compression specimen in this stage is different; the smaller the volume fraction, the weaker the ability to resist deformation. The specimen enters the failure stage of the stress–strain curve after the phenomenon of sudden change in stress. However, the densification stage did not occur during the compression of the specimen, which was caused by the fact that the tester stopped compressing the specimen when a local failure occurred in the porous structure during the compression performance test of the porous structure. The modulus of elasticity and compressive strength of different specimens were measured according to the stress–strain curves, as shown in Table 7.

Table 7. Compression experimental results.

Volume Fraction	10%	20%	30%	40%
Modulus of elasticity (GPa)	0.75	1.51	7.99	12.63
Compressive strength (MPa)	8	48	219	338

4. Friction and Wear Performance Analysis

In order to analyze the experimental results more intuitively, Origin software (Version number: 8.0) was used to realize the conversion from data to image. Figure 16 shows the graph of friction coefficient comparison between SLM-molded specimens and casting specimens under different lubrication conditions. From the Figure, it can be seen that the initial change in each curve is not regular, and it is guessed that the reason for this phenomenon is that when the TC4 alloy specimen and the UHWMPE specimen are experimented with, because the molding specimen is not identical, the initial contact state of the two specimens is different, which leads to a certain difference in the trend of the friction coefficient at the initial stage.

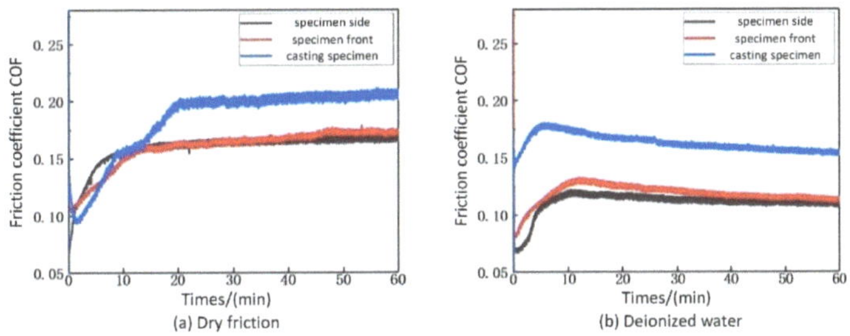

Figure 16. Friction coefficient of samples with different forming modes.

Figure 16a shows the comparison curve of the friction coefficients of the casting specimen and the SLM-molded specimen under dry friction conditions, from which it can be seen that the casting specimen maintains its friction coefficient basically unchanged and stabilized at about 0.2 after about 20 min of the break-in period, while the SLM-molded specimen maintains its friction coefficient at about 0.16 after about 10 min of the break-in period, which is obviously lower than the friction coefficient of the casting specimen. As can be seen from Figure 16b, under the condition of deionized water lubrication, the friction coefficient of the casting specimen is still higher than that of the SLM-molded specimen, and both of them have experienced the same break-in period (10 min); the friction coefficient of the casting specimen fluctuates up and down around 0.17 and the friction coefficient of the SLM-molded specimen fluctuates up and down around 0.11 after the break-in period.

Figure 17 shows the average coefficient of friction of different specimens under different lubrication conditions. Combined with Figure 16, it can be seen that the wear resistance of TC4 specimens of both molding methods is improved under lubrication conditions, and the coefficients of friction are reduced. The reason for this result is that under dry friction conditions, the repeated friction between the specimen and the counter-abrasive specimen produces fine wear particles, and due to the lack of lubricant, the wear particles cannot be discharged in time, so that the wear particles accumulate between the specimen and the counter-abrasive specimen, and the friction coefficient between the two is larger. Under the deionized water lubrication condition, there is water lubrication between the specimen and the counter-abrasive specimen, and the wear particles are discharged in time by the hydrodynamic effect of the lubricant, which reduces the degree of wear of the specimen and the counter-abrasive specimen and the coefficient of friction between the two.

Figure 18 shows the friction coefficient comparison curves of different surfaces of SLM-molded specimens, from which it can be seen that the friction coefficient difference between the side and the front of the SLM-molded specimens is not very large, and its tendency to change and the size of the value is almost the same. However, it can be seen from the figure that the friction coefficient of the front side of the molded specimen is slightly higher than that of the side; the reason for this result is related to the surface roughness of the SLM-molded specimen. The SLM molding layer-by-layer stacking process means that the melt channel and the real position achieve a certain degree of deviation, and in the direction of the stacking of the phenomenon there is sticking powder, so that the front side of the specimen is not flat, the specimen is rougher, and, therefore, the front side of the specimen has a slightly larger coefficient of friction.

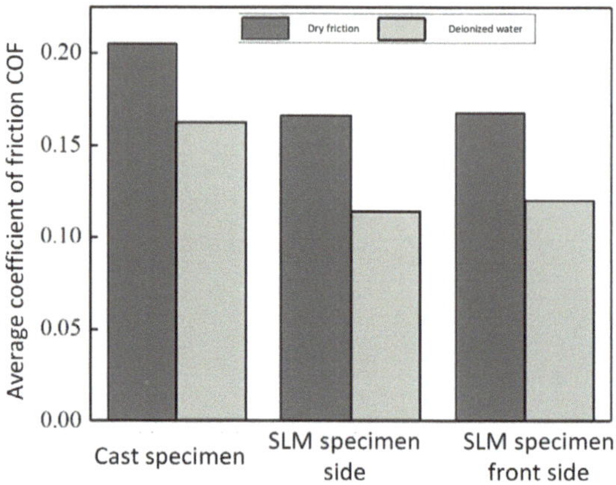

Figure 17. Average friction coefficient of samples under different lubrication conditions.

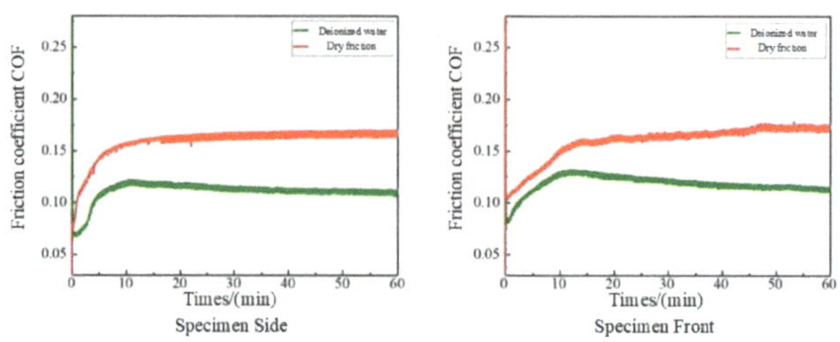

Figure 18. Friction coefficient of samples under different conditions.

4. Conclusions

1. The SLM process parameters were optimized through orthogonal tests, and for roughness, the optimized process parameter combinations were determined as follows: laser power of 300 W, scanning speed of 1250 mm/s, scanning spacing of 0.09 mm, and a powder layer thickness of 0.05 mm. When the density of the molded specimen is high, it has fewer internal pore defects, and, on the contrary, when the density is low, the internal pore defects are more obvious;
2. The microstructure of the cross-section of the molded specimen is mainly α' martensite, and there are oriented β columnar crystals in the microstructure of the elongated cross-section; the microhardness of the side of the molded specimen averaged 369.13 HV, and the microhardness of the front side has an average value of 364.06 HV, and the sides and fronts of the molded specimen are harder than those of the casting specimen;
3. The average yield strength of the molded specimens was 1217.40 MPa, the tensile strength was 1315.32 MPa, and the elongation at break was 9.6%. It was found that the yield strength and tensile strength of the SLM-molded specimens were higher than that of the ASTM F136 and casting standards but the elongation at break was slightly lower than that of the ASTM F136 standard; the elastic modulus of the molded specimens was in the range of 0.7~12.7 GPa and the compressive strength between 8~338 MPa, which is in line with the range of human bone requirements [41]. For the materials used to make femoral implants, Ti-6Al-4V femoral implants made by

SLM meet the requirements, but in practical orthopedics application, it is necessary to apply a coating that enables osseointegration to the alloy obtained by SLM;

4. Under the same experimental conditions, the friction coefficients of SLM-molded specimens are lower than those of casting specimens, and their wear resistance is higher than that of casting specimens. The friction and wear performance of the molded specimens on the side and the front side do not differ much, and there is basically no difference in the friction coefficient trends and sizes of the two, with the friction coefficient on the front side of the specimens being slightly higher.

Author Contributions: Conceptualization, W.Z. and Y.L.; methodology, W.Z.; software, Y.H.; validation, W.Z., H.L. and Z.L.; formal analysis, W.Z.; investigation, Y.L.; resources, Y.H.; data curation, W.Z. and Y.H.; writing—original draft preparation, W.Z.; writing—review and editing, W.Z., H.L. and Z.L.; visualization, W.Z. and Y.H.; supervision, Z.L. and H.L.; project administration, Z.L.; funding acquisition, Z.L. All authors have read and agreed to the published version of the manuscript.

Funding: This research received no external funding.

Data Availability Statement: The original contributions presented in the study are included in the article, further inquiries can be directed to the corresponding author.

Conflicts of Interest: The authors declare no conflicts of interest.

References

1. Shnaekel, A.; Morrison, J.C. High Rates of Aseptic Loosening in Modern Posterior-stabilized Femoral Components from a Single Manufacturer. *Arthroplast. Today* **2023**, *22*, 101132. [CrossRef]
2. Apostu, D.; Lucaciu, O.; Berce, C.; Lucaciu, D.; Cosma, D. Current methods of preventing aseptic loosening and improving osseointegration of titanium implants in cementless total hip arthroplasty: A review. *J. Int. Med. Res.* **2018**, *46*, 2104–2119. [CrossRef] [PubMed]
3. Munemoto, M.; Grammatopoulos, G.; Tanaka, Y.; Gibbons, M.; Athanasou, N.A. The pathology of failed McKee-Farrar implants: Correlation with modern metal-on-metal-implant failure, Journal of materials science. *Mater. Med.* **2017**, *28*, 66. [CrossRef] [PubMed]
4. Levai, J.P.; Descamps, S.; Roch, G.; Boisgard, S. Early aseptic loosening of cemented cup in metal-on-metal total hip arthroplasties. *Rev. Chir. Orthop. Reparatrice L'appareil Mot.* **2006**, *92*, 575–580. [CrossRef] [PubMed]
5. Doi, K.; Kobatake, R.; Makihara, Y.; Oki, Y.; Umehara, H.; Kubo, T.; Tsuga, K. Osseointegration Aspects of Implants at the Bone Reconstruction Site by a Novel Porous Titanium Scaffold. *J. Oral Maxillofac. Res.* **2021**, *12*, e4. [CrossRef] [PubMed]
6. Wu, S.; Xu, J.; Zou, L.; Luo, S.; Yao, R.; Zheng, B.; Liang, G.; Wu, D.; Li, Y. Long-lasting renewable antibacterial porous polymeric coatings enable titanium biomaterials to prevent and treat peri-implant infection. *Nat. Commun.* **2021**, *12*, 3303. [CrossRef]
7. Gao, B.; Zhao, H.; Peng, L.; Sun, Z. A Review of Research Progress in Selective Laser Melting (SLM). *Micromachines* **2022**, *14*, 57. [CrossRef] [PubMed]
8. Li, L.; Gao, H.; Wang, C.; Ji, P.; Huang, Y.; Wang, C. Assessment of Customized Alveolar Bone Augmentation Using Titanium Scaffolds vs Polyetheretherketone (PEEK) Scaffolds: A Comparative Study Based on 3D Printing Technology. *ACS Biomater. Sci. Eng.* **2022**, *8*, 2028–2039. [CrossRef] [PubMed]
9. Lim, H.-K.; Choi, Y.-J.; Choi, W.-C.; Song, I.-S.; Lee, U.-L. Reconstruction of maxillofacial bone defects using patient-specific long-lasting titanium implants. *Sci. Rep.* **2022**, *12*, 7538. [CrossRef]
10. Elshaer, R.N.; Ibrahim, K.M. Performance, Study of Microstructure, Mechanical Properties, and Corrosion Behavior of As-Cast Ni-Ti and Ti-6Al-4V Alloys. *J. Mater. Eng. Perform.* **2022**, *32*, 7831–7845. [CrossRef]
11. Sangali, M.; Cremasco, A.; Soyama, J.; Caram, R.; Contieri, R.J. Selective Laser Melting of Ti-6Al-4V Alloy: Correlation Between Processing Parameters, Microstructure and Corrosion Properties. *Mater. Res.* **2023**, *26*, e20230055. [CrossRef]
12. Feng, X.; Qiu, J.; Ma, Y.; Lei, J.; Cui, Y.; Wu, X.; Yang, R. Influence of Processing Conditions on Microstructure and Mechanical Properties of Large Thin-Wall Centrifugal Ti-6Al-4V Casting. *J. Mater. Sci. Technol.* **2016**, *32*, 362–371. [CrossRef]
13. Zhu, Y.; Zhao, Q.; Liu, X.; Geng, R.; Wang, B.; Jiang, Q.J.N. Improved Strength-Ductility of Ti-6Al-4V Casting Alloys with Trace Addition of TiC-TiB2 Nanoparticles. *Nanomaterials* **2020**, *10*, 2330. [CrossRef]
14. Lee, H.-W.; Kong, B.-O.; Kim, S.E.; Joo, Y.; Yoon, S.; Lee, J.H. Microstructure and Mechanical Properties of Vacuum Centrifugal Casted Ti-6Al-4V Alloy by Casting and Heat Treatment. *Korean J. Met. Mater.* **2022**, *60*, 282–290. [CrossRef]
15. Wen, F.; Zhao, J.-H.; Yuan, M.; Wang, J.; Zheng, D.; Zhang, J.; He, K.Z.; Shangguan, J.-J.; Guo, Y. Alloys, Influence of Ni interlayer on interfacial microstructure and mechanical properties of Ti-6Al-4V/AZ91D bimetals fabricated by a solid–liquid compound casting process. *J. Magnes. Alloys* **2021**, *9*, 1382–1395. [CrossRef]
16. Banu, A.; Preda, L.; Marcu, M.; Dinca, L.L.; Maxim, M.E.; Dobri, G. Electrochemical Behavior of SLM Ti–6Al–4V Alloy After Long Time of Immersion in Lactic Acid Environment. *Met. Mater. Trans. A* **2022**, *53*, 2060–2070. [CrossRef]

17. Cheng, Y.-J.; Hung, F.-Y.; Zhao, J.-R. Microstructural Characteristics and Material Failure Mechanism of SLM Ti-6Al-4V-Zn Alloy. *Materials* **2023**, *16*, 7341. [CrossRef] [PubMed]
18. Metalnikov, P.; Ben-Hamu, G.; Eliezer, D. Corrosion behavior of AM-Ti-6Al-4V: A comparison between EBM and SLM. *Prog. Addit. Manuf.* **2022**, *7*, 509–520. [CrossRef]
19. Naghavi, S.A.; Tamaddon, M.; Garcia-Souto, P.; Moazen, M.; Taylor, S.; Hua, J.; Liu, C. A novel hybrid design and modelling of a customised graded Ti-6Al-4V porous hip implant to reduce stress-shielding: An experimental and numerical analysis. *Front. Bioeng. Biotechnol.* **2023**, *11*, 1092361. [CrossRef]
20. Mondal, P.; Das, A.; Mondal, A.; Chowdhury, A.R.; Karmakar, A. Fabrication of Ti-6Al-4V Porous Scaffolds Using Selective Laser Melting (SLM) and Mechanical Compression Test for Biomedical Applications. *J. Inst. Eng. India Ser. D* **2022**, *103*, 181–190. [CrossRef]
21. Kutsukake, A.; Seki, A.; Yoshida, Y. Influence of Implantation Period on Bioactivity on the Surface of Ti-6Al-4V Implant Model Made by Selective Laser Melting. 2021. Available online: https://popups.uliege.be/esaform21/index.php?id=4350 (accessed on 5 March 2024).
22. *ASTM F136*; Standard Specification for Wrought Titanium-6Aluminum-4Vanadium ELI (Extra Low Interstitial) Alloy for Surgical Implant Applications (UNS R56401). ASTM International: West Conshohocken, PA, USA, 2021.
23. *GB/T 13810-2007*; Wrought Titanium and Titanium Alloy for Surgical Implants. Standardization Administration of China: Beijing, China, 2007.
24. Hao, Y.; Liu, Z.; Yao, C.; Liang, Y.; Li, Y. Topological design and high-performance additive manufacturing of porous femoral implants. *J. Phys. Conf. Ser.* **2023**, *2566*, 012086. [CrossRef]
25. Bormann, T.; Müller, B.; Schinhammer, M.; Kessler, A.; Thalmann, P.; De Wild, M. Microstructure of selective laser melted nickel-titanium. *Mater. Charact.* **2014**, *94*, 189–202. [CrossRef]
26. Greco, S.; Gutzeit, K.; Hotz, H.; Kirsch, B.; Aurich, J.C. Selective laser melting (SLM) of AISI 316L—Impact of laser power, layer thickness, and hatch spacing on roughness, density, and microhardness at constant input energy density. *Int. J. Adv. Manuf. Technol.* **2020**, *108*, 1551–1562. [CrossRef]
27. Li, X.; Ni, X.; Xin, M.; Kong, B.; Wu, J. Wear Behavior of Reinforced Modified UHMWPE Material under Water Lubrication. *China Plast. Ind.* **2019**, *47*, 23–26+57.
28. Stwora, A.; Skrabalak, G.; Maszybrocka, J. To enhance mechanical properties of parts produced from AlSi10Mg powder with use of SLS/SLM methods by densification of the product surface layer. *Mechanik* **2017**, *90*, 426–431. [CrossRef]
29. Chen, L.-Y.; Liang, S.-X.; Liu, Y.; Zhang, L.-C. Additive manufacturing of metallic lattice structures: Unconstrained design, accurate fabrication, fascinated performances, and challenges. *Mater. Sci. Eng. R Rep.* **2021**, *146*, 100648. [CrossRef]
30. Tardelli, J.D.C.; Firmino, A.C.D.; Ferreira, I.; dos Reis, A.C. Influence of the roughness of dental implants obtained by additive manufacturing on osteoblastic adhesion and proliferation: A systematic review. *Heliyon* **2022**, *8*, e12505. [CrossRef]
31. Matos, G.R.M. Surgery, Surface Roughness of Dental Implant and Osseointegration. *J. Maxillofac. Oral Surg.* **2021**, *20*, 1–4. [CrossRef]
32. Zhang, W. Investigation on Process and Performance of TC4 Alloy Fabricated by Selective Laser Melting. Master's Thesis, Guangdong University of Technology, Guangzhou, China, 2019.
33. Vasylyev, M.; Mordyuk, B.M.; Voloshko, S.M. Wire-Feeding Based Additive Manufacturing of the Ti–6Al–4V Alloy. Part II. Mechanical Properties. 2023. Available online: https://ufm.imp.kiev.ua/en/abstract/v24/i01/005.html (accessed on 10 March 2024).
34. Eskandari, H.; Lashgari, H.R.; Zangeneh, S.; Kong, C.; Ye, L.; Eizadjou, M.; Wang, H. Microstructural characterization and mechanical properties of SLM-printed Ti–6Al–4V alloy: Effect of build orientation. *J. Mater. Res.* **2022**, *37*, 2645–2660. [CrossRef]
35. Simonelli, M.; Tse, Y.Y.; Tuck, C. On the Texture Formation of Selective Laser Melted Ti-6Al-4V. *Met. Mater. Trans. A* **2014**, *45*, 2863–2872. [CrossRef]
36. Jiang, J. *Organization and Properties of Medical Ti-6Al-4V Alloy by Selective Zone Laser Melting and Forming*; Chongqing University: Chongqing, China, 2016.
37. *ASTME8/E8M-15a*; Standard Test Methods for Tension Testing of Metallic Materials. ASTM International: West Conshohocken, PA, USA, 2016.
38. Wang, D.; Yang, Y.Q.; Sun, X.B.; Chen, Y. Study on energy input and its influences on single-track, multi-track, and multi-layer in SLM. *Int. J. Adv. Manuf. Technol.* **2012**, *58*, 1189–1199.
39. Yang, T.; Zheng, J.; Yao, Z.; Ma, L.; Zhang, Y. Finite element analysis of the biomechanical properties of the walking femur. *Guangdong Med. J.* **2016**, *37*, 512–515. [CrossRef]
40. *ISO 13314*; Mechanical Testing of Metals—Ductility Testing—Compression Test for Porous and Cellular Metals. International Organization for Standardization: Geneva, Switzerland, 2011.
41. Gong, M.; Tan, L.; Yang, K. Bone tissue engineering scaffold materials and their mechanical properties. *Mater. Rep.* **2007**, *10*, 43–46+54.

metals MDPI

Article

Planning Mechanical Behavior of A356 Alloy Wheels by Using Distinct Heat Treatments

Adriano L. Tonetti [1], Wislei R. Osório [1,2], Ausdinir D. Bortolozo [1,2] and Giovana S. Padilha [1,*]

[1] Faculdade de Ciências Aplicadas, FCA, Centro de Pesquisa em Manufatura e Materiais Avançados (CPMMA), Campus II, Universidade de Campinas, UNICAMP, Limeira 13484-350, Brazil; a261621@dac.unicamp.br (A.L.T.); wislei1@unicamp.br (W.R.O.); ausdinir@unicamp.br (A.D.B.)
[2] Faculdade de Tecnologia, FT, Campus I, Universidade de Campinas, UNICAMP, Limeira 13484-332, Brazil
* Correspondence: giovanap@unicamp.br

Abstract: The aim of this investigation concerns evaluating the mechanical strength and microhardness values of A356 alloy samples in distinctive heat treatments, including those commonly applied to automotive wheels. It is recognized that A356 and Al-Si-based alloys exhibit considerable versatility across numerous industrial applications. The mechanical behavior obtained is intimately associated with different operational parameters (e.g., cooling rates, solution treatment, quenching, and artificial aging). In this study, a group of samples are quenched at 30, 60 and 80 °C. Another set is quenched and subsequently aged at three different temperatures, i.e., 180, 200, and 220 °C for 5 h, and mechanical responses are compared. Microstructural characterization, X-ray diffraction (XRD) analysis, tensile testing, and microhardness measurements are carried out. Using the Rietveld data and based on the modified Williamson–Hall method, the microstrains, crystallite size, and dislocation densities are calculated. Based on this, the resulting mechanical strengths from distinctive quenching and aging are understood. It was found that there exists a "quasi-optimal range" of operational parameters involving different A356 alloy treatments, which vary depending on the manufacturing route. Considering A356 alloy wheels, the planning of the powder coat treatment before or after T6 treating provides better mechanical properties and ductility.

Keywords: T6 treating; wheel manufacturing; mechanical strengths; powder coating; fractography; dislocation density

Citation: Tonetti, A.L.; Osório, W.R.; Bortolozo, A.D.; Padilha, G.S. Planning Mechanical Behavior of A356 Alloy Wheels by Using Distinct Heat Treatments. *Metals* **2024**, *14*, 461. https://doi.org/10.3390/met14040461

Academic Editors: Jingxiang Xu, Xingwei Zheng and Zhenhua Chu

Received: 5 March 2024
Revised: 8 April 2024
Accepted: 9 April 2024
Published: 13 April 2024

1. Introduction

The automotive sector has adopted light alloys as an efficient strategy for decreasing vehicle weight [1]. Aluminum and its alloys are replacing ferrous alloys, leading to a decrease in greenhouse gas emissions. This shift has resulted in the substitution of traditional steel frameworks in complex wheel configurations with lightweight alloys. The use of lightweight materials in wheel production not only enhances visual appeal but also reduces stress on the suspension system, requiring less effort for maneuvering. Furthermore, it facilitates the smoother braking and acceleration of vehicles [2–4]. Investigations concerning Al-Si-based alloys are widely reported [4–25]. These provide various discussions concerning selecting die-casting parameters (e.g., mold materials, pouring temperature, etc.) and addition elements such as Ti and Sr as grain refiners, and Si for eutectic morphology modification. Moreover, the literature also discusses solution treating, quenching, and artificial aging of the microstructure and on mechanical behavior [10–25]. These articles were probably written because Al-Si-based alloys exhibit excellent castability and affordability and have a low coefficient of thermal expansion and attractive mechanical properties, making them suitable for manufacturing wheels by casting [4–6].

It has also been reported that the microstructural characteristics are influenced by the presence of α-Al dendrites and lamellar eutectic Si [7]. It is widely recognized that a typical as-cast microstructural array affects the resulting properties [6–15]. When coarse

Al and interdendritic Si particles are constituted, the reached ductility is considerably decreased [6–11,15–23]. This is associated with a plate-like eutectic Si morphology formation [6–8]. On the other hand, when manufacturing routes or treatments provide an acicular-like morphology and/or spheroidal-like morphology, reasonable yield or gains are obtained for both the strength and ductility [6–11].

The automotive wheel industry demands specific alloying elements to reduce porosity and improve the casting quality. For this purpose, Al-Ti-B master alloys are commonly used [10–15]. Moreover, an increase in microstructural homogeneity in the casting is also obtained [10–15]. Al-B master alloys are effective grain refiners [14,15], and consequently, the resulting mechanical properties are improved. A refined particle size not only improves the flowability of the molten material into the mold, minimizing the entrapped air and preventing cracks, but it also facilitates a more manageable machining [15,16].

In the manufacture of aluminum wheels, low-pressure die-casting (LPDC) is the most widely applied method. It provides a finer microstructural array, leading to improved fatigue characteristics, and a significant reduction in the resulting shrinkage. It is also reported that LPDC decreases the required lead time for wheel production [11,17–19]. T6 heat treatment improves A356 mechanical behavior mainly in terms of ductility and elongation [11,19]. Conventional heat treatment involves a typical solution treatment, quenching, and artificial aging, which improves the resulting ductility [20]. It is recognized that eutectic Si morphology is affected by the temperature at which it dissolves, with grain growth being maximized at temperatures around 550 °C.

Jin et al. [21] demonstrated that hardness and tensile strength increase with the increase in solution temperature, i.e., close to 555 °C. These authors have also shown that mechanical behavior is increased. This is because heat treatment and aging affect Mg_2Si formation [21].

It is remarked that certain studies have demonstrated the effect of the cooling rate on Si morphology formation. Dang et al. [10] have shown that the dendrite arm spacings decrease with increased cooling rate. Also, eutectic Si morphology is modified from coarse plate-like to short-rod shape and even spherical shape [21–23]. Associated with this, they have also observed that microhardness is increased. Yang et al. [23] have also shown that the eutectic morphology of an as-cast A356 alloy is modified from large plate-like eutectic to fine plate-like eutectic, applying cooling rates of 3 and 96 K/s, respectively. It is also extensively reported [10,23–28] that the modification of eutectic silicon during solution heat treatment is affected by using distinctive solution heat treatment temperatures. A combination of solidification parameters with solution heat treatment also implies an efficient eutectic Si morphology modification [10]. Roy and Maijer [25] have reported that a wide range of solution treatments associated with water quenching at distinct temperatures and various artificial agings minimizes or decreases the precipitation driving force necessary for artificial aging [25].

In the field of material science and engineering, extensive literature addresses Al-Si-based and A356 alloys, scrutinizing various manufacturing conditions, Si morphology modifiers, and different heat treatments [4,8]. The literature reported [10–26,29–38] has shown a great gamma of studies. When the PRISMA concept (preferred reporting items for systematic reviews and meta-analyses), stringers "Al-Si-based alloys", and "eutectic" are utilized when searching articles, more than 1400 papers are obtained. When replacing the term "eutectic" with "solution" or T6 heat treatment", more than 90,000 objects are reached. This suggests that Al-Si-based alloys and A356 alloys have been intensively studied. However, certain future gaps concerning these Al-based alloys still need to be investigated. Recently, a review proposed by Dash and Chen [37], containing more than 320 references, reported that Al-Si-based alloys are still focused on a wide range of automotive components. They are manufactured to supply the premier auto giants (e.g., Audi, GM, Nissan, Volkswagen, Chevrolet, Tesla, etc.) [38]. Although there exists a promising and exponentially growing market for hybrid and fully electric vehicles or electrified automobiles, the forecast of the global Al-Si alloy market indicates an increasing trend up to 2030 [37]. These alloys are also focused on the complex structural Al-Si alloy components in aerospace industries, e.g., Boeing, Airbus, and Bombardier [37].

Since the 1990s, the literature concerning Al-Si-based and/or A356 alloys, specifically covering the effects of heat treatments upon microstructural array and mechanical behavior, is vast with numerous articles, as previously commented [10–38]. However, in this present investigation, the novelty provided has two distinct technological and scientific aspects. The first has an issue core involving technological aspects, independent of particularities or peculiarities of manufacturing stages. For the automotive wheel industry, the contribution favors planning the manufacturing route, including the paint curing temperature and time (e.g., powder polymeric coating). Another novelty concerns the use of the Rietveld method as a viable and cheaper tool to understand the microstructural modifications provided by different treatments. This permits the decline in more sophisticated and complex sample preparation techniques and more technologically robust equipment (e.g., transmission electron microscopy or high-resolution TEM). With this, it is possible to obtain, easily or conveniently, a "quasi-optimal operational range" of parameters considering different A356 alloy treatments.

2. Experimental Procedure

2.1. Die Casting Sample Preparation

An as-received commercial A356 (Al-7Si-0.3Mg) alloy ingot is produced at 760 (±5) °C by using an electric-induction furnace. The minimum and maximum contents of the main alloying elements are shown in Table 1. It is noted that, except for Si content, all other identified elements have their corresponding errors up to 10% from the reported values. This is because duplicates are considered, and the accuracy of the utilized equipment is also taken into account.

Table 1. The chemical composition of the commercial A356 alloy.

Alloying Elements	Si	Mg	Cu	Fe	Ti	Sr	Al
Min. content (%wt.)	6.5	0.2	----	---	0.1	0.01	Balance
Max. content (%wt.)	7.5	0.45	0.2	0.3	0.2	0.02	Balance

The molten alloy is poured into the permanent molds (AISI H13 steel), which are specifically designed for automotive wheels. A nitrogen gas flux (3 L.p.m) at 780 (±5) °C is applied to remove any impurities. A graphite rod is utilized for mechanical stirring of the molten. Both the grain refiners and modifiers, i.e., Al-5Ti-1B (0.12 wt.%) master alloy and Al-10Sr (0.004 wt.%) master alloy, are used. The molten alloy is subjected to a network of graphite pipes to the injector, operating under squeeze casting (~1 bar) at 680 (±5) °C, and the AISI H13 mold is filled. The sample solidifies with internal cooling channels in the top and bottom dies. This solidification is promoted by compressed air under pressures of 2 and 3 bars, for 4 min. Once the solidification process is finished, the mold parts are opened, and the casting is extracted. The average duration of the normal cycle is about 360 s. Triplicates are considered to obtain the specimens. The selected specimen is taken for all other complementary experimentations, i.e., microstructural characterization, heat treating, tensile tests, etc.

2.2. Heat Treatments

After the wheel design with the 360 s cycle, the samples are obtained by cutting the front surface of the wheel to proceed with the heat treating (Graphical Abstract). A solution treating is carried out with a heating rate of 20 °C/min, reaching a plateau at 540 (±3) °C. Subsequently, the samples are maintained at 540 (±3) °C for 5 hours. Sequentially, the quenching in three distinct water temperatures is carried out, i.e., 30, 60, and 80 (±2) °C. This provides three different quenching rates. For this mentioned purpose, a water-to-sample ratio of 320 mL H_2O per gram of the sample and a magnetic stirrer (Fisatom—752A,

São Paulo, Brazil) are utilized. For the next stage, the three artificial agings are adopted. After water quenching, three distinct artificial aging treatments (T6) are carried out at 180, 200, and 220 (±3) °C for 300 min. These parameters represent the commonly applied T6 in wheel manufacturing. In order to guarantee reproducibility, a duplicate is considered for each selected condition, and the average result is taken into account for analysis. Figure 1 depicts the representative temperature vs. time profile for the solution treating, quenching, and three distinctive T6 heat treatments for the examined A356 alloy samples.

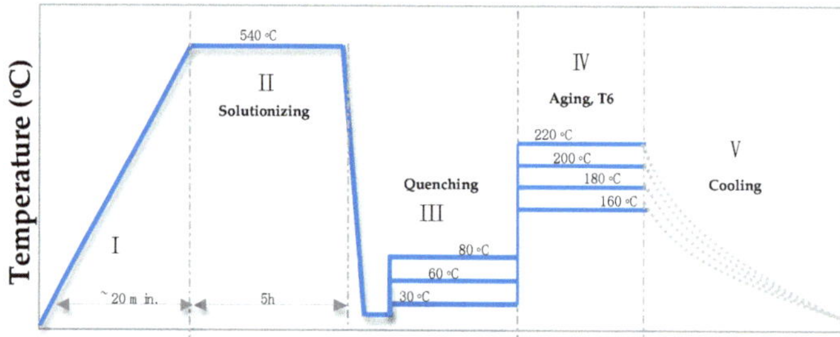

Figure 1. Representative temperature vs. time profile for the heating rate (I), solution treating (II), quenching (III), and three distinctive T6 heat treatments for the examined A356 alloy samples (IV). All adopted quenching are carried out in 120 s, and all aging treatments are performed for 180 min (3 h).

2.3. *The Sample Characterization*

A universal electro-hydraulic servo machine operating at environmental temperature (25 ± 2 °C) with a loading speed of approximately 25 mm/min and a strain rate of 2×10^{-4} s^{-1} is utilized. In order to guarantee reproducibility, triplicate is considered, and average values are depicted. Vickers hardness measurements (HV) (Buehler—Wilson VH1102, Lake Bluff, IL, USA) are conducted at environmental temperature (25 ± 2 °C). A least 20 measurements are performed for each one of the examined samples. A polished surface by applying a load of 9.8 N and a dwell time of 10 s is utilized to attain Vickers microhardness.

The experimental thermal curves comprising TG/DTG/DSC (TA Instruments—Q600, New Castle, DE, USA) are obtained between 50 and 700 °C. For this purpose, the samples with 18 (±0.8) inside α-alumina pan (30 μL), and applying a heating rate of 20 °C/min, are utilized. A nitrogen atmosphere of about 50 mL/min during examination is maintained. A duplicate is adopted and average curves are considered.

A scanning electron microscopy (SEM) (TESCAN® model VEGA3, Brno, Czech Republic) coupled with an energy-dispersive X-ray spectroscopy (EDS) is utilized for the microstructural characterization. Magnifications ranging between 500 and 1000 times are used. Typical SEM micrographs representing each selected condition are taken into account. To characterize the constituents and microstructural parameters, standard metallographic procedures are adopted. X-ray diffractometry (Panalytical diffractometer, X'Pert model, Malvern, UK), in Bragg–Brentano geometry is utilized. This is carried out at 25 (±2) °C with a Ni filter and Cu-Kα radiation (λ = 0.15406 nm, 45 kV, and 40 mA). Angles (2θ) between 10° and 120° with a scan rate of 0.008 s are utilized.

3. Results and Discussion

3.1. *Quenching and Aging Effects on Mechanical Behavior*

Three distinctive aging temperatures are evaluated, i.e., 30, 60, and 80 (±2) °C. It has previously been reported that a higher quenching rate provides increased mechanical strength [10,23]. These are attributed to finer eutectic Si morphology, which is modified from the coarse plate-like to spherical shape and/or fine plate-like eutectic. It has also been observed that the solution temperature affects the constituted eutectic silicon [10,16,23,33,35]. Based on this, solutioning at 540 °C is adopted. To correlate the mechanical strengths with heat treatments, the resulting mechanical behavior represented by conventional stress vs. strain curves and microhardness are carried out.

Figure 2a,b depict the experimental results of stress vs. strain for distinct quenching and aging temperatures, respectively. As previously reported, the highest ultimate tensile strength (UTS) is that of the A356 alloy sample quenched at 30 °C, i.e., 241 (±6) MPa. The intermediate and the lowest UTS values are those corresponding to the samples quenched at 60 and 80 °C, respectively. From a scientific and technological aspect, no intrinsic novelty was obtained, as previous investigations [10,16,23,33,35] have also reported similar UTS values.

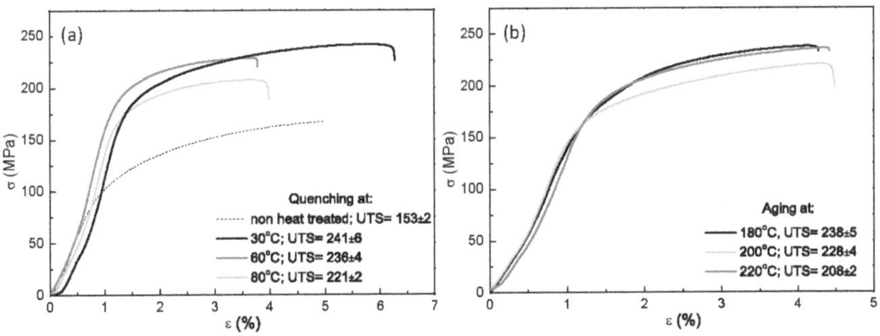

Figure 2. Experimental engineering stress vs. strain curves of A356 alloy after distinct (**a**) quenching and (**b**) aging. The as-cast (non-treated) results are also depicted.

It is recognized that Mg_2Si particles decrease mechanical behavior [21–23,38]. Additionally, it has been found that its dissolution occurs at solutioning, and this process is commonly rapidly obtained, i.e., between 2 and 4 min [38]. It is also understood that the optimal solutioning temperature and time is close to 540 °C for 3 and 6 h [10,35,38]. In the quenching stage, a supersaturated solid solution forms a significant number of vacancies. Dang et al. [11] demonstrated that after solutioning at 540 °C for 1 h, followed by water quenching (~2.6 K/s), a completely spheroidized Si morphology microstructural array is obtained, and the resulting microhardness is increased. It is also found that the highest ductility (represented by resulting elongation) is that of the sample quenched at 30 °C.

For the as-cast samples, the intermediate ductility values are confirmed, while the lowest are those of samples quenched at 60 and 80 °C. Ductility is a critical attribute for the performance of vehicle wheels, and the mechanical responses are affected by plastic deformation of the Al matrix and dislocation interactions. The yield strength (YS) and elongation (ε) values are determined from stress vs. strain at 0.2% proof strength, and these values are depicted in Figure 3.

Figure 3a shows that a similar decreasing trend of UTS and YS is obtained. It is remarkable that those mechanical strengths corresponding to the three different aging temperatures are obtained for quenching at 30 °C. Another interesting observation corresponds to the ductility, which is not substantially modified after the three different aging temperatures, as shown in Figure 3b. Moreover, the YS values corresponding to the aged and quenched samples are not substantially modified. This induces a very similar resulting microstructural array in terms of how the dendritic arm spacings and possible intermetallic phases are constituted. This will be confirmed when SEM micrographs are analyzed below. It is also noteworthy that decreasing trends are verified for the microhardness values obtained, i.e., with the increase in quenching and aging, the microhardness decreases, as depicted in Figure 4.

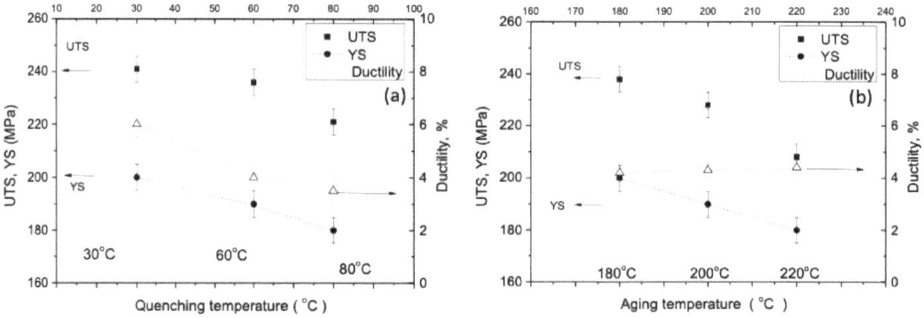

Figure 3. Relation among UTS, YS, and ductility in distinctive (**a**) quenching and (**b**) aging temperatures.

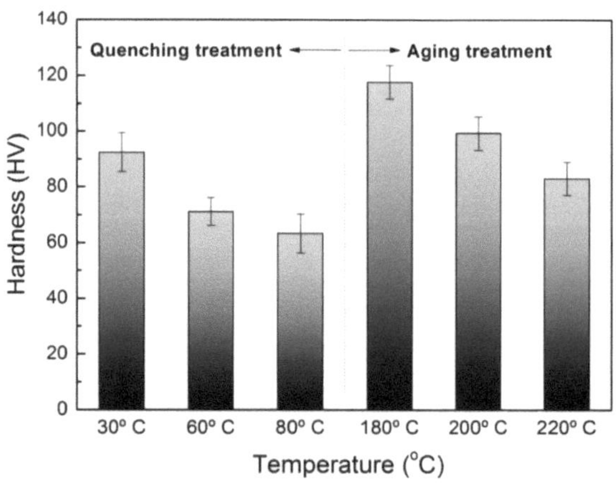

Figure 4. Microhardness obtained after three different quenching and three different aging temperatures.

Quenching temperature and time have important roles in the achieved hardness. When selecting or adopting appropriate quenching parameters, the hardness can be tailored to reach certain desirable requirements. Based on the obtained results, a low quenching temperature induces a hardness increase. On the other hand, high temperature promotes a decreasing trend in hardness. At 30 °C, the high hardness obtained seems to be attributed to the formation and transformation of the strengthening phases. These phases act as obstacles to dislocation movement, impeding deformation and contributing to the increase in the hardness.

When aging at 180 °C is carried out, an optimized mechanical response is achieved. This suggests that there is an optimal aging temperature without excessive particle growth. With an increase in the aging temperature to 200 °C and 220 °C, a decrease of about 10% in the tensile strength is verified. This seems to be associated with a possible occurrence of overaging, which leads to the dissolution of precipitated phases. Notably, the decrease in tensile strength does not coincide with the major alterations in the elongation. This suggests that similar ductility is obtained in all examined temperature ranges.

The correlations between UTS and YS and microhardness (HV) are proposed in order to compare with previous investigations [39–45], as shown in Figure 4. The problem of finding meaningful relationships between hardness and tensile strength has been widely reported [39–45].

Figure 5a depicts the relations between UTS and HV and UTS and YS and HV of the examined samples. These were quenched at 30, 60, and 80 °C, and aged at 180, 200, and 220 °C. Exponential equations with exponents 1/2 and 1/4 prescribe the relation for both the YS and UTS (i.e., 21.5 $HV^{\frac{1}{2}}$) for the quenched and aged samples, respectively, and between UTS and YS with HV for the quenched and aged samples, respectively.

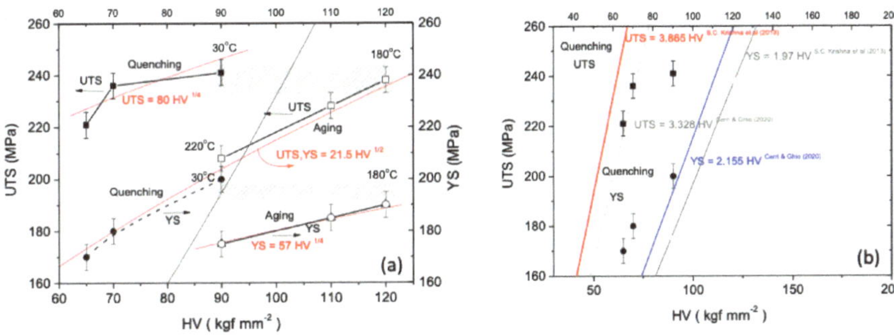

Figure 5. Correlation between yield strength (YS) and microhardness and ultimate tensile strength (UTS) and microhardness by using (**a**) exponential and (**b**) linear regression equations based on Krishna et al.' [43] and Cerri and Ghio's [42] equations.

However, a linear regression model previously proposed by Cerri and Ghio [42] and Krishna et al. [43] is evaluated in Figure 5b. Krishna et al. [43] have proposed linear equations, i.e., UTS = 3.865 HV and YS = 1.97 HV, which overestimate the experimental results obtained for the samples aged in the three different aging temperatures. On the other hand, when only the quenched samples are considered (closed point in Figure 5b), the experimental results are within the range prescribed by the linear equations proposed, i.e., UTS = 3.865 HV and YS = 1.97 HV.

Cerri and Ghio [42] proposed a simplified correlation between HV and UTS, modifying the Tabor and Cahoon equations [42]. In this sense, they also utilized linear equations adjusting the regression (constant) coefficients, i.e., to 3.328 and to 2.155 for UTS and YS correlations, respectively, as depicted in Figure 5b.

It is also clearly perceived that both the YS and UTS = f(HV) equations proposed by Cerri and Ghio mainly encompass or outline the experimental results. However, the experimental YS values corresponding to the aging treatment are overestimated by equation YS = 2.155 HV. To adjust these regression coefficients, the equations proposed by Cerri and Ghio [43] are modified, i.e., new equations prescribing the experimental results are UTS = 2.12 HV and YS = 1.65 HV, respectively, as shown in Figure 6a. It is worth noting that the coefficient (constant) described in both Krishna et al. and Cerri and Ghio's equations should range from 2.9 and 3.1, as suggested by Tabor and Cahoon [42]. This is intimately associated with Meyer's coefficient, which represents the "strain hardening" of the material, as reported by Cerri and Ghio [42].

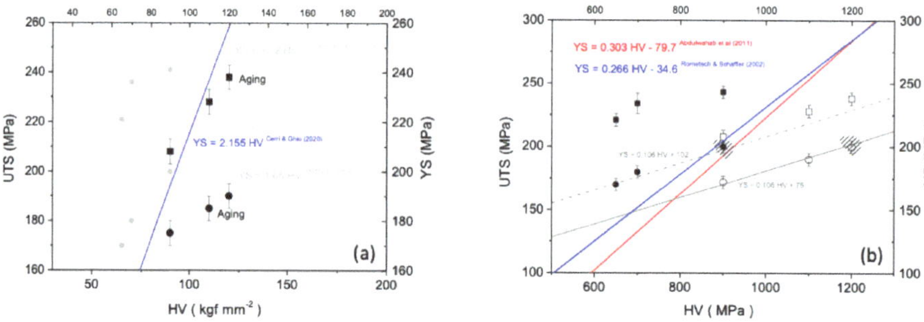

Figure 6. Correlation between YS and UTS with microhardness fitted by (**a**) Cerri and Ghio's [42] linear equation and (**b**) YS and UTS with microhardness (in MPa) considering the investigations of Abdulwahab et al. [40] and Rometsch and Schaffer [39].

Figure 6b shows a correlation between the yield strength (YS) and microhardness (in MPa) considering two other investigations. The first equation is based on Abdulwahab et al.'s [40] study with YS = 0.303 HV − 79.7, and another investigation, by Rometsch and Schaffer [39], describes YS = 0.266 HV − 34.6.

The negative portion highlighted in the referenced equations corresponds to the strain (during Vickers indentation) on the flow stress during indentation, which is less than the representative strain, as previously reported [35,39–41]. Considering all experimental data (both quenching and aging), it is confirmed that no match is obtained when both the Abdulwahab et al. [40] and Rometsch and Schaffer [39] equations are used.

These equations only prescribe the best results of the quenching, i.e., the value of YS of the sample quenched at 30 °C, and, similarly, the result of the UTS corresponding to the sample aged at 220 °C. Mathematically, adjustments should be provided in these equations to describe the results of the applied quenching and aging, i.e., $YS_{quenching}$ = 0.106 HV + 102 and YS_{aging} = 0.106 HV + 75. Although the same slope is used, this is lower than the range indicated by Abdulwahab et al. [40] and Rometsch and Schaffer [39]. Additionally, the positive portion is not commonly reported. With these considerations, it is observed that the adapted linear equations prescribed by Cerri and Ghio [43] and Krishna et al. [44] provide a more inclusive range that encompasses both the quenching and aging outcomes when assessing YS and UTS.

3.2. XRD and DSC Characterizations

The experimental X-ray diffractometry (XRD) patterns and Rietveld profiles of the A356 alloy samples in the as-cast, quenched at 30 °C, and aged at 180 °C are shown in Figure 7a–c, respectively. Based on the obtained mechanical behavior of the A356 alloy samples, only these aforementioned three distinct samples are selected. The as-cast is the initial condition and acts as a control or reference. The other two samples, i.e., quenched at 30 °C and aged at 180 °C, are those samples with "quasi-optimal" obtained mechanical responses. It is observed that all conceivable reflections of the α-Al and Si phase are identified. Major constituent phases, such α-aluminum and the Al-Si eutectic phase, are indexed. It is considered that Mg_2Si, β-Al_5FeSi, and many other Fe-containing non-stoichiometric phases have been fully dissolved in these samples. These mentioned phases are not identified. It is remarkable that the Fe-rich intermetallic phases are not detected in XRD patterns. This seems to be associated with few portions being detectable, which has its fraction lower than the detection limit of the diffractogram equipment.

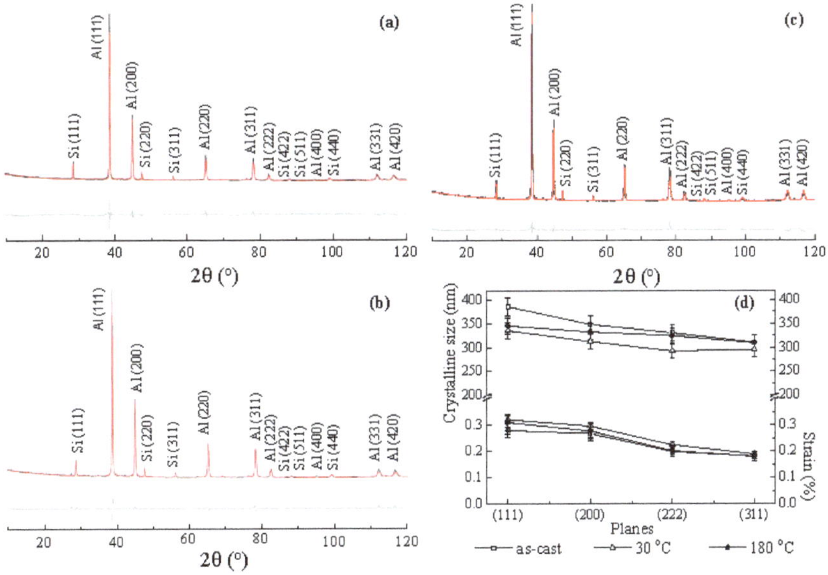

Figure 7. XRD patterns of the A356 alloy samples in condition: (**a**) as-cast, (**b**) water-quenched at 30 °C only, (**c**) artificially aged (T6) at 180 °C for 3 h; and (**d**) the calculated crystallite size and remaining strain determined by using Rietveld data (gray lines).

From Rietveld refinement of XRD data, the average crystallite sizes and lattice strains of the examined samples are determined, as shown in Figure 7d. The dislocation density (ρ) of the as-cast, quenched at 30 °C, and aged at 180 °C samples are determined by using the Equation (1) [45,46], as follows:

$$\rho = \frac{2\sqrt{3}\varepsilon}{Db} \tag{1}$$

where the term "ε" is the microstrain, D is the crystallite size, and the term "b" means the burger vector. Based on the modified Williamson–Hall method [45–49], the microstrain (ε) and crystallite size (D) are determined. For this purpose, XRD peaks obtained for each examined sample and an analytical software X'pert HighScore Plus (version 4.5), by PANalytical® are used. There are other models to calculate ε and D, e.g., the modified Warren–Averbach model [49] and Shear Lag and Nes models [50], which are not used in this present investigation.

The fundamental equation utilized in this methodology incorporates the phenomenon of XRD peak broadening resulting from both ε and D, as described in Equation (2) [46].

$$B\cos\theta_b = \frac{\lambda K}{D} + 4\varepsilon\sin\theta_b \tag{2}$$

where B is the width of the XRD peak measured at FWHM (full width at half-maximum); q_b is the Bragg's angle sample; K is a constant (commonly defined as 0.9); and l is the wavelength of the incident radiation (Cu Ka), i.e., 0.15405 nm. It is assumed that the crystallite strain is uniform in all directions of crystallography, as previously reported by Sutapa et al. [48].

Wang et al. [48] have also demonstrated that by employing the modified Williamson–Hall method, the FWHM is inversely proportional to the grain size and is proportional to the mean square strain, which is directly correlated with the dislocation density. Although it is recognized that TEM (Transmission Electron Microscopy) and EBSD (Electron Backscattering Diffraction) are techniques used to identify dislocations, XRD measurements offer more precision in quantifying dislocation density [48]. However, in this present study, no substantial apparent increases in the peak widths of these examined samples are verified, as shown in Figure 7.

Table 2 shows the calculated dislocation density (ρ) considering the (200) and (111) peaks of the as-cast, quenched at 30 °C, and aged at 180 °C. Although there exist limitations when using XRD techniques to precisely quantify the dislocation density of a material, qualitative assessments of the variations in the dislocation densities from different treatments are provided. The compensation for instrumental broadening in the peak width is achieved by estimating the Gaussian and Lorentz coefficients of the examined A356 alloy samples, and these are used as the default settings for line profile analysis. It is also recognized that the heat treatment potentially provokes a significant impact on dislocation density, which affects the resulting mechanical behavior.

Table 2. Calculated dislocation density (ρ) at (200) and (111) Al peaks obtained from XRD patterns of the as-cast, quenched at 30 °C, and aged at 180 °C samples.

	As-Cast	**Quenched at 30 °C**	**Aging at 180 °C**
		ρ ($m^{-2} \times 10^{15}$)	
(200)	6.2 (±0.2)	7.2 (±0.2)	7.9 (±0.2)
(111)	7.9 (±0.2)	8.9 (±0.2)	9.4 (±0.2)

Table 2 indicates that the sample quenched at 30 °C, and considering its (200) and (111) planes, is about 16 and 13% higher ρ than the as-cast sample (i.e., ρ of 7.2 × 10^{15} m^{-2} against 6.2 × 10^{15} m^{-2} and 8.9 against 7.9 × 10^{15} m^{-2}), respectively. It is evidenced that the quenching provokes internal stresses possibly associated with the differences in the thermal expansion among its constituent phases.

At (111) and (200) planes, both the quenched and aged alloy samples have slightly higher strains than the as-cast sample. Moreover, the crystallite sizes are considerably (12%) lower (~350 nm) than those of the as-cast sample (~390 nm). This occurred in the majority of all the planes examined when the heat-treated samples and the as-cast sample were compared. This is coherent with the dislocation density ρ shown in Table 2. Additionally, this also corroborates with our understanding of the UTS and YS values obtained, i.e., with the increase in the ρ, mechanical behavior also increases. However, when the quenched sample is compared with the sample artificially aged at 180 °C, the dislocation densities ρ in both (200) and (111) planes are higher than those observed for the sample quenched (i.e., 7.9 against 7.2 × 10^{15} m^{-2} and 9.4 against 8.9 × 10^{15} m^{-2}), respectively.

Considering these slight differences (~10 and 5%), the crystallite size and lattice strain reveal noteworthy disparities. It is observed that aging increases the dislocation density of all examined planes, accompanied by comparable strain values, as shown in Figure 7d. It is known that the dislocation density of a material is directly proportional to the lattice strain and inversely proportional to the crystallite size. Based on this, and associated with the fact that the aged sample has a similar microstrain level (from Rietveld treatment) and slightly higher crystallite size than the quenched sample, we deduce that the dislocations are responsible for the similar UTS and YS values when compared with the quenched sample, as shown in Figure 5a. Additionally, it is also clearly verified that it has substantially higher (~33%) microhardness (i.e., 120 against 90 HV), as depicted in Figure 4. These observations seem to be associated with the resulting microstructural arrays, not only at the micrometer scale but also at the nanometer scale. For example, at the microscale, there is a similarity in the dendrite arm spacings and the eutectic Si morphology and size. On the nanoscale, the characterization and observation of the dislocation densities and microstrains require TEM, EBSD, and/or HRTEM techniques. These advanced methods demand a more complex sample preparation and are more expensive than conventional microstructural characterization techniques.

From the metallurgical point of view, it is known that quenching forms a supersaturated solid solution containing numerous vacancies [38]. Associated with this, the sharp difference in the thermal expansion coefficient between the Al matrix and Si particle induces high thermal stress at the Al/Si interface. Consequently, high dislocation density is constituted, initiating from the interface inducing a mismatched lattice. Since Mg_2Si dissolution and homogenization are very fast processes (occurring from 2 up to 15 min to complete its homogenization [38]), the remaining possible particles are β-Al_5FeSi, as previously reported [38].

Conversely, the aging treatment achieves a more even distribution of precipitates within the material. To reach peak hardness, a specific duration of treatment is required, typically 10 h at 170 °C, although a period as brief as approximately 20 min at 210 °C can be sufficient [38,42]. From the calculated dislocation densities and microstrains results associated with the mechanical responses, it was found that the sample aged at 180 °C still exhibits a high dislocation level. This seems to be correlated with fine θ' precipitates, which commonly nucleate on dislocation sites. It has been reported [42] that these θ'' coherent precipitates on dislocations are responsible for the earlier and higher peak hardness.

Unfortunately, in this present investigation, to identify these dislocations, TEM and EBDS techniques are not carried out. However, SEM micrographs corresponding with the fractography of the examined samples are exploited and will be depicted and discussed below. Associated with the micrographs, mechanical results, and the calculated parameters from Rietveld treatments, the correlation between the mechanical behavior and the dislocation level is evaluated. This seems to constitute an important and complementary tool to prescribe and correlate the mechanical responses with applied heat treatments. Thus, a more complex and expensive microstructural characterization technique is not required.

The experimental DSC melting results show the cooling curves as a function of the thermodynamical transformation temperatures of the A356 alloy samples before (as-cast) and after aging (adopting T6 at 220 °C), as depicted in Figure 8. A first endothermic peak is associated with an inflexion and plateau characterized at ~620 °C, as shown in Figure 8a. This is correlated with α-Al dendritic matrix transformation, and from the Al-Si phase diagram, it is designated as liquid transformation (T_L) [10]. Dang et al. [10] also obtained this peak at 620 °C. Figure 8b shows a second one, which represents an endothermic peak occurring at ~590 °C. This corresponds with the Al-Si binary eutectic reaction, as reported in [10].

Dang et al. [10] also verified this transformation at a similar temperature range (i.e., ~580 °C). Considering a thermodynamic equilibrium condition, this transformation corresponds to the eutectic transformation temperature (T_E) occurring at ~570 °C [10]. Considering the solidification of an A356 alloy, depending on the Si and Mg contents, another reaction can occur, i.e., one associated with Al + Si + Mg_2Si ternary eutectic at 563 °C, as also demonstrated by Dang et al. [10]. As expected, after T6 treatment, the ternary eutectic transformation trends to be substantially dismissed or eliminated. This is suggested when DSC curves corresponding with the aged samples are observed.

Also, from the metallurgical point of view, depending on the Si and Mg contents, the Mg-containing phases during solidification are the Mg_2Si and π-Al_8Mg_3FeSi phases, [10,21,24,38–40]. Additionally, Mg_2Si particles tend to become dissolute and homogenized in a short period when heat treatment is carried out [38]. Based on the DSC results associated with the dislocation density and the microhardness and mechanical strength results, the as-cast sample is the sample exhibiting more remaining Mg_2Si particles, as shown in Figure 9c. This is also clearly confirmed in DSC curves.

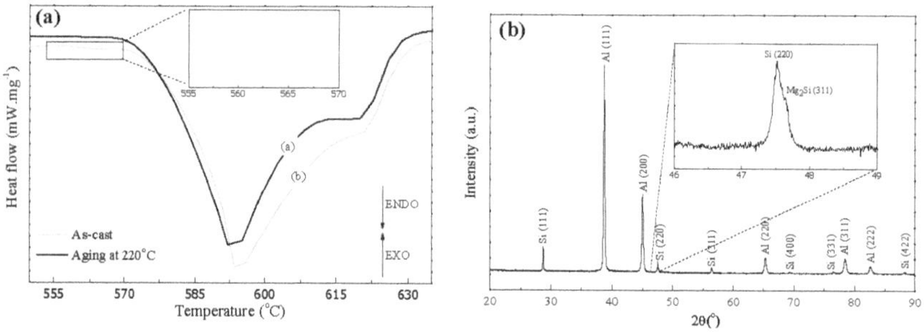

Figure 8. (**a**) Experimental DSC curves obtained by using 10 °C/min of the A356 alloy samples: (**a**) quenched at 30 °C and artificially aged at 220 °C and (**b**) the as-cast condition; (**b**) A typical XRD pattern corresponding with the as-cast A356 alloy sample, evidencing (inset) remaining Mg_2Si particles.

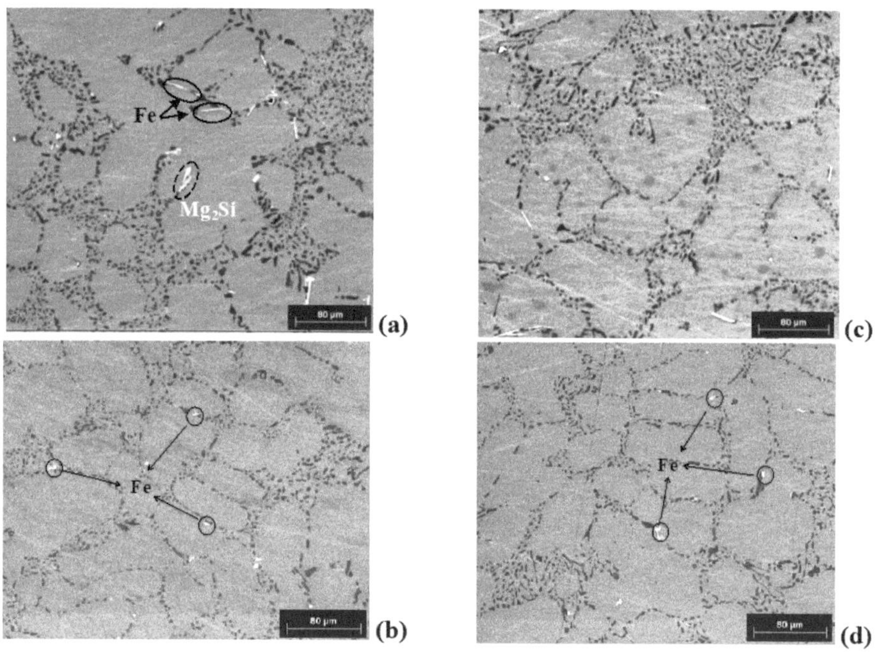

Figure 9. Typical SEM micrographs of the A356 alloy samples of the as-cast samples (**a**), and quenched at (**b**) 30 °C, (**c**) 60 °C, and (**d**) 80 °C, evidencing Fe-rich particles and Mg_2Si particles (as-cast sample).

Considering the aged sample, the eutectic ternary reaction is almost omitted or eliminated, as depicted in the obtained DSC curves. This is provided by a solution treatment and possible thermodynamic transformation of these Mg-rich phases in β-Al_5FeSi, mainly associated with more dilute Mg content (i.e., 0.3 wt.%), as reported by Chen et al [38]. Although the results of DSC and XRD have not substantially confirmed Mg_2Si particles, the dislocation density results reveal a high level, which permits us to understand the reason for the increasing microhardness. This seems to be associated with fine θ phases, which are nucleated at the dislocation sites.

3.3. Microstructural Array and Fractograph Characterizations

Figure 9a–d show typical microstructural arrays of the examined A356 alloy in the as-cast condition and after quenching at three distinctive temperatures, i.e., 30, 60, and 80 °C, respectively. It is known that Al-Si-Mg alloys are constituted by a lamellar Si morphology and primary α-Al phase constituting the dendritic Al-rich matrix. It is also known that Al-Sr master alloy has effective action to spheroidize Si particles. Since all examined samples contain portions of the mentioned Al-Sr master alloy, similar spheroidization levels are observed. This spheroidized Si morphology is associated with a finer eutectic microstructure and seems to be responsible for improving the resulting ductility of the casting, as previously reported [11,19,25–36]. Moreover, similar dendrite arm spacings (λ) are also characterized, i.e., λ = 38 ($\pm$6) μm. As expected, throughout the lattice of all examined samples, Fe-rich phases (commonly designated as β-Al_5FeSi) are homogeneously distributed. These resulting microstructural arrays are confirmed and are associated with the results previously discussed concerning DSC and XRD. Based only on these microstructural characterizations, it is not possible to conclude that a correlation exists between the results of the distinct quenching temperatures with the obtained mechanical behavior.

Although it is recognized that a finer dendrite spacing increases the mechanical response, this parameter is very similar for all quenched samples. The Si spheroidization is also very similar among the three quenched samples, including the as-cast sample. This differentiates only with Mg_2Si particles constituting its resulting microstructural morphology. These observations reasonably corroborate with the mechanical behavior reached, mainly when low mechanical strengths are compared. However, among the quenched samples, the microstructural array characterizations do not substantially corroborate to explain the distinctive results of the UTS, YS, and microhardness.

As previously discussed and reported [38], quenching induces high thermal stress at the Al/Si interface, which generates a higher dislocation density (~7.2×10^{15} m^{-2}) than the as-cast sample (6.2×10^{15} m^{-2}), as shown in Table 2. Additionally, due to the same thermal mechanism, when quenching is carried out at a slightly higher temperature, such as at 60 and 80 °C, it is expected that a lower dislocation density is obtained. This helps us understand the decreasing trends of UTS, YS, and microhardness including the ductility behavior corresponding with the samples quenched at high temperatures.

The typical microstructural array of the as-cast sample is similar to the quenched samples. However, differences concerning the Mg_2Si particles are observed, and it induces a distinct interpretation of their portions or volumes contributing to the decrease in mechanical strength. On the other hand, by analyzing the XRD and dislocation density results, the differences in the mechanical responses between the as-cast and the quenched samples are reasonably understood. The experimental XRD patterns characterize peaks corresponding with Mg_2Si particles into the as-cast lattice, which are dissolute and homogenized into the quenched lattice samples due to solution treatment and/or associated with subsequent aging, as previously demonstrated and discussed.

Figure 10a depicts a typical fractured surface and the corresponding SEM micrographs of the examined A356 samples after distinct quenching and aging treatments. An analysis of the physical features of fracture surfaces indicates the possible failure reason. With this, a correlation and understanding between these microstructural characteristics, fractured surface, and adopted heat treating can be provided.

Figure 10b,c correspond with the examined fractures of the quenched A356 alloy samples. The typical cleavage planes (white arrows) coexisting with typical dimples with certain tendon fossa sites (yellow dashed circles) are characterized. These characteristics have been previously reported when heat treatments of the A356 alloys are applied [50].

Additionally, the tearing edges (red dashed circles) are clearly characterized, interposing the dimple regions. These tearing edges suggest typical brittle fracture behavior [51].

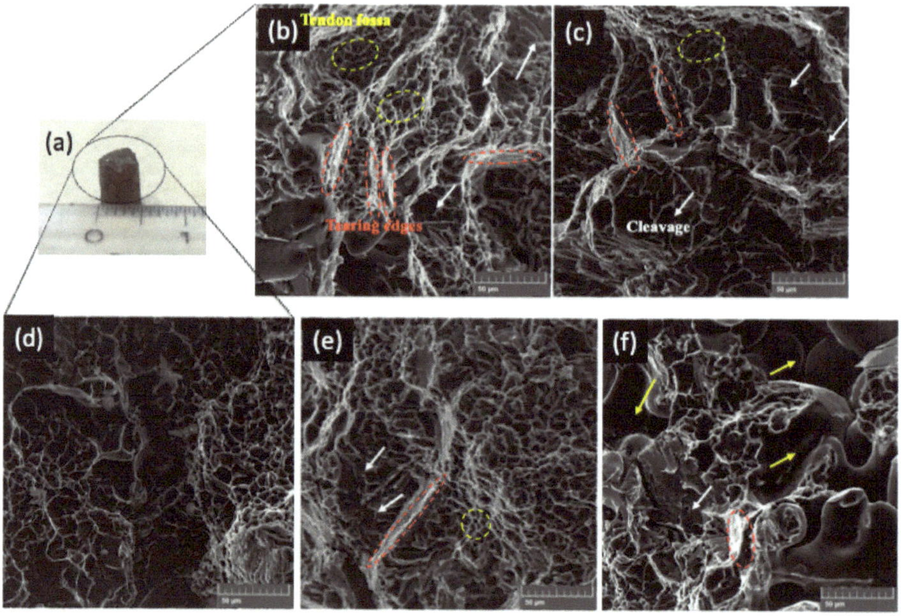

Figure 10. (**a**) A typical micrograph of the fractured surface of the specimens examined, SEM micrographs of the samples quenched at (**b**) 30 °C and (**c**) 60 °C; and the samples after aging at (**d**) 180 °C, (**e**) 200 °C, and (**f**) 220 °C. The white arrows indicate the cleavage planes and yellow and red dashed circles represent tendon fossa (dimples sites) [50] and tearing edges, respectively. Yellow arrows depict the constituted porosity/voids.

Figure 10d shows the fractured SEM micrograph of the aged sample at 180 °C. When this is compared with the micrographs shown in Figure 10e,f (after quenching), both the cleavage and dimples still coexist. However, it is perceived that the former fractography has both cleavage and tearing edge regions that decreased considerably when compared with the latter one. This observation corroborates with the obtained results of the UTS, YS, and dislocation density. Based on this, it is seen that aging decreases the brittle aspect, and UTS and YS are very similar, and no substantial decrease in the ductility has been verified. Additionally, the aged samples at 180 °C have their microhardness increased when compared with the quenched samples at 30 °C.

From both quenching and aging treatments, it can reasonably be concluded that the highest mechanical behavior is that of the sample treated at 30 °C (i.e., UTS of ~245 MPa and hardness of about 120 HV). Interestingly, the highest elongation (~6%) is also reached at this temperature. When samples are aged, the highest mechanical behavior is reached after aging at 180 °C, i.e., ~240 MPa, with elongation of about 4.5%. Interestingly, its hardness is about 50% lower than the aged sample. This means that if the highest UTS is desired, quenching and aging provide very similar mechanical behavior, based on the UTS results. However, if yield strength or deformation capacity is the crucial point of the design, then the sample quenched at 30 °C has YS ~ 16% higher than the sample aged at

180 °C. These assertions are corroborated by the calculated dislocation results. The samples aged at 180 °C achieved up to 10% more dislocations than those quenched at 30 °C. This is not characterized by using typical SEM micrographs. However, due to microstructural modifications at the nanoscale level, dislocation changes are calculated by using XRD patterns and then corroborated with the mechanical results observed.

With this, Al-based alloy wheel manufacturers can plan and control heating treatments, taking into account that curing coating treatments take up to 2 h at 160 and 180 °C to complete. Consequently, we found that a quasi-overaging of the wheels can be carried out, resulting in a deleterious effect on the mechanical response.

4. Conclusions

Based on the experimental results of the mechanical behavior and microstructural characterizations concerning the A356 alloy samples after different quenching and aging treatments, the following conclusions can be drawn:

- When the A356 alloy samples are quenched at 30, 60, and 80 °C, decreasing trends in both UTS and YS results are observed. Analog tendencies are also obtained when aging treatments at 180, 200, and 220 °C are applied. It was found that the UTS results corresponding with the aging at 180 °C and quenching at 30 °C are similar (i.e., ~240 MPa). Analog YS results are also similar in these same treatment temperature ranges, i.e., ~190 MPa.
- When comparing the experimental correlations between UTS and YS with microhardness with those previously reported (e.g., Krishna et al. [43] and Cerri and Ghio [42]), a good exponential adjustment prescribes the experimental data. On the other hand, when linear equations proposed by Abdulwahab et al. [40] and Rometsch and Schaffer [39] are utilized, no good adjustments are obtained.
- It was found that the calculated microstrains, crystallite sizes, and dislocation densities from the Rietveld treatments corroborate our understanding of the distinctive mechanical behavior of the A356 alloy samples subjected to different aging and quenching treatments. We found that the dislocations are responsible for similar UTS and YS values when compared with those samples quenched at 30 °C and aged at 180 °C. It was also found that by using the Rietveld treatment, the dislocation densities are calculated and correlation with mechanical responses can be constituted.
- It is important to remember that industrial practices commonly use artificial aging after a quenching procedure. However, for automotive wheels, it is noted that a complementary powder coating wheel treatment is often applied. This can promote a certain curing of coating, which prevents corrosion damage and provides a shiny aesthetic aspect. Although certain industrial secrets are required, this abovementioned curing coating treatment takes up to 2 h at 160 and 180 °C to complete. In this particular context, it seems that it is thermodynamically sufficient for a "preaging" or "overaging" to be conducted. We found that a "quasi-overaging" of the wheels results in a deleterious effect on mechanical behavior. This is attributed to "overaging" when associated with T6. With this, operational practices can be planned in order to obtain the required mechanical properties, which is particularly useful for manufacturers of A356 alloy wheels.

Author Contributions: Methodology, A.L.T., A.D.B. and G.S.P.; Formal analysis, A.L.T.; Investigation, A.L.T. and G.S.P.; Resources, W.R.O.; Data curation, W.R.O., A.D.B. and G.S.P.; Writing—original draft, W.R.O., A.D.B. and G.S.P. All authors have read and agreed to the published version of the manuscript.

Funding: The financial support provided by FAEPEX-UNICAMP (#2252/23), CAPES (Coordination for the Improvement of Higher Education Personnel), Ministry of Education, Brazil, Grant #1), and CNPq (The Brazilian Research Council) Grants, #407595/2022-8; #313272/2021-2, 305207/2023-7, and 310010/2020-9.

Data Availability Statement: The original contributions presented in the study are included in the article, further inquiries can be directed to the corresponding author.

Acknowledgments: Acknowledgments are also provided to Luiz Antonio Garcia (technician department) who contributed with technical aspects and equipment organization.

Conflicts of Interest: The authors declare no conflicts of interest.

References

1. Abdelaziz, M.H.; Samuel, A.M.; Doty, H.W.; Samuel, F.H. Effect of morphological changes of eutectic Si particles on the ambient and high temperature tensile properties of Zr containing Al–Si alloys. *J. Mater. Res. Tecnol.* **2020**, *9*, 5962–5981. [CrossRef]
2. Zhang, H.; Yu, S.; Yang, Z.; Zhang, C. The influence of porosity and precipitates on the corrosion behavior of A356 aluminum alloy. *J. Electroanal. Chem.* **2023**, *948*, 117796. [CrossRef]
3. Ortiz, E.L.; Osório, W.R.; Bortolozo, A.D.; Padilha, G.S. Alternative liquid-assisted sintering of Al/Cu composites using selected powders of As-Cast Al-Zn alloy. *Metals* **2022**, *12*, 962. [CrossRef]
4. Gu, G.; Li, R.; Xiang, L.; Zheng, H.; Lv, Y. Semi-solid Die Casting of Some Aluminum Alloys for Lightweight Automotive Components. In Proceedings of the International Conference on the Technology of Plasticity, Mandelieu-La Napoule, France, 24–29 September 2023; pp. 249–257.
5. Berlanga-Labari, C.; Biezma-Moraleda, M.V.; Rivero, P.J. Corrosion of cast aluminum alloys: A review. *Metals* **2020**, *10*, 1384. [CrossRef]
6. Samuel, E.; Samuel, A.M.; Songmene, V.; Samuel, F.H. A review on the analysis of thermal and thermodynamic aspects of grain refinement of aluminum-silicon-based alloys. *Materials* **2023**, *16*, 5639. [CrossRef] [PubMed]
7. Caceres, C.H.; Griffiths, J.R. Damage by the cracking of silicon particles in an Al-7Si-0.4Mg casting alloy. *Acta Mater.* **1996**, *44*, 25–33. [CrossRef]
8. Samuel, E.; Tahiri, H.; Samuel, A.M.; Songmene, V.; Samuel, F.H. A review on fundamentals of grain refining of Al-Si cast alloys. *Recent Adv. Alum. Alloys* **2024**. [CrossRef]
9. Chen, Z.; Wang, T.; Gao, L.; Fu, H.; Li, T. Grain refinement and tensile properties improvement of aluminum foundry alloys by inoculation with Al–B master alloy. *Mater. Sci. Eng. A* **2012**, *553*, 32–36. [CrossRef]
10. Dang, B.; Liu, C.-C.; Liu, F.; Liu, Y.-Z.; Li, Y.-B. Effect of as-solidified microstructure on subsequent solution-treatment process for A356 Al alloy. *Trans. Nonferrous Met. Soc. China* **2016**, *26*, 634–642. [CrossRef]
11. Samuel, A.M.; Samuel, E.; Songmene, V.; Samuel, F.H. A comparative study of grain refining of Al-(7–17%) Si cast alloys using Al-10% Ti and Al-4% B master alloys. *Materials* **2023**, *16*, 2867. [CrossRef]
12. Easton, M.A.; Prasad, A.; St John, D.H. The grain refinement of Al-Si alloys and the cause of Si poisoning: Insights revealed by the interdependence model. *Mater. Sci. Forum* **2014**, *6*, 161–166. [CrossRef]
13. Elias, A.L.P.; Koizumi, M.S.; Ortiz, E.L.; Rodrigues, J.F.Q.; Bortolozo, A.D.; Osório, W.R.; Padilha, G.S. Corrosion behavior of an Al-Si casting and a sintered Al/Si composite immersed into biodiesel and blends. *Fuel Process. Technol.* **2020**, *202*, 106360. [CrossRef]
14. El Haj, B.A.; Hamadellah, A.; Bouayad, A.; El Akili, C. Review of grain refinement performance of aluminium cast alloys. *Metall. Mater. Eng.* **2023**, *29*, 1–15. [CrossRef]
15. Jiang, D.; Yu, J. Simultaneous refinement and modification of the eutectic Si in hypoeutectic Al–Si alloys achieved via the addition of SiC nanoparticles. *J. Mater. Res. Technol.* **2019**, *8*, 2930–2943. [CrossRef]
16. Jin, B.-R.; Ha, D.-W.; Jeong, C.-Y. Effect of solution treatment on the hardness and tensile properties of Al–Mg–Si alloys for automotive chassis. *Mater. Trans.* **2019**, *60*, 815–823. [CrossRef]
17. Liu, G.; Gao, J.; Che, C.; Lu, Z.; Yi, W.; Zhang, L. Optimization of casting means and heat treatment routines for improving mechanical and corrosion resistance properties of A356-0.54Sc casting alloy. *Mater. Today Commun.* **2020**, *24*, 101227. [CrossRef]
18. Wei, B.; Li, S.; Jiang, T.; Zhang, Y.; Xu, G.; Li, Y.; Wang, Z. Optimization of Si content to inhibit inhomogeneous deformation in Al-Mg-Si alloy fabricated via twin-roll casting. *Metals* **2022**, *12*, 941. [CrossRef]
19. Xia, G.; Zhao, Q.; Ping, X.; Zhang, Y.; Yu, Q.; Li, Z.; Cai, Q. Effect of Al–5Ti–1B–La intermediate alloy on microstructure and mechanical properties of A356 aluminum alloy. *J. Mater. Res. Technol.* **2024**, *7*, 036535.
20. De Oliveira Santos, P.G.B.; Gomes, L.F.; Spinelli, J.E. Solidification and strength behavior of A356 Al alloy wheels. *Int. J. Met.* **2024**, 1–19. [CrossRef]
21. Miller, A.E.; Maijer, D.M. Investigation of erosive-corrosive wear in the low pressure die casting of aluminum A356. *Mater. Sci. Eng. A* **2006**, *435–436*, 100–111. [CrossRef]
22. Ozturk, F.; Sisman, A.; Toros, S.; Kilic, S.; Picu, R.C. Influence of aging treatment on mechanical properties of 6061 aluminum alloy. *Mater. Des.* **2010**, *31*, 972–975. [CrossRef]
23. Yang, C.-L.; Li, Y.-B.; Dang, B.; Lu, H.-B.; Liu, F. Effects of cooling rate on solution heat treatment of as-cast A356 alloy. *Trans. Nonferrous Met. Soc. China* **2015**, *25*, 3189–3196. [CrossRef]
24. Zhang, X.L.; Liu, J.H.; Li, W.S.; Zhou, H.T.; Wang, S.C. Effect of Sr–Ce modification on the microstructure and mechanical properties of A356 alloy. *Int. J. Met.* **2023**, *17*, 1008–1020. [CrossRef]

25. Roy, M.J.; Maijer, D.M. Response of A356 to warm rotary forming and subsequent T6 heat treatment. *Mater. Sci. Eng. A* **2014**, *611*, 223–233. [CrossRef]
26. Shabestari, S.G.; Shahri, F. Influence of modification, solidification conditions and heat treatment on the microstructure and mechanical properties of A356 aluminum alloy. *J. Mater. Sci.* **2004**, *39*, 2023–2032. [CrossRef]
27. Sjölander, E.; Seifeddine, S. The heat treatment of Al-Si-CuMg casting alloys. *J. Mater. Process. Technol.* **2010**, *210*, 1249–1259. [CrossRef]
28. Xu, C.L.; Wang, H.Y.; Qiu, F.; Yang, Y.F.; Jiang, Q.C. Cooling rate and microstructure of rapidly solidified Al-20 wt. % Si alloy. *Mater. Sci. Eng. A* **2006**, *417*, 275–280. [CrossRef]
29. Chen, R.; Shi, Y.-F.; Xu, Q.-Y.; Liu, B.-C. Effect of cooling rate and microstructure of Al-7Si-0.3Mg-0.15Fe alloys. *Trans. Nonferrous Met. Soc. China* **2014**, *24*, 1645–1652. [CrossRef]
30. Osório, W.R.; Cheung, N.; Spinelli, J.E.; Goulart, P.R.; Garcia, A. The effects of a eutectic modifier on microstructure and surface corrosion behavior of Al-Si hypoeutectic alloys. *J. Solid State Electrochem.* **2007**, *11*, 1421–1427. [CrossRef]
31. Osório, W.R.; Cheung, N.; Spinelli, J.E.; Cruz, K.S.; Garcia, A. Microstructural modification by laser surface remelting and its effect on the corrosion resistance of an Al–9 wt% Si casting alloy. *Appl. Surf. Sci.* **2008**, *254*, 2763–2770. [CrossRef]
32. Apelian, D.; Shivkumar, S.; Sigworth, G. Fundamental aspects of heat treatment of cast Al-Si-Mg alloys. *AFS Trans.* **1989**, *97*, 727–742.
33. Mahan, H.M.; Konovalov, S.V.; Panchenko, I. Effect of heat treatment on the mechanical properties of the aluminium alloys AA2024 with nanoparticles. *Int. J. Appl. Sci. Eng.* **2023**, *20*, 2022324. [CrossRef]
34. Wang, Q.G.; Cáceres, C.H. On the strain hardening behaviour of Al-Si-Mg casting alloys. *Mater. Sci. Eng. A* **1997**, *234–236*, 106–109. [CrossRef]
35. Tiryakioglu, M. The effect of solution treatment and artificial aging on the work hardening characteristics of a cast Al–7%Si–0.6%Mg alloy. *Mater. Sci. Eng. A* **2006**, *427*, 154–159. [CrossRef]
36. Campbell, J.; Tiryakioglu, M. Review of effect of P and Sr on modification and porosity development in Al–Si alloys. *Mater. Sci. Technol.* **2010**, *26*, 262–268. [CrossRef]
37. Dash, S.S.; Chen, D. A review on processing–microstructure–property relationships of Al-Si alloys: Recent advances in deformation behavior. *Metals* **2023**, *13*, 609. [CrossRef]
38. Chen, L.; Zhao, Y.; Wen, Z.; Tian, J.; Hou, H. Modelling and optimization for heat treatment of Al-Si-Mg alloy prepared by indirect squeeze casting based on response surface methodology. *Mater. Res.* **2017**, *20*, 1274–1281. [CrossRef]
39. Rometsch, P.A.; Schaffer, G.B. An age hardening model for Al–7Si–Mg casting alloys. *Mater. Sci. Eng. A* **2002**, *325*, 424–434. [CrossRef]
40. Abdulwahab, M.; Madugu, I.A.; Yaro, S.A.; Hassan, S.B.; Popoola, A.P.I. Effects of multiple-step thermal ageing treatment on the hardness characteristics of A356.0-type Al-Si-Mg alloy. *Mater. Des.* **2011**, *32*, 1159–1166. [CrossRef]
41. Sekhar, A.P.; Nandy, S.; Ray, K.K.; Das, D. Hardness- yield strength realtion of Al-Mg-Si alloys. *IOP Conf. Ser. Mater. Sci. Eng.* **2018**, *338*, 012011. [CrossRef]
42. Cerri, E.; Ghio, E. AlSi10Mg alloy produced by selective laser melting: Relationships between Vickers microhardness, Rockwell hardness and mechanical properties. *Metall. Ital.* **2020**, *7–8*, 5–17.
43. Krishna, S.C.; Gangwar, N.K.; Jha, A.K.; Pant, B. On the prediction of strength from hardness for copper alloys. *J. Mater.* **2013**, *9–10*, 1–6. [CrossRef]
44. Ungár, T. Dislocation densities, arrangements and character from X-ray diffraction experiments. *Mater. Sci. Eng. A* **2001**, *309–310*, 14–22. [CrossRef]
45. Ungár, T.; Ott, S.; Sanders, P.G.; Borbély, A.; Weertman, J.R. Dislocations, grain size and planar faults in nanostructured copper determined by high resolution X-ray diffraction and a new procedure of peak profile analysis. *Acta Mater.* **1998**, *46*, 3693–3699. [CrossRef]
46. Sutapa, I.W.; Wahab, A.W.; Taba, P.; Nafie, N.L. Dislocation, crystallite size distribution and lattice strain og magnesium oxide nanoparticles. *J. Phys. Conf. Ser.* **2018**, *979*, 01221. [CrossRef]
47. Wang, C.; Wang, Z.; Xu, H.; Zhang, G. Decreased dislocation density as an origin for the quench sensitivity of the Al-Si-Mg alloys with high Si content. *J. Alloys Compd.* **2022**, *910*, 165011. [CrossRef]
48. Muiruri, A.; Maring, M.; Preez, W. Evaluation of dislocation densities in various microstructures of additively manufactured Ti6Al4V (eli) by method of X-ray diffraction. *Materials* **2022**, *13*, 5355. [CrossRef]
49. Kavosi, J.; Saei, M.; Kazeminezhad, M.; Dodangeh, A. Modeling of dislocation density and strength on rheoforged A356 alloy during multi-directional forging. *Comput. Mater. Sci.* **2014**, *81*, 284–289. [CrossRef]
50. Teng, D.; Zhang, G.; Zhang, S.; Li, J.; Guan, R. Mechanical properties of refined A356 alloy in response to continuous rheological extruded Al-5Ti-0.6C-1.0Ce alloy prepared at different temperatures. *Metals* **2023**, *13*, 1344. [CrossRef]
51. Dewangan, S.; Selvaraj, S.K.; Karthikeyan, B.; Chadha, U.; Singhal, P.; Sarma, P.P.; Raju, P.V.; Kumar, U.; Kumar, M.P. Fractography analysis into low-C steel undergone through various destructive mechanical tests. *Mater. Res. Express* **2022**, *9*, 126502. [CrossRef]

metals

Article

In Situ Pure Shear Tests on Textured Magnesium AZ31B Sheets

Lawrence Whitmore [1,*], Anton Nischler [2], Holger Saage [2] and Otto Huber [2]

[1] Department of Chemistry and Physics of Materials, University of Salzburg, 5020 Salzburg, Austria
[2] Competence Centre for Lightweight Design (LLK), University of Applied Sciences Landshut, Am Lurzenhof 1, 84036 Landshut, Germany; anton.nischler@haw-landshut.de (A.N.); holger.saage@haw-landshut.de (H.S.); otto.huber@haw-landshut.de (O.H.)
* Correspondence: lawrence.whitmore@plus.ac.at; Tel.: +43-680-2260446

Abstract: Pure shear tests of textured magnesium AZ31B sheet samples were carried out using a 5 kN Kammrath & Weiss in situ tension-compression stage in a scanning electron microscope in combination with real-time electron backscatter diffraction lattice orientation mapping. The sample design was optimized to produce a pure shear stress in the central gauge zone. Distributions of the deformation twins were correlated with finite element simulations using a linear-elastic constitutive law considering large deformations to show that twins form in areas where the principal compressive stress σ_3^* is a maximum and that they form normal to the trajectories of that minor principal stress. Mappings of the same area at different load values revealed the formation and growth of individual twins and their relationship to the internal elastic strain of individual grains as indicated by the internal grain disorientation. All twins observed were of the extension type, with an 86.3° disorientation with respect to the parent grains. A more detailed study was conducted using transmission electron microscopy to correlate with the EBSD observations and to further elucidate the twin structures within samples.

Keywords: in situ testing; pure shear; electron backscatter diffraction; magnesium AZ31B; deformation twinning; twin growth; bands of twinned grains

Citation: Whitmore, L.; Nischler, A.; Saage, H.; Huber, O. In Situ Pure Shear Tests on Textured Magnesium AZ31B Sheets. *Metals* **2024**, *14*, 404. https://doi.org/10.3390/met14040404

Academic Editors: Jingxiang Xu, Zhenhua Chu and Xingwei Zheng

Received: 27 February 2024
Revised: 19 March 2024
Accepted: 20 March 2024
Published: 29 March 2024

1. Introduction

Magnesium alloys have significant value as lightweight engineering materials [1,2], with magnesium recently being heralded as the metal of the future [3,4]. Their high specific strength, low density (1.74 g·cm^{-3} compared with aluminium (2.7 g·cm^{-3})), high stiffness, ductility, and good machinability give them a wide range of applications, especially in the automotive and aerospace industries [5,6]. With a melting point of 650 °C, magnesium forms a polycrystalline metal with a hexagonal crystal structure that has a c/a ratio of 1.62 [7,8]. Through alloying, thermal treatment, and strain hardening, the mechanical properties of these alloys can be optimized for specific applications [9–11]. However, in extruded or sheet form, their strong basal texture, in which a majority of grains are aligned with the hexagonal c-axis normal to the sheet, gives them strongly anisotropic yield behaviour and asymmetric yield stresses [12], which complicates their machinability and practical use [13,14].

Figure 1 shows the basic unit cell of magnesium and their planes for dislocation slip and twinning [15]. Because of the hexagonal crystal system, basal slip and extension twinning are the main modes of plastic deformation in magnesium at room temperature, both with a critical resolved shear stress ($CRSS_{RT}$) of approximately 0.5 MPa [16–21]. At higher temperatures, prismatic and pyramidal slip are possible with $CRSS_{RT}$s of approximately 40 and 60 MPa, respectively, as well as contraction twinning with a $CRSS_{RT}$ of approximately 200 MPa [16,22]. Extension twinning extends the lattice in the c-direction by reorientation across the [10–12] planes [23], while basal slip extends the lattice parallel to the basal plane. The basal texture in an AZ31B sheet has a large influence on the activation of deformation modes during mechanical testing, and it is reported that prismatic slip and extension twinning can activate preferentially to basal slip according to the direction of the applied force [24].

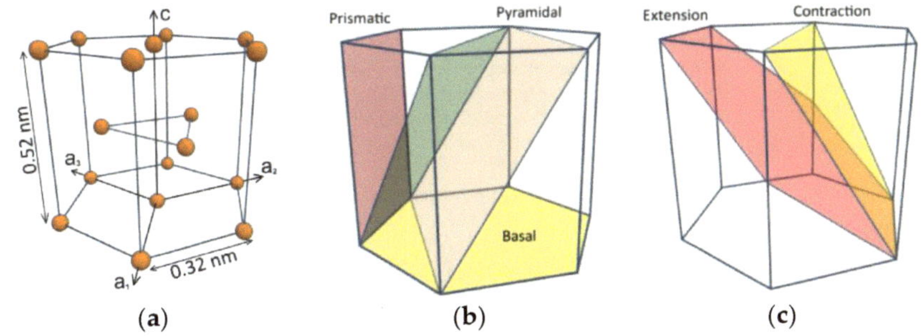

Figure 1. The hexagonal unit cell of magnesium (**a**) with atom sites, axes, and dimensions indicated, (**b**) the slip planes for basal, prismatic, and pyramidal slip, and (**c**) the twinning planes for extension and contraction twinning.

In this study, the plastic deformation mechanisms of twin-roll cast AZ31B alloy sheets (weight percentage composition [25]: 2.75% Al, 1.08% Zn, 0.368% Mn, 0.0262% Cu, 0.0187% Si, 0.00282% Fe, 0.00038% Ni, and 0.00041% Ca) are investigated when subjected to pure shear stress [26,27]. By adapting a recently developed sample geometry [28] and using an in situ mechanical tester, the samples' behavior was studied inside a scanning electron microscope (SEM). Real-time electron backscatter diffraction (EBSD) lattice orientation mapping [29,30] was used to reveal the initiation and growth of deformation structures within the samples when they were subjected to incrementally increasing pure shear stress.

In situ mechanical testing was initially developed in the late 1960s [31] and, since that time, has undergone significant further development due to its high value for investigating material behavior under controlled conditions of stress combined with real-time imaging at the micron and sub-micron scales. The initiation and growth of deformation structures can be studied with increasing or decreasing stress states or cyclic stress to examine how plastic strain appears in the microstructure. This knowledge is especially relevant to the development of materials for engineering applications, as well as the design of individual stress-bearing components, because the macroscopic properties of bulk materials are derived from their behavior at the microscopic scale. In combination with EBSD mapping, changes in the lattice orientation can be studied in situ in relation to specific applied stress states, which further elucidates the localized distribution of elastic and plastic strain that accompanies and precedes macroscopic plastic flow.

Shear testing is a form of mechanical stress test that applies offset opposing forces across a volume of material rather than directly opposing forces, which are otherwise used for uniaxial compression and tension tests. The shear test emulates real-world conditions that involve shearing forces such as those encountered through friction when components are displaced relative to each other while maintaining mechanical contact. Also, in a sheet loaded with in-plane shear forces or in a torsionally loaded plate, shear stresses occur within the bulk of the material. Knowledge of how a material behaves under shear stress can then be used to design components to withstand these stresses in real world applications. In the case of sheet materials, shear tests can be made parallel or inclined to the sheet plane with appropriate sample and experiment design. Recent studies have shown the extent to which shear studies have been conducted on magnesium alloys [32,33], while developments in sample design [28] permit the application of pure shear stress to an optimized gauge area as a result of applied tensile force.

Earlier studies of magnesium AZ31B subjected to uniaxial compressive stress observed plastic deformation in the form of bands of twinned grains (BTG) [25,34]. These bands formed normal to the maximal principal compressive stress σ_3^* and individual twins inside the bands were aligned approximately parallel to those bands. Recent in situ studies of magnesium subjected to uniaxial compressive stress confirmed this and investigated in more detail the initiation and growth of individual twins [15] under this stress state. The current study applies this approach of in situ testing with pure shear stress, an area that has not yet been explored in the literature. The goal of this study is to investigate the nature of plastic flow at the microscopic level and to observe its evolution with increasing stress.

2. Materials and Methods

Textured AZ31B magnesium sheets were obtained from Magnesium Flachprodukte GmbH in Freiberg [35] and test samples were milled from these sheets using a PFE1000-PX CNC milling machine from BZT Maschinenbau GmbH (Leopoldshöhe, Germany) mill fitted with a 202515-2 carbide cutter from Garant Productions GmbH (Hoffmann Group, Munich, Germany) operated at 5000 min^{-1} with a cutting speed of 100 mm s^{-1} and a 10 µm depth of cut and cooled using compressed air. Sample geometry was adapted from the studies of [29] and surface roughness after milling was nominally Ra = 0.6 µm. Mechanical tests were carried out in a Zeiss Merlin SEM with an accelerating voltage of 20 kV. A 5 kN in situ tension-compression stage from Kammrath & Weiss GmbH (Schwerte, Germany) (referred to herein as the in situ stage) was used to incrementally increase the shear stress on the sample.

Figure 2 shows details of the pure shear test sample. The experiment used the in situ stage to apply a tensile force (F_1) to the sample in the x_1 direction. Dimensions of the sample were chosen so that it would fit into the in situ stage and have the same length (60 mm) and thickness (3 mm) as the earlier uniaxial compression samples [15]. The geometry, based on [28], was optimized with the help of numerical simulations to produce an almost pure shear stress state in the area between the 2 notches. This is shown using the triaxiality factor $\eta = \sigma_m/\sigma_{v.Mises}$ in Figure 3a. Samples were finely ground and polished on one side before in situ testing so that the analysed surface was suitable for imaging and EBSD mapping [36,37]. Figure 2c shows the 3D-printed (3DP) plastic sample holder designed to fit a RotoPol 11 polishing machine from Struers GmbH (Ballerup, Denmark). Using this holder, three samples could be ground and polished simultaneously with an even surface finish.

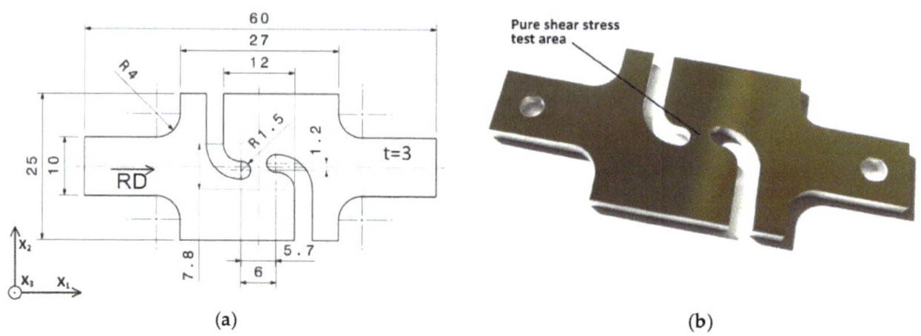

(a) (b)

Figure 2. *Cont.*

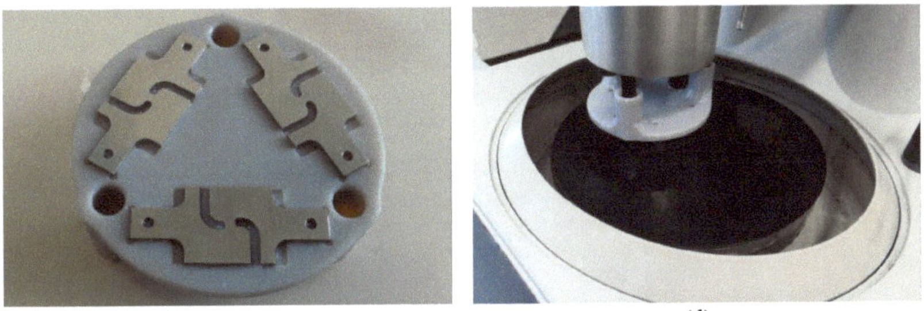

(c) (d)

Figure 2. The pure shear sample: (**a**) technical drawing with dimensions in mm showing the coordinate system (x_1, x_2, and x_3), (**b**) photograph indicating the central pure shear gauge zone, (**c**) sample mounted in a 3DP plastic holder for polishing, and (**d**) sample being polished on a neoprene cloth.

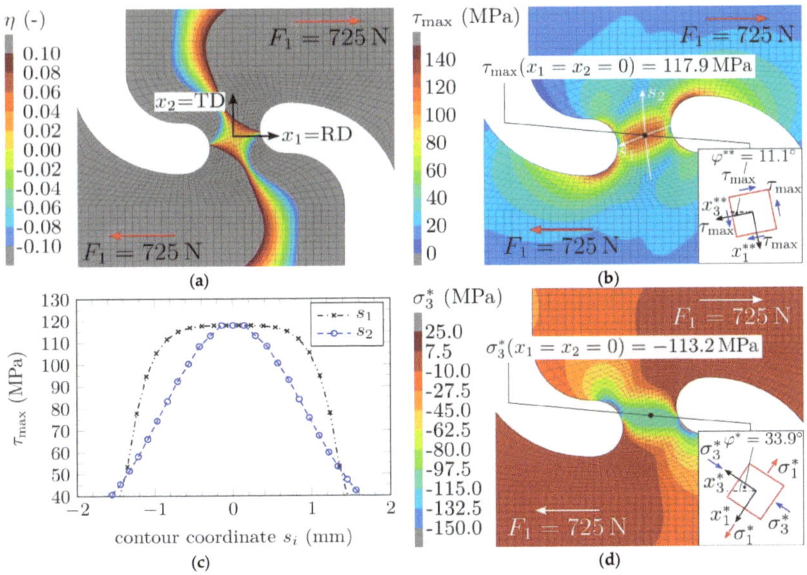

Figure 3. Finite element analysis of the pure shear sample under 725 N load showing (**a**) the scalar field $\eta(x_1, x_2)$ of the triaxiality factor in the range of $-0.1 \leq \eta \leq 0.1$ from the FEM simulation, (**b**) distribution of the maximum shear stress $\tau_{max}(x_1, x_2)$ from the FEM simulation, (**c**) plot of the maximum shear stress τ_{max} along the contour coordinates s_1 and s_2, and (**d**) distribution of the minor principal stress $\sigma_3^*(x_1, x_2)$ and the principal axis system, which is valid for the central gauge area.

Grinding was carried out using 2500- and 4000-grit silicon carbide papers, polishing using a colloidal suspension of 40 nm silicon dioxide powder solution in deionized water on a neoprene cloth, vibrational polishing using 30 nm aluminum oxide powder suspension on a neoprene cloth in a Buehler Vibromet, and cleaning in an ultrasonic bath. By using deionized water for each stage and absolute ethanol for final cleaning, it was possible to produce a surface without a hydroxide layer [38], which would otherwise form with water-containing solutions and diminish the EBSD signal which is derived from the upper 100 nm or so of the surface [39]. This high surface quality allowed the same area of samples to be studied and mapped multiple times; only the physical deformation of grains at the surface by the mechanical test eventually obstructed the electron signal reaching the CCD sensor of the EBSD detector.

The geometry of the pure shear sample, based on [28], was optimized with the help of numerical simulations to produce an almost pure shear stress state in the area between the 2 notches. Figure 3 shows the results of the finite element simulation of the pure shear sample based on a linear-elastic constitutive law considering large deformations. The dimensions of the shear sample were optimized by an isotropic elastic FEM (finite element method) simulation considering large deformations. A Young's modulus of 42,866.5 MPa and a Poisson's ratio of 0.32 [25] were used for the isotropic elastic material model. The geometry, shown in Figure 3a, was discretized using 94,500 hexahedral isoparametric elements (C3D20R) with quadratic shape functions and a reduced integration scheme. The boundary conditions specified on the left surface of the clamping area (x_1–x_3 surface) are such that the displacements u_1, u_2, and u_3 are set to zero. In addition, a homogenous distributed load of $F_1 = 725$ N was applied on the right-hand side of the right clamping surface.

The main purpose of using the FEM simulation was to optimize the geometry of the sample to achieve a homogeneous pure shear stress state in the gauge area. A defining characteristic of the pure shear stress state is that the major principal stress σ_1^* (positive value), the minor principal stress σ_3^* (negative value), and the maximum shear stress τ_{max} are equal in magnitude. It should be noted that the principal stresses and the principal axis system are marked with $\blacksquare^*$ in this study. In addition, the principal stresses are ordered according to the convention $\sigma_1^* \geq \sigma_2^* \geq \sigma_3^*$. It is well established that the x_1^{**}–x_3^{**} axis system, where the maximum shear stress τ_{max} occurs, is rotated by 45° relative to the principal x_1^*–x_3^* axis system. The corresponding values and the corresponding axis system are denoted as $\blacksquare^{**}$.

To verify whether a pure shear stress state has been achieved in the gauge region, the triaxiality factor η is used. The triaxiality factor is calculated using the formula

$$\eta = \frac{\sigma_{\text{m}}}{\sigma_{\text{v.Mises}}} = \frac{I_1}{3\sigma_{\text{v.Mises}}}, \quad (1)$$

where σ_{m} is the arithmetic mean of the normal stresses, $I_1 = \sigma_{11} + \sigma_{22} + \sigma_{33}$ the first stress invariant, and $\sigma_{\text{v.Mises}}$ the von Mises equivalent stress. For a pure shear stress state, $\eta = 0$. The triaxiality factor converts a multiaxial stress tensor into a simple scalar quantity, making it easy to evaluate the multiaxiality of the stress state. Figure 3a displays the plot of the scalar field $\eta(x_1, x_2)$ in the range of $-0.1 \leq \eta \leq 0.1$. The plot confirms the attainment of an almost perfect pure shear stress state within the central gauge area.

Figure 3b shows the distribution of the maximum shear stress $\tau_{\text{max}}(x_1, x_2)$ and the orientation of the corresponding x_1^{**}–x_3^{**} axis system. The axis system is rotated by the angle $\varphi^{**} = 11.1°$ in relation to the global x_1–x_2 axis system presented in Figure 3a. The shear stress should be the maximum and as homogeneous as possible in the central gauge area. To verify this, τ_{max} is plotted in Figure 3c along the paths s_1 and s_2, which are illustrated in Figure 3b. Figure 3c confirms that the maximum value of τ_{max} is attained at $x_1 = 0$ and $x_2 = 0$. Furthermore, τ_{max} remains constant along the paths s_1 and s_2 for a certain length. This is particularly valid for the path s_1, which verifies the presence of an acceptable homogeneous stress distribution in the central gauge zone.

In addition to the triaxiality factor η and the maximum shear stress τ_{max}, the principal stresses σ_1^* and σ_3^* and the orientation of the corresponding principal x_1^*–x_3^* axis system were analyzed for the central gauge area. This is of crucial importance, as one of the main objectives of this study is to investigate whether the twins align perpendicular to the minor principal stress σ_3^* as in the uniaxial experiments in [15], although a pure shear stress state prevails. Figure 3d shows the distribution of the minor principal stress σ_3^* for an external load of $F_1 = 725$ N. Additionally, Figure 3d shows the orientation of the principal axis system and the associated principal stresses. The major principal stress in the central gauge area is $\sigma_1^*(x_1 = x_2 = 0) = 123.0$ MPa, which is a tensile stress. Conversely, the minor principal stress of $\sigma_3^*(x_1 = x_2 = 0) = -113.2$ MPa is a compressive stress. Consequently, the criterion for the pure shear stress state $\sigma_1^* \approx |\sigma_3^*|$ is approximately valid with a deviation of 8.66%. The principal x_1^*–x_3^* axis system is rotated by the angle $\varphi^* = 33.9°$ with respect to the global x_1–x_2 axis system, introduced in Figure 3a. The alignment of the principal stress directions is crucial for comprehending the microscopic deformation mechanisms later on.

Figure 4 shows details of the experimental set-up. The in situ stage was attached to the microscope stage using an electrically conductive steel adapter provided by Kammrath & Weiss GmbH. The adapter plate tilted the sample by 50° to the horizontal, and when combined with the 20° tilt within the in situ stage, the required 70° tilt was achieved for EBSD measurement. The horizontal and vertical positions of the stage were precisely set in order to avoid contacting the SEM objective pole piece or the sensors when the in situ stage was inserted into the chamber. Figure 3b shows the polished sample fixed using hardened steel gripping plates in the in situ stage. Figure 4c illustrates the EBSD system, where electrons backscattered from atoms within approximately 100 nm of the sample surface are diffracted by the upper layers of the sample lattice, collected on the CCD sensor of an Oxford Instruments Nordlys detector, and processed in the AZtecHKL software (version 3.3 SP1) into phase and orientation of the scattering site. By scanning the electron beam raster-style across an area of the sample, a map of phase and lattice orientation, called the Inverse Pole Figure or IPF, was produced. These maps were recorded at different values of load and later processed using the ATEX analysis software (version 4.05) [40].

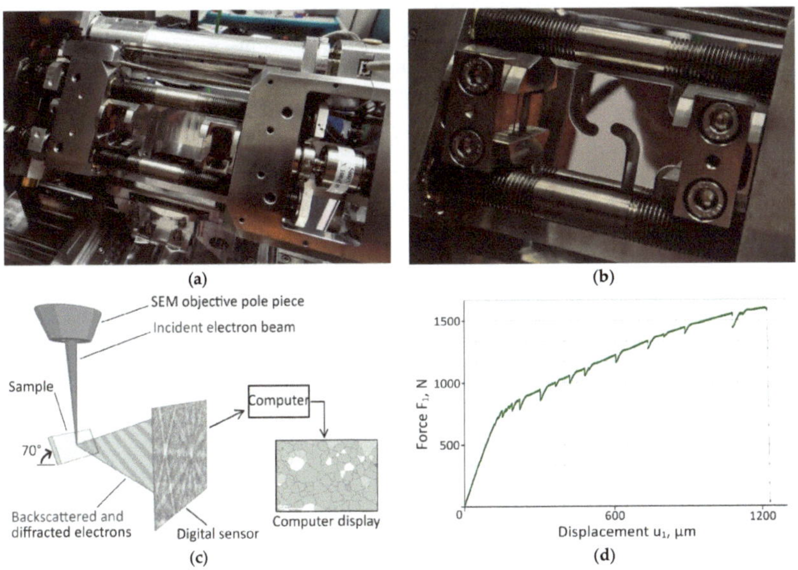

Figure 4. Experimental set-up showing: (**a**) the in situ stage mounted in the SEM, (**b**) the pure shear sample within the in situ stage, (**c**) a schematic of the EBSD measurement system, and (**d**) a typical experimental loading curve of force F_1 vs displacement u_1 with breakpoints for EBSD data acquisition.

Figure 4d shows a typical experimental load–displacement curve. As the stress and strain within the pure shear gauge zone could not be measured directly due to the complex geometry of the sample, the curve displays measured values of applied force F_1 vs. macroscopic displacement u_1 in the x_1 direction. Force was increased incrementally by driving the DC motor at a speed of 2 μm per second to specific pre-set breakpoints. At each breakpoint, power to the motor was held constant while SEM images and EBSD lattice orientation maps were collected. Maps covered a typical area of 100 × 80 μm^2, with a measurement spacing of 0.1 or 0.2 μm. This allows a compromise between resolution and collection time while including an area of some 100 grains. The clean surface produced a strong clear EBSD signal, which allowed for rapid processing times so that each map required around 1 h to collect and multiple mappings could be made of the same area. During collection times, system relaxation gave rise to a slight reduction in force; consequently, these measurement periods appear as inverted spikes in the loading curve.

Figure 5 shows the method used to prepare samples for investigation using TEM. A sample that had been loaded to 950 N was mounted on the table of a 3D-printed core drill and a 3 mm cylinder was removed from the central zone using a stainless-steel tube with 4.0 mm outer diameter and 3.25 mm inner diameter. The core drill, a simplified version of that developed earlier [41], was operated at 200 min^{-1} with a slurry of 15 μm diamond particle paste under a pressure of approximately 5 N which was applied using a 50 mm long 10 mm-diameter stainless-steel compression spring with 0.35 Nmm^{-1} spring constant. Approximately 5 h was required for the drill to grind through the 3 mm-thick sheet. By grinding the annular channel slightly off-center, a slight impression on the side remaining from the sample geometry made it possible to maintain knowledge of the orientation relative to the larger test sample during subsequent studies. After cleaning, the 3 mm cylinder was mounted on a brass rod and sliced into three discs, each approximately 250 μm thick, using a fine-grit diamond wafering blade in a Struers Minitom. The discs were then ground using a self-made 3D-printed manual grinding tool on a 12 μm alumina abrasive sheet on a glass plate to obtain an even surface finish and thickness of approximately 100 μm.

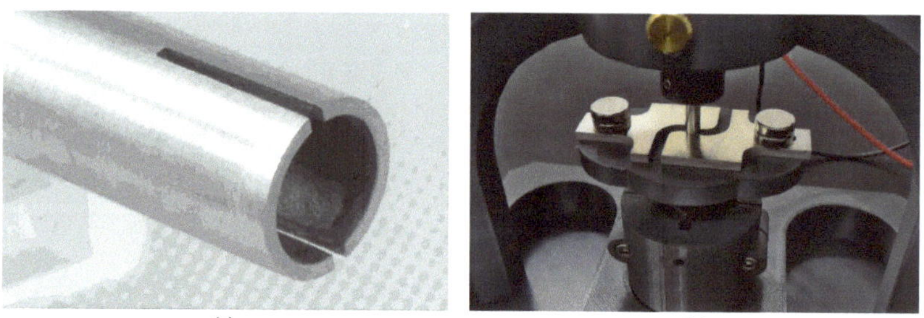

(a) (b)

Figure 5. *Cont.*

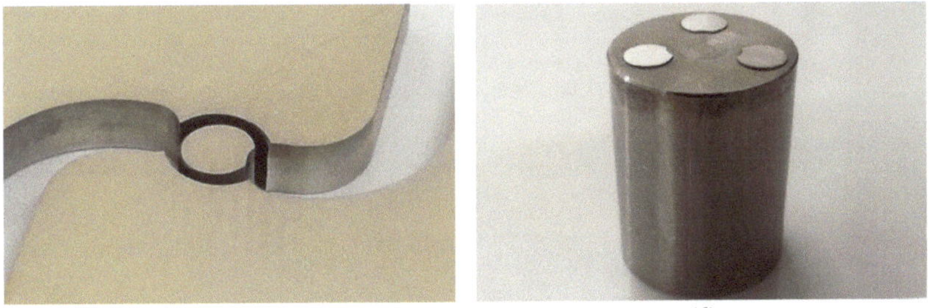

(c) (d)

Figure 5. Extraction of the 3 mm-diameter cylindrical sample from the central pure shear area using a 3D-printed core drill showing (**a**) the stainless-steel grinding tube, (**b**) the sample mounted on the spring-loaded holder with magnets, (**c**) the sample after 5 h drilling, and (**d**) three sections of the sample on a stainless-steel holder ground to 100 μm thickness.

Figure 6 shows the dimpling process that thins the central area of the round sample to <30 μm thickness before ion milling. The dimple grinder-polisher was a 3D-printed device [42] with a horizontally-mounted DC motor for the grinding/polishing disc and a vertically-mounted DC motor to rotate the sample. The sample was mounted on a 2 mm-thick 8 mm-diameter glass disc mounted in a 3DP plastic holder and rotated at 10 min^{-1}. A plastic grinding disc and 3 μm diamond abrasive in ethylene glycol required 20 min to form the initial dimple with a vertical force of approximately 60 gf or 0.6 N and rotation speed of approximately 100 min^{-1}. Polishing was carried out using a strip of flocked polishing cloth mounted on a 3D-printed plastic disc with 1 μm diamond in ethylene glycol and deionized water with the same load as for grinding. A small hydroxide layer would inevitably form on the surface, but by using deionized water the thickness of the layer was minimized, and it would be removed later by ion milling. The sample was dimple-ground and polished on both sides.

(a) (b)

Figure 6. Dimpling of samples for TEM using (**a**) a 3D-printed plastic grinding disc with a slurry of 3 μm diamond and glycol and (**b**) polished using a 3D-printed holder with a strip of polishing cloth fixed with silicone paste and a slurry of 1 μm diamond in glycol.

Figure 7a shows the 3 mm-diameter sample before dimple grinding and Figure 7b after dimple grinding: the dimple is well-centered and shows the expected fine uniform grinding lay of 3 μm diamond combined with a plastic grinding disc. The central area at this stage was 40 μm thick as measured using the calibrated depth control of the Zeiss Axio Imager II optical microscope. Figure 7c shows the dimple after polishing: the grinding lay have been removed and only very fine marks remain, typical of 1 μm diamond polishing.

At this stage, the central area was approximately 30 μm thick and any subsurface damage remaining would be expected to be less than 1 μm deep [43,44]; ion-beam milling would remove this damage as it removes 15 μm from both sides. Figure 7d shows the sample after ion-beam milling using a Gatan PIPS II. The sample holder was cooled using liquid nitrogen to −50 °C, beam energy was 6 keV until perforation by 2 keV for 5 min and 1 keV for one minute to remove amorphization, gun angles were 6° from above and 6° from below, and a rotation speed of 3 min^{-1} was used. With these parameters, perforation was achieved in approximately 25 min.

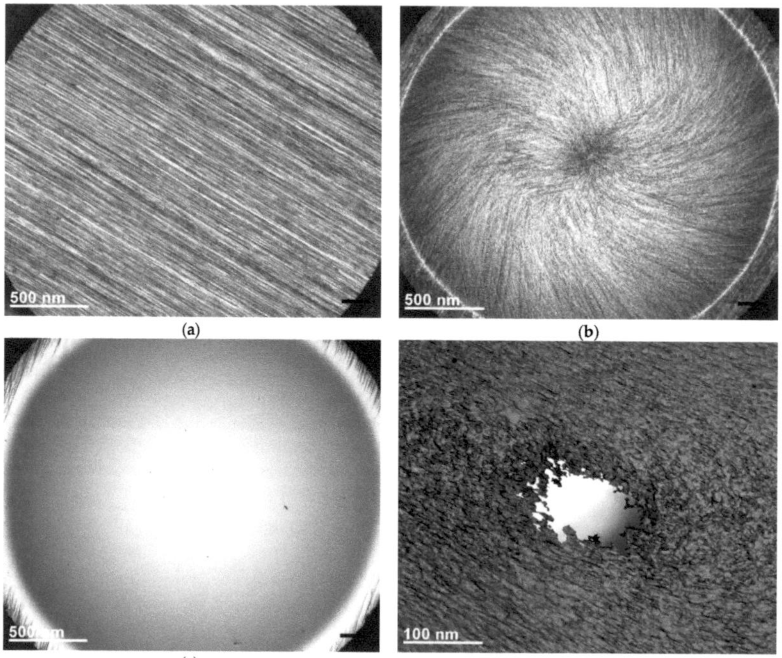

(a) (b) (c) (d)

Figure 7. The sample (**a**) after drilling, (**b**) after dimple grinding, (**c**) after dimple polishing showing very fine surface, and (**d**) after ion-beam milling showing the thinned central area.

3. Results and Discussion

Figure 8 shows SEM images of a sample at six different pure shear-stress states corresponding to applied force F_1 of: (a) 600 N, (b) 780 N, (c) 850 N, and (d) 1500 N. The dark curve in each image is the focal band resulting from the 70° tilt of the sample. Plastic deformation first appears in the central pure shear zone at around a 700 N, load forming an elliptical shape, and grows in size with an increasing load until it extends across the gauge zone. This is consistent with the results of the FEM simulation in Figure 3. Bands of twinned grains at the edges of the notch roots first appear at around 600 N. These regions exhibit a large negative minor principal stress σ_3^*, as shown in Figure 3d. Along with this, the stress state is almost uniaxial with $\sigma_1^* = \sigma_2^* = 0$, which favors the formation of BTGs. These deformation microstructures, together with the surface disturbance, grew in size and became more pronounced as the sample distorted macroscopically until, at approximately a

1500 N force, the sample failed. The dark rectangular area in the zone of plastic deformation is the area studied using EBSD to observe changes in the microstructure with an increasing applied load. The rectangle itself is caused by carbon deposition along the electron beam, as this area was mapped for several hours and collected carbon contamination during the scans.

Figure 9 shows details of the sample surface during the pure shear test before loading and at increasing values of force F_1 of (b) 700 N, (c) 780 N, (d) 850 N, (e) 950 N, and (f) 1400 N. Clearly, the plastic flow has formed deformation structures slanted from the lower left to upper right, normal to the principal compressive stress σ_3^*. The contaminated area is useful for reference of the area used for EBSD mapping.

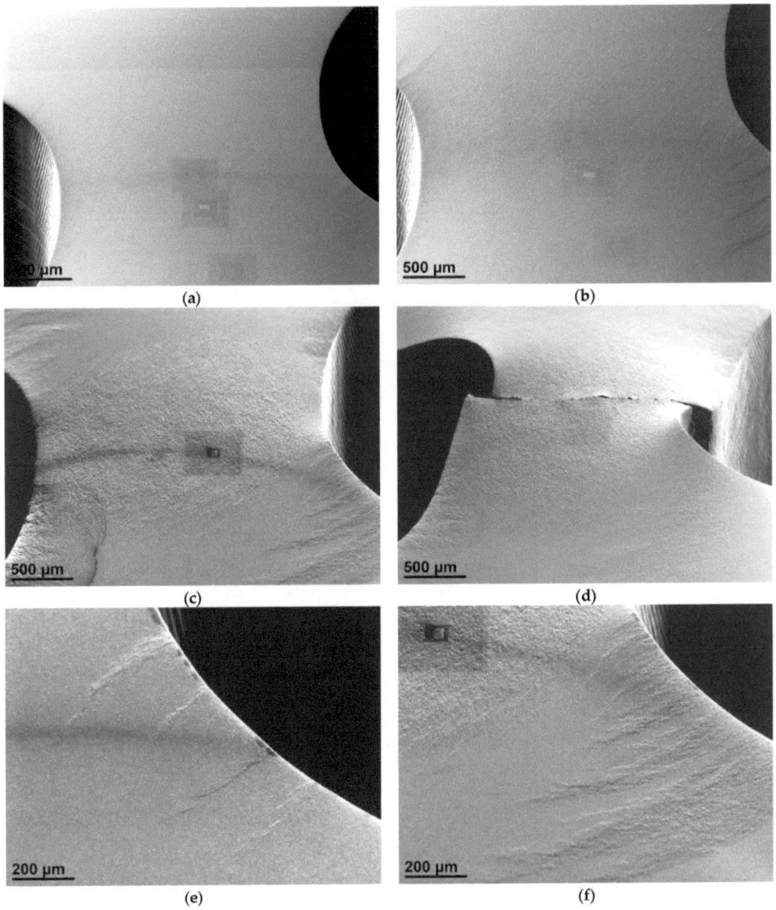

Figure 8. A typical sequence of SEM images of the central pure shear zone of the sample at applied forces of F_1, (**a**) 600 N, (**b**) 780 N, (**c**) 850 N, and (**d**) 1500 N, and details of the edge twin bands at the notch root at (**e**) 650 N and (**f**) 850 N.

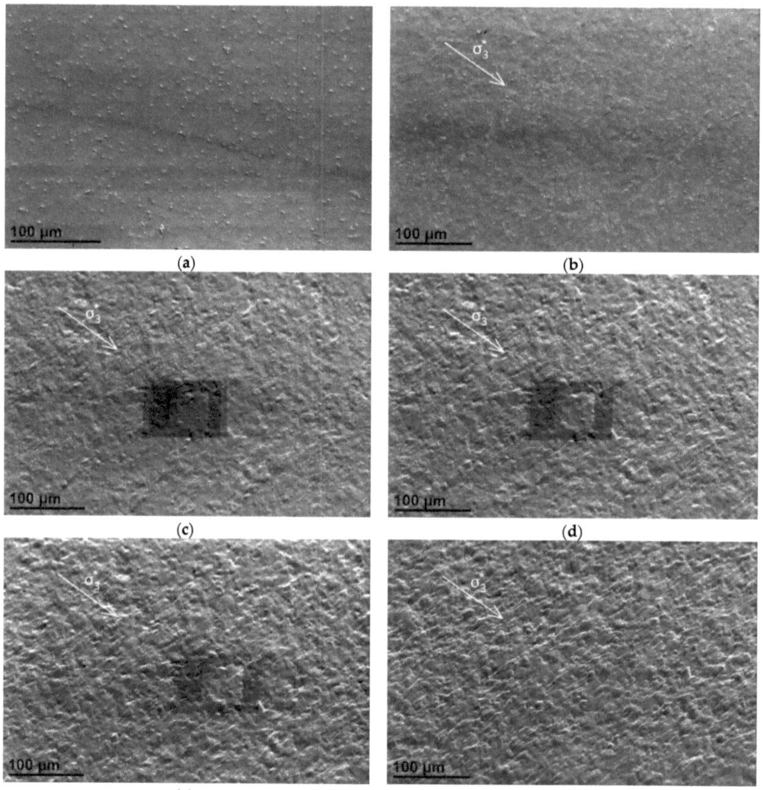

Figure 9. SEM images of the sample showing gradual evolution of plastic deformation at loads of (**a**) 0 N, (**b**) 700 N, (**c**) 780 N, (**d**) 850 N, (**e**) 950 N, and (**f**) 1400 N, where an angular alignment of twins normal to the σ_3 principal stress is clearly visible.

Figure 10 shows higher magnification details of the pure shear zone at (a) 950 N and (b) 1400 N of applied force adjacent to the contaminated rectangle areas. Throughout the deformation zone, dislocation slip is found to be more prevalent than twinning. In Figure 10b, deep slip steps are seen in a number of grains. An in-plane compressive force will result in an expansion normal to the sheet in accordance with the Poission's ratio [45], while a tensile force will have the opposite effect; it is reasonable to expect shear stress to have a combined effect, with some grains expanding out of the sheet while others retract from the surface. This could explain the type of appearance in these images. Twinned grains under compression show a plastic Poisson's ratio ν_{3n} (the first index denotes the σ_3^* direction, the second denotes the normal direction of the sheet surface plane) of approximately 1.0 and ν_{31} of approximately 0.0 [45], whereas the macroscopic material under tension show a plastic Poisson's ratio of approximately 0.5 [45]. Therefore, because of compatibility reasons, dislocation slip and the irregular surfaces appear.

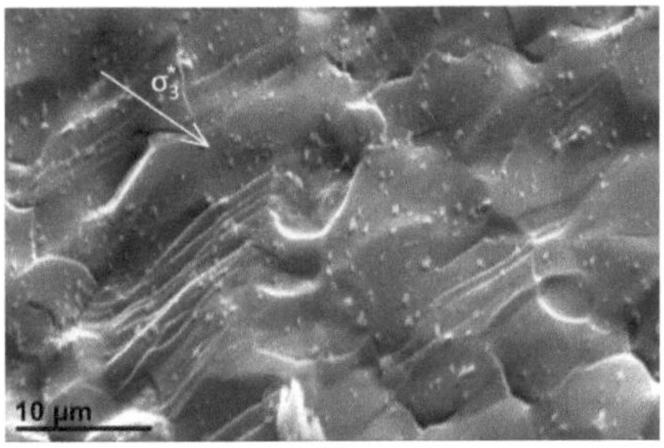

(**a**) (**b**)

Figure 10. SEM images of the sample (**a**) with 950 N and (**b**) 1400 N force applied showing the irregular surface and slip steps as grains were deformed plastically at high levels of shear stress.

Figure 11 shows the marked rectangular area of the sample at four different loading conditions: before loading and at 700 N, 900 N, and 1000 N of applied force. The inset colour code that relates colour to the lattice orientation normal to the sheet is applicable to all subsequent orientation maps and is not reproduced in each map. The mostly red colour denotes the basal texture with the c-axis of the grain lattice normal to the sheet. The few non-aligned grains are typical of AZ31B sheets. Grain identification is made using a disorientation angle of 5°. At 750 N, the first few twins have appeared, visible in the top left, but otherwise there is little change. At 900 N, several grains have become twinned throughout and appear as regular grains tilted away from normal. An inclination of twins is clearly appearing from the lower left to upper right. At 1000 N, surface irregularity has reduced the EBSD signal in many areas and many points are zero solutions (cannot be indexed by the software due to poor signal) and appear black. With new developments in EBSD detectors [46], these types of low signa andl high noise areas might be more measurable. However, it is clear that a high density of twins has formed up with the same inclination.

Figure 12 shows the growth of a band of extension twins that extends across eight grains in the middle of the pure shear zone. The band is oriented normal to the principal compressive stress σ_3^*. The four images correspond to applied forces of 700 N, 730 N, 780 N, and 850 N. The twins have expanded laterally and become wider as a result of the increasing stress, while additional smaller twins parallel to the main band have appeared and grown. The growth of the main band has been in both directions, left and right, but more significantly to the right, filling out the parent grains. The shearing of the area can be seen in the image, with the upper right grains moving to the right and the lower grains progressively shifting to the left. The pole figures for each map show the growth of the twinned lattice. Additionally, the x_1, x_2 and x_3 Kearn's factors [47] for each map are (0.10, 0.10, 0.8), (0.11, 0.10, 0.79), (0.12, 0.10, 0.78), and (0.14, 0.10, 0.76), which shows the realignment of the lattice away from x_3 mainly into x_1. This means that an increase of 150 N in applied force corresponds to an increase of approximately 5% in the twinned lattice area. The two SEM images show the changes in the surface appearance between 700 and 850 N. It is clear that the band of twins is inclined to the sheet surface, as this is where the grains have expanded outwards from the bulk sample under stress.

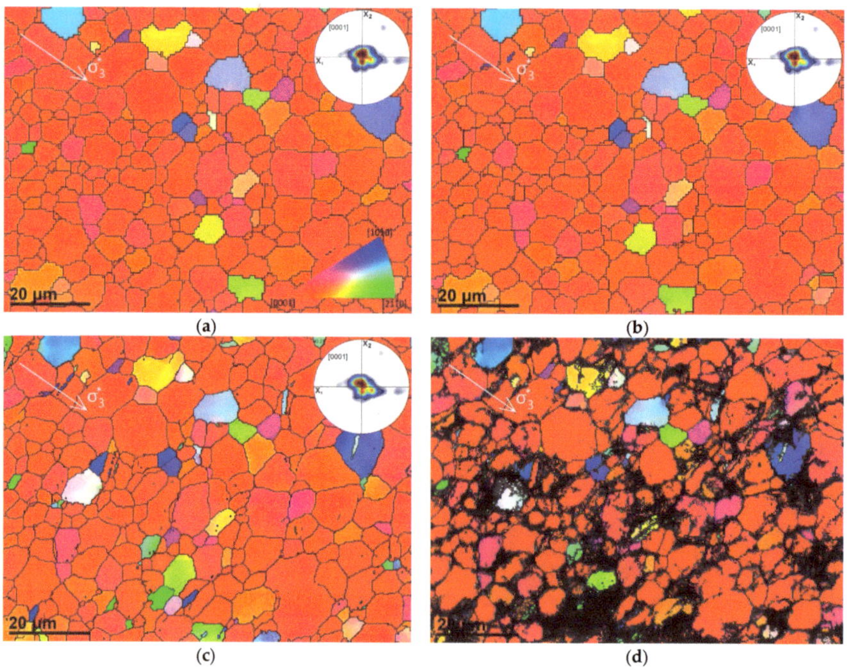

Figure 11. Lattice orientation maps of the same area of the sample (**a**) before loading and at (**b**) 750 N, (**c**) 900 N, and (**d**) 1000 N.

Figure 13 shows detailed lattice orientation maps of the large twin in the top left of Figure 12 at 700 and 850 N of applied force. The lower right area of this grain has become progressively twinned and an area in the top left of the grain has begun to twin. Maps of internal grain disorientation can, to some degree, be identified with elastic strain in grains, as inhomogeneous bending and torsion will give rise to disorientation away from the grain average. Clearly, the 850 N sample contains more elastic strain in the grains, although the twins themselves contain very little elastic strain; this suggests that twinning has the effect of releasing elastic strain in the lattice. To process the internal grain disorientation data, it was necessary to reduce the grain tolerance angle to 1°, otherwise separate grains with small-angle boundaries appear as single grains and can give false disorientation values; this has produced a large number of dark spots in the images. Figure 13d represents an orientation linescan across the grain from A to B, and it is clear how the strain has increased with increasing stress, including the top left area which has developed a discontinuity from the main grain that appears to be the early stages of a twin. The maximum disorientation is 10.9°, which is consistent with previous observations that indicate that this could be a threshold for twinning to occur.

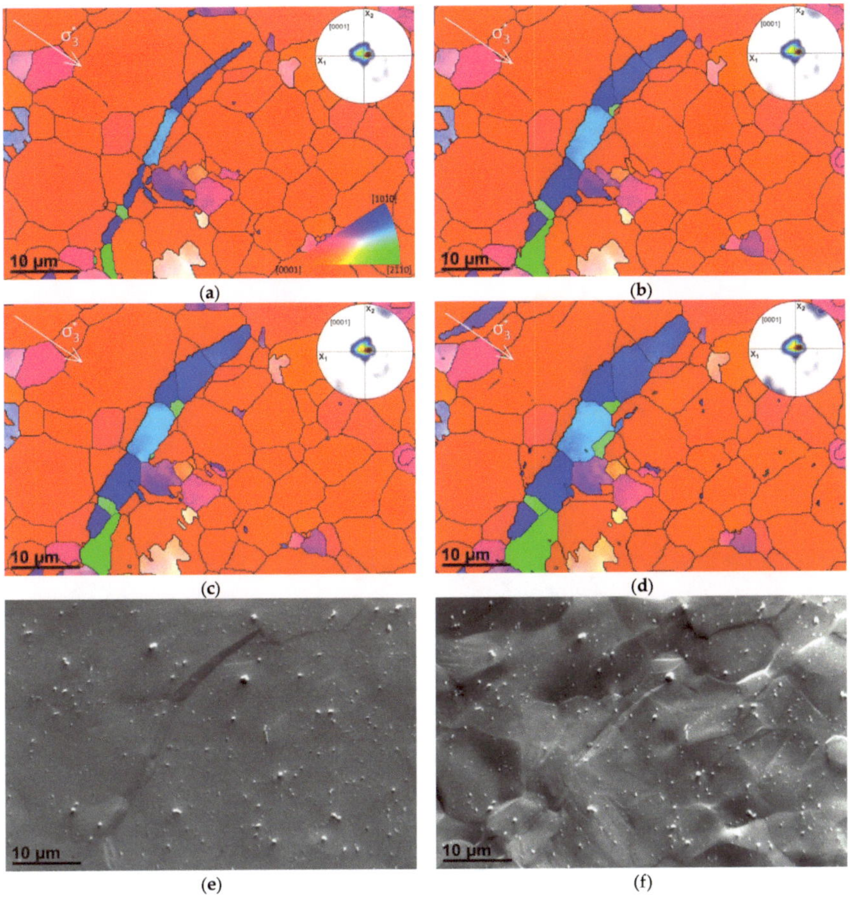

Figure 12. Lattice orientation maps and corresponding pole figures of the magnesium AZ31B sample at increasing values of applied force F_1 of (**a**) 700 N, showing direction of principal compressive stress σ_3^*, (**b**) 730 N, (**c**) 780 N, and (**d**) 850 N, showing the initiation and growth of a band of extension twins that cross several grains; (**e**,**f**) are SEM images of the same area corresponding to (**a**,**d**).

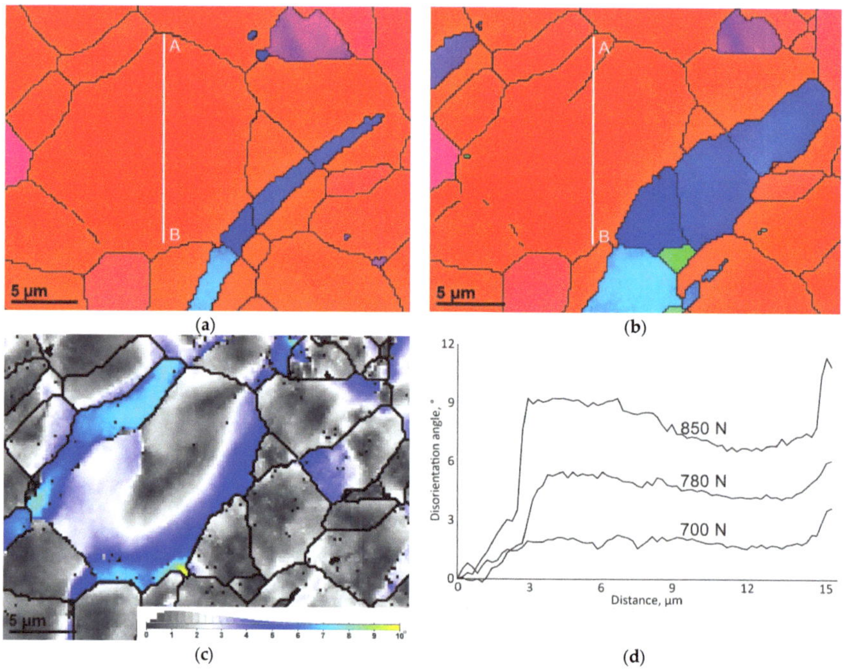

Figure 13. SEM images from the deformed area at (**a**) 700 N and (**b**) 850 N showing the band of twins and (**c**) corresponding internal grain disorientation map at 850 N, and (**d**) internal disorientation within the large grain along the line A-B for loads of 700, 780, and 800 N.

Figure 14 shows EBSD maps from within the pure-stress zone after an applied load of 950 N and after fresh surface preparation. In Figure 14a, the area contains around 120 grains and the twin density is very high, many grains are fully twinned, and the angle of the twins with respect to σ_3^* is approximately 90°. All twins are extension type with an 86.3° disorientation. Many of the grains contain internal disorientation, which can be identified with elastic strain, with an average of 2.23° and a maximum of 9°. The Kearns factors for this condition are $x_1 = 0.2$, $x_2 = 0.48$, and $x_3 = 0.32$, indicating significant loss of texture as grains are reoriented through twinning. Figure 14c, showing a higher magnification, reveals multiple twins in several grains. Figure 14d shows grain disorientation along the line A-B across the larger pink grain and a disorientation of approx. 86.3° characteristic of extension twins, and approx. 5° disorientation on the right side, indicating that, with increasing stress, the twin would likely grow in that direction.

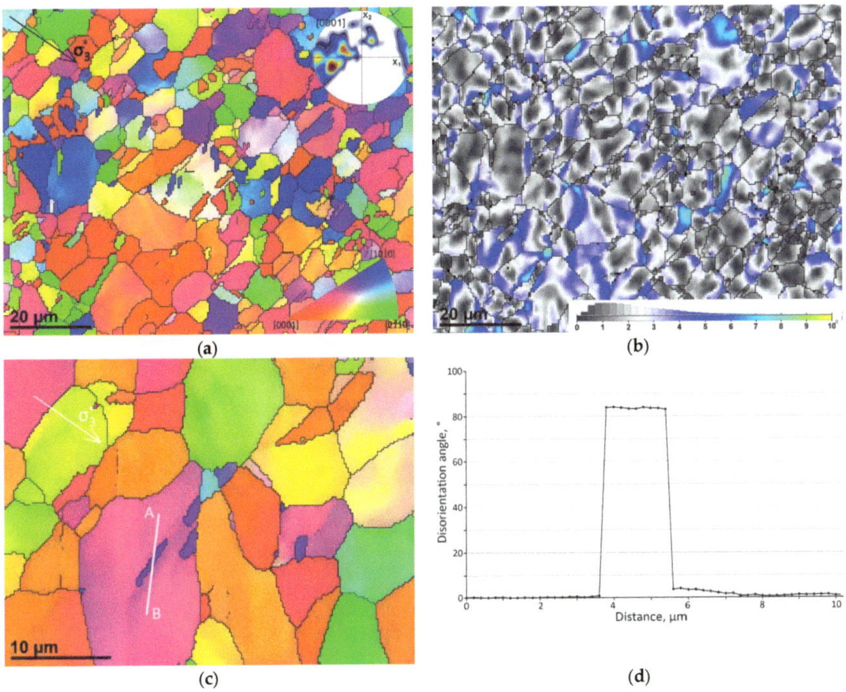

Figure 14. EBSD orientation map from the central area of the pure shear sample after a 950 N applied load and further surface preparation showing (**a**) unidirectional twinning at 90° to σ_3^* with (**b**) internal grain disorientation of up to 9°, (**c**) an adjacent area at higher magnification with inclined twins in several grains, and (**d**) disorientation angle line scan along A-B in (**c**) across the twin showing the characteristic 86.3° disorientation and 5° disorientation right of the twin.

Figure 15 shows a higher magnification lattice orientation map of the 950 N-loaded sample at a grain containing a large twin that has grown from the lower left to upper right. Things to note in this image are: the higher strain is in the parent grain next to the base of the orange twin, suggesting that this is where twining is growing; the maximum measured disorientation within the grains is 9°; most of the very small spots are extension twins; the lower end of the grain where the strain is higher is where the twin would grow; the strain on the right of the twin in the parent grain is greater than on the left, suggesting that the grain would grow to the right; the band of elastic strain across grains (line 4) seems to be derived from the grain on the right; the red twin appears to have grown into the yellow grain. The dark orange grain in the lower left is not a twin of the pink grain, and the interface between the two is the grain boundary of the pink grain.

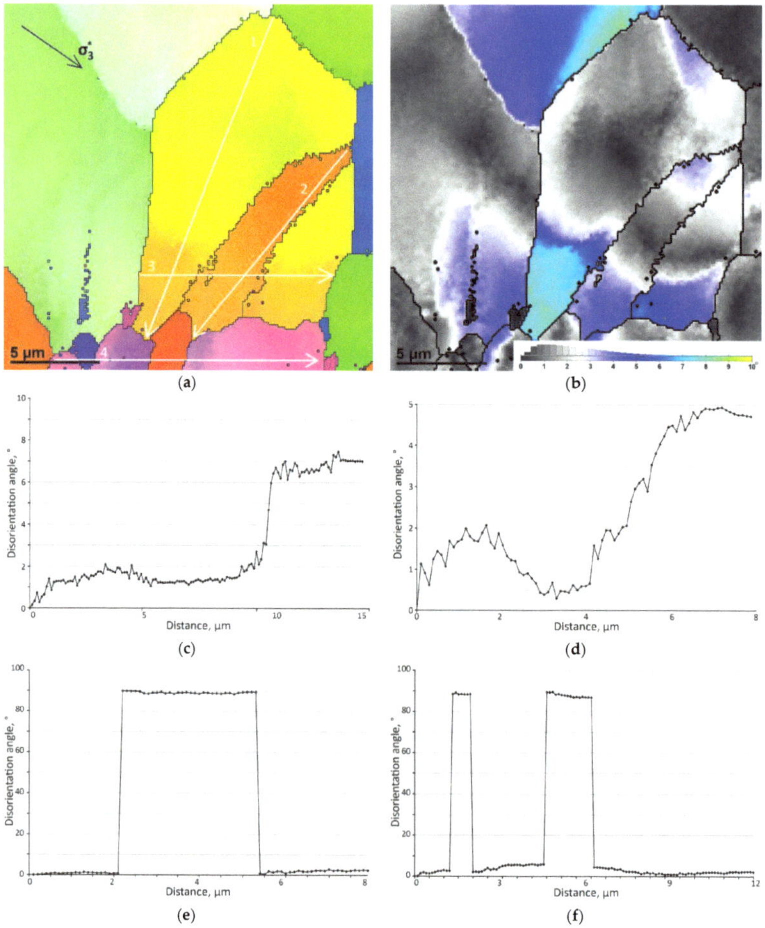

Figure 15. EBSD lattice orientation map at higher magnification of the 950 N-loaded sample showing (**a**) large extension twins (orange) in the yellow grain and red and blue in the pink grain, (**b**) a map of internal grain disorientation, (**c**,**d**) disorientation line scans through grains (lines 1 and 2), and (**e**,**f**) line scans across the twins (lines 3 and 4) showing the characteristic 86.3° disorientation angle.

Figure 16 shows bright field TEM images of a typical area from the gauge zone of the sample subjected to a 950 N force. The diffraction pattern shows the [100] or [2-1-10] zone axis of the grain, while the hexagonal unit cell (with side length ratio 1.62) corresponds to the lattice inside the twins and indicates the [0001] direction (c-axis), the basal plane (0001), and the (10–12) extension twinning plane. The direction of the minor principal stress σ_3^* is indicated by the arrow, and this direction applies to all subsequent images. The image contains three parallel twins of extension type, which is consistent with the EBSD maps that show a high concentration of twins mostly parallel to each other and approximately perpendicular to the direction of the compressive stress σ_3^*. Figure 16c shows more details of the middle twin; features inclined to the length of the twin are stacking faults on the basal plane, which form as atoms rearrange themselves during the twinning process. Figure 16d shows details of the end of the twins where they intersect the grain boundary. This area corresponds to the white square in Figure 16a below the diffraction pattern. In both cases, the stacking faults are normal to the grain boundary, which shows that the twin at this location has the lattice with the c-axis parallel to the grain boundary. This suggests that a favorable location for twinning to initiate is at grain boundaries. TEM examines very small volumes of material at a high resolution. It is suited to revealing details that are not present in SEM or EBSD images, but macroscopic generalisms should not be extrapolated from the TEM study unless it is extensive and examines a larger number of sample areas. There is a periodicity of 106 nm in the spacing of the parallel stacking faults, their presence adding to the mechanical properties of the alloy [48,49].

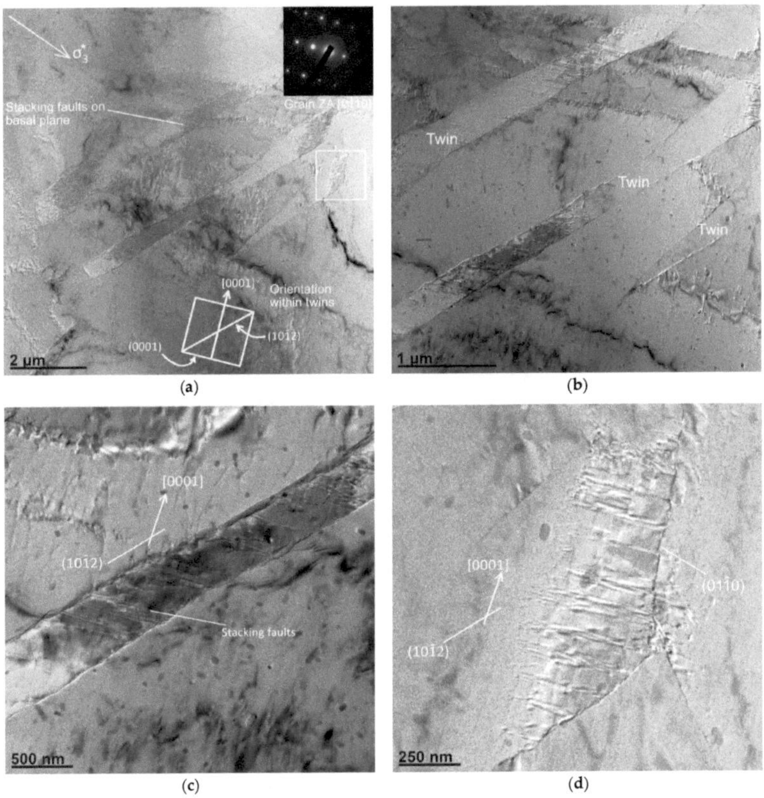

(a) (b) (c) (d)

Figure 16. TEM bright field images of a typical area of the sample containing three parallel twins (**a**) at low magnification, showing an orientation diagram (hexagonal unit cell sideways on) to illustrate key planes and directions within the twins and direction of σ_3^*, (**b**) more detail of the twinned structures, (**c**) detail of the central twin showing parallel stacking faults on the basal plane and (**d**) detail of the twin shown in (**a**) with a white square at the grain boundary showing stacking faults and steps at the boundary.

Figure 17 shows TEM bright field images of intersecting twins. These images and those in Figure 18 are made using multibeam conditions to optimize the contrast. In addition to parallel twins, twins were also found to cross and intersect with each other. The number of these was very low compared with that of parallel twins, but still consistent with the EBSD maps from the gauge zone that do show occasional crossing twins (e.g., Figure 12c). The faint lines in the image from the top left to bottom right are tracks from ion-beam milling, and can be ignored. The two twins in Figure 17a share the same orientation relation with the parent grain, as illustrated by the parallel stacking faults. The location of intersection reveals some complexity, and an area that could be considered as a transition zone. In Figure 17b, there is clearly a dominant twin (top left to lower right) that is intact through the intersection and one that crosses it but is not connected. Figure 17c,d shows the propagating tip of a twin and a lenticular extension twin with a high aspect ratio.

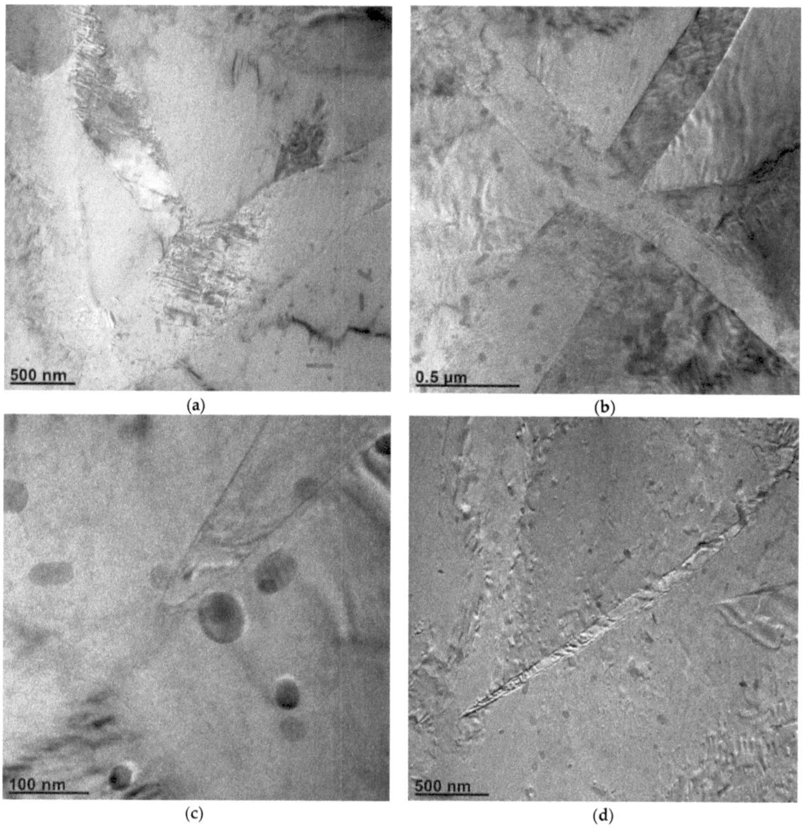

Figure 17. TEM bright field images with multibeam conditions of features relating to twins: (**a**) one-side intersecting twins, (**b**) crossing twins, (**c**) propagating twin tip, and (**d**) high aspect twin.

Figure 18 shows TEM bright field images of other features observed during investigation that are evidence of the high levels of strain taking place in the lattice. Figure 18a shows the dense networks of dislocations that are present, which is expected at high levels of strain and also observed in the SEM images [50–52]. Figure 18b shows the typical distribution of a smaller population of precipitates, consistent with an earlier study [34]. Further studies on dislocation slip and deformation twinning in magnesium alloys can be found to add context to these investigations and give an indication of the direction of research in this field [53–56].

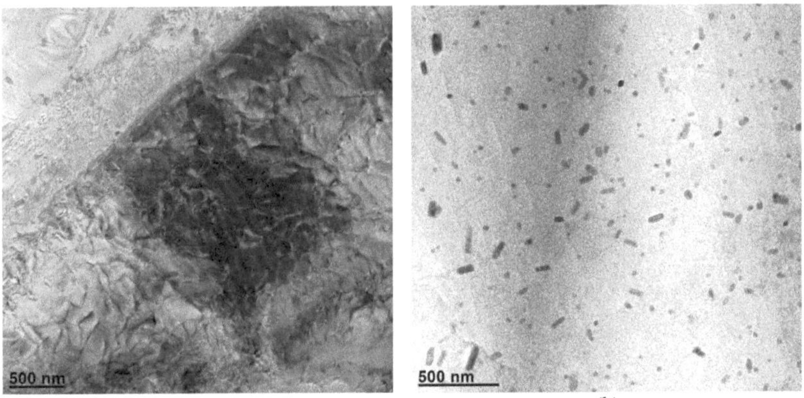

(a) (b)

Figure 18. Additional bright field TEM images that show (**a**) dense networks of dislocations present in the sample and (**b**) typical distribution of the population of smaller precipitates.

4. Conclusions

By adapting a recently developed sample design for pure shear testing and using an in situ mechanical tester, the elastic and plastic deformation of magnesium AZ31B sheets were investigated in a scanning electron microscope in real time with an incrementally increasing load until the point of failure. Within the pure shear zone as well as in notch roots, bands of deformation twins were found to form approximately normal to the local direction of the macroscopic maximum principal compressive stress σ_3^*. Maps of the internal grain lattice disorientation, identified with the elastic strain within grains, revealed several features: the average elastic strain within twins was around a 2–3° disorientation, the elastic strain around a twin in the parent grain is almost always higher on one side, which suggests the growth direction of the twin, the elastic strain tends to cross grain boundaries, and the maximum elastic strain within any single grain was found to be approximately 10°. TEM investigation confirmed the general distribution of deformation structures observed with EBSD, but additionally observed detailed structures at the edges of twins as well as a small percentage of crossing and intersecting twins.

This tells us that, for the uniaxial compressive stress state as well as for the biaxial shear stress state, the twins form approximately normal to the minor principal stress σ_3^*. Compared to uniaxial compression, where twinning is the major plastic deformation mechanism, within the biaxial shear zone, twinning and dislocation slip occur. The proposed shear specimen is also useful for the development of elasto-plastic constitutive equations for FEM analysis. To determine the initial principal yield loci for the pure shear stress state, a mixed numerical-experimental method must be used after determining the initial force for yielding. The required FEM simulation can be performed using a linear elastic material law. To gain the complete yield surface, more tests with different biaxial stress states should be performed and the stress–strain behavior has to be evaluated.

Author Contributions: Conception, L.W., H.S. and O.H.; methodology, O.H., L.W., H.S. and A.N.; software, L.W. and A.N.; validation, L.W., A.N., O.H. and H.S.; formal analysis, L.W., O.H. and A.N.; investigation, L.W. and A.N.; resources, H.S. and L.W.; data curation, L.W.; writing—original draft preparation, L.W.; writing—review and editing, L.W. and O.H.; visualization, L.W. and O.H.; supervision, L.W. and H.S.; project administration, L.W.; funding acquisition, L.W. and O.H. All authors have read and agreed to the published version of the manuscript.

Funding: This research was funded by the Austrian Science Fund (FWF) under grant number I-4782-N and the Deutsche Forschungsgemeinschaft (DFG) under project number 438040004. Open Access Funding by the Austrian Science Fund (FWF).

Data Availability Statement: The raw data supporting the conclusions of this article will be made available by the authors on request.

Conflicts of Interest: The authors declare no conflicts of interest.

References

1. Tan, J.; Ramakrishna, S. Applications of Magnesium and Its Alloys: A Review. *Appl. Sci.* **2021**, *11*, 6861. [CrossRef]
2. Kaya, A.A. A Review on Developments in Magnesium Alloys. *Front. Mater.* **2020**, *7*, 198. [CrossRef]
3. Magnesium: Metal of the Future. Available online: www.financialtribune.com/articles/economy-business-and-markets/56890/magnesium-metal-of-the-future (accessed on 23 February 2024).
4. Dieringa, H.; St. John, D.; Pérez Prado, M.T.; Kainer, K.U. Editorial: Latest developments in the field of magnesium alloys and their applications. *Front. Mater.* **2021**, *8*, 726297. [CrossRef]
5. Jayasathyakawin, S.; Ravichandran, M.; Baskar, N.; Anand Chairman, C.; Balasundaram, R. Mechanical properties and applications of Magnesium alloy—Review. *Mater. Today Proc.* **2020**, *27*, 909–913. [CrossRef]
6. Polmear, I.J. Magnesium alloys and applications. *J. Mater. Sci. Technol.* **1994**, *10*, 1–16. [CrossRef]
7. Gahr, K.-H.Z. (Ed.) *Chapter 2 Microstructure and Mechanical Properties of Materials*; Tribology Series; Elsevier: Amsterdam, The Netherlands, 1987; Volume 10, pp. 8–47. [CrossRef]
8. Shackelford, J.F. *Introduction to Materials Science for Engineers*, 8th ed.; Pearson: Hoboken, NJ, USA, 2014.
9. Cahn, R.W.; Haasen, P. (Eds.) *Physical Metallurgy*, 4th ed.; North-Holland: Amsterdam, The Netherlands, 1996.
10. Loukil, N. Alloying elements of magnesium alloys: A literature review. In *Magnesium Alloys Structure and Properties*; IntechOpen: London, UK, 2022.
11. Askeland, D.R. Strain hardening and annealing. In *The Science and Engineering of Materials*; Springer: New York, NY, USA, 1996; pp. 188–215.
12. Wagoner, R.H.; Lou, X.Y.; Li, M.; Agnew, S.R. Forming behavior of magnesium sheets. *J. Mater. Process. Technol.* **2006**, *177*, 483–485. [CrossRef]
13. Wu, J.; Jin, L.; Dong, J.; Wang, F.; Dong, S. The texture and its optimization in magnesium alloy. *J. Mater. Sci. Technol.* **2020**, *42*, 175–189. [CrossRef]
14. Wang, H.; Wu, P.D.; Gharghouri, M.A. Effects of basal texture on mechanical behaviour of magnesium alloy AZ31B sheet. *Mat. Sci. Eng. A-Struct.* **2010**, *527*, 3588–3594. [CrossRef]
15. Whitmore, L.; Nischler, A.; Saage, H.; Huber, O. In situ uniaxial compression of textured magnesium AZ31B. *Metals* **2023**, *14*, 20. [CrossRef]
16. Nie, J.F.; Shin, K.S.; Zeng, Z.R. Microstructure, deformation, and property of wrought magnesium alloys. *Metall. Mater. Trans. A* **2020**, *51*, 6045–6109. [CrossRef]
17. Barnett, M.R. Twinning and the ductility of magnesium alloys. *Mat. Sci. Eng. A-Struct.* **2007**, *464*, 1–7. [CrossRef]
18. Serra, A.; Bacon, D.J.; Osetsky, Y.N. Strengthening and microstructure modification associated with moving twin boundaries in hcp metals. *Philos. Mag. Lett.* **2007**, *87*, 451–459. [CrossRef]
19. Christian, J.W.; Mahajan, S. Deformation twinning. *Prog. Mater. Sci.* **1995**, *39*, 1–157. [CrossRef]
20. Agnew, S.R. Advances in wrought magnesium alloys. In *Fundamentals of Processing, Properties and Applications*; Bettles, C., Barnett, M., Eds.; Woodhead Publishing: Pittsburgh, PA, USA, 2012; pp. 63–104.
21. Partridge, P.G. The crystallography and deformation modes of hexagonal close-packed metals. *Metall. Rev.* **1967**, *12*, 169–194. [CrossRef]
22. Liu, T.; Yang, Q.; Guo, N.; Lu, Y.; Song, B. Stability of twins in Mg alloys—A short review. *J. Magnes. Alloy.* **2020**, *8*, 66–77. [CrossRef]
23. Beyerlein, I.J.; Arul Kumar, M. The Stochastic Nature of Deformation Twinning: Application to HCP Materials. In *Handbook of Materials Modeling*; Andreoni, W., Yip, S., Eds.; Springer: Cham, Switzerland, 2018. [CrossRef]
24. Yu, H.; Li, C.; Xin, Y.; Chapuis, A.; Huang, X.; Liu, Q. The mechanism for the high dependence of the Hall-Petch slope for twinning/slip on texture in Mg alloys. *Acta. Materialia.* **2017**, *128*, 313–326. [CrossRef]

25. Denk, J.; Nischler, A.; Whitmore, L.; Huber, O.; Saage, H. Discontinuous and inhomogeneous strain distributions under monotonic and cyclic loading in textured wrought magnesium alloys. *Mat. Sci. Eng. A-Struct.* **2019**, *764*, 138182. [CrossRef]
26. Portella, P.D.; Frenz, H. Mechanical Properties: Static Tests. In *Encyclopedia of Materials: Science and Technology*; Elsevier: Amsterdam, The Netherlands, 2001; pp. 5251–5253. [CrossRef]
27. Thiel, C.; Voss, J.; Martin, R.J.; Neff, P. Shear, pure and simple. *Int. J. Non-Linear Mech.* **2018**, *112*, 57–72. [CrossRef]
28. Peirs, J.; Verleysen, P.; Degrieck, J. Novel technique for static and dynamic shear testing of Ti6Al4V sheet. *Exp. Mech.* **2012**, *52*, 729–741. [CrossRef]
29. Dingley, D.J.; Meaden, G.; Dingley, D.J.; Day, A.P. A review of EBSD: From rudimentary on line orientation measurements to high resolution elastic strain measurements over the past 30 years. *IOP Conf. Series. Mater. Sci. Eng.* **2018**, *375*, 012003. [CrossRef]
30. Zaefferer, S. A critical review of orientation microscopy in SEM and TEM. *Cryst. Res. Technol.* **2011**, *46*, 607–628. [CrossRef]
31. Dingley, D.J. A simple straining stage for the scanning electron microscope. *Micron* **1969**, *1*, 206–210. [CrossRef]
32. Seok, M.-Y.; Gopalan, H.; Nandy, S.; Zaefferer, S.; Raabe, D.; Kirchlechner, C.; Dehm, G. Microscale plastic anisotropy of basal and pyramidal I slip in pure magnesium tested in shear. *Materialia* **2020**, *14*, 100932. [CrossRef]
33. Panchal, M.; Kaushik, L.; Ravi, K.R.; Khatirkar, R.; Choi, S.-H.; Singh, J. Recent advances in the in-plane shear testing of Mg alloy sheets. *JMA* **2023**, *11*, 405–424. [CrossRef]
34. Whitmore, L.C.; Denk, J.; Zickler, G.A.; Bourret, G.R.; Huber, O.; Huesing, N.; Diwald, O. Macro to nano: A microscopy study of a wrought magnesium alloy after deformation. *EJP* **2019**, *40*, 045501. [CrossRef]
35. Kawalla, R.; Oswald, M.; Schmidt, C. New technology for the production of magnesium strips and sheets. *J. Metall.* **2008**, *47*, 195–198.
36. Nowell, M.M.; Witt, R.A.; True, B. EBSD sample preparation: Techniques, tips, and tricks. *Microsc. Microanal.* **2005**, *11*, 504–505. [CrossRef]
37. Voort, G.V.; Geertruyden, W.V.; Dillon, S.; Manilova, E. Metallographic Preparation for Electron Backscattered Diffraction. *Microsc. Microanal.* **2006**, *12*, 1610–1611. [CrossRef]
38. Taheri, M.; Phillips, R.C.; Kish, J.R.; Botton, G.A. Analysis of the surface film formed on Mg by exposure to water using a FIB cross-section and STEM–EDS. *Corros. Sci.* **2012**, *59*, 222–228. [CrossRef]
39. Humphreys, F.J. Grain and subgrain characterisation by electron backscatter diffraction. *J. Mat. Sci.* **2001**, *36*, 3833–3854. [CrossRef]
40. Beausir, B.; Fundenberger, J.J. *Analysis Tools for Electron and X-Ray Diffraction, ATEX-Software*; Université de Lorraine—Metz: Metz, France, 2017; Available online: www.atex-software.eu (accessed on 20 March 2024).
41. Whitmore, L. A precision core drill for transmission electron microscopy sample preparation produced by 3D printing. *Ultramicroscopy* **2022**, *241*, 113613. [CrossRef] [PubMed]
42. Whitmore, L. A precision dimple grinder-polisher produced by 3D printing. *Ultramicroscopy* **2023**, *253*, 113813. [CrossRef] [PubMed]
43. Karci, O.; Beldek, T. Quantitative investigation of abrasive grit size dependency of subsurface damages for the metal-bonded abrasives on Zerodur glass-ceramic. *Appl. Opt.* **2021**, *60*, 2624–2632. [CrossRef]
44. Yang, W.; Li, Y. The influence of crystal orientation on subsurface damage of mono-crystalline silicon by bound-abrasive grinding. *Micromachines* **2021**, *12*, 365. [CrossRef] [PubMed]
45. Nischler, A.; Huber, O. Entwicklung eines Materialmodells zur Simulation des inhomogenen 3D-Dehnungsfelds von texturierten Magnesium-Knetlegierungen mittels der FEM. In *11. Landshuter Leichtbau-Colloquium*; Huber, O., Bicker, M., Patzelt, P., Eds.; LC-Verlag: Landshut, Germany, 2023; ISBN 978-3-9818439-7-2.
46. DeRonja, J.; Nowell, M.; Wright, S.; Kacher, J. Generational assessment of EBSD detectors for cross-correlation-based analysis: From scintillators to direct detection. *Ultramicroscopy* **2024**, *257*, 113913. [CrossRef] [PubMed]
47. Gruber, J.A.; Brown, S.A.; Lucadamo, G.A. Generalized Kearns texture factors and orientation texture measurement. *J. Nucl. Mats.* **2011**, *408*, 176–182. [CrossRef]
48. Jian, W.W.; Cheng, G.M.; Xu, W.Z.; Koch, C.C.; Wang, Q.D.; Zhu, Y.T.; Mathaudhu, S.N. Physics and model of strengthening by parallel stacking faults. *App. Phys. Letts.* **2013**, *103*, 133108. [CrossRef]
49. Su, R.; Neffati, D.; Zhang, Y.; Cho, J.; Li, J.; Wang, H.; Kulkarni, Y.; Zhang, X. The influence of stacking faults on mechanical behavior of advanced materials. *Mat. Sci. Eng. A* **2021**, *803*, 140696. [CrossRef]
50. Voyiadjis, G.; Yaghoobi, M. *Size Effects in Plasticity: From Macro to Nano*; Academic Press: Cambridge, MA, USA, 2019.
51. Xu, Y.; Joseph, S.; Karamched, P.; Fox, K.; Rugg, D.; Dunne, F.; Dye, D. Predicting dwell fatigue life in titanium alloys using modelling and experiment. *Nat. Commun.* **2020**, *11*, 5868. [CrossRef]
52. Hull, D.; Bacon, D.J. *Introduction to Dislocations*, 5th ed.; Butterworth-Heinemann: Oxford, UK, 2011.
53. Shi, Q.; Allison, J. An experimental study on twinning behavior in mg alloys with different solute elements. In *Magnesium Technology*; Springer Nature Switzerland: Cham, Switzerland, 2024; pp. 69–71.
54. Frydrych, K.; Libura, T.; Kowalewski, Z.; Maj, M.; Kowalczyk-Gajewska, K. On the role of slip, twinning and detwinning in magnesium alloy AZ31B sheet. *Mat. Sci. Eng. A* **2021**, *813*, 141152. [CrossRef]

55. Lin, H.; Lv, L.; Shao, L.; Jin, T. Deformation responses of AZ31B magnesium alloy under combined shear-compression loading. *Mat. Sci. Tech.* **2023**, *39*, 321–326. [CrossRef]
56. Zhu, Y.; Hou, D.; Li, Q. Quasi in-situ EBSD analysis of twinning-detwinning and slip behaviors in textured AZ31 magnesium alloy subjected to compressive-tensile loading. *J. Mag. Alloy.* **2022**, *10*, 956–964. [CrossRef]

metals

Article

The Effects of Strain Rate and Anisotropy on the Formability and Mechanical Behaviour of Aluminium Alloy 2024-T3

Martin Harant [1,*], Patricia Verleysen [2,*], Milan Forejt [1] and Stepan Kolomy [1]

1 Faculty of Mechanical Engineering, Brno University of Technology, Technická 2896/2, 616 69 Brno, Czech Republic; mforejt@vutbr.cz (M.F.); stepan.kolomy@vutbr.cz (S.K.)
2 MST-DyMaLab, Department of Electromechanical, Systems and Metal Engineering, Ghent University, Technologiepark 46, 9052 Zwijnaarde, Belgium
* Correspondence: 170977@vutbr.cz (M.H.); patricia.verleysen@ugent.be (P.V.)

Abstract: The present study focuses on the mechanical behaviour and formability of the aluminium alloy 2024-T3 in sheet form with a thickness of 0.8 mm. For this purpose, tensile tests at quasi-static and intermediate strain rates were performed using a universal testing machine, and high strain rate experiments were performed using a split Hopkinson tension bar (SHTB) facility. The material's anisotropy was investigated by considering seven different specimen orientations relative to the rolling direction. Digital image correlation (DIC) was used to measure specimen deformation. Based on the true stress–strain curves, the alloy exhibited negative strain rate sensitivity (NSRS). Dynamic strain aging (DSA) was investigated as a possible cause. However, neither the strain distribution nor the stress–strain curves gave further indications of the occurrence of DSA. A higher deformation capacity was observed in the high strain rate experiments. The alloy displayed anisotropic mechanical properties. Values of the Lankford coefficient lower than 1, more specifically, varying between 0.45 and 0.87 depending on specimen orientations and strain rate, were found. The hardening exponent was not significantly dependent on specimen orientation and only moderately affected by strain rate. An average value of 0.183 was observed for specimens tested at a quasi-static strain rate. Scanning electron microscopy (SEM) revealed a typical ductile fracture morphology with fine dimples. Dimple sizes were hardly affected by specimen orientation and strain rate.

Keywords: 2024-T3; aluminium alloy; high strain rates; anisotropy; mechanical behaviour

Citation: Harant, M.; Verleysen, P.; Forejt, M.; Kolomy, S. The Effects of Strain Rate and Anisotropy on the Formability and Mechanical Behaviour of Aluminium Alloy 2024-T3. *Metals* **2024**, *14*, 98. https://doi.org/10.3390/met14010098

Academic Editors: Jingxiang Xu, Zhenhua Chu and Xingwei Zheng

Received: 2 December 2023
Revised: 9 January 2024
Accepted: 11 January 2024
Published: 13 January 2024

1. Introduction

Aluminium is one of the most frequently used metals. However, its application in pure form is limited because of its low strength. By adding alloying elements such as copper, magnesium, or manganese, aluminium alloys with high strength-to-weight ratios can be obtained. Additionally, several heat treatments can be imposed to achieve specific properties. The aluminium alloy 2024-T3 of the present study has been used in the automotive and aerospace industry thanks to its excellent resistance to fatigue crack propagation and its high damage tolerance. Aluminium 2024-T3 is primarily alloyed with copper and magnesium, which form Al_2Cu and Al_2CuMg phases responsible for its hardening. The presence of copper, however, has a negative effect on the corrosion resistance. The T3 condition refers to solution heat treatment, cold working, and natural aging [1–4].

During manufacturing, processing, and use, materials are subjected to various loading conditions and a wide range of strain rates and temperatures, which affect their properties. Moreover, the materials' behaviour can also be anisotropic, which can have major implications for forming operations. The strain rate and directional dependence of the mechanical behaviour of aluminium alloy 2024-T3 have been studied by several authors. Li et al. [5] analysed the compressive behaviour of specimens machined from a plate at three different strain rates, i.e., 0.001, 10, and 3500 s^{-1}. The alloy showed a low, though positive, strain rate sensitivity. This resulted in a 3% increase in engineering stress at an engineering strain of 0.05% and an 8% increase at a strain of 0.2% at the highest strain rate compared to the lowest quasi-static strain rate. For compression tests at strain rates ranging from 0.006 to 11,000 s^{-1} on specimens cut out of rods, on the other hand, a strain rate-insensitive response was observed by Abotula and Chalivendra [6]. Rodríguez-Martínez et al. [7] investigated the influence of strain rate and temperature by tensile tests. A reduced strain rate sensitivity was reported at room temperature for strain rates ranging from 0.001 to 200 s^{-1}. A decreasing strain hardening was observed with increasing temperatures in the range from −50 to 200 °C. Shamchi et al. [8] studied the mechanical behaviour of 2 mm thick 2024-T3 plates cladded at both sides by pure aluminium with a thickness of 0.05 mm. Tension experiments were performed at strain rates ranging from 0.001 to 1150 s^{-1}, at temperatures from 24 up to 250 °C for the dynamic and up to 350 °C for the quasi-static experiments. The temperature had a significant effect on the plastic flow: at temperatures higher than 300 °C, no more hardening was observed at quasi-static rates. Comparing the dynamic flow stress levels with the quasi-static values, an NSRS was observed at 24 °C. Additionally, the failure strain increased by 28%, moving from quasi-static to dynamic tension.

Low to negative strain rate sensitivity is characteristic of many aluminium alloys. It is often attributed to DSA and/or thermal softening. Next to an NSRS, DSA can also result in serrated flow, known as the Portevin–Le Chatelier effect, accompanied by a localisation of the deformation in bands which can be detected using full-field DIC deformation measurements [9,10]. Thermal softening might result from the adiabatic temperature rise at dynamic strain rates. Indeed, a major part of plastic work is converted into heat. At high strain rates, the heat is not dissipated to the environment and results in a temperature rise in the specimen. Shamchi et al. [8] used an infrared camera and observed a significant temperature rise within the experiment at the strain rate of 500 s^{-1}. The lower flow stress at this strain rate is considered to be a result of thermal softening.

The anisotropy of aluminium 2024-T3 was studied by Barlat et al. [11], who tested specimens from a 0.3 mm thick sheet, which were machined in seven different orientations relative to the rolling direction (0°—rolling direction; 15°, 30°, 45°, 60°, 75°, and 90°—transverse direction). The results of tensile tests at a quasi-static strain rate clearly showed anisotropic properties. Lesuer [12] examined the alloy in the form of a 4 mm thick plate under compression at a strain rate of 4000 s^{-1} (specimen orientations: 0°, 90°, and through thickness) and tension at a strain rate of 8000 s^{-1} (specimen orientations: 0° and 90°). Only small deviations between true stress–strain curves for the different specimen orientations were observed, and it was concluded that the material was isotropic. Houria et al. [13] performed tensile tests at a strain rate of 0.0005 s^{-1} on specimens cut out along the 0°, 45°, and 90° directions of a 0.6 mm thick plate. The tests showed anisotropic properties. A mild anisotropy was also observed by Shamchi et al. [8], who considered two test orientations, 0° and 90°, at a strain rate of 0.001 s^{-1}.

The above-mentioned investigations reveal inconsistencies in the influence of strain rate on the mechanical properties of aluminium alloy 2024-T3. Furthermore, there is a lack of studies focusing on the material's anisotropy and formability indicators at different strain rates. Hence, in the present work, the alloy in sheet form with a thickness of 0.8 mm was subjected to tensile loading at quasi-static (0.001 s^{-1}), intermediate (1 s^{-1}), and high strain rates (≈800 s^{-1}). The anisotropy at different strain rate regimes was examined using specimens cut out in seven different orientations relative to the rolling direction. The mechanical behaviour and formability were evaluated based on the true stress–strain curves, hardening exponents, and Lankford coefficients [14]. Moreover, the fracture behaviour was examined using scanning electron microscopy.

2. Materials and Methods

2.1. Material and Geometry of Specimen

The tensile specimens were cut using electrodischarge machining from a 0.8 mm thick aluminium alloy 2024-T3 sheet. The chemical composition of the material is provided in Table 1. Figure 1 depicts the microstructure of the alloy in the as-received state in the rolling plane observed by a Zeiss Axio Observer Z1m light microscope (Carl Zeiss AG, Oberkochen, Germany). The metallographic specimens were ground using emery papers and finely polished with 1 μm diamond paste. To reveal the microstructure, specimens were etched with Keller's reagent (2.5 mL HNO_3, 1.5 mL HCl, 1 mL HF, and 95 mL distilled water). The average grain size was measured using Stream Motion software version 2.4 (Olympus Corporation, Tokyo, Japan) with a built-in module for the "Grain Intercept method". The measurement was performed according to ASTM E112 applying the cross-and-circles pattern [15].

Table 1. Chemical composition of 2024-T3 aluminium alloy (% of weight).

Si	Fe	Cu	Mn	Mg	Cr	Ni	Zn	Ti	Other Elements	Al
max. 0.50	max. 0.50	3.80–4.90	0.30–0.90	1.20–1.80	max. 0.10	-	max. 0.25	max. 0.15	0.05 each, 0.15 in total	rest

Figure 1. Microstructure in the as-received state of the 2024-T3 alloy in the rolling plane.

To study material anisotropy, specimens were taken along 0°, 15°, 30°, 45°, 60°, 75°, and 90° relative to the rolling direction. The same specimen geometry with a pin connection presented in Figure 2 was used for all tests, regardless of the strain rate. In this way, the effects of both geometry and boundary conditions on the results was suppressed.

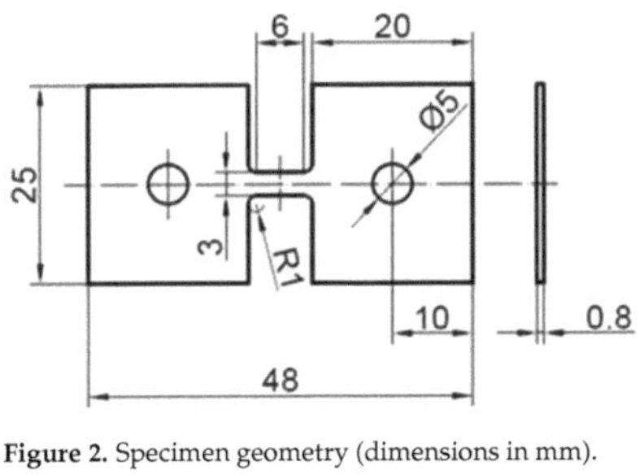

Figure 2. Specimen geometry (dimensions in mm).

2.2. Experimental Methods

The tensile tests were carried out at ambient temperature using setups available at DyMaLab at Ghent University. Two experiments were performed for each test condition, i.e., the seven specimen orientations and the three strain rate regimes. For testing at quasi-static and intermediate strain rates, an Instron 5569 (Instron GmbH, Darmstadt, Germany) machine equipped with a 50 kN load cell was used. Crosshead speeds of 0.36 mm·min^{-1} and 360 mm·min^{-1} were used to obtain strain rates of 0.001 s^{-1} and 1 s^{-1}, respectively. Specimens were fixed to the setup using pins.

An SHTB setup was used for the experiments at high strain rates ($\approx$800 s^{-1}). The setup is schematically presented in Figure 3 and consisted of an input and output bar between which the specimen was fixed. The input and output bars were made of aluminium alloy and had a diameter of 25 mm and lengths of 6 m and 3 m, respectively. A tubular striker made of ertalon was put around the input bar and was accelerated by a pneumatic mechanism towards a flange at the end of the input bar. The impact with the flange generated an incident tensile wave that propagated to the specimen. The wave interacted with the specimen and was partly reflected as a compressive wave and partly transmitted to the output bar as a tensile wave. The waves were measured using strain gages wired in full Wheatstone bridge configuration attached to the bars. As elaborated below, from the acquired data, it was possible to evaluate the stress–strain curve and to calculate the strain rate. Indeed, as soon as force equilibrium is established in the specimen and during uniform deformation, stress and strain rate in the specimen can be calculated from the transmitted and reflected wave [16]. The strain rate can be adjusted by the striker impact velocity, which was chosen to be 13 m·s^{-1} [17].

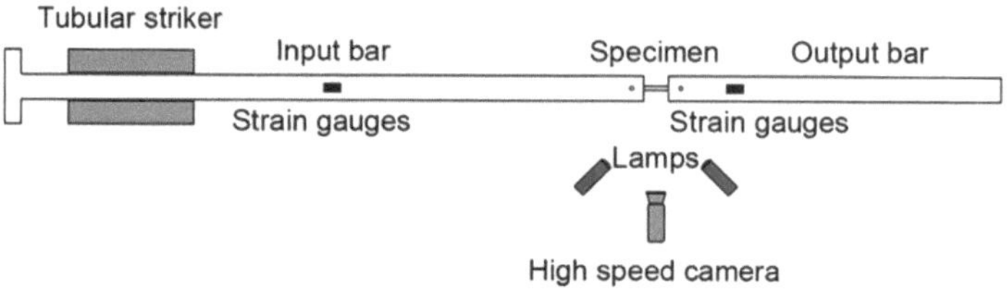

Figure 3. Schematic representation of the SHTB facility.

Analysis of SHTB tests relies on the one-dimensional wave propagation theory [18]. The amplitudes of the waves are kept below the elastic limit of the input and output bars, allowing the use of Hooke's law to calculate the forces $F_1(t)$ and $F_2(t)$ at the interfaces between the specimen and the input bar and output bar, respectively:

$$F_1(t) = E_B \cdot A_B \cdot (\varepsilon_I(t) + \varepsilon_R(t)), \tag{1}$$

$$F_2(t) = E_B \cdot A_B \cdot \varepsilon_T(t), \tag{2}$$

where E_B is the Young's modulus of the bars, A_B is the bar's cross-section area, $\varepsilon_I(t)$ is the incident strain wave, $\varepsilon_R(t)$ is the reflected strain wave, and $\varepsilon_T(t)$ is the transmitted strain wave.

Assuming force equilibrium in the specimen, i.e., $F_1(t) = F_2(t)$, the following equation can be applied to calculate the engineering stress and strain rate [18]:

$$\sigma(t) = \frac{A_B}{A_S} \cdot E_B \cdot \varepsilon_T(t), \tag{3}$$

$$\dot{\varepsilon}(t) = -2 \cdot \frac{C_B}{L_S} \cdot \varepsilon_R(t), \tag{4}$$

where C_B is the velocity of wave propagation in the bars, and L_S is the specimen gauge length.

Figure 4 shows incident, reflected, and transmitted stress waves shifted in time to the specimen/bar interfaces for test in the rolling direction. The sign of the reflected compression wave was reversed.

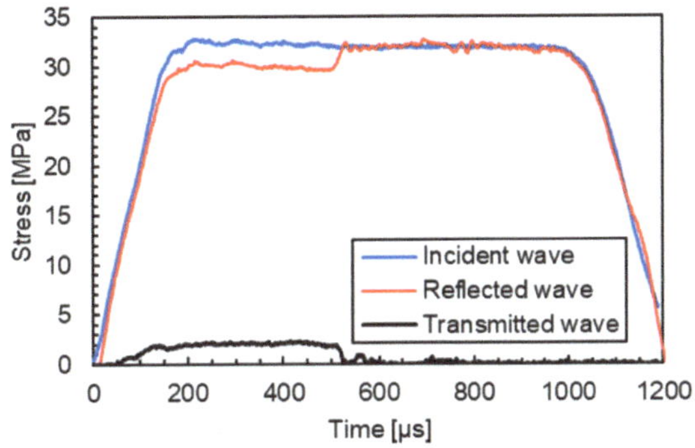

Figure 4. Incident, reflected, and transmitted stress waves during SHTB shifted in time to the specimen/bar interfaces.

The true stress was calculated as

$$\sigma_{true} = \frac{F}{A_{Si}}, \tag{5}$$

where F is the actual tensile force (for experiments at dynamic strain rate, it was calculated using Equation (2)), and A_{Si} is the actual specimen cross section area:

$$A_{Si} = A_S \cdot e^{-\varphi}, \tag{6}$$

where φ is the true longitudinal strain obtained from optical deformation measurements.

For all tests, the specimen deformation was measured using DIC [19]. The method is non-contact and provides a complex full-field deformation analysis. The measurement process consists of four stages, i.e., specimen surface preparation, calibration of devices, image recording (before and during the experiment), and image processing [20]. The accuracy of the technique is affected by factors that can be grouped as errors related to the specimen, loading, and imaging (size of the speckle pattern, image resolution, illumination intensity, etc.) and errors related to the correlation algorithm (inappropriate subset size, correlation function, shape function, etc.) [21,22]. The speckle pattern (Figure 5) was applied to the gauge section prior to the experiment. The pattern was made by spraying black paint on a white background. To measure in-plane deformations at quasi-static strain rates, a 5Mpxl F-504B Allied Vision Stingray camera (Allied Vision Technologies GmbH, Stadtroda, Germany) was used. The specimen was illuminated by two LED lamps. The experiments at intermediate and high strain rates were captured by a high-speed Photron Mini AX200 camera (Photron, Tokyo, Japan) and illuminated with two Dedocool lamps (see Figure 5).

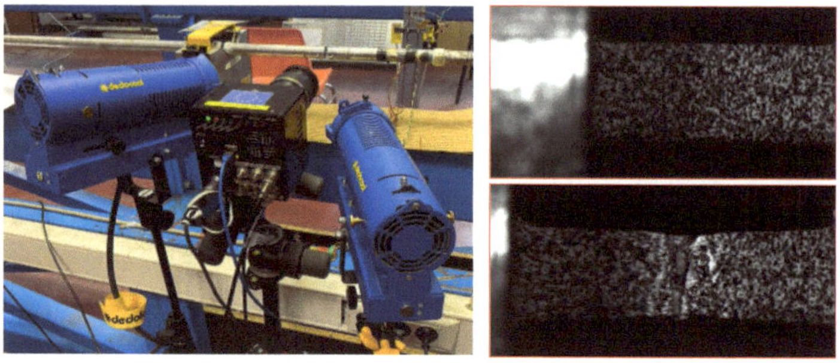

Figure 5. Experimental setup for the tensile test at high strain rates (**left**), and specimen prior to and after testing (**right**).

The images were analysed by the commercial software MatchID 2D version 2021.2.1. The parameters of the optical deformation measurement are presented in Table 2.

Table 2. DIC analysis parameters.

	Quasi-Static Strain Rate	Intermediate Strain Rate	High Strain Rate
Resolution	2452 × 2056 pixel^2	256 × 272 pixel^2	256 × 128 pixel^2
Scale	1 pixel = 0.0053 mm	1 pixel = 0.0564 mm	1 pixel = 0.0407 mm
Frame rate	1/3 Hz	10,000 Hz	120,000 Hz
Image filtering	Gaussian	Gaussian	Gaussian
Subset size	45 pixel^2	9 pixel^2	7 pixel^2
Step size	11 pixel	1 pixel	2 pixel
Shape function	Quadratic	Quadratic	Quadratic
Matching criterion	Zero-normalized sum of square differences (ZNSSD)	ZNSSD	ZNSSD
Strain window	17 data points	15 data points	13 data points
Virtual strain gage	221 pixel^2	23 pixel^2	31 pixel^2

Crystallographic texture resulting from material processing steps, such as rolling, gives rise to anisotropic material properties, which significantly affect the formability. To assess the anisotropy of a sheet, the Lankford coefficient [14] was used and determined by the following equation:

$$r = \frac{\varphi_w}{\varphi_t}, \tag{7}$$

where φ_w and φ_t are the true strain in the width and thickness direction, respectively. The thickness strain was obtained using the volume conservation principle: $\varphi_t = -(\varphi + \varphi_w)$. The DIC technique allowed for the evaluation of the Lankford coefficients for each frame from the onset of yielding to necking, from which the average value was calculated. Typical values of the Lankford coefficient for aluminium alloys are close to the values obtained for isotropic materials, i.e., 1 [23].

The formability of a material is also affected by strain hardening, commonly quantified by the hardening exponent n. The n-value is derived from the slope of the true ln(stress)–ln(strain) curve from yielding to necking. For materials following Hollomon's law, the following equation was used:

$$\sigma_{true} = K \cdot \varphi^n, \tag{8}$$

where K is the strength coefficient, and the n-value is equal to the true necking strain. Higher n-values, thus, result in better formability.

3. Results and Discussion

Grain size influences material properties, such as ductility, machinability, and mechanical properties. Decreasing grain size results in an increase in hardness, yield stress, or strength [24]. The etched microstructure of 2024-T3 (Figure 1) is composed of grains with an average size of 17.4 μm and coarse inclusions. Following [1,25,26], the microstructure can, thus, be considered fine-grained.

Representative true stress–strain curves up to necking from tensile tests at quasi-static, intermediate, and high strain rates for different specimen orientations are shown in Figure 6. The curves show a moderate NSRS in the considered strain rate regime. Indeed, the flow stress decreases when the strain rate increases from 0.001 s^{-1} to 1 s^{-1} and then further to 800 s^{-1}.

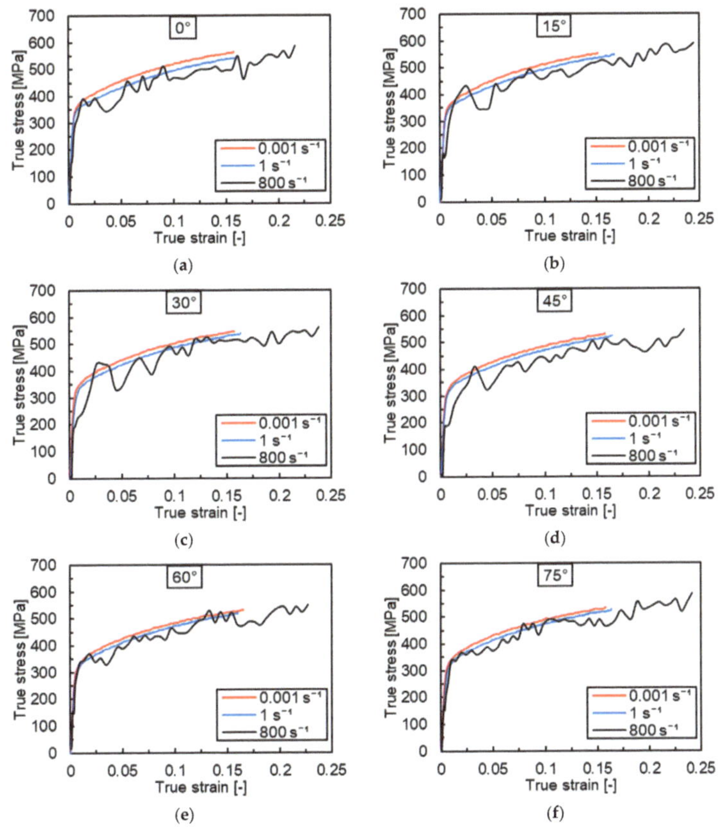

Figure 6. *Cont.*

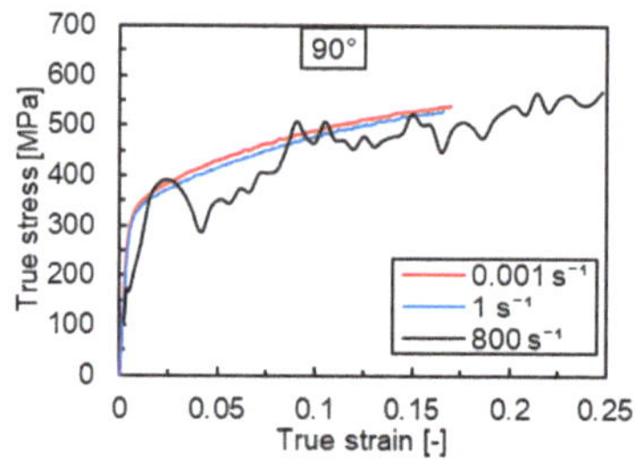

(g)

Figure 6. True stress–strain curves for different strain rates and specimen orientations relative to the rolling direction of (**a**) 0°, (**b**) 15°, (**c**) 30°, (**d**) 45°, (**e**) 60°, (**f**) 75°, and (**g**) 90°.

Figure 7a shows the 0.2% value of yield stress versus specimen orientation for quasi-static and intermediate strain rates. Results from SHTB were not included, as oscillations prevented the calculation of reliable values. NSRS was observed for yield stress values at all specimen orientations. The maximum and minimum yield stresses at the strain rate of 0.001 s^{-1} were 355 MPa (specimen orientation: 15°) and 314 MPa (specimen orientation: 90°), respectively. The values at the strain rate of 1 s^{-1} ranged from 294 MPa to 342 MPa. Figure 7b depicts flow stresses extracted at a true strain of 0.05 and 0.1 for the different specimen orientations. To exclude the effect of oscillations, values at the highest strain rate were calculated from curves fitted to the Hollomon law (see Equation (8)). The flow stress values confirm the NSRS. The evolution of the flow stresses at different specimen orientations followed the trend seen for the yield stress in Figure 7a.

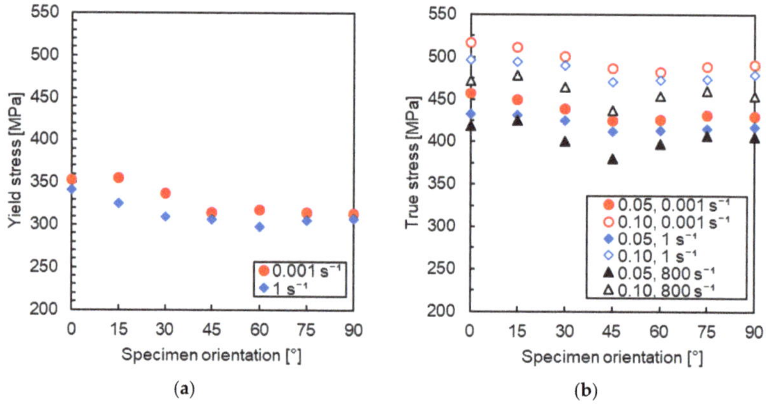

Figure 7. (**a**) Yield stress and (**b**) true stress at 0.05 and 0.1 true strain versus specimen orientation at different strain rates.

The strain rate sensitivity can be quantified using the semi-logarithmic parameter S:

$$S = \frac{d\sigma}{d(ln\dot{\varepsilon})}, \tag{9}$$

For the stress σ in the equation, the yield stress, the stress at a specific strain level, or the ultimate stress can be taken. The S parameter can be determined as the slope of σ–$ln\dot{\varepsilon}$ curves.

Based on the values represented in Figure 7, the S parameter at different specimen orientations was calculated as the slope of σ–$ln\dot{\varepsilon}$ curves for the yield stress, as well as for true stresses at strain levels of 0.05 and 0.1. An overview of corresponding S-values is presented in Table 3.

Table 3. Overview of S-values for different specimen orientations.

Specimen Orientation	S Yield Stress ($MPa \cdot s^{-1}$)	S Stress at 0.05 Strain ($MPa \cdot s^{-1}$)	S Stress at 0.1 Strain ($MPa \cdot s^{-1}$)
0°	−1.62	−2.82	−3.32
15°	−4.26	−1.76	−2.46
30°	−4.02	−2.77	−2.66
45°	−3.04	−2.12	−2.07
60°	−3.04	−2.12	−2.07
75°	−1.40	−1.78	−2.12
90°	−0.88	−1.80	−2.69

For all specimen orientations and considered stresses, S-values were negative. The NSRS might be attributed to DSA [27,28]. However, the strain fields obtained from the tests did not show any indication of strain bands. Also, no Portevin–Le Chatelier (PLC) serrations were observed in the true stress–strain curves. The oscillations in the high strain rate curves are believed to have a mechanical origin. Thermal softening might also result in a negative SRS. Indeed, the adiabatic temperature rise in specimens tested at high strain rates might have a softening effect on the flow stress. It is worth noting that the strain rate limit above which adiabatic conditions prevail is relatively low, i.e., typically around 1 s^{-1} for metals [29]. Consequently, dynamic tests reflect the combined influence of strain rate and temperature. However, adiabatic heating does not occur at the onset of plastic deformation. Therefore, changes in stress levels with a strain rate at yielding or low strain levels can be attributed solely to the effect of the strain rate, and the corresponding SRS parameters provide the most reliable insight into strain rate sensitivity. The negative S-values for the yield stress indicate that the NSRS cannot be attributed to thermal softening. The fact that S-values do not systematically decrease with increasing deformation further points towards a non-dominant role of thermal softening.

For high strain rate experiments, the elongation strain at fracture was significantly higher compared to the values at other strain rates (see Figure 8a), thus confirming the findings in [8]. Moreover, the uniform elongation calculated from the gauge length at maximum force (Figure 8b) exhibited the same trend. Compared to the flow stresses, the effect of specimen orientation was less pronounced at all strain rates considered.

The plastic anisotropy exhibited by the sheet metal resulted in Lankford coefficients being significantly lower than 1 and the variation of the Lankford coefficient with respect to the rolling direction, as can be seen in Figure 9a. The maximum values for all strain rates were located at specimen orientations of 30° and 40°. Coefficients lower than 1 indicate higher strains along the thickness direction compared to the width direction, which is critical for forming applications. A similar evolution of the Lankford coefficient with testing direction is presented in Barlat et al. [11]. The hardening exponent (Figure 9b), on the contrary, showed a negligible dependence on the specimen orientation. The deviations of dynamic strain rates could be due to the oscillations of the stress–strain curves from which the hardening exponent was extracted. As was the case for the Lankford coefficient, the hardening exponent was only moderately influenced by the strain rate.

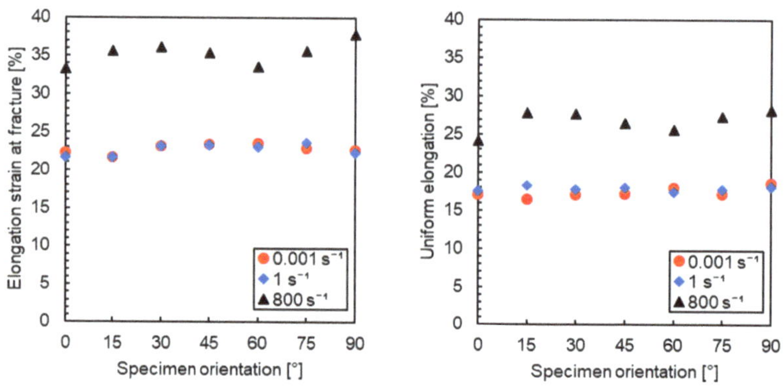

(**a**) (**b**)

Figure 8. (**a**) Elongation strain at fracture and (**b**) uniform elongation for the different specimen orientations considered.

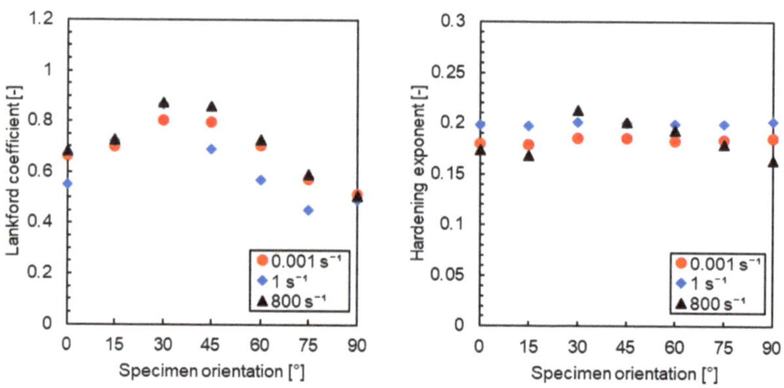

(**a**) (**b**)

Figure 9. (**a**) Lankford coefficient and (**b**) hardening exponent versus specimen orientation.

Figure 10 shows the fracture surfaces of specimens cut along the 45° (Figure 10a,c,e) and 90° (Figure 10b,d,f) directions relative to the rolling direction. The micrographs were taken at a high magnification (5000×) to comprehensively identify the fine dimple morphology typical of a ductile fracture process resulting from the nucleation, propagation, and coalescence of voids. A combination of very shallow and deeper dimples is a typical type of fracture for aluminium alloys and other metals tested at ambient temperature [30]. The aluminium alloy 2024-T3 was subjected to natural aging, which resulted in the precipitation of homogeneously distributed submicron particles, causing strengthening and acting as initiation points for the creation of very fine dimples on the fracture surfaces (Figure 10c). No clear relationship between fracture morphology and strain rate was observed. Indeed, no noticeable changes were observed in the size or shape of the dimples when the strain rate was increased.

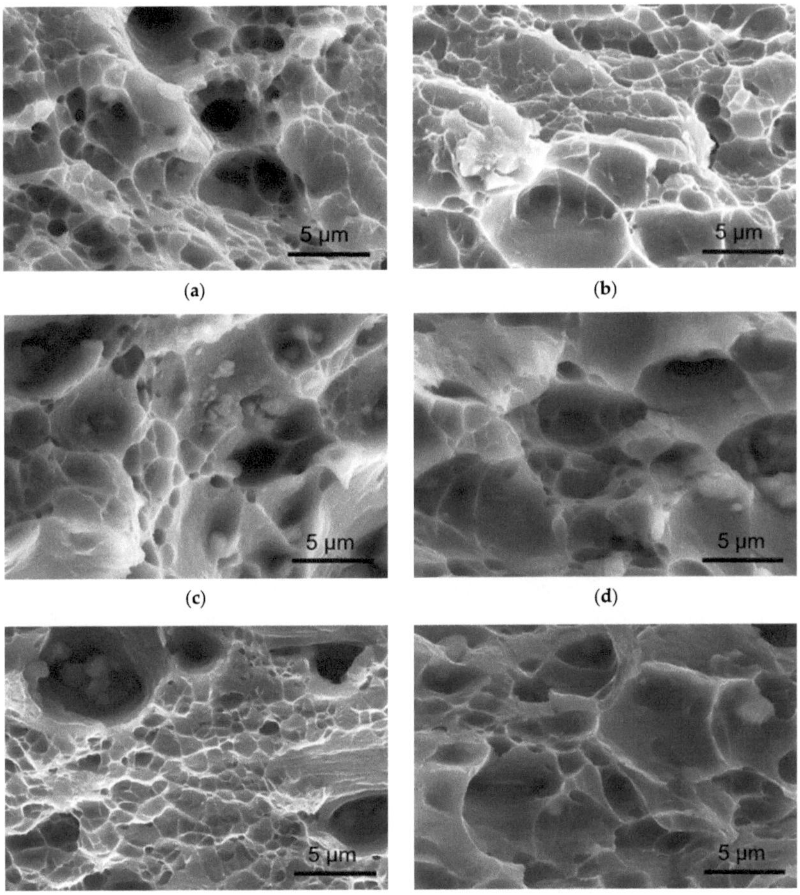

(**a**) (**b**) (**c**) (**d**) (**e**) (**f**)

Figure 10. Fracture surfaces of selected specimens after tensile testing: (**a**) at 0.001 s^{-1} for the 45° direction, (**b**) at 0.001 s^{-1} for the 90° direction, (**c**) at 1 s^{-1} for the 45° direction, (**d**) at 1 s^{-1} for the 90° direction, (**e**) at 800 s^{-1} for the 45° direction, and (**f**) at 800 s^{-1} for the 90° direction.

4. Conclusions

The mechanical behaviour and formability of aluminium alloy 2024-T3 in sheet form with a thickness of 0.8 mm were investigated under tensile loading from quasi-static to dynamic strain rates for seven different specimen orientations. The DIC technique was used to measure the evolution of the deformation fields in the specimens. An NSRS was observed from the onset of yielding over the entire strain rate regime and in all directions. The NSRS might have been induced by DSA. However, the DIC strain fields and true stress–strain curves did not show further indications of DSA, such as propagating strain bands or PLC serrations. The values of uniform elongation and fracture elongation showed a significant increase in deformation capacity at high strain rates. The specimens cut at a 15° angle against the rolling direction showed the highest increase in both uniform and fracture elongation, with values of 9.8% and 11.5%, respectively, moving from quasi-static to dynamic strain rates. The Lankford coefficients revealed a pronounced anisotropy over the entire strain rate regime. Indeed, strongly directional-dependent coefficients, varying between 0.45 and 0.87, were found. The underlying dominant thinning in the thickness direction limited the materials' formability. The influence of the strain rate on the Lankford coefficient was negligible. No noticeable dependency of the hardening exponent on the specimen orientation or the strain rate was seen. An average value of 0.183 was obtained at quasi-static strain rates. The fracture mode proved to be ductile within the entire analysed range of strain rates. Variations in the size and depth of dimples between the considered orientations were minimal.

Author Contributions: Conceptualisation, M.H. and P.V.; methodology, M.H. and P.V.; validation, P.V.; formal analysis, M.H.; investigation, M.H.; resources, P.V.; data curation, M.H. and P.V.; writing—original draft preparation, M.H. and S.K.; writing—review and editing, M.H., P.V. and M.F.; visualisation, M.H. and S.K.; supervision, P.V. and M.F. All authors have read and agreed to the published version of the manuscript.

Funding: This research was supported by the project "Development of an apparatus for testing of flat specimens in tension" no. FV 22-06 within the projects of the Faculty of Mechanical Engineering, Brno University of Technology.

Data Availability Statement: The data that support the findings of this study are available from the corresponding authors. The data are not publicly available due to privacy.

Acknowledgments: We thank Sarath Chandran from the Department of Electromechanical, Systems and Metal Engineering, MST-DyMaLab Research Group, Ghent University, for assistance with tensile testing. We also express our gratitude to Marek Doubrava and Libor Pantělejev from the Institute of Materials Science and Engineering, Faculty of Mechanical Engineering, Brno University of Technology, for providing the fractographic images of tensile specimens.

Conflicts of Interest: The authors declare no conflicts of interest.

References

1. Huda, Z.; Taib, I.; Zaharinie, T. Characterization of 2024-T3: An aerospace aluminum alloy. *Mater. Chem. Phys.* **2009**, *113*, 515–517. [CrossRef]
2. Moy, C.K.S.; Weiss, M.; Xia, J.; Sha, G.; Ringer, S.P.; Ranzi, G. Influence of heat treatment on the microstructure, texture and formability of 2024 aluminium alloy. *Mat. Sci. Eng. A* **2012**, *552*, 48–60. [CrossRef]
3. Lin, Y.C.; Xia, Y.; Jiang, Y.; Zhou, H.; Li, L. Precipitation hardening of 2024-T3 aluminum alloy during creep aging. *Mat. Sci. Eng. A* **2013**, *565*, 420–429. [CrossRef]
4. Dursun, T.; Soutis, C. Recent developments in advanced aircraft aluminium alloys. *Mater. Des.* **2014**, *56*, 862–871. [CrossRef]
5. Li, P.; Siviour, C.R.; Petrinic, N. The Effect of Strain Rate, Specimen Geometry and Lubrication on Responses of Aluminium AA2024 in Uniaxial Compression Experiments. *Exp. Mech.* **2009**, *49*, 587–593. [CrossRef]
6. Abotula, S.; Chalivendra, V.B. An experimental and numerical investigation of the static and dynamic constitutive behaviour of aluminium alloys. *J. Strain Anal. Eng.* **2010**, *45*, 555–565. [CrossRef]
7. Rodríguez-Martínez, J.A.; Rusinek, A.; Arias, A. Thermo-viscoplastic behaviour of 2024-T3 aluminium sheets subjected to low velocity perforation at different temperatures. *Thin Wall Struct.* **2011**, *49*, 819–832. [CrossRef]
8. Shamchi, S.P.; Queirós de Melo, F.J.M.; Tavares, P.J.; Moreira, P.M.G.P. Thermomechanical characterization of Alclad AA2024-T3 aluminum alloy using split Hopkinson tension bar. *Mech. Mater.* **2019**, *139*, 103198. [CrossRef]
9. Lee, S.; Takushima, C.; Hamada, J.; Nakada, N. Macroscopic and microscopic characterizations of Portevin-LeChatelier effect in austenitic stainless steel using high-temperature digital image correlation analysis. *Acta Mater.* **2021**, *205*, 116560. [CrossRef]
10. Xu, J.; Holmedal, B.; Hopperstad, O.S.; Mánik, T.; Marthinsen, K. Dynamic strain ageing in an AlMg alloy at different strain rates and temperatures: Experiments and constitutive modelling. *Int. J. Plast.* **2022**, *151*, 103215. [CrossRef]
11. Barlat, F.; Lege, D.J.; Brem, J.C. A six-component yield function for anisotropic materials. *Int. J. Plast.* **1991**, *7*, 693–712. [CrossRef]
12. Lesuer, D.R. *Experimental Investigations of Material Models for Ti-6Al-4V Titanium and 2024-T3 Aluminum: Final Report*; Report No.—DOT/FAA/AR-00/25 20591; Office of Aviation Research: Washington, DC, USA, 2000.
13. Houria, M.; Matougui, N.; Mehdi, B.; Kherrouba, N.; Jahazi, M. Effect of Plastic Anisotropy on the Kinetics of Static Softening in AA2024–T3 Aluminum Alloy. *Met. Mater. Int.* **2022**, *28*, 2042–2058. [CrossRef]

14. Lankford, W.T.; Snyder, S.C.; Bausher, J.A. New criteria for predicting the press performance of deep drawing sheets. *Trans. Am. Soc. Met.* **1950**, *42*, 1197–1232.
15. *ASTM E112-13*; Standard Test Method for Determining Average Grain Size. ASTM International: West Conshohocken, PA, USA, 2021.
16. Kuhn, H.; Medlin, D. (Eds.) *Mechanical Testing and Evaluation, ASM Handbook*; ASM International: Materials Park, OH, USA, 2000.
17. Chen, W.W.; Song, B. *Split Hopkinson (Kolsky) Bar: Design, Testing and Applications*; Springer: New York, NY, USA, 2011.
18. Meyers, M.A. *Dynamic Behavior of Materials*; John Wiley and Sons: New York, NY, USA, 1994.
19. Chu, T.C.; Ranson, W.F.; Sutton, M.A. Applications of digital-image-correlation techniques to experimental mechanics. *Exp. Mech.* **1985**, *25*, 232–244. [CrossRef]
20. Górszczyk, J.; Malicki, K.; Zych, T. Application of Digital Image Correlation (DIC) Method for Road Material Testing. *Materials* **2019**, *12*, 2349. [CrossRef] [PubMed]
21. Pan, B.; Qian, K.; Xie, H.; Asundi, A. Two-dimensional digital image correlation for in-plane displacement and strain measurement: A review. *Meas. Sci. Technol.* **2009**, *20*, 062001. [CrossRef]
22. Janeliukstis, R.; Chen, X. Review of digital image correlation application to large-scale composite structure testing. *Compos. Struct.* **2021**, *271*, 114143. [CrossRef]
23. Buschow, K.H.J.; Cahn, R.W.; Flemings, M.C.; Ilschner, B.; Kramer, E.J.; Mahajan, S.; Veyssière, P. (Eds.) *Encyclopedia of Materials: Science and Technology*; Elsevier: Amsterdam, The Netherlands, 2001.
24. De Sousa Araujo, J.V.; Milagre, M.X.; Ferreira, R.O.; de Souza Carvalho Machado, C.; de Abreau, C.P.; Costa, I. Microstructural Characterizatics of the Al Alloys: The Dissimilarities Among the 2XXX Alloys Series used in Aircraft Structures. *Metallogr. Microstruct. Anal.* **2020**, *9*, 744–758. [CrossRef]
25. Khatami, R.; Fattah-alhosseini, A.; Mazaheri, Y.; Keshavarz, M.K.; Haghshenas, M. Microstructural evolution and mechanical properties of ultrafine grained AA2024 processed by accumulative roll bonding. *Int. J. Adv. Manuf. Technol.* **2017**, *93*, 681–689. [CrossRef]
26. Goli, F.; Jamaati, R. Effect of strain path during cold rolling on the microstructure, texture, and mechanical properties of AA2024 aluminum alloy. *Mater. Res. Express* **2019**, *6*, 066514. [CrossRef]
27. Chandran, S.; Liu, W.; Lian, J.; Münstermann, S.; Verleysen, P. Dynamic strain aging in DP1000: Effect of temperature and strain rate. *Mat. Sci. Eng. A* **2022**, *832*, 142509. [CrossRef]
28. Noh, Y.; Lee, M.S.; Chaudry, U.M.; Jun, T.S. Effect of strain rate on the deformation of 6061-T6 aluminum alloy at cryogenic temperature. *Mater. Charact.* **2023**, *206*, 113403. [CrossRef]
29. Hokka, M. (Ed.) *Dynamic Behavior of Materials*; Elsevier: Amsterdam, The Netherlands, 2023.
30. Alvand, M.; Naseri, M.; Borhani, E.; Abdollah-Pour, H. Nano/ultrafine grained AA2024 alloy processed by accumulative roll bonding: A study of microstructure, deformation texture and mechanical properties. *J. Alloys Compd.* **2017**, *712*, 517–525. [CrossRef]

coatings

MDPI

Article

Effect of the Molybdenum Content on Wear and Corrosion Behavior of Fe-B-Based Surface-Alloyed Layer

Engin Kocaman

Faculty of Engineering, Department of Aerospace Engineering, Zonguldak Bulent Ecevit University, Incivez, Zonguldak 67100, Turkey; enginkocaman@beun.edu.tr

Abstract: In this study, Fe-Mo-B-based hardfacing electrodes containing different amounts of Mo were coated on an AISI 1020 steel substrate using the electric arc welding method. The findings show that molybdenum is highly effective on the microstructure and minor changes in the coating composition affect the phases and morphological properties. In the hardness tests, an increase of 73% was achieved in the $Fe_{14}Mo_2B_4$-based hardfacing coating, compared to the base material, and a 30% increase was achieved, compared to the $Fe_{16}B_4$-based coating. The highest hardness value was measured as 56.4 HRC and the highest phase hardness was measured as 3228 HV in the $FeMo_2B_4$ phase. The lowest wear rate was measured in the $Fe_{14}Mo_2B_4$-based coating. The wear rate of the $Fe_{14}Mo_2B_4$-based coating was 8.1 times lower than that of the substrate material and 4.7 times lower than that of the $Fe_{16}B_4$-based coating. According to corrosion test results, the highest corrosion resistance was obtained in the $Fe_{16}B_4$-based coating. The current density value of the $Fe_{16}B_4$-based coating was measured to be 13.6 times lower than that of the substrate material.

Keywords: hardfacing; surface alloying; hardness; wear; corrosion

Citation: Kocaman, E. Effect of the Molybdenum Content on Wear and Corrosion Behavior of Fe-B-Based Surface-Alloyed Layer. *Coatings* **2023**, *13*, 2050. https://doi.org/10.3390/coatings13122050

Academic Editors: Jingxiang Xu, Xingwei Zheng and Zhenhua Chu

Received: 13 November 2023
Revised: 29 November 2023
Accepted: 4 December 2023
Published: 6 December 2023

1. Introduction

Steel is one of the most important metals, owing to its wide range of uses from structural applications to automotive, aerospace, and shipbuilding industries [1–4]. The reason why steel is an important metal is because of its mechanical properties such as high strength, high toughness, and relatively easy production and processing [5]. However, most steels, particularly carbon steels, cannot exhibit sufficient performance under service conditions where high wear resistance and corrosion resistance are required [6,7]. In today's industrial conditions, interest in materials that exhibit high performance is increasing day by day. At this point, researchers are making intense efforts to improve the properties of the material. In service conditions where the use of steel is inevitable, improving the surface properties instead of improving the properties of the entire material appears to be a very economical engineering approach [8–11].

Although the surface properties of steel materials can be improved via various methods, hardfacing coatings are a good alternative, especially in aggressive working conditions such as mining and excavation, where the material is exposed to high wear and corrosion. In hardfacing coatings, the surface of the material is coated with a material that provides high performance using a suitable welding method [12]. Various welding methods, such as shielded manual arc welding (SMAW) [13], flux cored arc welding (FCAW) [14], plasma transfer arc welding (PTA) [15], gas tungsten arc welding (GTAW) [16], and laser welding (LW) [17–19], are used in hardfacing applications. Although these methods have advantages and disadvantages compared to each other, the SMAW is one of the most preferred methods due to being economical and simple. Many alloy compositions are used as coating materials in the hardfacing process. Nowadays, it is seen that coating materials containing carbides and borides formed via transition metals such as Cr, Ti, V, Nb, and Mn are frequently used [20]. One of the transition metals, molybdenum, is an element that can contribute to the improvement of material surface properties via in situ-forming boride and carbide phases on the coating surface [21,22]. There are various studies in the literature where molybdenum is used as a coating material [23,24]. In these studies, it is generally seen that the wear or corrosion properties on the surface are improved after hardfacing [25–29]. On the other hand, some studies have reported that molybdenum increases corrosion resistance in the presence of an element such as chromium or nickel, but its effect alone is still ambiguous [30,31]. It is seen in the literature that hardfacing coating studies carried out with molybdenum-containing compounds focus mostly on wear. The studies performed with Fe-Mo-B-based hardfacing coatings are very limited, especially the studies examining the microstructures formed in varying molybdenum ratios and the corrosion behavior of the coating. The effect of molybdenum is unclear in some studies, which reveals that there are not enough studies on Fe-Mo-B-based coatings. In this study, Fe-Mo-B-based hardfacing electrodes containing different amounts of molybdenum were produced and coated on an AISI 1020 steel substrate using the electric arc welding method. Microstructural investigations, hardness test, and wear and corrosion resistance of the coated samples were comparatively examined with the substrate metal and the Fe-B-based coating.

2. Experimental Procedure

In the experimental studies, hardfacing electrodes containing different ratios of Fe-Mo-B were produced. In the production of hardfacing electrodes, a H08A low carbon steel wire rod was used as the core metal, and ferromolybdenum, ferroboron, flux material (5%), and binder potassium silicate (17.5%) were used in the cover composition. The chemical compositions of ferro alloy powders and core metal are given in Table 1.

Table 1. Compositions of powders used in experimental studies (wt.%).

	C	Cr	Mo	B	Mn	Si	P	S	Fe
H08A	<0.1	0.064	-	-	0.35–0.40	0.10	<0.02	<0.02	Bal.
AISI 1020	0.22	0.025	0.02	-	0.52	0.17	0.023	0.019	Bal.
Ferro-Mo	-	-	60	-	-	1.5	0.050	0.10	38.35
Ferro-B	0.312	-		18.58		0.39	0.029	0.003	80.602

Covered electrodes were weighed by calculating their atomic ratios according to the compositions given in Table 2, and the powders were first mixed wet by adding a binder and then mixed as dry. The covering paste in the viscose form was pressed in a PTFE-coated aluminum mold and the core metal was passed through it. Atomic calculation of the electrode compositions and details of the production process can be found in previous studies carried by Kocaman et al. [32,33]. The coating process was carried out using an electric arc welding machine at a 125 A (amperes) and 25 V (volts) constant potential on the AISI 1020 steel substrate, of which its chemical composition is given in Table 1.

Table 2. Chemical composition of electrodes cover (at.%).

Compound	B	Mo	Fe
$Fe_{16}B_4$	20	-	Bal.
$Fe_{15}MoB_4$	20	5	Bal.
$Fe_{14}Mo_2B_4$	20	10	Bal.
$Fe_{14}Mo_4B_2$	10	20	Bal.

The coated samples were cut and subjected to standard metallographic operations as can be seen in Figure 1. Firstly, the samples' surfaces were sanded with 240, 320, 600, 800, 1200, and 2500 mesh SiC paper, respectively. Then, the samples were polished with 6, 1, and

0.05 μm Al_2O_3 paste, respectively. The samples to be used for microstructural examinations were etched using 4% Nital solution. The surface morphology and elemental distribution of the samples were analyzed using the JEOL-JSM-6060 scanning electron microscope (SEM) and X-ray spectroscopy (EDS) device coupled to the scanning electron microscope. The phases in the microstructure were analyzed with Rigaku brand X-ray diffraction analysis (XRD) and defined using the X-pert High score program.

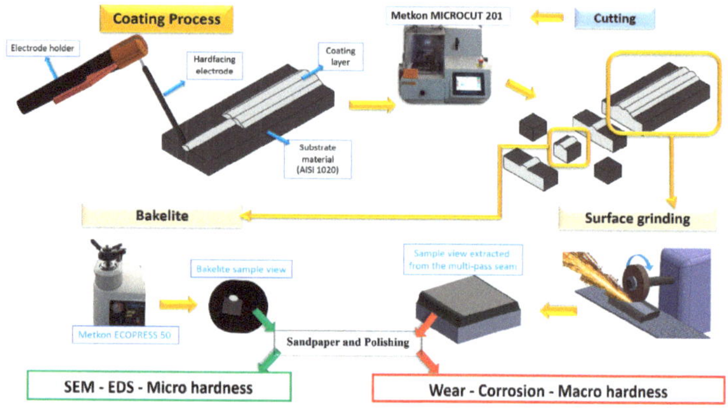

Figure 1. Schematic illustration of production processes.

Hardness measurements of hardfacing coatings were carried out with Bulut Makina DIGIROCK RB model hardness device with Hardness Rockwell A Scale (HRA) and Hardness Rockwell C Scale (HRC). Rockwell hardness measurements were carried out using a conical diamond indenter under a load of 60 kgf for HRA and 150 kgf for HRC. Shimadzu HMV-G21D Hardness Vickers (HV) microhardness device was used for microhardness measurements. A dry sliding wear test was performed using a TriboTechnic wear device with reciprocating wear mode against an Al_2O_3 (2050 HV) ball with 10 mm dimensions. The average surface roughness of wear samples was measured as Ra = 0.035 (±0.003) μm before the wear test. Wear tests were carried out at two different loads, 4 N and 8 N, and at two different wear distances, 400 m and 800 m. Following the wear tests, the area of wear tracks was measured using a Taylor Hobson 2D profilometer. All wear tests were performed at a constant humidity (31%–34%) and room temperature (23–25 °C). Friction coefficient (*FC*) data were provided via a computer software connected to the wear device and was calculated according to the following formula.

$$FC = \frac{F_S}{F_N} \quad (1)$$

In the formula, F_S represents the lateral load (*N*) and F_N the normal load.

The wear rate was calculated according to the formula [34] below, using the data obtained from the wear tracks' measurement.

$$WR = V/S \quad (2)$$

where *V* is the wear trace volume and *S* is the sliding distance.

The wear track was calculated with Equation (3).

$$V = A * E \quad (3)$$

where A is the cross-sectional surface area of the wear track and E is the eccenter distance which was chosen during all tests as 10 mm.

The Gamry 1010 E potansiyotat electrochemical work station equipped with a classic three-electrode system was used during electrochemical tests. A saturated calomel electrode (SCE)'s reference electrode and graphite counter electrode was used in the corrosion tests. Open circuit potential (OCP) was measured during 1200 s. The tests were performed in a 0.5 M NaCl solution with a potential range of −0.5 V, +1.5 V and the scanning rate was chosen as 1 mV/s. E_{cor} (corrosion potential) and I_{cor} (corrosion current) were calculated by using Gamry Echem software (https://www.gamry.com/). Electrochemical impedance spectroscopy (EIS) tests were performed with an alternative current signal of 10 mV in the range of 100 kHz to 10 mHz.

3. Results and Discussion

3.1. Microstructure and Phases Analysis

Figure 2 shows the substrate–coating transition zone of hardfacing coatings. One of the most important features that distinguish hard coating from other coating methods is that their coating thickness is quite high and they create strong metallurgical bonds with the substrate. These features enable hardfacing coatings to be used for a long time in aggressive environments. On the other hand, porosities that will occur in the transition zone and coating layer after coating will significantly reduce the performance of the coating. SEM images show that the alloys selected as the coating material and the process are compatible and a metallurgical bond is formed between the substrate material and the coating layer. In this study, the average thickness of coatings was measured as 5–5.5 mm. Additionally, no significant porosity formation was observed in either the transition zone or the coating layer. In the SEM images, during the coating process carried out at room temperature, columnar grains were formed at the interface, extending from the substrate material towards the coating area. During repeated melting and solidification, the temperature increases and the columnar grains grow from the colder substrate material towards the coating region. The interface, which develops depending on the dynamic solidification conditions of the welding process, has been reported in similar studies [25,32].

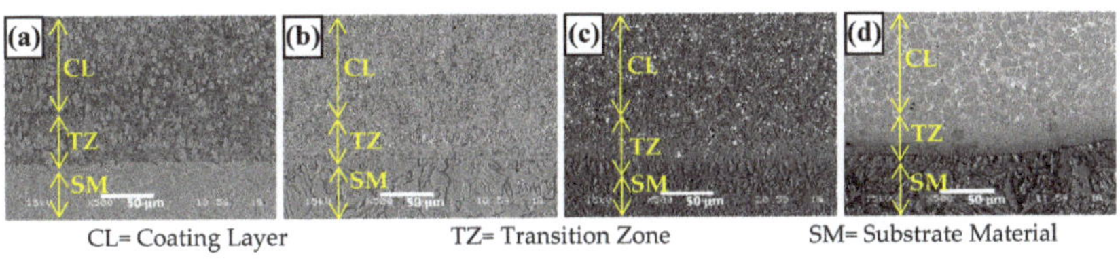

Figure 2. SEM images of transition zone (**a**) $Fe_{16}B_4$, (**b**) $Fe_{15}MoB_4$, (**c**) $Fe_{14}Mo_2B_4$, and (**d**) $Fe_{14}Mo_4B_2$.

XRD analysis of hardfacing coatings is given in Figure 3. According to XRD analysis, the changing alloying element in the coating composition was affected by the phases in the microstructure. In the XRD analysis, α-Fe, $FeMo_2B_2$, M_2B-type boride, R-$Fe_{63}Mo_{37}$, and trace amounts of the Fe_3B and μ-Fe_7Mo_6-phase were detected. Normally, the R-$Fe_{63}Mo_{37}$ and Fe_3B phase is a non-equilibrium phase formed under rapid cooling conditions [35]. However, it can be seen in trace amounts in a solidified microstructure under dynamic solidification conditions such as the welding process.

SEM images of hardfacing coatings are given in Figure 4. The $Fe_{16}B_4$-based hardfacing coating composition contains 20% atomic boron. During solidification, the Fe_2B phase is expected to solidify first as seen in the Fe-B phase diagram given in Figure 5. As the eutectic temperature drops below the eutectic temperature, which is approximately ~1177 °C, the liquid phase will solidify eutectically. For this reason, primary Fe_2B phases and the eutectic α-Fe + Fe_2B structure can be found in the coating microstructure. However, since the electrode composition passes very close to the eutectic point, the amount of primary phases is expected to be low. In EDS analysis, α-Fe (point 1) represents Fe_2B with a partial block structure (point 2) and a α-Fe + Fe_2B eutectic structure (point 3). Normally, the boron element cannot be detected quantitatively via EDS analysis. However, with EDS analysis, signals can be received from points containing boron. As expected, there were trace amounts of block-structured Fe_2B phases that solidify primarily, and the microstructure largely solidifies as a eutectic structure.

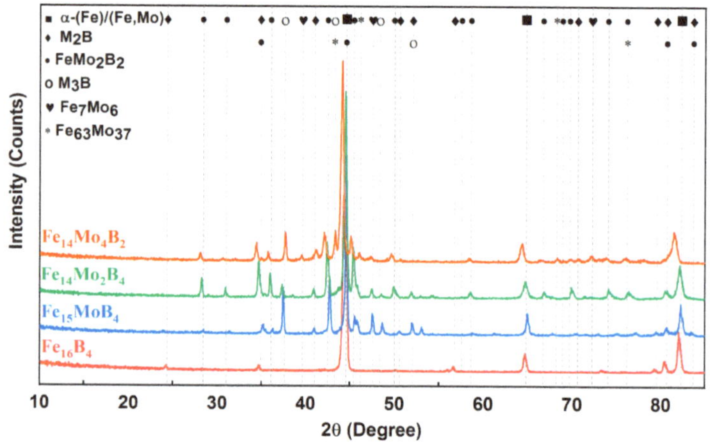

Figure 3. XRD analysis of hardfacing coatings.

It is seen that significant changes occur in the microstructure of the $Fe_{15}MoB_4$-based coating with the addition of molybdenum in the electrode cover composition. According to EDS and MAP analysis, a molybdenum signal is received from these block-structured phases. It is also understood that these block-structured phases contain boron. It is known that for a fixed boron ratio in the Fe-Mo-B system, increasing the amount of molybdenum shifts the solidification to the hyper-eutectic side [36]. On the other hand, the high melting temperature and thermodynamically stable phase in the $Fe_{15}MoB_4$-based coating is the $FeMo_2B_2$ compound [37,38]. For this reason, solidification begins with the formation of $FeMo_2B_2$ solid crystals and this phase grows with a block structure. After the formation of the $FeMo_2B_2$ phase, theoretically, molybdenum is consumed in the melt, but there is still 15% atomic boron. Solidification continues with the formation of α-Fe, similar to a sub-eutectic composition according to the Fe-B phase diagram. For this reason, there are α-Fe islands around the $FeMo_2B_2$ phase. When the temperature drops below the eutectic line, the remaining liquid solidified as a α-Fe + Fe_2B eutectic structure and continued to grow. It indicates that the phase detected as M_2B in the XRD analysis is taken from the α-Fe + M_2B structure, which solidifies eutectically around the block-structured phases. Although it was assumed that this system solidifies according to the Fe-B phase diagram, the eutectic formed is morphologically different from the eutectic structure formed in the $Fe_{16}B_4$-based coating. It is known that molybdenum can be dissolved in the Fe_2B phase [39]. Possibly, molybdenum dissolved in the Fe_2B phase causes this phase to solidify as $(Fe,Mo)_2B$ and the morphology of the eutectic structure changes. A similar microstructure has been reported as a reticular eutectic structure in the literature [36]. Also, a molybdenum signal was received from EDS region number 2. Although the solubility of molybdenum in α-Fe is quite low under normal conditions, its solubility has been reported to be ~24% at ~1450 °C [40,41]. In a process where solidification is dynamic, such as welding, rapid regional or structural cooling might have occurred [33]. For this reason, it was normal for the molybdenum solubility to increase regionally in α-Fe, and the dark gray regions are called α-(Fe,Mo). As a result, block-structured $FeMo_2B_2$, α-(Fe,Mo), and α-Fe + $(Fe,Mo)_2B$ eutectics were expected to form in the final microstructure.

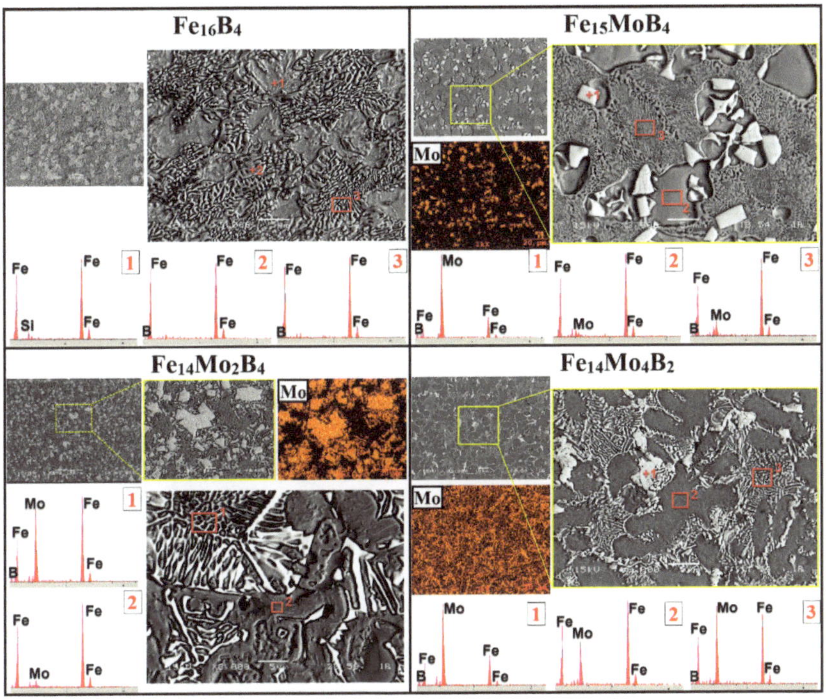

Figure 4. SEM images, EDS, and MAP analysis of coatings layer.

There is a significant increase in the ratio of block-structured $FeMo_2B_2$ phases in the SEM image of the $Fe_{15}MoB_4$-based coating given in Figure 4. The thermodynamic analysis performed with PANDAT software (https://computherm.com/) (Figure 5) show similar phases occurring in the $Fe_{15}MoB_4$- and $Fe_{14}Mo_2B_4$-based coatings. In XRD analysis, it was determined that the $Fe_{14}Mo_2B_4$-based hardfacing coating contained α-Fe, $FeMo_2B_2$ phase, M_2B-type boride, and trace amounts of Fe_3B borides in its microstructure. Solidification is expected to begin with the formation of the $FeMo_2B_2$ phase, similar to the $Fe_{15}MoB_4$-based coating. However, in the microstructure of the $Fe_{14}Mo_2B_4$-based coating, it is seen that the block-structured phases and the eutectic structure change morphologically. It also has been reported in the literature that the $FeMo_2B_2$ phase formed in the Fe-Mo-B system can solidify in different forms and the morphology of this phase can change depending on the Mo/B ratio [38]. In similar studies, it has been reported that a eutectic structure consisting of a $FeMo_2B_2$ and Fe,Mo solid solution can be found around the $FeMo_2B_2$ phase [39]. However, this eutectic structure solidifies into a morphologically floriform structure. Another phase detected in the XRD analysis, the M_2B type phase, indicates the eutectic-solidified Fe_2B phase. As stated in previous studies, Fe_2B can dissolve some molybdenum during solidification and form the $(Fe,Mo)_2B$ phase [39]. The Fe_3B phase detected in trace amounts in the XRD analysis was not visible in the microstructure in a distinguishable way. Yin et al. [36] stated that the Fe_3B phase was distributed within the Fe-M_2B eutectic structure. In the final microstructure, which consists of $FeMo_2B_2$, α-(Fe/Mo) + $FeMo_2B_2$ eutectic (floriform structure that solidifies around $FeMo_2B_2$), α-(Fe/Mo), and α-(Fe/Mo) + $(Fe,Mo)_2B$ (reticular eutectic structure).

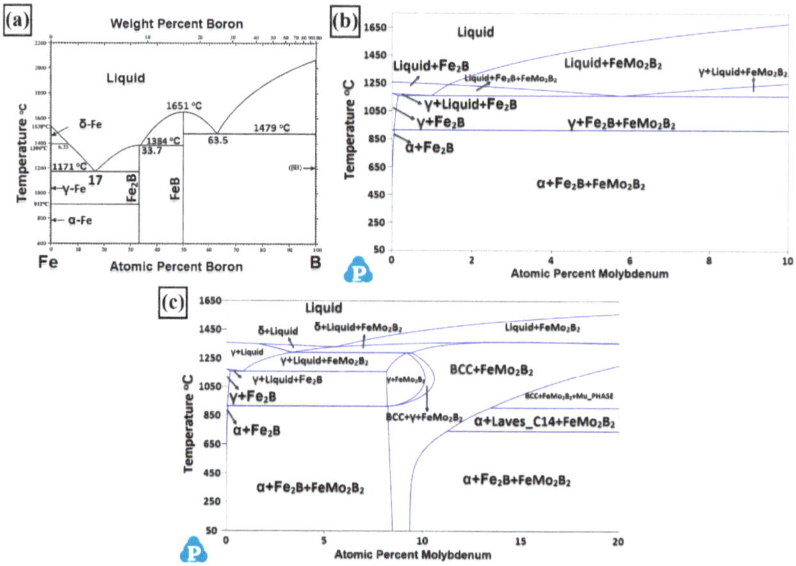

Figure 5. (**a**) Fe-B phase diagram; (**b**,**c**) equilibrium phase diagram of Fe-Mo-B system.

A structure consisting of light-colored phases, eutectic-like structures with the same contrast as these phases, and a matrix phase can be seen in the SEM image of the $Fe_{14}Mo_4B_2$-based coating. According to the thermodynamic analysis performed with PANDAT software (https://computherm.com/), the α-Fe, $FeMo_2B_2$, and Fe_2B phases were expected to be present in the microstructure under stable solidification conditions. However, in XRD analysis, in addition to the α-Fe and $FeMo_2B_2$ phases, μ-Fe_7Mo_6 and R-$Fe_{63}Mo_{37}$ phases were detected. According to the Fe-Mo phase diagram, the R-$Fe_{63}Mo_{37}$ phase is an unstable phase below 1200 °C and indicates solidification at a high cooling rate [25]. A molybdenum signal was detected from white island-shaped phases in the EDS analysis. The first phase expected to form under balanced conditions was the $FeMo_2B_2$ phase. However, the surface of the island-shaped phase shown at EDS point number 1 has a faceted structure, unlike the $FeMo_2B_2$ phase. Possibly, this phase is the R-$Fe_{63}Mo_{37}$ phase, which solidifies out of equilibrium, as stated in the literature [25]. The μ-Fe_7Mo_6 phase, which was detected in trace amounts in the XRD analysis, could not be clearly detected in the microstructure. Previous studies also show that this phase is not clearly distinguishable [25,42]. Eutectic regions indicate the $FeMo_2B_2$ phase detected in XRD analysis. The amount of molybdenum affects not only the morphological properties of $FeMo_2B_2$ phases, but also the morphology of the eutectic structure (α-(Fe,Mo) + $FeMo_2B_2$). For this reason, increasing the amount of molybdenum and decreasing the boron rate affects the distance between the lamellar in the eutectic structure in the $Fe_{14}Mo_4B_2$-based coating.

3.2. Hardness and Wear Tests

Figure 6 shows the hardness test results of hardfacing coatings. In the $Fe_{16}B_4$-, $Fe_{15}MoB_2$-, $Fe_{14}Mo_2B_2$-, $Fe_{14}Mo_4B_2$-based coatings carried out on the AISI 1020 steel substrate, there was an increase of 58%, 70%, 73%, and 66% according to the substrate material, respectively. Additionally, it was observed that an increase in hardness as the amount of molybdenum added to the Fe-B-based hardfacing electrode composition increases. With 5% molybdenum added to the electrode cover composition, an increase in hardness of 22.8% was observed, and with the addition of 10% molybdenum, an increase in HRC hardness of 30.48% was observed compared to the $Fe_{16}B_4$-based coating. On the other hand, the hardness value of the $Fe_{14}Mo_2B_4$-based hardfacing coating was measured as 56.5 HRC, while the hardness value of the $Fe_{14}Mo_4B_2$-based hardfacing coating was measured as 49.9 HRC. In other words, the hardness value of $Fe_{14}Mo_2B_4$-based hardfacing was approximately ~13% higher than that of $Fe_{14}Mo_4B_2$-based hardfacing. This shows that although the hardness increases with molybdenum for Fe-Mo-B-based coatings, the amount of boron in the coating composition has a significant effect on the increase in hardness. In addition, the top values of hardness were related to the phases in the microstructure, and especially the $FeMo_2B_2$ phase plays an important role in increasing the hardness.

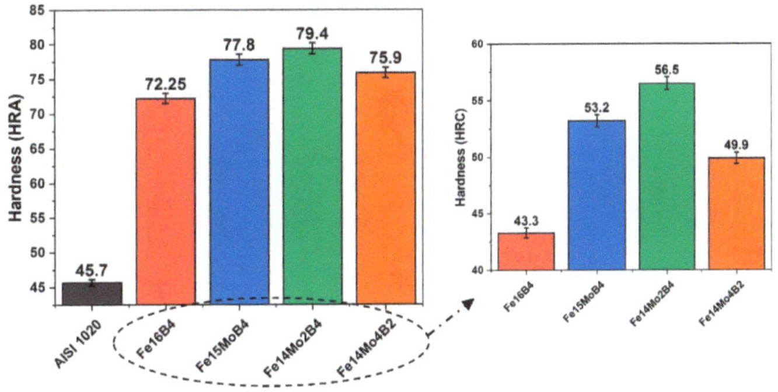

Figure 6. Hardness results of hardfacing coatings.

In the study, the lowest microhardness value was measured from the substrate material as seen in Table 3. Among the coated samples, the lowest matrix phase hardness was measured in the $Fe_{16}B_4$-based hardfacing coating. The matrix phase hardness of the $Fe_{16}B_4$-based coating gives almost the same results as the substrate material. The solubility of boron, an interstitial element, in the α-Fe matrix was quite low. For this reason, the boron in the composition was found in Fe_2B or the eutectic phase formed by this phase. With the addition of atomic 5% molybdenum into the alloy, there was a significant increase in the hardness of the matrix phase. Additionally, it was observed that this increase was not only in the matrix phase, but also in the eutectic structure. Due to the nature of the welding process, molybdenum may dissolve to some extent in the matrix phase and the Fe_2B phase. This situation causes an increase in hardness in the matrix phase as a result of solid solution hardening [43]. Similarly, molybdenum dissolved in the Fe_2B phase causes the hardness of this phase to increase. Although the hardness of the other phases was similar, trace changes were observed depending on the composition, especially in the matrix phase. On the other hand, the matrix hardness of the $Fe_{14}Mo_4B_2$-based coating was less than that of other coatings containing molybdenum. The hardness of the R-$Fe_{63}Mo_{37}$ phase in the microstructure of the same coating was also measured to be lower than the boron-containing phases. As can be seen from previous studies, changes in both microstructural and atomic dimensions change the hardness of the coating [44,45]. The difference in the matrix phase between the compositions might cause distortion in the lattice, causing the hardness to be slightly different. Microstructurally, each phase prevents dislocation movement, but the physical properties of the phase determine the limit values of plastic deformation. As stated in the literature, boron-containing phases play a decisive role on the hardness of the coating [19]. The hardness values measured from the phases in the microstructure have an effect on the hardness. Although the hardness values measured in the study depend on the composition and production method for Fe-Mo-B-based alloys, they appear to be compatible with various studies [46–48].

Table 3. Microhardness results of hardfacing coatings.

Sample	Matrix ($HV_{0.01}$)	Eutectic (α-Fe-M_2B) ($HV_{0.01}$)	Eutectic (α-Fe-$FeMo_2B_2$) ($HV_{0.01}$)	$FeMo_2B_2$ ($HV_{0.01}$)	R-($Fe_{63}Mo_{37}$) ($HV_{0.01}$)
AISI 1020	143–147	-	-	-	-
$Fe_{16}B_4$	173–180	478–542	-	-	-
$Fe_{15}MoB_4$	424–542	996–1200	-	1953–2973	-
$Fe_{14}Mo_2B_4$	459–573	1053–1242	642–956	1970–3228	-
$Fe_{14}Mo_4B_2$	368–379	-	520–754	-	718–840

Figure 7 shows the wear rate graph of hardfacing coatings depending on load and distance. According to the graph, it is understood that the wear rate increases with both the increasing distance and increasing load. In the study, the highest wear rate was measured from the substrate material. Among the coated samples, the highest wear rate was measured in $Fe_{16}B_4$ and the lowest wear rate in the $Fe_{14}Mo_2B4$-based coating. Wear rates exhibit a similar behavior to the hardness results. As Archard stated [49], the wear rate varies inversely with the hardness. The basic philosophy of hardfacing coatings is to cover the surface with a harder layer than the substrate material. As mentioned above, each hard phase in the microstructure of this hard layer formed on the substrate surface changes the effect of the coating on wear. In addition, not only the hardness of these phases, but also their compatibility with the surface, morphology, etc., properties also affect the wear behavior of the coating. In this study, the hard eutectic α-Fe + Fe_2B structure, which forms the microstructure of the $Fe_{16}B_4$-based coating, increases the wear resistance of the substrate material. The high-hardness $FeMo_2B_2$ phase formed in the microstructure with the addition of molybdenum and the different morphologies of this phase cause further increase in wear resistance. On the other hand, although the wear resistance increases with the increasing molybdenum amount, the wear resistance of the $Fe_{14}Mo_2B_4$-based coating was measured to be higher at all loads and distances than the wear resistance of the $Fe_{14}Mo_4B_2$-based coating. Boron is a critical element in the formation of hard phases, and the volumetric fraction of the phases formed by boron affects the wear resistance of the coating. In addition, although boron provides the formation of the hard phase, the matrix phase is critical in terms of both hardness and wear resistance. The fact that molybdenum is soluble in the matrix phase and increases its hardness resultingly has an effect on wear. Likewise, both the hardness and wear resistance of the $Fe_{16}B_4$-based coating, where the matrix phase is soft, were measured to be lower than the $Fe_{14}Mo_4B_2$-based coating.

Figure 8 shows the friction coefficient graphs obtained after the wear test at different loads and distances applied to hardfacing coatings. Friction coefficient values vary between ~0.4 and 0.7 μ. According to the friction coefficient data of the coated samples, it can be said that the highest friction coefficient value was obtained in $Fe_{16}B_4$-based hardfacing and the lowest friction coefficient values were obtained in $Fe_{14}Mo_4B_2$-based hardfacing. The overall friction coefficient is closely related to the surface profile and phases in the microstructure [50–52]. In composite materials such as hardfacing, each phase has a different effect on wear. On the other hand, the hardness, morphology, volumetric ratio, and distribution of these phases are other parameters that affect the friction coefficient. The friction coefficients given in Figure 8 act independently of the surface hardness. However, it can be said that the smoothest surface among the samples is on the $Fe_{14}Mo_4B_2$-based hardfacing coating after the substrate material. On the other hand, the composite structure seen in the microstructures of other coatings is likely to create resistance during sliding. These composite structures might cause an increase in the coefficient of friction. Similar studies in the literature have stated that the coefficient of friction increases with the increase in the amount of hard phase in the metal matrix composite [53]. In the study, no correlation was found in the change in friction coefficients according to load and distance.

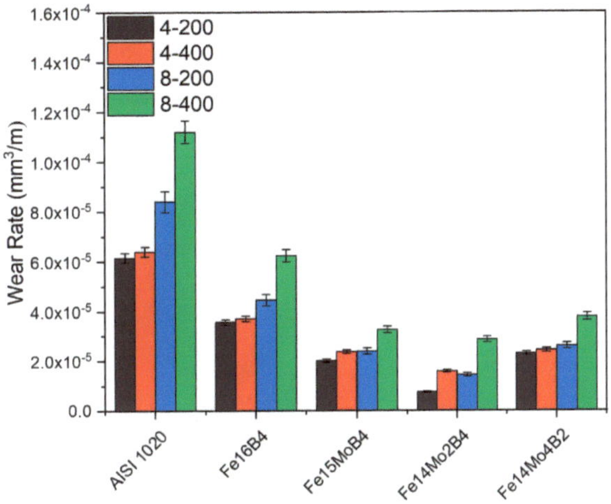

Figure 7. Wear rate of samples.

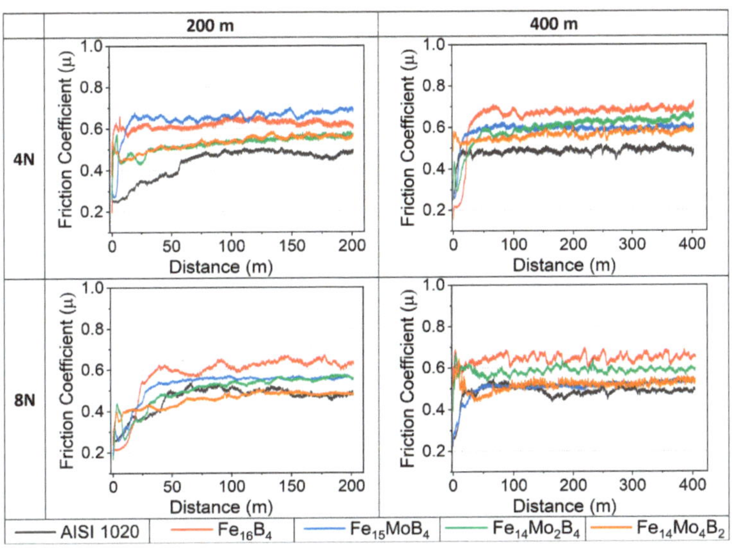

Figure 8. Friction coefficient graph of hardfacing coatings.

In the EDS analysis of the substrate material shown in Figure 9, an intense amount of oxygen signal was received from the dark areas. The formation of an FeO-based oxide layer during wear on the carbon steel surface has been reported in similar studies [54]. The microstructure of the substrate material did not contain factors such as grain boundaries, secondary phases, and eutectic structure that would create resistance during wear compared to the coated samples. For this reason, friction occurs on a smoother surface. In the friction coefficient graph given in Figure 8, it reached the steady-state wear phase after a short running-in period. The homogeneous oxide layer formed on the surface might acted as a solid lubricant, causing the friction coefficient to be relatively lower than other samples. It is seen that this oxide layer forms debris at some points. Local adhesions may occur as a result of micro weld in softer and easily oxidized substrate material. Increasing heat during friction and the change in surface topography due to the oxide layer formed may also cause fluctuations in the friction coefficient [55]. On the other hand, oxide-based debris formed as a result of the delamination of the oxide layer was seen in the SEM image of the substrate material. It was understood that wear was dominated by oxidative and adhesive wear mechanism. The worn surface appearance of the hardfacing-coated samples shown in Figure 9 shows that the oxidized parts remained more localized compared to the substrate material.

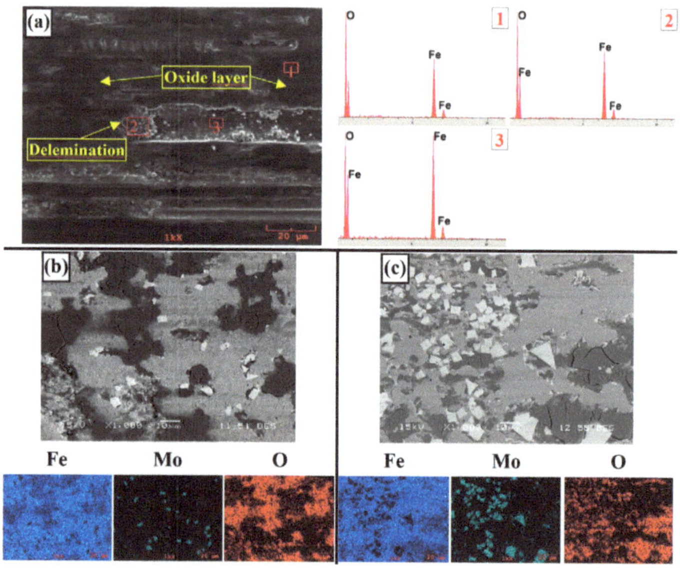

Figure 9. Worn surfaces. (**a**) SEM images and EDS analysis of AISI 1020 steel substrate; (**b**) SEM image and MAP analysis of $Fe_{15}MoB_4$-based coating; and (**c**) SEM image and MAP analysis of $Fe_{14}Mo_2B_4$-based coating.

As a result of the SEM analysis performed on the coated samples, different mechanisms played an active role during wear. It can be seen that debris and delaminations were formed on the surface of the $Fe_{16}B_4$-based hardfacing coating shown in Figure 10a after wear. Debris might have occurred due to the sawing effect created by the eutectic phase during friction. It was also observed that delamination occurs in the soft (matrix) phase during friction in the eutectic structure. This indicates that it might be pulled out due to adhesion formed in the soft matrix phase with the increasing heat in the later stages of friction. With the addition of molybdenum, the delamination of hard phases formed in the microstructure decreased by reducing the oxide detachment on the surface. However, the abrasive wear increases with the increasing hard phase friction and micro-abrasive lines appear on the surface. One of the basic mechanisms expected to develop during the wear of hardfacing coatings is micro-abrasive wear, which was expected to occur due to a three-body wear mechanism. As a result of the micro-abrasive wear mechanism, small pieces pulled out from the friction surfaces were plastered on the surface, causing wear called smeared [56–58]. The SEM image and MAP analysis shown in Figure 9 indicate this situation, where cracks occur directly in oxidized areas. Similarly, the cracks formed in Figure 10b and c indicate removal from the surface after smearing. On the other hand, block-structured phases formed via the addition of molybdenum caused a large hardness difference between the matrix and the particle. The difference in hardness between the matrix and the hard phase might have caused fatigue in the matrix phase. Fatigue occurring during wear might cause fatigue cracks and subsequent ruptures after a while, as shown in the SEM image of the $Fe_{15}MoB_2$-based coating given in Figure 9b. As a result, hard phases such as M_2B and $FeMo_2B_2$ play an active role during friction. These hard phases and eutectic structures were in direct contact with the abrasive. For this reason, the increasing amount of block-structured or eutectic hard phase acts in the direction of the increasing wear resistance. Similar results have been reported in previous studies [27]. In the study, it was determined that micro-abrasive, adhesive, and oxidative wear were the dominant wear mechanism.

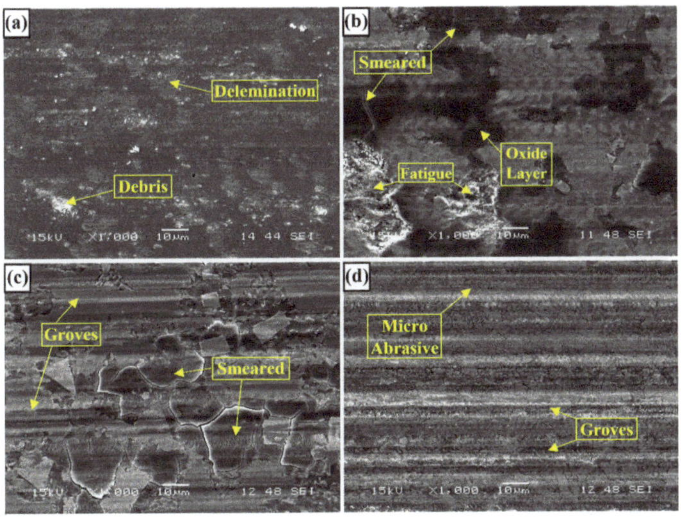

Figure 10. Wear surface of hardfacing coatings (**a**) Fe16B4, (**b**)$Fe_{15}MoB_4$, (**c**) $Fe_{14}Mo_2B_4$, and (**d**) $Fe_{14}Mo_4B_2$.

3.3. Corrosion Behavior

OCP curves are given in Figure 11a. The OCP curves become stable in a short time. As stated in the literature, when the open circuit potential reaches equilibrium, it indicates that the film formation and dissolution on the surface are in balance [59]. In all OCP curves, there was first a slight movement towards the anodic side and then the curves became stable. The OCP value of the substrate material was measured to be slightly more positive than the coated samples. This might be due to the oxide layer that forms more easily on the surface of the substrate material. According to the OCP test, the anodic corrosion tendency decreased with the increasing molybdenum amount in the coating composition. The OCP value of the $Fe_{16}B_4$-based coating was measured to be similar to molybdenum-containing coatings.

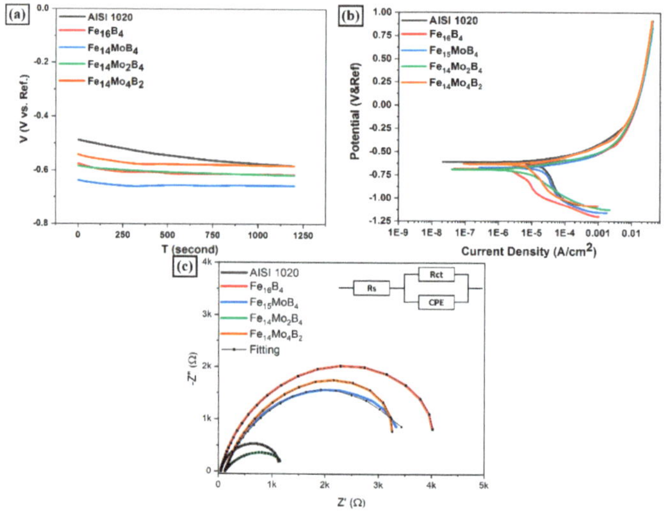

Figure 11. (**a**) OCP diagram and (**b**) Tafel curves of coating layers; (**c**) EIS diagram.

Tafel curves obtained from the samples subjected to the electrochemical corrosion test in the 0.5 M NaCl solution are given in Figure 11b. The corrosion potential obtained from these curves gives information about the direction of corrosion [27,60]. According to the curves, there was no big difference in corrosion potentials. However, the most noble sample in terms of corrosion potential is the AISI 1020 steel substrate. This shows the sample substrate material with the lowest corrosion tendency according to its corrosion potential. According to the corrosion parameters given in Table 4, it is seen that the addition of molybdenum to the molybdenum Fe-B-based coating composition does not cause a linear change on the corrosion potential. Among the coated samples, the noblest corrosion potential was measured in the $Fe_{14}Mo_4B_2$-based coating. On the other hand, current density, which is a critical parameter in the kinetic interpretation of corrosion, is directly related to the rate of corrosion. According to corrosion current density values, the lowest current density was measured in the $Fe_{16}B_4$-based coating. Many researchers have reported that Fe-B alloys near the eutectic point have very good corrosion resistance [61–64]. There was a decrease in the current density values with the addition of molybdenum into the $Fe_{16}B_4$-based coating. In other words, the corrosion resistance of the coating has increased.

Table 4. Tafel parameters of hardfacing coatings.

Sample	E_{cor} (mV)	I_{cor} (μA/cm^2)	Cr (mpy)
AISI 1020	−609.315	28.331	16.291
$Fe_{16}B_4$	−683.455	2.078	1.544
$Fe_{15}MoB_4$	−669.678	15.287	11.245
$FeMo_2B_4$	−688.666	6.601	4.808
$FeMo_4B_2$	−632.627	5.650	3.603

Another method used to understand corrosion in kinetic terms is the EIS method. The similarity of the EIS curves seen in Figure 11c indicates that similar corrosion mechanisms have developed in the coating layers. On the other hand, the diameter of the capacitive loop varies in direct proportion to the corrosion resistance. The corrosion resistance of the coating with a large capacitive loop diameter is high [65,66]. EIS test parameters obtained by equivalent circuit modeling are given in Table 5. According to the model, Rs is a measure of the solution resistance and Rct is a measure of the polarization resistance. A higher Rct value indicates higher corrosion resistance. The fact that both the capacitive loop diameter and the Rct value are large indicates that the corrosion resistance of the $Fe_{16}B_4$-based coating is higher. This can be associated with the microstructure of the $Fe_{16}B_4$-based coating. Factors such as phases and grain boundaries within the microstructure increase the polarization resistance [67]. Rct resistance increases especially in $Fe_{16}B_4$-based coatings that contain a high fraction of eutectic structure. According to the EIS data, the AISI 1020 steel substrate, which does not contain grain boundaries in the microstructure, has low corrosion resistance. Although the microstructure of the $Fe_{14}Mo_2B_4$-based coating is complex, the Rct value was low. This may be related to the morphology and composition of the eutectic structure.

Table 5. EIS test parameters.

Sample	Rs (Ω)	CPE-1 (μF.cm^{-2})	Rct-1 (kΩ)
AISI 1020	42.78	2.2×10^{-4}	1.29
$Fe_{16}B_4$	40.41	2.29×10^{-4}	4.87
$Fe_{15}MoB_4$	126.3	1.08×10^{-4}	4.25
$Fe_{14}Mo_2$B4	134.4	2.43×10^{-4}	1.4
$Fe_{14}Mo_4B_2$	129	1.11×10^{-4}	4.38

In the post-corrosion SEM image of the substrate material shown in Figure 12, it is seen that a cotton-like layer is covering the surface. In the EDS analysis carried out in this region, intense signals belonging to Fe and Cl elements were received. This indicates that the iron on the surface is dissolved as a result of the anodic reaction and ferrous salts are formed on the surface.

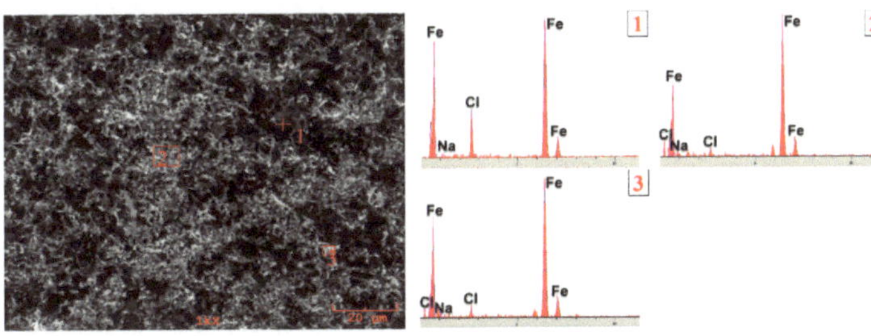

Figure 12. SEM image and EDS analysis of corroded surface of AISI 1020 steel substrate.

The reaction of the anodic dissolution occurring on the substrate surface is the oxidation reaction of α-Fe and is expected to occur according to the reaction given below. In the SEM image, it is understood that this reaction develops homogeneously on the surface. This type of homogeneous corrosion is a less dangerous corrosion mechanism, and generally, the homogeneous oxide layer formed on the surface polarizes the surface. However, the breakdown of this layer accelerates corrosion.

Anodic reaction [68,69]:

$$Fe \rightarrow Fe^{+2} + 2e^{-}$$

Cathodic Reaction:

$$1/2O_2 + H_2O + 2e^- \rightarrow 2OH^-$$

direct oxygen reduction of the cathodic reaction

$$2Fe + 2H_2O + O_2 \rightarrow 2Fe^{+2} + 4OH^- \rightarrow 2Fe(OH)_2$$

Figure 13 shows SEM images of hardfacing coatings after corrosion. In the images, there was no corrosion damage in the parts formed by $FeMo_2B_2$, R-$Fe_{63}Mo_{37}$ and these phases within this eutectic, but there was a significant amount of dissolution in the α-Fe/(Fe,Mo) regions between the matrix phase and the eutectic phase. This situation indicates the formation of galvanic corrosion caused by the potential difference between the boride phase and α-Fe. Also, this results in selective corrosion of the matrix phase.

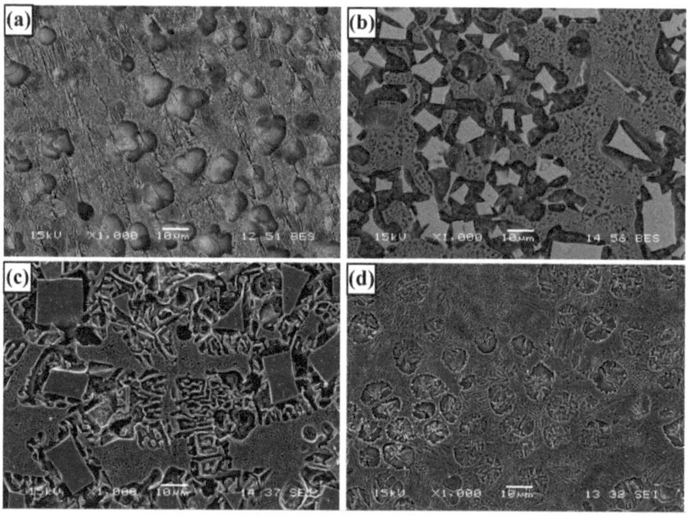

Figure 13. Corroded surface of (**a**) $Fe_{16}B_4$-, (**b**) $Fe_{15}MoB_4$-, (**c**) $Fe_{14}Mo_2B_4$-, and (**d**) $Fe_{14}Mo_4B_2$-based hardfacing coatings.

In the $Fe_{16}B_4$-based coating, the matrix phase consists of α-Fe and dissolution can occur according to the reactions given above. The molybdenum added to the $Fe_{16}B_4$-based coating might have increased the corrosion resistance of the matrix by dissolving it in the matrix phase. Similarly, there was a significant dissolution in the phases around the $FeMo_2B_2$ phase. As a result of the potential difference between the $FeMo_2B_2$ phase and the surrounding phases, it may have an accelerating effect on dissolution. Similarly, it has been reported in the literature that the low molybdenum-containing phase in the region adjacent to the high molybdenum-containing phase in the eutectic structure will have corroded preferentially and caused intergranular corrosion [70]. It was also clearly seen in the SEM images. In the SEM image and MAP analysis given in Figure 14, the iron signal received from the eutectic region shows that there was no dissolution in these regions. That is, both M_2B and $FeMo_2B_2$ phases behave as noble, with respect to the matrix phase.

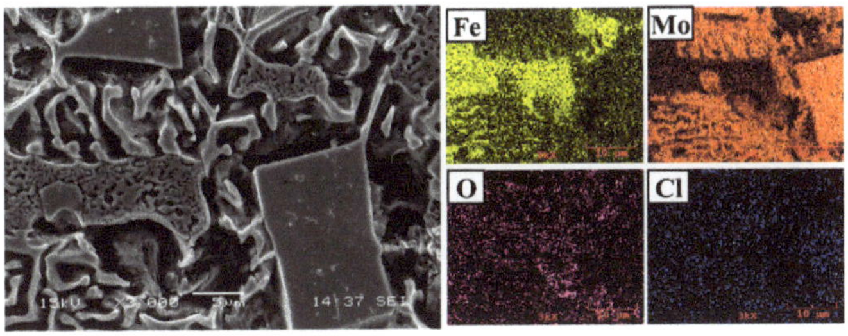

Figure 14. Corroded surface of $Fe_{14}Mo_2B_4$-based coating.

4. Conclusions

In this study, it was aimed to investigate the effect of molybdenum on an Fe-B-based hardfacing coating. For this purpose, Fe-Mo-B-based hardfacing electrodes with different compositions were produced. These produced electrodes were observed microstructurally and hardness, wear, and corrosion tests were performed. Additionally, the results were compared with an AISI 1020 steel substrate and Fe-B-based hardfacing. The findings obtained in the study are as follows;

- It has been determined that changes made in the electrode cover composition change the phases in the microstructure and that even trace changes have a significant morphological effect on some phases. In the study, α-Fe, $FeMo_2B_2$, Fe_2B, and R-$Fe_{63}Mo_{37}$ phases were detected as major phases, and Fe_3B and Fe_7Mo_6 phases were detected as minor phases. It has also been determined that molybdenum can dissolve in the α-Fe and Fe_2B phase and can affect the morphological and mechanical properties of both these phases and the eutectic structures formed by these phases.
- Neither molybdenum nor boron could provide the effect of both on the hardness and wear resistance of the coating. In addition, this effect reaches its maximum level for mixtures made in certain proportions. For this reason, optimizing the compositions in hardfacing coating works was critical. In the study, the highest hardness value was obtained in the $Fe_{14}Mo_2B_4$-based coating as 56.4 HRC. It was observed that the hardness of this coating was ~73% higher than the substrate material and ~30.5% higher than the $Fe_{16}B_4$-based coating. According to microhardness measurements, although the hardness of the phases in the microstructure varies over a wide range, the highest phase hardness was measured as 3228 HV in the $FeMo_2B_2$ phase. In the study, the highest wear resistance was obtained in the $Fe_{14}Mo_2B_4$-based coating. According to the wear rate values, up to ~8.1 times higher wear resistance was obtained in the $Fe_{14}Mo_2B_4$-based coating compared to the AISI 1020 substrate material and up to ~4.7 times higher than the $Fe_{16}B_4$-based coating.
- According to the corrosion test results, it was observed that there was no significant difference between the corrosion potentials of the substrate material and the coated samples. However, a significant difference was detected between the corrosion current density values. The current density of all samples with hardfacing coating is lower than the base material, meaning their corrosion resistance was better. In the study, the lowest current density value was measured as 2.078 $\mu A/cm^2$ in the $Fe_{16}B_4$-based coating and it was found to be ~13.6 times more resistant to corrosion than the substrate material. Although the corrosion resistance of the $Fe_{16}B_4$-based coating, that is, molybdenum-free, was high, it has been determined that the corrosion resistance increases with the increasing molybdenum amount in molybdenum-containing hardfacing coatings.
- From this study, it can be concluded that molybdenum is a good alternative that can be added to Fe-B-based hardfacing alloys to improve the properties of the coating.

Furthermore, extending this study to reveal the effect of high temperature wear behavior on the coating will be beneficial for potential applications.

Funding: The study has been supported by the Scientific and Technological Research Council of Türkiye (TÜBİTAK) project number 219M195.

Institutional Review Board Statement: Not applicable.

Informed Consent Statement: Not applicable.

Data Availability Statement: The data cannot be made publicly available upon publication because there is no suitable repository that exists for hosting data in this field of study. The data that support the findings of this study are available upon reasonable request from the authors.

Acknowledgments: The author would like to thank Zonguldak Bulent Ecevit University and Sakarya University, whose laboratory facilities were used to carry out the study.

Conflicts of Interest: The author declares no conflict of interest.

References

1. Zhang, K.; Zheng, Z.; Zhang, L.; Liu, Y.; Chen, S. Method for Dynamic Prediction of Oxygen Demand in Steelmaking Process Based on BOF Technology. *Processes* **2023**, *11*, 2404. [CrossRef]
2. Steel Pipe Coatings Market to Hit $14.0 Billion by 2033. *Focus Powder Coat.* **2023**, *2023*, 1–6. [CrossRef]
3. Pelser, W.A.; Marais, J.H.; van Laar, J.H.; Mathews, E.H. Development and Application of an Integrated Approach to Reduce Costs in Steel Production Planning. *Process. Integr. Optim. Sustain.* **2022**, *6*, 819–836. [CrossRef]
4. Schubert, E.; Klassen, M.; Zerner, I.; Walz, C.; Sepold, G. Light-Weight Structures Produced by Laser Beam Joining for Future Applications in Automobile and Aerospace İndustry. *J. Mater. Process. Technol.* **2001**, *115*, 2–8. [CrossRef]
5. Dong, Q.; Chen, X.; Gao, Y.; Hu, J.; Chen, X.; Xu, G. *Steel BT—Civil Engineering Materials for Transportation Infrastructure*; Dong, Q., Chen, X., Gao, Y., Hu, J., Chen, X., Xu, G., Eds.; Springer Nature: Singapore, 2023; pp. 267–296; ISBN 978-981-99-1300-8.
6. Ren, X.; Fu, H.; Xing, J.; Yi, Y. Research on High-Temperature Dry Sliding Friction Wear Behavior of Ca Ti Modified High Boron High Speed Steel. *Tribol. Int.* **2019**, *132*, 165–176. [CrossRef]
7. Qi, Y.; Luo, H.; Zheng, S.; Chen, C.; Lv, Z.; Xiong, M. Effect of Temperature on the Corrosion Behavior of Carbon Steel in Hydrogen Sulphide Environments. *Int. J. Electrochem. Sci.* **2014**, *9*, 2101–2112. [CrossRef]
8. Liu, D.S.; Liu, R.P.; Wei, Y.H.; Pan, P. Properties of Cobalt Based Hardfacing Deposits with Various Carbon Contents. *Surf. Eng.* **2013**, *29*, 627–632. [CrossRef]
9. Hutchings, I.; Shipway, P. *7—Surface Engineering*, 2nd ed.; Hutchings, I., Shipway, P.B.T.-T., Eds.; Butter-worth-Heinemann: Oxford, UK, 2017; pp. 237–281. ISBN 978-0-08-100910-9.
10. Burkov, A.A.; Kulik, M.A. Wear-Resistant and Anticorrosive Coatings Based on Chrome Carbide Cr_7C_3 Obtained by Electric Spark Deposition. *Prot. Metall. Phys. Chem. Surf.* **2020**, *56*, 1217–1221. [CrossRef]
11. Bazhin, P.; Titov, N.; Zhidovich, A.; Avdeeva, V.; Kolomeichenko, A.; Stolin, A. Features of the Carbo-Vibroarc Surfacing in the Development of Multicomponent Cermet Wear-Resistant Coatings. *Surf. Coat. Technol.* **2022**, *429*, 127952. [CrossRef]
12. Vasilescu, M.; Dobrescu, M. Hardfacing Corrosion and Wear Resistant Alloys. *Adv. Mater. Res.* **2015**, *1114*, 196–205. [CrossRef]
13. Kocaman, E.; Kılınç, B.; Şen, Ş.; Şen, U. Development of Surface Properties with In Situ TiB2 Intermetallic-Assisted Coating by Fe(18-X)Ti2BX (x = 3,4,5)-Based Electrodes. *Arab. J. Sci. Eng.* **2021**, *47*, 8485–8501. [CrossRef]
14. Trembach, B. Comparative Studies of the Three-Body Abrasion Wear Resistance of Hardfacing Fe-Cr-C-B-Ti Alloy. *IOP Conf. Ser. Mater. Sci. Eng.* **2023**, *1277*, 012016. [CrossRef]
15. Ardigo-Besnard, M.; Tellier, A.; Besnard, A.; Chateau-Cornu, J.-P. Effect of the Microstructure on the Tribological Properties of HIPed and PTA-Welded Fe-Based Hardfacing Alloy. *Surf. Coat. Technol.* **2021**, *425*, 127691. [CrossRef]
16. Kılınç, B.; Kocaman, E.; Şen, Ş.; Şen, U. Effect of Vanadium Content on the Microstructure and Wear Behavior of Fe(13-x)VxB7 (x = 0–5) Based Hard Surface Alloy Layers. *Mater. Charact.* **2021**, *179*, 111324. [CrossRef]
17. Chen, C.; Wang, J.; Ge, Y.; Zhuang, M.; Ma, Z. Microstructure and Wear Resistance of High-Chromium Cast Iron with Multicomponent Carbide Coating via Laser Cladding. *Coatings* **2023**, *13*, 1474. [CrossRef]
18. Durmuş, H.; Çömez, N.; Gül, C.; Yuddaşkal, M.; Uzun, R.O. Ferromolibden ve Ferrobor Takviyeli Lazer Kaplamaların Aşınma Karakteristiği ve Mikroyapısı. *DÜMF Mühendislik Derg.* **2019**, *10*, 1009–1017. [CrossRef]
19. Zhang, H.; Pan, Y.; Zhang, Y.; Lian, G.; Cao, Q.; Yang, J. Sensitivity Analysis for Process Parameters in Mo2FeB2 Ternary Boride Coating by Laser Cladding. *Coatings* **2022**, *12*, 1420. [CrossRef]
20. Belov, D.S.; Blinkov, I.V.; Sergevnin, V.S.; Chernogor, A.V.; Demirov, A.P.; Polyanskii, A.M. Structure and Phase Formation in Arc PVD Zr–B–Si–C–Ti–(N) Coatings. *Inorg. Mater.* **2023**, *59*, 157–163. [CrossRef]
21. Prysyazhnyuk, P.; Shlapak, L.; Ivanov, O.; Korniy, S.; Lutsak, L.; Burda, M.; Hnatenko, I.; Yurkiv, V. In Situ Formation of Molybdenum Borides at Hardfacing by Arc Welding with Flux-Cored Wires Containing a Reaction Mixture of B4C/Mo. *Eastern-Eur. J. Enterp. Technol.* **2020**, *4*, 46–51. [CrossRef]

22. Wang, Y.-W.; Yang, Y.; Wang, X.-Y.; Li, W.; Tian, W. Microstructure and Wear Properties of İn-Situ Molybdenum Composite Coatings by Plasma Spraying Mo-B4C and MoO3-Al-B4C Composite Powders. *Surf. Coat. Technol.* **2023**, *469*, 129769. [CrossRef]
23. Ye, F.; Zhang, Y.; Lou, Z.; Wang, Y. Enhanced Tribological Performance of Micro-Beam Plasma-Cladded Ni60 Coatings with Addition of Mo and Ag Lubricants in a Wide Temperature Range. *Coatings* **2023**, *13*, 1996. [CrossRef]
24. Sun, M.; Pang, M. Defect Formation Mechanism and Performance Study of Laser Cladding Ni/Mo Composite Coating. *Coatings* **2021**, *11*, 1460. [CrossRef]
25. Deng, X.; Zhang, G.; Wang, T.; Ren, S.; Bai, Z.; Cao, Q. Investigations on Microstructure and Wear Resistance of Fe-Mo Alloy Coating Fabricated by Plasma Transferred arc Cladding. *Surf. Coat. Technol.* **2018**, *350*, 480–487. [CrossRef]
26. Dilawary, S.A.A.; Motallebzadeh, A.; Paksoy, H.A.; Akhter, R.; Atar, E.; Cimenoglu, H. LSM of Stellite 12 + 10wt % Mo Hardfacing Alloy for Enhancement in Wear Resistance. In Proceedings of the XXXI International Conference on Surface Modification Technologies (SMT31), Mons, Belgium, 5–7 July 2017.
27. Li, Q.; Wang, Q.; Zhang, L.; Chen, D.X.; Jin, H.; Li, J.D.; Zhang, J.W. Microstructure, Wear and Electrochemical Behaviors of Laser Cladding Fe-Based Coatings with Various Molybdenum Contents. *Mater. Res. Express* **2022**, *9*, 026504. [CrossRef]
28. Jiang, R.S.; Wang, Y.T.; Hu, L.; Xu, G.; De Liu, Z. Effect of Mo Content on the Corrosion Resistance of Fe-Based Amorphous Composite Coating. *Mater. Sci. Forum* **2016**, *849*, 636–641. [CrossRef]
29. Bahmaeia, M.; Pourzarea, R.; Ghezloo, A. The Effect of Cerium Pre-Treatment on the Corrosion Resistance of Steel Sheets. *Russ. J. Gen. Chem.* **2013**, *83*, 2386–2391. [CrossRef]
30. Hazza, M.; El-Dahshan, M. The Effect of Molybdenum on the Corrosion Behaviour of Some Steel Alloys. *Desalination* **1994**, *95*, 199–209. [CrossRef]
31. Rockel, M.B. The Effect of Molybdenum on the Corrosion Behavior of Iron-Chromium Alloys. *Corrosion* **2013**, *29*, 393–396. [CrossRef]
32. Kocaman, E.; Kılınç, B.; Şen, Ş.; Şen, U. In-situ TiB2 and Fe2Ti İntermetallic Assisted Hard Coatings by Fe-Ti-B Based Hardfacing electrodes. *J. Alloys Compd.* **2022**, *900*, 163478. [CrossRef]
33. Kocaman, E.; Kilinç, B.; Şen, S.; Şen, U. Effect of Chromium Content on $Fe_{(18-x)}Cr_xB_2$(X=3,4,5) Hardfacing Electrode on Microstructure, Abrasion and Corrosion Behavior. *J. Fac. Eng. Archit. Gazi Univ.* **2020**, *36*, 177–190. [CrossRef]
34. Duraisamy, R.; Kumar, S.M.; Kannan, A.R.; Shanmugam, N.S.; Sankaranarayanasamy, K.; Ramesh, M. Tribological Performance of Wire Arc Additive Manufactured 347 Austenitic Stainless Steel under Unlubricated Conditions at Elevated Temperatures. *J. Manuf. Process.* **2020**, *56*, 306–321. [CrossRef]
35. Sunbul, S.E.; Akyol, S.; Onal, S.; Ozturk, S.; Sozeri, H.; Icin, K. Effect of Co, Cu, and Mo Alloying Metals on Electrochemical and Magnetic Properties of Fe-B Alloy. *J. Alloys Compd.* **2023**, *947*, 169652. [CrossRef]
36. Ouyang, X.; Chen, G.; Yin, F.; Liu, Y.; Zhao, M. Effect of Molybdenum on the Microstructures of As-Cast Fe-B Alloys and Their Corrosion Resistance in Molten Zinc. *Corrosion* **2017**, *73*, 942–952. [CrossRef]
37. Beardsley, M.B.; Sebright, J.L. *Structurally Integrated Coatings for Wear and Corrosion*; Caterpillar Inc.: Peoria, IL, USA, 2018.
38. Wang, H.Q.; Sun, J.S.; Li, C.N.; Geng, S.N.; Sun, H.G.; Wang, G.L. Microstructure and Mechanical Properties of Molybdenum–İron–Boron–Chromium Cladding Using Argon arc Welding. *Mater. Sci. Technol.* **2016**, *32*, 1694–1701. [CrossRef]
39. Sarasola, M.; Gómez-Acebo, T.; Castro, F. Microstructural Development during Liquid Phase Sintering of Fe and Fe–Mo Alloys Containing Elemental Boron Additions. *Powder Metall.* **2005**, *48*, 59–67. [CrossRef]
40. Guillermet, A.F. The Fe−Mo (Iron−Molybdenum) System. *Bull. Alloy. Phase Diagr.* **1982**, *3*, 359–367. [CrossRef]
41. Cho, I.-S.; Savelyev, K.D.; Golod, V.M. Development of Thermophysical Calculator for Stainless Steel Casting Alloys by Using CALPHAD Approach. *China Foundry* **2017**, *14*, 353–358. [CrossRef]
42. Utkin, S.; Bondar, A.; Kublii, V.; Kapitanchuk, L.; Tikhonova, I. Solidus Surface of the Mo–Fe–B System. *Powder Metall. Met. Ceram.* **2020**, *59*, 89–105. [CrossRef]
43. Chen, K.; Yang, X.; Li, W.; Xia, G.; Wang, S.; Wang, K. Study on the Wear and Corrosion Resistance of Fe–Mo Coatings on 65Mn Steel Ploughshares by Laser Cladding. *Appl. Phys. A* **2022**, *128*, 795. [CrossRef]
44. Huang, Z.; Xing, J.; Guo, C. Improving Fracture Toughness and Hardness of Fe2B in High Boron White Cast İron by Chromium Addition. *Mater. Des.* **2010**, *31*, 3084–3089. [CrossRef]
45. Lentz, J.; Röttger, A.; Theisen, W. Hardness and Modulus of Fe_2B, Fe_3(C,B), and $Fe_{23}(C,B)_6$ Borides and Carboborides in the Fe-C-B System. *Mater. Charact.* **2018**, *135*, 192–202. [CrossRef]
46. Azakli, Y.; Cengiz, S.; Tarakci, M.; Gencer, Y. Characterisation of Boride Layer Formed on Fe–Mo Binary Alloys. *Surf. Eng.* **2016**, *32*, 589–595. [CrossRef]
47. Hocaoğlu, R. *Fe-Mo-Ti-B-C Esaslı Sert Dolgu Alaşımlı Örtülü Elektrotların Üretimi ve Özelliklerinin Incelenmesi = The Production and Investigation of the Properties of Fe-Mo-Ti-B-C Based Hardfacing Alloy Covered Electrodes*; Sakarya University: Serdivan, Turkey, 2021.
48. Márquez-Herrera, A.; Fernandez-Muñoz, J.; Zapata-Torres, M.; Melendez-Lira, M.; Cruz-Alcantar, P. Fe_2B coating on ASTM A-36 Steel Surfaces and İts Evaluation of Hardness and Corrosion Resistance. *Surf. Coat. Technol.* **2014**, *254*, 433–439. [CrossRef]
49. Archard, J.F. Contact and Rubbing of Flat Surfaces. *J. Appl. Phys.* **1953**, *24*, 981–988. [CrossRef]
50. Sedlaček, M.; Podgornik, B.; Vižintin, J. Influence of Surface Preparation on Roughness Parameters, Friction and Wear. *Wear* **2009**, *266*, 482–487. [CrossRef]
51. Chapter 8 Lubrication of Solid Surfaces. In *Surface Effects in Adhesion, FrictiON, Wear, and Lubrication*; Buckley, D.H.B.T.-T.S. (Ed.) Elsevier: Amsterdam, The Netherlands, 1981; Volume 5, pp. 511–552; ISBN 0167-8922.

52. Szeri, A.Z. *Tribology*, 3rd ed.; Meyers, R.A., Ed.; Academic Press: New York, NY, USA, 2003; pp. 127–152. ISBN 978-0-12-227410-7.
53. Kalel, S.M.; Patil, S.R. Ceramic Reinforced Metal Matrix Composite (MMC)—Processing. *Int. J. Adv. Res. Sci. Eng.* **2018**, *7*, 1047–1058.
54. Lee, H.-Y. Effect of Changing Sliding Speed on Wear Behavior of Mild Carbon Steel. *Metall. Mater. Int.* **2020**, *26*, 1749–1756. [CrossRef]
55. Menezes, P.L.; Kishore; Kailas, S.V.; Lovell, M.R. Role of Surface Texture, Roughness, and Hardness on Friction During Unidirectional Sliding. *Tribol. Lett.* **2011**, *41*, 1–15. [CrossRef]
56. Sen, U. Wear Properties of Niobium Carbide Coatings Performed by Pack Method on AISI 1040 Steel. *Thin Solid Films* **2005**, *483*, 152–157. [CrossRef]
57. Sen, U. Friction and Wear Properties of Thermo-Reactive Diffusion Coatings against Titanium Nitride Coated Steels. *Mater. Des.* **2005**, *26*, 167–174. [CrossRef]
58. Sen, S.; Sen, U. Sliding Wear Behavior of Niobium Carbide Coated AISI 1040 Steel. *Wear* **2008**, *264*, 219–225. [CrossRef]
59. Zhao, H.; Xie, L.; Xin, C.; Li, N.; Zhao, B.; Li, L. Effect of Molybdenum Content on Corrosion Resistance and Corrosion Behavior of Ti-Mo Titanium Alloy in Hydrochloric Acid. *Mater. Today Commun.* **2023**, *34*, 105032. [CrossRef]
60. Guo, R.; Zhang, C.; Chen, Q.; Yang, Y.; Li, N.; Liu, L. Study of Structure and Corrosion Resistance of Fe-Based Amorphous Coatings Prepared by HVAF and HVOF. *Corros. Sci.* **2011**, *53*, 2351–2356. [CrossRef]
61. Zhang, J.; Liu, J.; Liao, H.; Zeng, M.; Ma, S. A Review on Relationship between Morphology of Boride of Fe-B Alloys and the Wear/Corrosion Resistant Properties and Mechanisms. *J. Mater. Res. Technol.* **2019**, *8*, 6308–6320. [CrossRef]
62. Ma, S.; Xing, J.; Fu, H.; Yi, D.; Li, Y.; Zhang, J.; Zhu, B.; Gao, Y. Microstructure and İnterface Characteristics of Fe–B Alloy in Liquid 0.25wt.% Al–Zn at Various Bath Temperatures. *Mater. Chem. Phys.* **2012**, *132*, 977–986. [CrossRef]
63. Ma, S.; Xing, J.; Fu, H.; Yi, D.; Zhang, J.; Li, Y.; Zhang, Z.; Zhu, B.; Ma, S. Interfacial Morphology and Corrosion Resistance of Fe–B Cast Steel Containing Chromium and Nickel in Liquid Zinc. *Corros. Sci.* **2011**, *53*, 2826–2834. [CrossRef]
64. Ma, S.; Xing, J.; Fu, H.; Yi, D.; Zhi, X.; Li, Y. Effects of Boron Concentration on the Corrosion Resistance of Fe–B Alloys İmmersed in 460 °C Molten Zinc Bath. *Surf. Coat. Technol.* **2010**, *204*, 2208–2214. [CrossRef]
65. Abakay, E.; Şen, U. Effect of Morphological and Microstructural Variations on the Properties of Electroless Nickel Boron Coatings. *Trans. Indian Inst. Metall.* **2023**, *76*, 657–664. [CrossRef]
66. Gu, Z.; Mao, P.; Gou, Y.; Chao, Y.; Xi, S. Microstructure and Properties of $MgMoNbFeTi_2Y_x$ High Entropy Alloy Coatings by Laser Cladding. *Surf. Coat. Technol.* **2020**, *402*, 126303. [CrossRef]
67. Qin, L.-Y.; Lian, J.-S.; Jiang, Q. Effect of Grain Size on Corrosion Behavior of Electrodeposited Bulk Nanocrystalline Ni. *Trans. Nonferrous Metall. Soc. China* **2010**, *20*, 82–89. [CrossRef]
68. Lu, B.; Luo, J.; Chiovelli, S. Corrosion and Wear Resistance of Chrome White İrons—A Correlation to Their Composition and Microstructure. *Metall. Mater. Trans. A* **2006**, *37*, 3029–3038. [CrossRef]
69. Kocaman, E.; Kılınç, B.; Durmaz, M.; Şen, Ş.; Şen, U. The İnfluence of Chromium content on Wear and Corrosion Behavior of Surface Alloyed Steel with Fe(16−x)Crx(B,C)4 Electrode. *Eng. Sci. Technol. Int. J.* **2021**, *24*, 533–542. [CrossRef]
70. Liu, C.; Liu, Z.; Gao, Y.; Wang, X.; Zheng, C. Effect of Cr Content on Corrosion Resistance of Ni-XCr-Mo Laser-Cladding Coatings under H2S-Induced High-Temperature Corrosion Atmosphere. *Materials* **2022**, *15*, 1885. [CrossRef] [PubMed]

materials

Article

Role of Cr Element in Highly Dense Passivation of Fe-Based Amorphous Alloy

Ziqi Song [1], Zhaoxuan Wang [1], Qi Chen [1], Zhigang Qi [1], Ki Buem Kim [2] and Weimin Wang [1,*]

[1] Key Laboratory for Liquid-Solid Structural Evolution and Processing of Materials (Ministry of Education), School of Materials Science and Engineering, Shandong University, Jinan 250061, China; 202234218@mail.sdu.edu.cn (Z.S.); 202114121@mail.sdu.edu.cn (Z.W.); 202120538@mail.sdu.edu.cn (Q.C.); 202034203@mail.sdu.edu.cn (Z.Q.)

[2] Department of Nanotechnology and Advanced Materials Engineering, Sejong University, Seoul 05006, Republic of Korea; kbkim@sejong.ac.kr

* Correspondence: weiminw@sdu.edu.cn

Abstract: The effect of the Cr element on the corrosion behavior of as-spun $Fe_{72-x}Cr_xB_{19.2}Si_{4.8}Nb_4$ ribbons with x = 0, 7.2, 21.6, and 36 in 3.5% NaCl solution were investigated in this work. The results show that the glass formability of the alloys can be increased as Cr content (c_{Cr}) is added up to 21.6 at.%. When c_{Cr} reaches 36 at.%, some nanocrystals appear in the as-spun ribbon. With increasing c_{Cr} content, the corrosion resistances of as-spun Fe-based ribbons are continually improved as well as their hardness properties; during the polarization test, their passive film shows an increase first and then a decrease, with the highest pitting potential as c_{Cr} = 7.2 at.%, which is confirmed by an XPS test. The dense passivation film, composed of Cr_2O_3 and $[CrO_x(OH)_{3-2x}, nH_2O]$, can reduce the number of corrosion pits on the sample surface due to chloride corrosion and possibly be deteriorated by the overdosed CrFeB phase. This work can help us to design and prepare the highly corrosion-resistant Fe-based alloys.

Keywords: Cr element; passivation; Fe based; amorphous alloys

Citation: Song, Z.; Wang, Z.; Chen, Q.; Qi, Z.; Kim, K.B.; Wang, W. Role of Cr Element in Highly Dense Passivation of Fe-Based Amorphous Alloy. *Materials* **2023**, *16*, 6630. https://doi.org/10.3390/ma16206630

Academic Editor: Tony Spassov

Received: 10 September 2023
Revised: 2 October 2023
Accepted: 4 October 2023
Published: 10 October 2023

1. Introduction

Since firstly fabricated in the last century, the amorphous alloys have a wide range of applications in many industrial fields, due to their special atomic structures, high strengths, and high hardness properties; Fe-based metallic amorphous alloys have especially excellent soft magnetic properties [1]. Meanwhile, Fe-based amorphous alloys are also used in various fields due to their excellent corrosion/wear resistances, such as drill bits [2] and artificial bone joints [3]. In addition, Fe-based metallic glasses contain major components with much lower costs than Co, Zr, Ni, Cu, and Ti [4]. Nowadays, in the rapid and high-level development of industry, it is necessary to solve the problems of corrosion and wear of metallic parts working in harsh environments to increase the corrosion resistances of Fe-based amorphous alloys is in line with the current industrial needs [5–7].

In the known stainless steels, Cr is one of the important and cheap elements to improve the corrosion resistance of the alloys [8]. Adding Cr can improve the corrosion/wear resistance of Fe-based amorphous alloys drastically [9–12]. At high temperatures, a high Cr content can ensure the mechanical strength and corrosion resistance of the alloys simultaneously, so a high-Cr alloy with high temperature corrosion resistance is widely used in the manufacturing of turbochargers, furnace fixtures, etc. [13,14]. The corrosion resistance performance of Fe-based amorphous alloys is attributed to three main factors: (1) amorphous alloys have a uniform passivation film [15]; (2) they have a high reactivity to diffuse the passive components to the surface [16]; and (3) they have no crystal defects. In biomedicine, the Cr-contained Fe-based amorphous alloys have excellent corrosion resistance and a superior biocompatibility, suitable for preparing surgical scissors. The (Fe,

Cr)-based glass exhibits better corrosion resistance than 316L steel and Ti-based commercial alloys in the human body-like environment [17,18]. The Fe-based amorphous alloys with Cr show high corrosion resistance in Cl^- environments, giving new ideas for the design of marine anti-corrosion alloys [2,19,20].

Mo is another important element to be often added to steel and cast iron to improve their corrosion resistance. According to the argument on the pitting resistance equivalent number (also denoted as PREM), the element Mo can increase the resistance to local pit corrosion of the alloy [21,22]. In the real preparation and machining process, some Mo is added to most stainless steels to improve their resistances to pitting corrosion. Generally, Mo has a high cost; other low cost elements are substituted and then reduce the cost of the alloys, keeping enough corrosion resistance [23]. In the alloys containing both Cr and Mo, Mo has a strong interaction with Cr to increase corrosion resistance, forming a Cr-rich passive oxide film, rather than forming a Mo-oxide film [24]. Thus, increasing the c_{Cr}, replacing Mo with other low-cost corrosion resistance elements in the alloy, is a practical idea [25]. It is necessary to be pointed out that Fe-based amorphous alloys can have a stable passivation rather than the active dissoluble film only when the Cr content is beyond a certain critical value [26]. Therefore, it is valuable to explore the role of the composition in the mechanism of formation and destruction of passivation films of Fe-based amorphous alloys in this work.

The marine environment contains lot of Cl^- and serious natural corrosion conditions. In a series of industrial circumstances, such as off-shore oil/gas exploitation and underwater operations, the selection of alloy types with high corrosion resistances is an important approach to solve the Cl^--induced degradation of parts [27]. Recently, the effect of Cr mainly involves low Cr steel in case of CO_2 corrosion [28]; however, its anti-corrosion mechanism in different solutions is different. In this paper, we mainly studied the corrosion behavior of Fe-based amorphous alloys with Cr in a simulated marine environment, i.e., Cl^--contained solution. At the same time, Mo in traditional Fe-based amorphous alloys containing Cr is replaced by Nb, Si, and other elements to improve the corrosion resistance and GFA of amorphous alloys. We discuss the effect of Cr on the corrosion behaviors of Fe-based amorphous alloys in neutral solutions. Meanwhile, the composition and destruction process of amorphous passivated films during electrochemical tests have been analyzed. The results of this work not only provide a new way to understand the effect of Cr on the corrosion resistance of Fe-based metallic glasses but also provide a new method for supplying the needed passivation film of Fe-based amorphous alloys.

2. Experimental

2.1. Sample Preparation

The alloy ingots with nominal compositions of $Fe_{72-x}Cr_xB_{19.2}Si_{4.8}Nb_4$ (x = 0, 7.2, 21.6, and 36) were smelted with pure Fe (99.99 wt.%), pure Si (99.999 wt.%), pure Nb (99.99 wt.%), pure Cr (99.99 wt.%), and Fe-B ingots in an arc melting furnace (MAM-1 Edmund Buhler, Berlin, Germany) protected by high purity argon (99.999%). Each prepared ingot was 5 g and remelted at least four times during the smelting process to ensure the uniformity of the composition. The obtained ingots were made into amorphous ribbons at a roller speed of 41.8 $m{\cdot}s^{-1}$ under the protection of purified argon (99.999%) by a single roll melt spinning system (SD500 SKY, Shenyang, China).

2.2. Structural Analysis

The microstructures of the ingots and the as-spun ribbons as well as the ground ribbon powders were characterized by X-ray diffraction (XRD, Bruker D8 Discover, Beijing Technology Co., Ltd., Beijing, China). The thermodynamic behaviors of the ribbons were measured by differential scanning calorimetry (DSC, NETZSCH-404, Netzsch-404, Netzsch, Bavaria, Germany) under the protection of high purity argon at a heating rate of 20 K/min. The ribbon microstructures were observed by transmission electron microscopy (TEM, JEM-2100F, Japan Electronics Co., Ltd., Beijing, China), and the surface morphologies and passivation film compositions were analyzed by a scanning electron microscope (SEM, JSM-7800 F Japan Electronics Co., Ltd., Beijing, China) and X-ray photoelectron spectroscopy (XPS, AXIS Supra, Manchester, UK).

2.3. Electrochemical Tests

An electrochemical workstation (CHI660E, Shanghai Chenhua Instrument Co., Ltd., Shanghai, China) was used to test and characterize the corrosion behavior of the ribbons at room temperature (300 K). The polarization curve and electrochemical impedance spectroscopy (EIS) of the ribbons were tested in 0.6 M NaCl neutral solution with a three-electrode system. One side of the ribbons was covered with silicone rubber 12 h in advance before the test to ensure the accuracy of the experiment. In addition, the ribbons were immersed in hydrochloric acid solution at pH = 3 for 215 h to calculate the weight loss rate, which was calculated by the following formula:

$$v_{\text{weight}} = \frac{m_0 - m_1}{S \times t} \tag{1}$$

m_0 (g) is the weight of the ribbon before immersion, m_1 (g) is the weight of the ribbon after immersion, S (cm^2) is the immersion area of the ribbon, and t (h) is the immersion time.

3. Results and Discussion

Microstructure

Figure 1a–d shows the SEM images of the $Fe_{72-x}Cr_xB_{19.2}Si_{4.8}Nb_4$ ingots with x = 0, 7.2, 21.6, and 36, denoted by Cr0, Cr7, Cr21, and Cr36, respectively. The corresponding XRD patterns are shown in Figure 1e–h. Based on the XRD patterns, the four ingots had a certain similarity; there were α-Fe phases. With the addition of Cr, three strong peaks of α-Fe gradually shifted to the left, indicating that the average atomic distance increased, probably because Cr atoms with larger atomic radii replaced some of the Fe atoms. In addition, with the addition of the Cr element, CrFeB phases appeared in the ingots. Interestingly, from the ingots, the tendency of CrFeB phase dendritic growth was more and more evident. It was related to the migration and diffusion of solute during solidification, which was in line with the classical theory of "constitutional supercooling" proposed by Thiller [29]. In Cr0 ingot, the FeB phase presented in large plate-like shape (Figure 1a). With the addition of the Cr element, an elongated CrFeB phase began to appear in the ingot, indicating a strong binding force between Cr and Fe (Figure 1b,d). Some elliptical protrusions of Nb-Si intermetallic compounds in the ingots were formed, indicating that the strengths of Nb-Si bonds were stronger than those the of Nb-Nb bonds (Figure 1c,d).

The microstructures of the Cr0, Cr7, Cr21, and Cr36 as-spun ribbons and the as-crushed Cr36 ribbon were investigated by XRD. The diffraction patterns of the as-spun Cr0, Cr7, and Cr21 ribbons exhibited a typical diffuse peak within the range of 44.2 $\pm$ 0.4, without evident sharp crystallization peaks, indicating the amorphous properties of these samples (Figure 2a–c). This indicated that the FeSiBNb alloy with a small amount of Cr could form amorphous ribbons. As the Cr content increased, the position of the diffuse emission peak decreased from 44.6° to 43.2°, indicating an increase in the average atomic spacing, and the free volume within the ribbons also increased [30]. At the same time, the ribbons became increasingly brittle, and the XRD of the Cr36 ribbon showed evident crystal peaks. Cr36 ribbons became shorter and more brittle, which may have been related to the existence of internal stress during the ribbons production process. As shown in Figure 2d,e, there is a set of crystal peaks on the XRD curve of Cr36 ribbon; the α-Fe phase is not evident in the as-spun Cr36 ribbon, but the α-Fe peak becomes very evident after the sample has been crushed into pieces, similar to Ref. [31], indicating that the orientation of CrFeB and α-Fe are developed in the as-spun ribbon and that the α-Fe phase in the ingot also obtained a certain inheritance in the ribbons. These results suggested the structural heredity between the melts/glasses and ingots of present alloys.

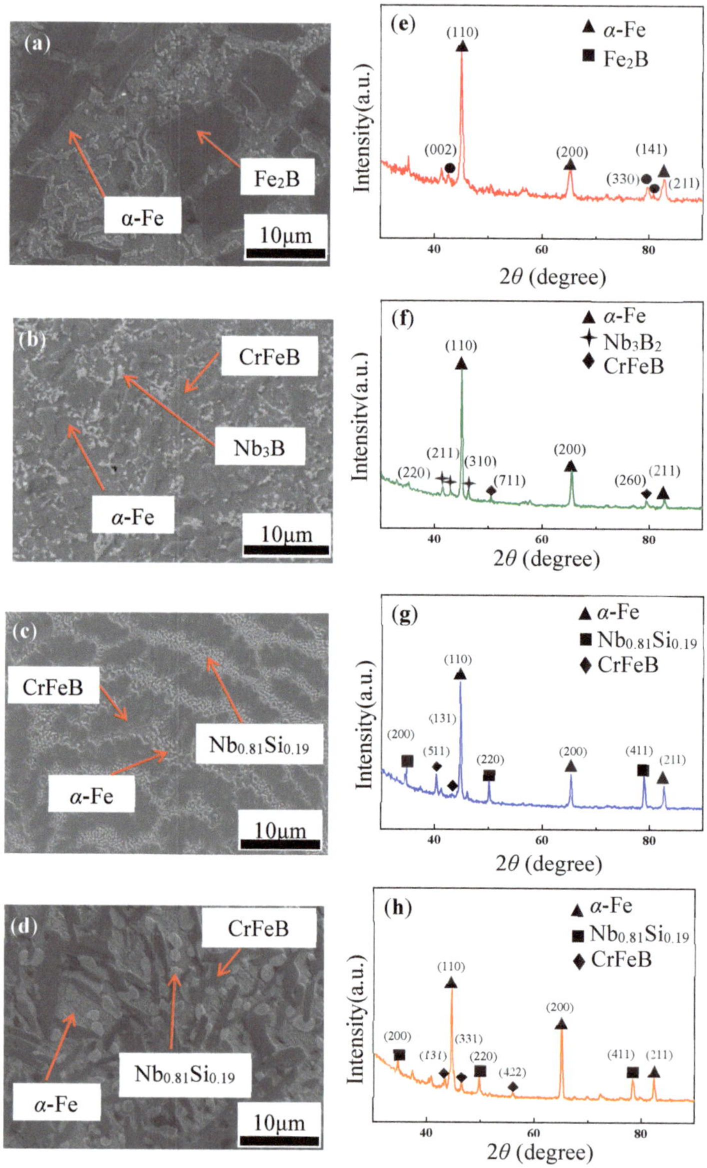

Figure 1. SEM micrographs of the $Fe_{72-x}Cr_xB_{19.2}Si_{4.8}Nb_4$. (**a**) $x = 0$ (Cr0), (**b**) $x = 7.2$ (Cr7), (**c**) $x = 21.6$ (Cr21), and (**d**) $x = 36$ (Cr36) ingots and XRD patterns of (**e**) Cr0, (**f**) Cr7, (**g**) Cr21, and (**h**) Cr36.

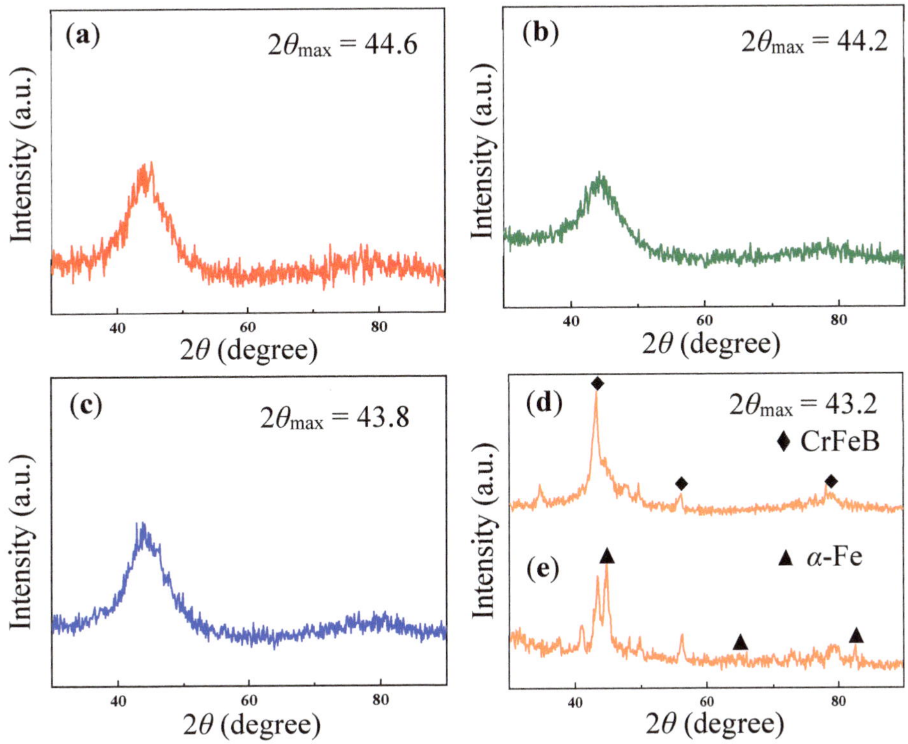

Figure 2. XRD curves of as-spun $Fe_{72-x}Cr_xB_{19.2}Si_{4.8}Nb_4$ ribbons together with a crushed sample. (**a**) $x = 0$ (Cr0) ribbon, (**b**) $x = 7.2$ (Cr7) ribbon, (**c**) $x = 21.6$ (Cr21) ribbon, (**d**) $x = 36$ (Cr36) ribbon, and (**e**) crushed Cr36 sample.

In order to further characterize the microstructure of the as-spun Cr36 ribbon, TEM tests were performed. The TEM images and selected area electron diffraction (SAED) patterns of Cr36 ribbon are shown in Figure 3. They present many nanocrystals in the Cr36 ribbon. As shown in Figure 3b,c, the grain diameter is about 100 nm, and there are some dislocation structures in the rod-shaped grains, which are distributed vertically along the head and tail of the rod-shaped grains. At the same time, EDS shows that the rod-shaped grains have oxygen enrichment, indicating that higher activation energy is displayed at the dislocations, and the metal oxide reaction preferentially occurs in the regions with high dislocation density [32]. Figure 3(d_1,d_2) are SAED corresponding to HRTEM (d) (High Resolution Transmission Electron Microscope), corresponding to the CrFeB and α-Fe phases, respectively.

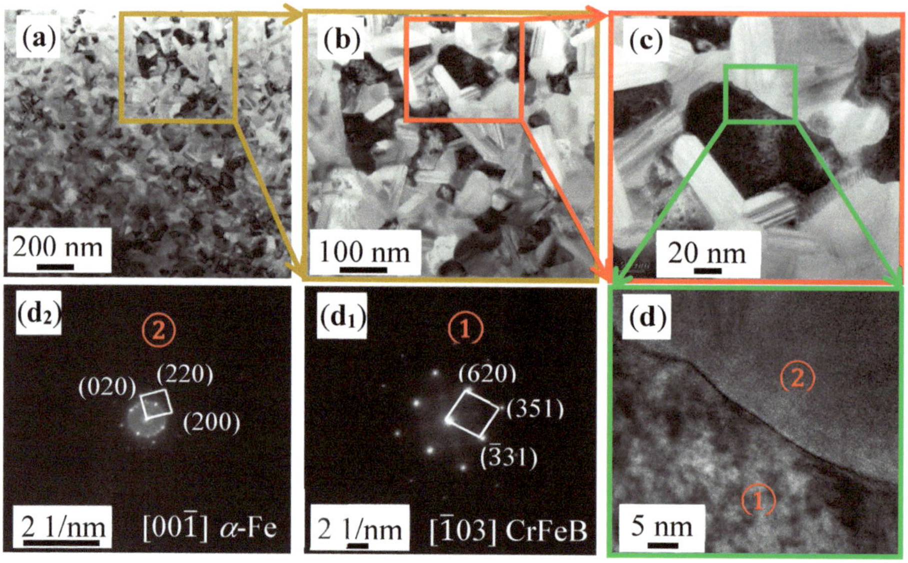

Figure 3. (**a**) LRTEM (Low Resolution Transmission Electron Microscope) image of $Fe_{36}Cr_{36}B_{19.2}Si_{4.8}Nb_4$ (Cr36). (**b**,**c**) are the enlarged part of (**a**); (**d**) is the HRTEM (High Resolution Transmission Electron Microscope) image of Cr36 (① and ② represent different phases); and (**d$_1$**) and (**d$_2$**) are the corresponding FFT (Fast Fourier Transform) patterns.

The DSC curves of the Cr0, Cr7, Cr21, and Cr36 ribbons with heating/cooling rates of 20 K/min are shown in Figure 4. The characteristic thermodynamic temperatures are listed in Table 1. The amorphous Cr0, Cr7, and Cr21 ribbons exhibit similar exothermic behaviors, and all the amorphous samples have only one exothermic crystallization peak, usually corresponding to eutectic crystallization [33]. The height of the crystallization peak increases with the increase in Cr content. As the Cr content increases up to 21.6 at.%, the reduced crystallization temperature T_{rx} (=T_x/T_l) and ΔH_{rc} of amorphous ribbons also gradually increases, indicating that the addition of 21.6 at.% Cr element can still improve the amorphous stability and amorphous formation ability of Fe-based ribbons [34]. When the addition of the Cr element reaches 36 at.%, there is no exothermic peak in the sample in Figure 4a, indicating that the ribbons has crystallized. In addition, the onset temperature of melting (T_m) of the amorphous ribbons shows an increasing trend, and the end temperature of melting (T_l) does not tends to decrease with increasing x (Figure 4b). In other words, the solidification temperature range ΔT_l (T_l–T_m) of ribbons does not change monotonically. The results show that Cr21 ribbon is the closest component to the eutectic point [30].

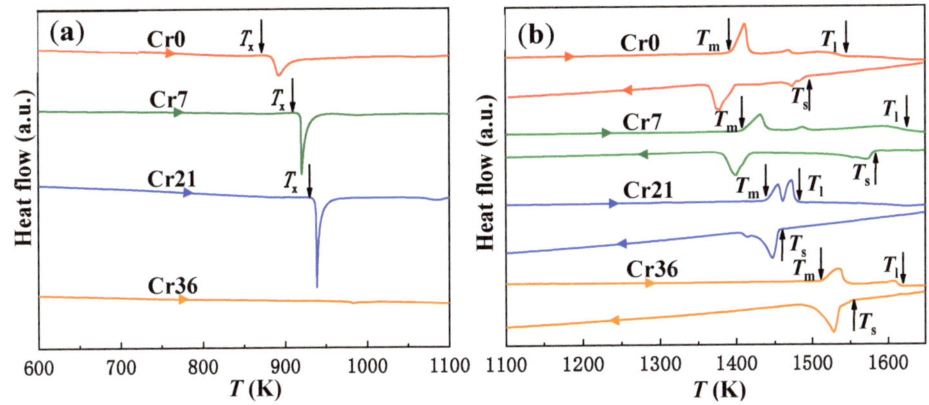

Figure 4. DSC curves of $Fe_{72-x}Cr_xB_{19.2}Si_{4.8}Nb_4$ with $x = 0$ (Cr0), $x = 7.2$ (Cr7), $x = 21.6$ (Cr21), and $x = 36$ (Cr36) ribbons at different temperature intervals. (**a**) Heating curves in 600–1100 K with 20 K/min, and (**b**) heating and cooling curves in 1100–1650 K.

Table 1. Thermal parameters deduced from the DSC curves, including the onset crystallization temperature T_x, the solidification temperature T_s, the onset melting temperature T_m, the liquidus temperature T_l, the solidification temperature range ΔT_l (=T_l–T_m), the reduced crystallization T_{rx} (=T_x/T_l), the degree of supercooling ΔT (=T_l–T_s), and the reduced crystallization heat ΔH_{rc} (=$\Delta H_c/\Delta H_m$). Here, ΔH_c and ΔH_m are heats of crystallization and melting, respectively.

Ribbon	T_x (K)	T_s (K)	T_m (K)	T_l (K)	ΔT_l (K)	ΔT (K)	T_{rx}	ΔH_{rc} ($\Delta H_c/\Delta H_m$)
$x = 0$	882	1492	1394	1541	147	49	0.57	0.18
$x = 7.2$	916	1580	1406	1619	213	39	0.57	0.23
$x = 21.6$	933	1459	1438	1482	44	23	0.63	0.52
$x = 36$	-	1554	1512	1615	103	61	-	-

In general, the smaller the residual penetration depth, the higher the nanohardness. The steeper the unloading curve, the larger the Young's modulus [35–37]. The nanoindentation images of Cr0, Cr7, Cr21, and Cr36 ribbons are shown in Figure 5, with the increase in Cr content, the hardness and Young's modulus of the ribbons both increased. The hardness and Young's modulus of the Cr36 ribbon as a crystal increase significantly, indicating that Cr-rich ordered clusters in the ribbon can improve the hardness of the ribbon.

Figure 6 shows the potentiodynamic polarization plots of the Cr0, Cr7, Cr21, and Cr36 as-spun ribbons and the corrosion rate R_{im} deduced from the weight loss in the HCl solution with a pH = 3. The corrosion potential E_{corr} (vs. SCE (saturated calomel (reference) electrode)), corrosion current density i_{corr} ($\mu A \cdot cm^{-2}$), pitting potential E_{pit} (V vs. SCE), and deduced corrosion rate CR ($\mu m \cdot y^{-1}$) with Equation (2) are summarized in Table 2 [38,39].

$$CR = 3.28\, i_{corr} \times M/nd\ (\mu m \times y^{-1}) \quad (2)$$

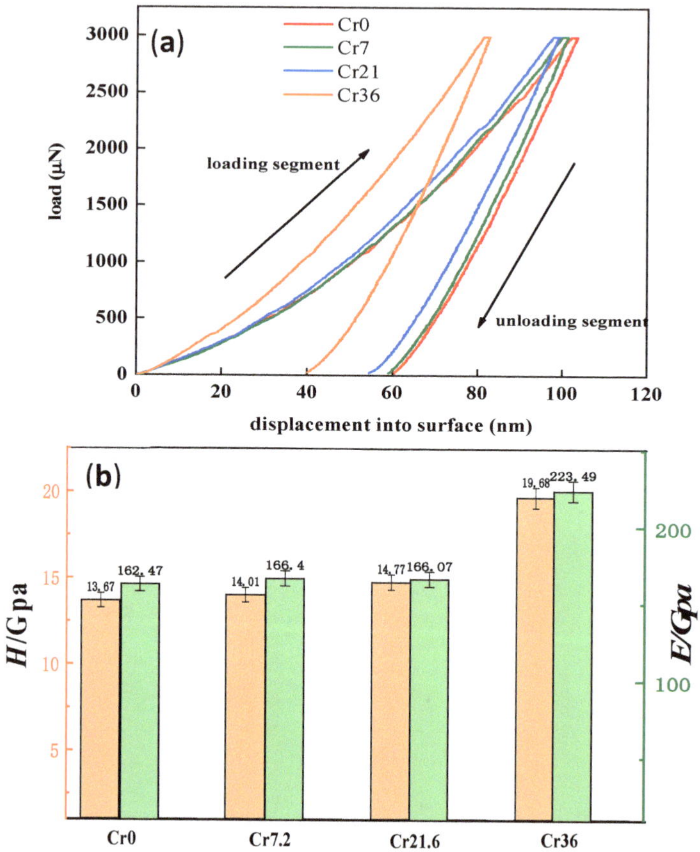

Figure 5. Nanoindentation of $Fe_{72-x}Cr_xB_{19.2}Si_{4.8}Nb_4$ with x = 0 (Cr0), x = 7.2 (Cr7), x = 21.6 (Cr21), and x = 36 (Cr36) ribbons: (**a**) Curve of the load and indentation depth; (**b**) calculated results of the nanoindentation, including hardness and modulus.

Table 2. Parameters deduced from potentiodynamic polarization plots in Figure 6, such as corrosion potential E_{corr} (V vs. SCE), corrosion current density i_{corr} (μA·cm^{-2}), pitting potential E_{pit} (V vs. SCE), and deduced corrosion rate CR (μm·y^{-1}).

Ribbon	E_{corr} (V)	i_{corr} (μA·cm^{-2})	E_{pit} (V)	CR (μm·y^{-1})
x = 0	−0.69	40.72	−0.60	473
x = 7.2	−0.56	14.83	1.21	172
x = 21.6	−0.55	8.43	1.04	98
x = 36	−0.50	5.28	0.68	61

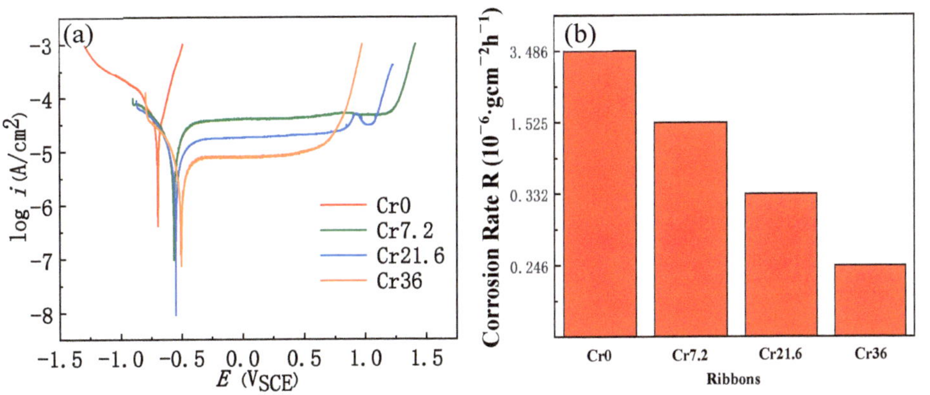

Figure 6. (**a**) Potentiodynamic polarization plots of as-spun ribbons $Fe_{72-x}Cr_xB_{19.2}Si_{4.8}Nb_4$ with $x = 0$ (Cr0), $x = 7.2$ (Cr7), $x = 21.6$ (Cr21), and $x = 36$ (Cr36) in neutral solutions. (**b**) Corrosion rate R_{im} derived from weight loss obtained by immersing in acid solution.

Here, M is the atomic weight of Fe (55.85 g), n is the number of electrons transferred in the corrosion reaction ($n = 2$), and d is the density of Fe (7.88 g·cm^{-3}). For the sake of similarity, the corrosion rate is mainly ascribed to the dissolution of Fe atoms.

From the potentiodynamic polarization curves (Figure 6a), the amorphous ribbon Cr0 has no evident passivation platform, the anode polarization curves of Cr7, Cr21, and Cr36 have significantly wide passivation platforms, and there is no evident anodic Tafel region. The corrosion current density was determined by cathodic polarization curves and corrosion potential. In addition, the addition of Cr increased the corrosion potential of the ribbons and decreased the corrosion current density of the ribbons (Table 2). As can be seen from Figure 6b, with the increase in Cr content, the corrosion rate of the ribbons after immersing in HCl solution with pH = 3 for 215 h decreased, which was consistent with the Tafel curves. At a high potential of about 0.7 V, Cr transpassive dissolution occurred in Cr36 ribbon, according to Ref. [40], resulting in a gradual increase in current. When the potential was less than 0, with the increase in c_{Cr} (Table 2), the value of i_{corr} and E_{corr} of the ribbons decreased, which was consistent with Figure 6b, indicating that the corrosion resistance of the ribbons was getting better. When the potential was increased until the ribbons were pitted, it can be seen that the E_{pit} of the ribbons reached the maximum value in Cr7, indicating that Cr7 had the best pitting corrosion resistance. On the one hand, the disorder of amorphous ribbons reduced the existence of crystal defects and prevented the formation of galvanic corrosion. On the other hand, the composition contained more easily passivated elements (such as Cr), which was conducive to improving corrosion resistance [41].

In comparison, we collect the date of 51 Fe-based glassy alloys, as shown in Figure 7. The i_{corr} and E_{corr} of Fe-based glassy alloys are listed in Table S1. Apparently, Cr36 has a very low i_{corr} and a certain positive E_{corr}, showing a super corrosion resistance.

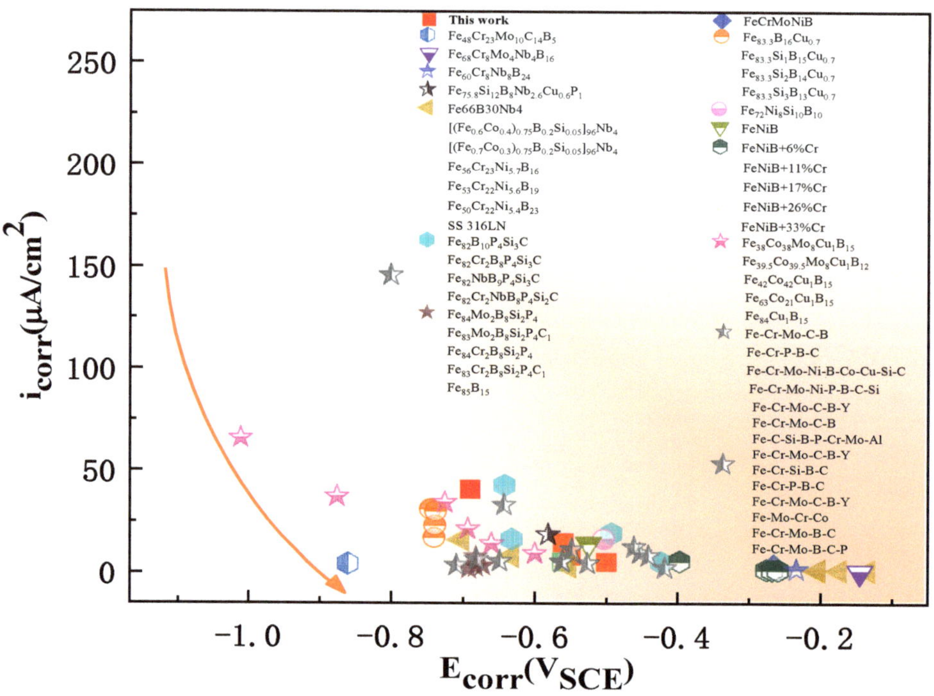

Figure 7. The comparison diagram of corrosion potential (E_{corr}) and corrosion current density (i_{corr}) of various Fe-based alloys in 3.5 wt.% NaCl solution; details are listed in Table S1 [42–53].

Figure 8 shows the surface schematic graph morphologies and cross-sections of the tested samples after the polarization corrosion experiment in the neutral solution. In the case of Cr0, a corrosion scale in the thickness of 1–2 μm (Fe_2O_3 and SiO_2) was mostly detached, with only small amounts remaining, and evident stress cracking was observed on the surface, exposing the matrix under the scale. In contrast, very little corrosion product was observed over the smooth surface of Cr7 (Figure 8b). It is thought that a relatively Cr-rich passive oxide may be produced on the surface, indicating that adding a small amount of Cr element can reduce the occurrence of local corrosion pits [54]. However, corrosion pits on the Cr21 ribbon's surface are more evident than Cr7. At this time, small pits are easily formed on the bottom of the early formed large pit. On the surface of the Cr36 ribbon, it can be observed that the ribbon surface is rougher than Cr21 (Figure 8d), and the surface passivation film caused by pitting corrosion is in a chain-like shape. These results show that the addition of Cr element can firstly reduce stress cracking and pitting by corrosion and then increase them, show a maximal pitting resistance at c_{Cr} = 7 at.%.

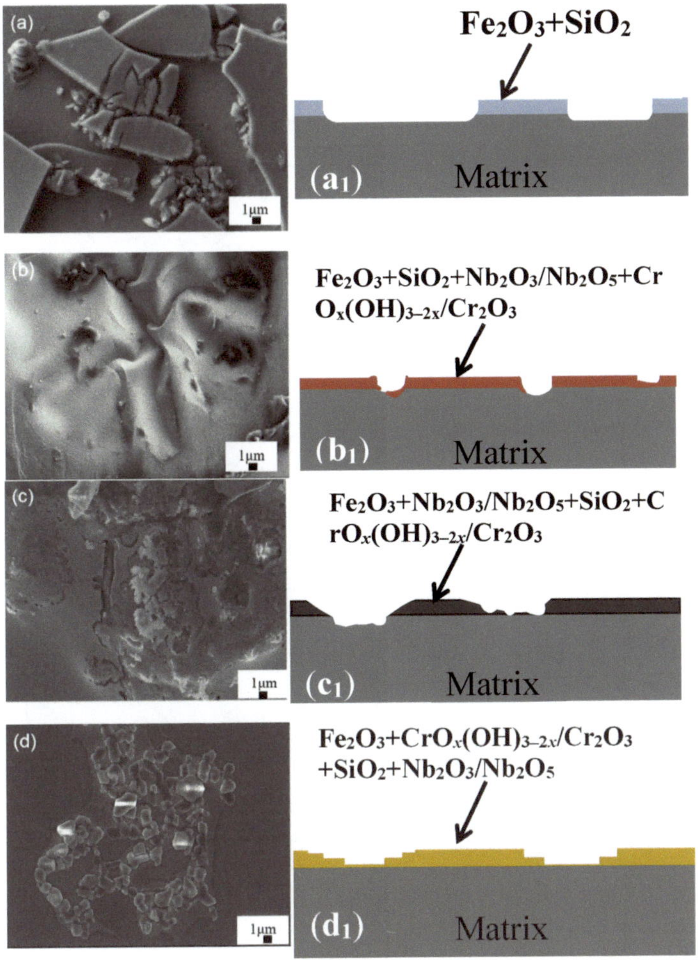

Figure 8. SEM images and simplified cross-sectional schematics of corroded $Fe_{72-x}Cr_xB_{19.2}Si_{4.8}Nb_4$ samples. (**a**) x = 0 (Cr0). (**b**) x = 7.2 (Cr7). (**c**) x = 21.6 (Cr21). (**d**) x = 36 (Cr36). (**a₁**–**d₁**) Simplified cross-section schematics showing their damaged characteristics.

To further understand the surface property changes, Figure 9 displays the XPS etching analysis on the surfaces of the $Fe_{72-x}Cr_xB_{19.2}Si_{4.8}Nb_4$ ribbons with x = 0, 7.2, 21.6, and 36 after the potentiodynamic polarization in neutral solution. On the Fe 2p spectra (Figure 9(a_0–a_3)), the XPS curves of four ribbons can be decomposed into two peaks at about 707 and 709/710 eV, being identified as Fe^0 and Fe^{2+}/Fe^{3+}, respectively [55]. As can be seen from Figure 8(a_0,a_1), the Fe^{3+} peak at 709.9 eV in Cr0 and Cr7 tend to decrease with the increase in etching time, whereas the Fe^{3+}/Fe^{2+} peak at 709.3/710 eV in Cr21 and Cr36 gradually increases, which is because the passivation films of the Cr21 and Cr36 are thicker and denser. Under the same etching time, the passivation films of Cr36 are still Fe^{2+}; there is no conversion of Fe^{2+} to Fe^{3+}, which is consistent with the low corrosion current density and high corrosion potential shown in the Tafel curve (Figure 6a), indicating that a dense passivated film containing Fe is formed on its surface and that the Fe element finally has experienced interactive oxidation in the passivation film.

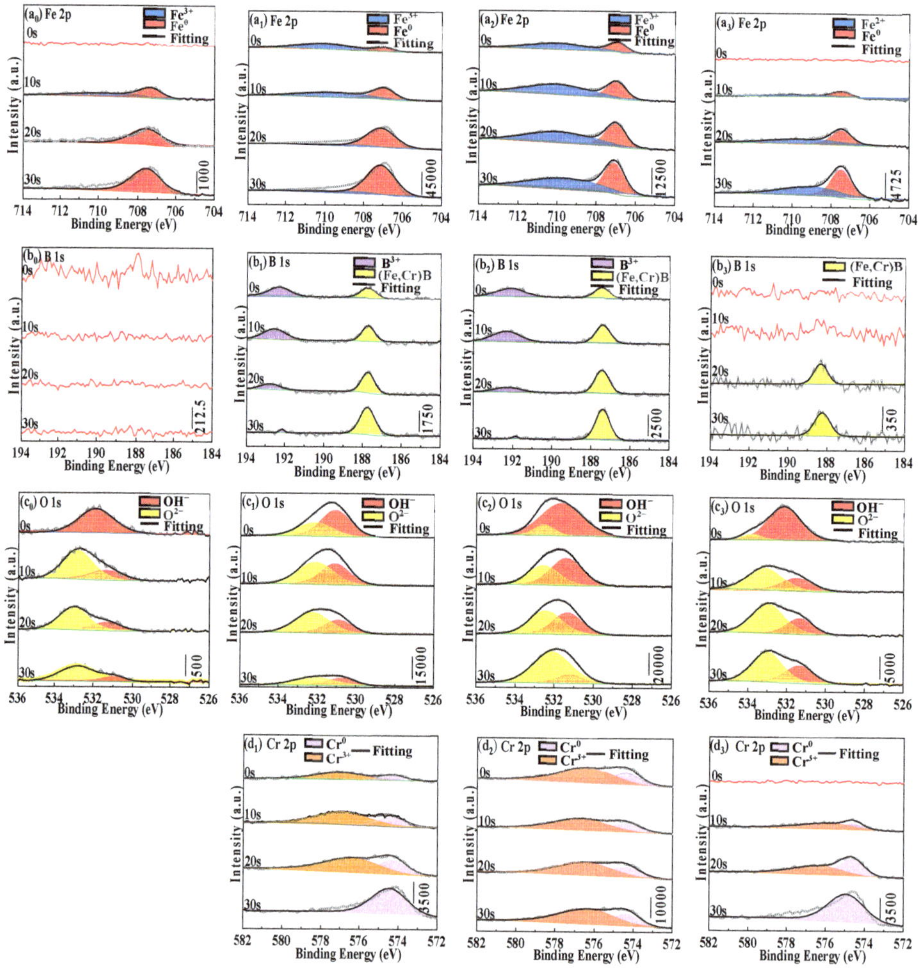

Figure 9. XPS spectra of Fe 2p (**a₀–a₃**), B 1s (**b₀–b₃**), O 1s (**c₀–c₃**) and Cr 2p (**d₁–d₃**) in binding energy regions for the as-spun $Fe_{72-x}Cr_xB_{19.2}Si_{4.8}Nb_4$ ribbons with $x = 0$ (Cr0), $x = 7.2$ (Cr7), $x = 21.6$ (Cr21), and $x = 36$ (Cr36) after corrosion in neutral solutions.

On most B 1s spectra (Figure 9(b_0–b_3)), we can decompose the XPS curves into two peaks at 192 and 188 eV, being indexed as B^{3+} and (Fe, Cr)B, respectively. It can be seen from Figure 9(b_0–b_3) that the oxides of the B element only exist on the surfaces of Cr7 and Cr21, and the contents of oxides are small; the (Fe, Cr)B compounds peak are stronger, and exist in Cr7, Cr21 and Cr36. Figure 9(b_0–b_3) shows that most of the element B oxides are corroded away during the corrosion process, indicating that element B is not the main factor in improving the corrosion resistance of the ribbons.

On the O 1s spectra, the XPS curves can be decomposed two peaks at about 531 and 532 eV, identified as OH^- (531.3 eV) and O^{2-} (532.9 eV), respectively [56]. With the increase in etching time, the reduction rate of OH^- of the four ribbons is greater than the reduction rate of O^{2-}, indicating that some O^{2-} still exists at a certain depth of etching and passivation film composed of Fe_2O_3 and Cr_2O_3, and a certain amount of Si, Nb, and B element oxides is formed on the surface of the sample. Meanwhile, the oxide thickness of Cr7 and Cr21 is much higher than Cr0 and Cr36, due to their higher amorphous fraction f_a (Figure 4a).

On the Cr 2p spectra, we can divide the XPS curve into two peaks, identified as Cr^0 (574.1 and 583.5 eV) and Cr^{3+} (577 and 586.4 eV), respectively, according to Refs. [57,58]. For Fe-based amorphous ribbons containing corrosion-resistant elements such as Cr, the surface of the amorphous ribbons will form a passivation film containing corrosion resistant elements, which will improve the corrosion resistance [59]. Cr is easy to form passivation films of chromium oxide and chromium hydroxide on the surfaces of amorphous ribbons, thus protecting the amorphous phase and inhibiting the active dissolution of the amorphous phase [60]. The formation mechanism can be explained by the following equation:

$$2Cr + 3H_2O = Cr_2O_3 + 6H^+ + 6e^- \quad (3)$$

Then, part of the Cr oxide will further form H_2O and Cr hydroxide [$CrO_x(OH)_{3-2x}$, nH_2O]; both oxide films can improve the resistance of the ribbons to chloride ion corrosion [61].

As can be seen from Figure 9, Figure S1 and S2, the passivation film thicknesses of Cr7 and Cr21 ribbons are slightly larger than those of a Cr36 ribbon; among them, Cr21 has the best amorphous formation ability, so a thicker passive layer is formed on the surface. As can be seen from Figure 6a, Cr7 has a strong pitting corrosion resistance and is more stable, indicating that the oxide thickness is not the sole factor for its stability. However, it can be seen from the i_{corr} and E_{corr} of Cr36 in Table 2 that the corrosion resistance of Cr36 is still the best, indicating that a dense passivation film is formed on the surface of Cr36 when the potential is lower than 0 V.

Figure 10 shows the Nyquist and Bode plots with fitting results of $Fe_{72-x}Cr_xB_{19.2}Si_{4.8}Nb_4$ glassy ribbons with x = 0, 7.2, 21.6, and 36 in a neutral solution. Among several equivalent circuit candidates, the equivalent circuit R(Q(R(CR))) is suitable for fitting the EIS data. The non-ideal capacitive behavior due to local inhomogeneity is represented by the constant phase element CPE (Q). The impedance of a CPE is defined as [62–64]:

$$Q = (j\omega)^{-n}/Y_0 \quad (4)$$

Here, Y_0 is the frequency independent parameter ($\Omega^{-1}cm^{-2}s^n$), j is the imaginary number, ω is the angular frequency (rad s^{-1}), and n indicates the CPE power, which is between 0.5 and 1. When n = 1, the Q describes a pure capacitor. For 0.5 < n < 1, the Q represents a distribution of dielectric relaxation times in the frequency domain, and Q represents a Warburg impedance with diffusion character (n = 0.5).

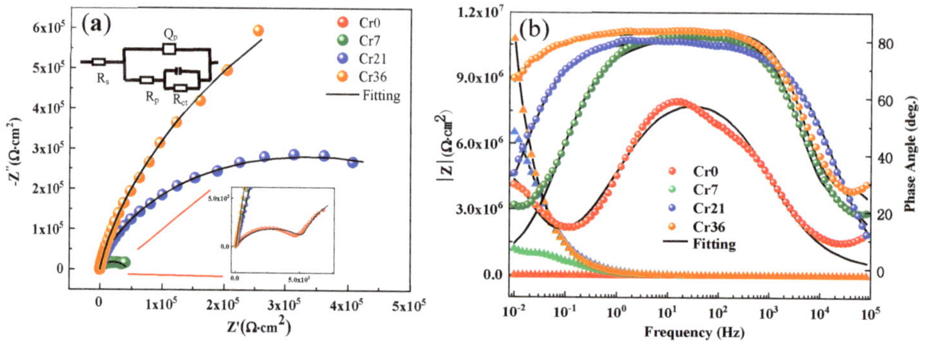

Figure 10. (**a**) Nyquist plots of as spun $Fe_{72-x}Cr_xB_{19.2}Si_{4.8}Nb_4$ ribbons with x = 0 (Cr0), 7.2 (Cr7), 21.6 (Cr21), and 36 (Cr36) in neutral solution with equivalent circuits R(Q(R(CR))) and (**b**) the corresponding Bode plots. Symbols show the experimental data while solid lines are fitting results.

The fitting results are summarized in Table 3. Here, the change transfer resistances (R_{ct}) of the ribbons increase from 1.1 to 2410 kΩ·cm² with the increase in c_{Cr}, indicating that the passivation film becomes more stable and dense, which is consistent with Ref. [65]. Meanwhile, we can see that the value of n ranges from 0.83 to 0.88. The n value of the Cr0 ribbon is low, which indicates a higher dispersion at the alloy/electrolyte interface, probably due to the defects on the passive film formed in this hypersaline solution [66,67]. With increasing c_{Cr}, the diameter of the Nyquist semicircle of Fe-based glassy ribbons significantly expands to the larger size (Figure 10a). The Bode plots is mainly divided into three different regions: (i) the low frequency region (10^{-2}–1 Hz) mainly reflects the capacitance and resistance at the interface between the ribbons and the passivation film; (ii) the intermediate frequency region (1–10^3 Hz) mainly reflects the resistance and capacitance of the passivation film; and (iii) the high frequency region (10^2–10^5 Hz) in this work mainly reflects the capacitances and resistances of the corrosion products [68–70]. In the bode plots, the closer the phase angle is to 90°, the more stable and compact the passivation film on the corrosion surface of the ribbons will be; it can be seen that the phase angle of the ribbons in the intermediate frequency region is closer to 90° when the element Cr is added [71]. Therefore, it can be seen that the increase in Cr content can make the passivation film more dense and thick, which is consistent with the passivation platform and the corrosion current density i_{corr} on the dynamic potential polarization curve (Figure 6a). The total resistance after fitting increases with the increase in Cr content, which is consistent with the Tafel results (Figure 6a).

Table 3. Simulated parameters of the $Fe_{72-x}Cr_xB_{19.2}Si_{4.8}Nb_4$ ribbons with x = 0, 7.2, 21.6, and 36, derived from EIS curves. R_s: the solution resistance, Q_p: the passive film capacitance, R_p: the passive film resistance, and R_{ct}: the charge transfer resistance.

Ribbon	R_s(Ω·cm²)	Q_p		R_p (kΩ·cm²)	C (μF·cm²)	R_{ct} (kΩ·cm²)	R_{total} (kΩ·cm²)
		Y(10^{-5} $Ω^{-1}s^ncm^{-2}$)	n				
x = 0	3.5	50.6	0.72	0.5	31,790	1.1	1.7
x = 7.2	1.1	3.3	0.84	0.0	9.8	43.0	43.0
x = 21.6	3.0	1.2	0.88	0.1	0.4	678.6	678.7
x = 36	1.3	1.1	0.86	3.2×10^{-3}	6.0	2410	2410

4. Conclusions

In this work, we prepared $Fe_{72-x}Cr_xB_{19.2}Si_{4.8}Nb_4$ ingots with x = 0, 7.2, 21.6, and 36 by arc melting, and ribbons with melt spinning in a vacuum. The microstructures of the ingots were investigated; microstructure investigations, hardness tests, electrochemical experiments, and XPS analyses on ribbons were conducted; and the surface morphologies of the ribbons after testing were analyzed. Combining the heredity between the ingots and ribbons, the effect of the Cr element on the properties of the Fe-based amorphous alloys could be found, as follows:

(1) With increasing Cr content c_{Cr} (x), the stability and glass formability (GFA) of Fe-based alloys firstly increased until x = 21 and then decreased drastically as x = 36. Simultaneously, the CrFeB phase and clusters in the samples tended to form with strong orientations in the ingots and as-spun ribbons, which deteriorated the GFA of the alloy, as its abundance was high enough.

(2) With increasing c_{Cr}, the corrosion rate R_{im}, corrosion current i_{corr}, and deduced corrosion rate CR of Fe-based ribbons increased monotonically, as well as their hardness, whereas their pitting resistance increased until x = 7 and decreased as x = 21, indicating that the most stable passive film formed at a proper c_{Cr}, which was confirmed by XPS data. This work can give a new clue to design and prepare highly corrosion-resistant Fe-based alloys.

Supplementary Materials: The following supporting information can be downloaded at: https://www.mdpi.com/article/10.3390/ma16206630/s1. Figure S1: XPS spectra of Si 2p and Nb 3d in binding energy regions for the as-spun Fe72−xCrxB19.2Si4.8Nb4 ribbons with x = 0 (Cr0), x = 7.2 (Cr7), x = 21.6 (Cr21), and x = 36 (Cr36) after corrosion in neutral solutions; Figure S2: XPS spectra of Fe 2p, Si 2p, B 1s, Nb 3d O 1s and Cr 2p in binding energy regions for the as-spun Fe72−xCrxB19.2Si4.8Nb4 ribbons with x = 0 (Cr0), x = 7.2 (Cr7), x = 21.6 (Cr21), and x = 36 (Cr36); Table S1: The comparison data of corrosion potential (E_{corr}) and corrosion current density (i_{corr}) of various Fe-based alloys in 3.5 wt.% NaCl solution.

Author Contributions: Z.S.: Conceptualization, Methodology, Validation, Data curation, Investigation, Formal analysis, Writing—original draft, Writing—review & editing. Z.W.: Methodology, Formal analysis. Q.C.: Methodology, Validation. Z.Q.: Methodology, Supervision. K.B.K.: Software, Formal analysis. W.W.: Conceptualization, Formal analysis, Writing—review & editing, Supervision, Funding acquisition. All authors have read and agreed to the published version of the manuscript.

Funding: This work was financially supported by the key research and development program of China (Grant No. 2022YFB2404102), the National Natural Science Foundation of China (51971093, 52171158, and 52101196), the Open Project Program of Shandong Marine Aerospace Equipment Technological Innovation Center (Ludong University) (Grant No. MAETIC2021-11), and the Key R&D Program of Shandong Province, China (Grant No. 2021ZLGX01 and 022CXGC020308).

Institutional Review Board Statement: Not applicable.

Informed Consent Statement: Not applicable.

Data Availability Statement: Data available on request due to restrictions eg privacy or ethical. The data presented in this study are available on request from the corresponding author. The data are not publicly available due to follow-up research on the work.

Conflicts of Interest: The authors declare no conflict of interest.

References

1. Lee, C.-Y.; Lin, T.-J.; Sheu, H.-H.; Lee, H.-B. A study on corrosion and corrosion-wear behavior of Fe-based amorphous alloy coating prepared by high velocity oxygen fuel method. *J. Mater. Res. Technol.* **2021**, *15*, 4880–4895. [CrossRef]
2. Farmer, J.; Choi, J.-S.; Saw, C.; Haslam, J.; Day, D.; Hailey, P.; Lian, T.; Rebak, R.; Perepezko, J.; Payer, J.; et al. Iron-Based Amorphous Metals: High-Performance Corrosion-Resistant Material Development. *Metall. Mater. Trans. A* **2009**, *40*, 1289–1305. [CrossRef]

3. Ibrahim, M.Z.; Sarhan, A.A.D.; Kuo, T.Y.; Yusof, F.; Hamdi, M.; Lee, T.M. Developing a new laser cladded FeCrMoCB metallic glass layer on nickel-free stainless-steel as a potential superior wear-resistant coating for joint replacement implants. *Surf. Coat. Technol.* **2020**, *392*, 125755. [CrossRef]
4. Souza, C.A.C.; Ribeiro, D.V.; Kiminami, C.S. Corrosion resistance of Fe-Cr-based amorphous alloys: An overview. *J. Non-Cryst. Solids* **2016**, *442*, 56–66. [CrossRef]
5. Montemor, M.F.; Simões, A.m.p.; Ferreira, M.G.S.; Belo, M.D.C. The role of Mo in the chemical composition and semiconductive behaviour of oxide films formed on stainless steels. *Corros. Sci.* **1999**, *41*, 17–34. [CrossRef]
6. Chen, Q.; Yan, Z.; Guo, L.; Zhang, H.; Zhang, L.; Kim, K.; Li, X.; Wang, W. Enhancing the acid orange dye degradation efficiency of Mg-based glassy alloys with introducing porous structure and zinc oxide. *J. Alloys Compd.* **2020**, *831*, 154817. [CrossRef]
7. Stokłosa, Z.; Rasek, J.; Kwapuliński, P.; Badura, G.; Haneczok, G.; Pająk, L.; Lelątko, J.; Kolano-Burian, A. Magnetic, electrical and plastic properties of Fe76Nb2Si13B9, Fe75Ag1Nb2Si13B9 and Fe75Cu1Nb2Si13B9 amorphous alloys. *J. Alloys Compd.* **2011**, *509*, 9050–9054. [CrossRef]
8. Li, L.Y.; Sun, C.H.; Ruan, Y.; Wei, B. Duplex corrosion mechanisms correlated to α/γ phase selection in containerlessly processed Fe-Cr-Ni-Mo alloy. *Corros. Sci.* **2023**, *221*, 111311. [CrossRef]
9. Kobayashi, A.; Yano, S.; Kimura, H.; Inoue, A. Mechanical property of Fe-base metallic glass coating formed by gas tunnel type plasma spraying. *Surf. Coat. Technol.* **2008**, *202*, 2513–2518. [CrossRef]
10. Yoon, S.; Kim, J.; Bae, G.; Kim, B.; Lee, C. Formation of coating and tribological behavior of kinetic sprayed Fe-based bulk metallic glass. *J. Alloys Compd.* **2011**, *509*, 347–353. [CrossRef]
11. Zenebe, D.; Yi, S.; Kim, S.S. Sliding friction and wear behavior of Fe-based bulk metallic glass in 3.5% NaCl solution. *J. Mater. Sci.* **2011**, *47*, 1446–1451. [CrossRef]
12. Zhang, C.; Liu, L.; Chan, K.C.; Chen, Q.; Tang, C.Y. Wear behavior of HVOF-sprayed Fe-based amorphous coatings. *Intermetallics* **2012**, *29*, 80–85. [CrossRef]
13. Lekakh, S.N.; Buchely, M.; Li, M.; Godlewski, L. Effect of Cr and Ni concentrations on resilience of cast Nb-alloyed heat resistant austenitic steels at extreme high temperatures. *Mater. Sci. Eng. A* **2023**, *873*, 145027. [CrossRef]
14. Xiang, S.; Fan, Z.; Chen, T.; Lian, X.; Guo, Y. Microstructure evolution and creep behavior of nitrogen-bearing austenitic Fe–Cr–Ni heat-resistant alloys with various carbon contents. *J. Mater. Res. Technol.* **2023**, *23*, 316–330. [CrossRef]
15. Long, Z.L.; Chang, C.T.; Ding, Y.H.; Shao, Y.; Zhang, P.; Shen, B.L.; Inoue, A. Corrosion behavior of Fe-based ferromagnetic (Fe, Ni)–B–Si–Nb bulk glassy alloys in aqueous electrolytes. *J. Non-Cryst. Solids* **2008**, *354*, 4609–4613. [CrossRef]
16. Hashimoto, K. What we have learned from studies on chemical properties of amorphous alloys? *Appl. Surf. Sci.* **2011**, *257*, 8141–8150. [CrossRef]
17. Tsai, P.H.; Xiao, A.C.; Li, J.B.; Jang, J.S.C.; Chu, J.P.; Huang, J.C. Prominent Fe-based bulk amorphous steel alloy with large supercooled liquid region and superior corrosion resistance. *J. Alloys Compd.* **2014**, *586*, 94–98. [CrossRef]
18. Zohdi, H.; Shahverdi, H.R.; Hadavi, S.M.M. Effect of Nb addition on corrosion behavior of Fe-based metallic glasses in Ringer's solution for biomedical applications. *Electrochem. Commun.* **2011**, *13*, 840–843. [CrossRef]
19. Ni, H.S.; Liu, X.H.; Chang, X.C.; Hou, W.L.; Liu, W.; Wang, J.Q. High performance amorphous steel coating prepared by HVOF thermal spraying. *J. Alloys Compd.* **2009**, *467*, 163–167. [CrossRef]
20. Otsubo, F.; Kishitake, K. Corrosion Resistance of Fe-16%Cr-30%Mo-(C,B,P) Amorphous Coatings Sprayed by HVOF and APS Processes. *Mater. Trans.* **2005**, *46*, 80–83. [CrossRef]
21. Garfias-Mesias, L.F.; Sykes, J.M.; Tuck, C.D.S. The effect of phase compositions on the pitting corrosion of 25 Cr duplex stainless steel in chloride solutions. *Corros. Sci.* **1996**, *38*, 1319–1330. [CrossRef]
22. Cleland, J.H. What does the pitting resistance equivalent really tell us? *Eng. Fail. Anal.* **1996**, *3*, 65–69. [CrossRef]
23. Liljas, M.; Johansson, P.; Liu, H.-P.; Olsson, C.-O.A. Development of a Lean Duplex Stainless Steel. *Steel Res. Int.* **2008**, *79*, 466–473. [CrossRef]
24. Tan, M.-W.; Akiyama, E.; Kawashima, A.; Asami, K.; Hashimoto, K. The effect of air exposure on the corrosion behavior of amorphous Fe-8Cr-Mo-13P-7C alloys in 1 M HCl. *Corros. Sci.* **1995**, *37*, 1289–1301. [CrossRef]
25. Jung, K.; Ahn, S.; Kim, Y.; Oh, S.; Ryu, W.-H.; Kwon, H. Alloy design employing high Cr concentrations for Mo-free stainless steels with enhanced corrosion resistance. *Corros. Sci.* **2018**, *140*, 61–72. [CrossRef]
26. Zhang, C.; Li, Q.; Xie, L.; Zhang, G.; Mu, B.; Chang, C.; Li, H.; Ma, X. Development of novel Fe-based bulk metallic glasses with excellent wear and corrosion resistance by adjusting the Cr and Mo contents. *Intermetallics* **2023**, *153*, 107801. [CrossRef]
27. Wang, Y.; Yu, H.; Wang, L.; Li, M.; Si, R.; Sun, D. Research on the structure-activity relationship of Fe-Cr alloy in marine environment based on synchrotron radiation: Effect of Cr content. *Surf. Interfaces* **2021**, *26*, 101370. [CrossRef]
28. Hua, Y.; Mohammed, S.; Barker, R.; Neville, A. Comparisons of corrosion behaviour for X65 and low Cr steels in high pressure CO2-saturated brine. *J. Mater. Sci. Technol.* **2020**, *41*, 21–32. [CrossRef]
29. Yan, J.; Liu, T.; Wang, M.; Sun, J.; Dong, S.; Zhao, L.; Wang, Q. Constitutional supercooling and corresponding microstructure transition triggered by high magnetic field gradient during directional solidification of Al-Fe eutectic alloy. *Mater. Charact.* **2022**, *188*, 111920. [CrossRef]
30. Qi, Z.G.; Chen, Q.; Feng, Y.; Liu, H.Z.; Wang, Z.X.; Song, Z.Q.; Yan, Z.C.; Guo, L.Y.; Zhang, X.H.; Wang, W.M. Light activated methylene blue degradation by magnetic Fe80P C20-(x = 0, 4, 7 and 13) glassy ribbons. *J. Alloys Compd.* **2023**, *953*, 170135. [CrossRef]

31. Meng, L.L.; Li, X.Y.; Pang, J.; Wang, L.; An, B.; Yin, L.J.; Song, K.K.; Wang, W.M. Casting Atmosphere Effects on the Precipitates, Magnetism, and Corrosion Resistance of Fe78Si9B13 Glassy Alloys. *Metall. Mater. Trans. A* **2013**, *44*, 5122–5133. [CrossRef]
32. Ma, J.; Zhang, B.; Fu, Y.; Hu, X.; Cao, X.; Pan, Z.; Wei, Y.; Luo, H.; Li, X. Effect of cold deformation on corrosion behavior of selective laser melted 316L stainless steel bipolar plates in a simulated environment for proton exchange membrane fuel cells. *Corros. Sci.* **2022**, *201*, 110257. [CrossRef]
33. Lu, K.; Sui, M.L.; Wang, J.T. Eutectic crystallization products and their orientation relationship in amorphous TM-M alloys. *J. Mater. Sci. Lett.* **1990**, 630–632. [CrossRef]
34. Lu, Z.P.; Liu, C.T. A new glass-forming ability criterion for bulk metallic glasses. *Acta Mater.* **2002**, *50*, 3501–3512. [CrossRef]
35. Han, B.; Chen, Y.; Tao, W.; Lei, Z.; Li, H.; Guo, S.; Li, P. Nano-indentation investigation on the local softening of equiaxed zone in 2060-T8/2099-T83 aluminum-lithium alloys T-joints welded by double-sided laser beam welding. *J. Alloys Compd.* **2018**, *756*, 145–162. [CrossRef]
36. Hatada, R.; Flege, S.; Ashraf, M.N.; Timmermann, A.; Schmid, C.; Ensinger, W. The Influence of Preparation Conditions on the Structural Properties and Hardness of Diamond-Like Carbon Films, Prepared by Plasma Source Ion Implantation. *Coatings* **2020**, *10*, 360. [CrossRef]
37. Jiang, B.; Doi, K.; Tsuchiya, K.; Kawano, Y.; Kori, A.; Ikushima, K. Micromechanical properties of steel corrosion products in concrete studied by nano-indentation technique. *Corros. Sci.* **2020**, *163*, 108304. [CrossRef]
38. Amin, M.A.; Abd El Rehim, S.S.; Abdel-Fatah, H.T.M. Electrochemical frequency modulation and inductively coupled plasma atomic emission spectroscopy methods for monitoring corrosion rates and inhibition of low alloy steel corrosion in HCl solutions and a test for validity of the Tafel extrapolation method. *Corros. Sci.* **2009**, *51*, 882–894. [CrossRef]
39. Amin, M.A.; Khaled, K.F.; Fadl-Allah, S.A. Testing validity of the Tafel extrapolation method for monitoring corrosion of cold rolled steel in HCl solutions – Experimental and theoretical studies. *Corros. Sci.* **2010**, *52*, 140–151. [CrossRef]
40. Zhang, S.D.; Zhang, W.L.; Wang, S.G.; Gu, X.J.; Wang, J.Q. Characterisation of three-dimensional porosity in an Fe-based amorphous coating and its correlation with corrosion behaviour. *Corros. Sci.* **2015**, *93*, 211–221. [CrossRef]
41. Chong, K.; Gao, Y.; Zhang, Z.; Zou, Y.; Liang, X. Thermal stability and corrosion behavior of a novel Zr22.5Ti22.5Hf22.5Ni22.5Ta10 high-entropy amorphous alloy. *Corros. Sci.* **2023**, *213*, 110979. [CrossRef]
42. Berger, J.E.; Jorge, A.M.; Koga, G.Y.; Roche, V.; Kiminami, C.S.; Bolfarini, C.; Botta, W.J. Influence of chromium concentration and partial crystallization on the corrosion resistance of FeCrNiB amorphous alloys. *Mater. Charact.* **2021**, *179*, 111369. [CrossRef]
43. Botta, W.J.; Berger, J.E.; Kiminami, C.S.; Roche, V.; Nogueira, R.P.; Bolfarini, C. Corrosion resistance of Fe-based amorphous alloys. *J. Alloys Compd.* **2014**, *586*, S105–S110. [CrossRef]
44. Coimbrão, D.D.; Zepon, G.; Koga, G.Y.; Godoy Pérez, D.A.; Paes de Almeida, F.H.; Roche, V.; Lepretre, J.C.; Jorge, A.M.; Kiminami, C.S.; Bolfarini, C.; et al. Corrosion properties of amorphous, partially, and fully crystallized Fe68Cr8Mo4Nb4B16 alloy. *J. Alloys Compd.* **2020**, *826*, 154123. [CrossRef]
45. Fan, Y.; Zhang, S.; Xu, X.; Miao, J.; Zhang, W.; Wang, T.; Chen, C.; Wei, R.; Li, F. Effect of the substitution of Si for B on thermal stability, magnetic properties and corrosion resistance in novel Fe-rich amorphous soft magnetic alloy. *Intermetallics* **2021**, *138*, 107306. [CrossRef]
46. Han, Y.; Chang, C.T.; Zhu, S.L.; Inoue, A.; Louzguine-Luzgin, D.V.; Shalaan, E.; Al-Marzouki, F. Fe-based soft magnetic amorphous alloys with high saturation magnetization above 1.5 T and high corrosion resistance. *Intermetallics* **2014**, *54*, 169–175. [CrossRef]
47. Han, Y.; Kong, F.L.; Han, F.F.; Inoue, A.; Zhu, S.L.; Shalaan, E.; Al-Marzouki, F. New Fe-based soft magnetic amorphous alloys with high saturation magnetization and good corrosion resistance for dust core application. *Intermetallics* **2016**, *76*, 18–25. [CrossRef]
48. Koga, G.Y.; Nogueira, R.P.; Roche, V.; Yavari, A.R.; Melle, A.K.; Gallego, J.; Bolfarini, C.; Kiminami, C.S.; Botta, W.J. Corrosion properties of Fe–Cr–Nb–B amorphous alloys and coatings. *Surf. Coat. Technol.* **2014**, *254*, 238–243. [CrossRef]
49. Liu, Y.; Li, J.; Sun, Y.; He, A.; Dong, Y.; Wang, Y. Effect of annealing temperature on magnetic properties and corrosion resistance of Fe75.8Si12B8Nb2.6Cu0.6P1 alloy. *J. Mater. Res. Technol.* **2021**, *15*, 3880–3894. [CrossRef]
50. Meghwal, A.; Pinches, S.; King, H.J.; Schulz, C.; Stanford, N.; Hall, C.; Berndt, C.C.; Ang, A.S.M. Fe-based amorphous coating for high-temperature wear, marine and low pH environments. *Materialia* **2022**, *25*, 101549. [CrossRef]
51. Sunbul, S.E.; Akyol, S.; Onal, S.; Ozturk, S.; Sozeri, H.; Icin, K. Effect of Co, Cu, and Mo alloying metals on electrochemical and magnetic properties of Fe-B alloy. *J. Alloys Compd.* **2023**, *947*, 169652. [CrossRef]
52. Vasić, M.M.; Simatović, I.S.; Radović, L.; Minić, D.M. Influence of microstructure of composite amorphous/nanocrystalline Fe72Ni8Si10B10 alloy on the corrosion behavior in various environments. *Corros. Sci.* **2022**, *204*, 110403. [CrossRef]
53. Wu, L.; Zhou, Z.; Zhang, K.; Zhang, X.; Wang, G. Electrochemical and passive film evaluation on the corrosion resistance variation of Fe-based amorphous coating affected by high temperature. *J. Non-Cryst. Solids* **2022**, *597*, 121892. [CrossRef]
54. Wang, P.; Qi, W.; Yang, K.; Qiao, Y.; Wang, X.; Zheng, T.; Bai, C.; Liu, Z.; Zhang, X. Systematic investigation of the oxidation behavior of Fe-Cr-Al cladding alloys in high-temperature steam. *Corros. Sci.* **2022**, *207*, 110595. [CrossRef]
55. Qin, C.; Hu, Q.; Li, Y.; Wang, Z.; Zhao, W.; Louzguine-Luzgin, D.V.; Inoue, A. Novel bioactive Fe-based metallic glasses with excellent apatite-forming ability. *Mater. Sci. Eng. C* **2016**, *69*, 513–521. [CrossRef]
56. Wu, G.; Tan, X.; Li, G.; Hu, C. Effect of preparation method on the physical and catalytic property of nanocrystalline Fe2O3. *J. Alloys Compd.* **2010**, *504*, 371–376. [CrossRef]
57. Tian, W.-P.; Yang, H.-W.; Zhang, S.-D. Synergistic Effect of Mo, W, Mn and Cr on the Passivation Behavior of a Fe-Based Amorphous Alloy Coating. *Acta Metall. Sin. (Engl. Lett.)* **2017**, *31*, 308–320. [CrossRef]

58. Wang, Y.; Jiang, S.L.; Zheng, Y.G.; Ke, W.; Sun, W.H.; Wang, J.Q. Electrochemical behaviour of Fe-based metallic glasses in acidic and neutral solutions. *Corros. Sci.* **2012**, *63*, 159–173. [CrossRef]
59. Wu, J.; Zhang, S.D.; Sun, W.H.; Gao, Y.; Wang, J.Q. Enhanced corrosion resistance in Fe-based amorphous coatings through eliminating Cr-depleted zones. *Corros. Sci.* **2018**, *136*, 161–173. [CrossRef]
60. Xu, D.D.; Zhou, B.L.; Wang, Q.Q.; Zhou, J.; Yang, W.M.; Yuan, C.C.; Xue, L.; Fan, X.D.; Ma, L.Q.; Shen, B.L. Effects of Cr addition on thermal stability, soft magnetic properties and corrosion resistance of FeSiB amorphous alloys. *Corros. Sci.* **2018**, *138*, 20–27. [CrossRef]
61. Zhai, H.; Yuan, H.; Li, W.; Zhang, X.; Li, X.; Cai, A. Corrosion resistance mechanisms of detonation sprayed Fe-based amorphous coating on AZ31B magnesium alloy. *J. Non-Cryst. Solids* **2022**, *576*, 121276. [CrossRef]
62. Čekerevac, M.; Simičić, M.; Bujanović, L.N.; Popović, N. The influence of silicate and sulphate anions on the anodic corrosion and the transpassivity of iron and silicon-rich steel in concentrated KOH solution. *Corros. Sci.* **2012**, *64*, 204–212. [CrossRef]
63. Luo, H.; Su, H.; Dong, C.; Li, X. Passivation and electrochemical behavior of 316L stainless steel in chlorinated simulated concrete pore solution. *Appl. Surf. Sci.* **2017**, *400*, 38–48. [CrossRef]
64. Zhang, Y.; Song, B.; Ming, J.; Yan, Q.; Wang, M.; Cai, C.; Zhang, C.; Shi, Y. Corrosion mechanism of amorphous alloy strengthened stainless steel composite fabricated by selective laser melting. *Corros. Sci.* **2020**, *163*, 108241. [CrossRef]
65. Kroll, R.; Kearns, P.; Usman, B.J.; Zhou, X.; Engelberg, D.L. A novel approach to determine cathodic passivation characteristics and semiconducting properties of pure aluminium 99.5 wt% and aluminium alloy 7075-T6 with an electrochemical pen electrode. *Corros. Sci.* **2023**, *211*, 110898. [CrossRef]
66. Liu, M.; Cheng, X.; Li, X.; Pan, Y.; Li, J. Effect of Cr on the passive film formation mechanism of steel rebar in saturated calcium hydroxide solution. *Appl. Surf. Sci.* **2016**, *389*, 1182–1191. [CrossRef]
67. Shi, J.; Ming, J.; Wu, M. Electrochemical behavior and corrosion products of Cr-modified reinforcing steels in saturated $Ca(OH)_2$ solution with chlorides. *Cem. Concr. Compos.* **2020**, *110*, 103587. [CrossRef]
68. Song, G.; Atrens, A.; John, D.S.; Wu, X.; Nairn, J. The anodic dissolution of magnesium in chloride and sulphate solutions. *Corros. Sci.* **1997**, *39*, 1981–2004. [CrossRef]
69. Lamaka, S.V.; Shchukin, D.G.; Andreeva, D.V.; Zheludkevich, M.L.; Möhwald, H.; Ferreira, M.G.S. Sol-Gel/Polyelectrolyte Active Corrosion Protection System. *Adv. Funct. Mater.* **2008**, *18*, 3137–3147. [CrossRef]
70. Song, G.-L.; Unocic, K.A. The anodic surface film and hydrogen evolution on Mg. *Corros. Sci.* **2015**, *98*, 758–765. [CrossRef]
71. Xu, J.; Liu, L.; Li, Z.; Munroe, P.; Xie, Z.H. Niobium addition enhancing the corrosion resistance of nanocrystalline Ti_5Si_3 coating in H_2SO_4 solution. *Acta Mater.* **2014**, *63*, 245–260. [CrossRef]

materials

Article

Enhancing the SCC Resistance of the Anchor Steel with Microalloying in a Simulated Mine Environment

Hailong Du [1,2], Na An [3], Xiyan Wang [3], Yongliang Li [4], Zhiyong Liu [3,5,*], Aibing Jin [1], Renshu Yang [1], Yue Pan [3,*] and Xiaogang Li [3]

[1] School of Civil and Resource Engineering, University of Science and Technology Beijing, Beijing 100083, China; 13703561606@163.com (H.D.)
[2] Technology Research Institute of Shanxi Jincheng Coal Group Co., Ltd., Jincheng 048000, China
[3] National Materials Corrosion and Protection Scientific Data Center, Key Laboratory for Corrosion and Protection (MOE), University of Science and Technology Beijing, Beijing 100083, China; b20150517@xs.ustb.edu.cn (X.W.)
[4] School of Energy and Mining Engineering, China University of Mining and Technology-Beijing, Beijing 100083, China
[5] Hebei Special Equipment Supervision and Inspection Institute, Key Laboratory of Safety Evaluation of Steel Pipes and Fittings for State Market Regulation, Shijiazhuang 050061, China
* Correspondence: liuzhiyong7804@126.com (Z.L.); panyue0130@126.com (Y.P.)

Abstract: This work explored a new idea for enhancing the resistance to stress corrosion cracking (SCC) of mining anchor steel through microalloying. Microalloyed anchor steels with Nb, Cu, Ni, Sb, and C were prepared through vacuum smelting and hot rolling. Electrochemical measurements, slow strain rate tensile (SSRT) tests, and fracture morphology observations were used to study the electrochemical and SCC behavior in the simulated mine environment. The results proved that the microstructure of microalloyed steels varies slightly. Adding Ni, Cu, and Sb can improve the mechanical properties of the anchor steel, while reducing C content decreases tensile strength as a result of loss of the solution-strengthening effect. The addition of Sb, Cu, Ni, and reducing the content of C enhances the resistance to corrosion and SCC by mitigating anodic dissolution (AD), while adding Nb improves SCC resistance by inhibiting hydrogen embrittlement (HE). The combined addition of 1% Ni, 0.5% Cu, 0.05% Nb, 0.1% Sb, and 0.5% C presented the highest SCC resistance, which is a promising prospect for the development of high-performance, low-alloy anchor steels. The combined addition of 1% Ni, 0.5% Cu, 0.05% Nb, and 0.1% Sb resulted in the inhibition of electrochemical reactions and corrosion. As a result of the synergistic effect of the microalloy, both AD and HE mechanisms were simultaneously inhibited, which greatly enhanced SCC resistance.

Keywords: microalloying; mining anchor bolts; SCC; hydrogen embrittlement

Citation: Du, H.; An, N.; Wang, X.; Li, Y.; Liu, Z.; Jin, A.; Yang, R.; Pan, Y.; Li, X. Enhancing the SCC Resistance of the Anchor Steel with Microalloying in a Simulated Mine Environment. *Materials* **2023**, *16*, 5965. https://doi.org/10.3390/ma16175965

Academic Editor: Oldrich Sucharda

Received: 1 August 2023
Revised: 25 August 2023
Accepted: 29 August 2023
Published: 31 August 2023

1. Introduction

Anchoring technology has been widely used in all types of mines, but the service conditions of the anchor cables and rods have been deteriorating year by year due to the increasing mining depth and corrosion problems in various aggressive mine environments. According to reported statistics [1,2], 29% of anchor bolts and 25% of anchor cables in the Australian mining industries are subjected to corrosion failures, and corrosion has been recognized as one of the biggest threats to the life of anchor bolt structures buried in rock and soil layers. In addition, at external load or residual stress, stress corrosion cracking (SCC) and hydrogen embrittlement (HE) happen frequently and lead to abrupt failures [3–5]. Therefore, SCC and HE problems pose a serious threat to the safe service of anchor bolt structures and severely affect the development and application of anchor steels.

Studies have shown that high-strength anchor steels are vulnerable to SCC in chloride-containing environments [6,7]. The SCC mechanism is considered a mixture of anodic dissolution (AD) and hydrogen embrittlement (HE). Microalloying has become the mainstream method for optimizing corrosion and SCC resistance by inhibiting AD and HE. It has been reported that Ni, Nb, Cu, Sb, and C have a great impact on corrosion and SCC. In general, Ni, Cu, and Sb can mitigate AD, while adding Nb prevents HE. It has been reported that Ni can greatly enhance the corrosion resistance of steels by improving the compactness of the rust layer [8,9]. Adding Cu can improve corrosion resistance because it facilitates CuO formation and enhances the protective capacity of the corrosion product layer [8]. In addition, in Ni-bearing steels, a small amount of Cu addition can be beneficial to the function of Ni in improving the rust layer protectiveness [10]. Sb is an acid-resistant element and has a synergetic effect with Cu and Ni, which can inhibit AD and improve corrosion resistance [11]. Previous work has proved that Sb induces the enrichment of Cu and Ni and improves the resistance of the rust layer to corrosive chloride ions [11]. Therefore, the combined addition of Ni, Cu, and Sb should be effective at improving the SCC resistance of anchor steels by mitigating AD. Nb can mitigate hydrogen-induced cracking problems by inhibiting crack initiation and propagation [12] because Nb addition can facilitate the formation of nanoprecipitated Nb(C, N), which can act as hydrogen traps and prevent hydrogen diffusion [13]. Adding C can refine grains and improve the strength of steels by the solution-strengthening mechanism but has an opposite effect on corrosion behavior. C addition can facilitate carbide formation at grain boundaries and cause intergranular corrosion [14]. Hence, the addition of Nb and adjusting the content of C are also expected to enhance the resistance to corrosion and SCC. The above research has demonstrated that microalloying is feasible in mitigating the SCC of low-alloy steels. However, few studies have paid attention to the microalloy effect on the SCC behavior and mechanism of anchor steels.

Therefore, based on the above progress, in this work, a series of anchor cable steels were prepared by adding Nb, Sb, Cu, Ni, and reducing C content. The effect of microalloying on the SCC behavior and mechanism was analyzed systematically, which is anticipated to propose new ideas and theoretical support for the application and safe service of anchor steels.

2. Experimental Procedures

2.1. Materials and Solution

Four types of low-alloy, ultra-high-strength anchor steels with different chemical compositions were prepared, and the chemical composition is listed in Table 1. Commercial YL82B steel was selected as the as-received material, which is labeled as the BM in this work. Three other kinds of steels were added to alloy elements compared to the BM. 0.5NiCu denotes the addition of 0.5% Ni with 0.5% Cu and other trace alloying elements (0.046% Nb + 0.11% Sb). 1NiCu represents the adjustment of Ni elements to 1% and the decrease in carbon content to 0.53% compared to 0.5NiCu. LNiCu represents the decrease in carbon content to 0.34% in comparison to 1NiCu as the decarburized version of 1NiCu. The ingots were prepared by vacuum smelting in a vacuum induction furnace, and the procedures are illustrated in Figure 1. As shown, the ingots were homogenized at 1200 °C for 2 h, furnace-cooled at 1010 °C, and then hot-rolled at the final temperature of 880 °C followed by air cooling. All the test specimens in this work were taken from the center thickness of the steel sheet to avoid the possible errors caused by surface decarburization during hot rolling.

Table 1. The chemical compositions (wt.%) of the tested anchor steels.

	C	Si	Mn	Cr	Ni	Cu	Nb	Sb	P	S
BM	0.71	0.21	0.80	0.20	0	0	0	0	0.018	0.002
0.5NiCu	0.71	0.21	0.80	0.20	0.49	0.51	0.046	0.11	0.021	0.016
1NiCu	0.53	0.21	0.80	0.20	0.98	0.51	0.046	0.11	0.022	0.012
LNiCu	0.34	0.21	0.80	0.20	0.98	0.51	0.046	0.11	0.015	0.014

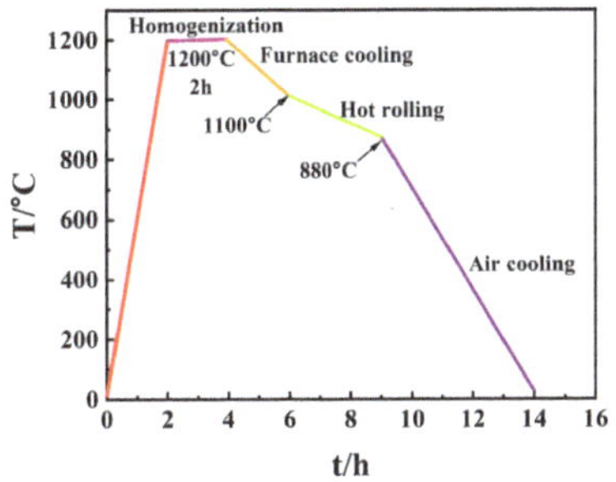

Figure 1. Schematic diagram of the heat treatment process for the prepared low-alloy anchor steels.

The corrosive medium in this work was the simulated mine environment, and the chemical composition is shown in Table 2 according to the analysis of the chemical composition of the groundwater and soluble gas contents in underground coal mines. The solution pH was adjusted to 5.0 with acetic acid. All the ingredients were analytically pure for the preparation of the simulated solution, and all the tests were performed at ambient temperature.

Table 2. The chemical composition of the simulated mine environment (g/L).

NaCl	KNO_3	Na_2SO_4	$NaHCO_3$	$NaHSO_3$
0.25	0.1	0.50	1.00	1.00

2.2. Microstructural Analysis

Metallographic samples with size of 10 mm × 10 mm × 3 mm were cut from the rolling and transverse (RD–TD) planes. Then, all samples were ground to 3000 grits with SiC paper, mechanically polished to 1 μm, and chemically etched with 4% nital solution. The microstructure was observed with an FEI Quanta 250 scanning electron microscope (Eindhoven, The Netherlands) (SEM).

2.3. Mechanical Property Test

Tensile tests were carried out to study the mechanical properties of the test steels. The test specimens were cut from the rolling–transverse (TD–RD) plane along the rolling direction, and the dimensions are shown in Figure 2 in accordance with the Chinese National Standard GB/T228.1-2021 [15]. Prior to the test, the gauge length of the specimens was sequentially ground with SiC paper to 3000#, and the final grinding direction was parallel to the gauge length. The tensile tests were performed with an MTS-Landmark-730 hydraulic servo fatigue testing machine (Eden Prairie, MN, USA) with the tensile rate of 10^{-3} s^{-1}. After fracture, the elongation and cross-sectional area were measured with a vernier caliper. All the tests were performed at least three times to ensure the repeatability and reliability of the results.

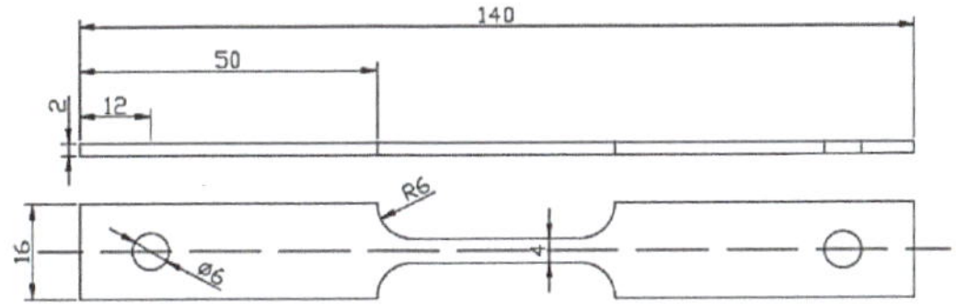

Figure 2. Schematic diagram of tensile specimens.

2.4. Electrochemical Measurements

The electrochemical properties of the test steels were measured with a VersaSTAT3 electrochemical workstation produced by Princeton Applied Research (Trenton, NJ, USA). The three-electrode system was applied, in which the saturated calomel electrode (SCE) was used as the reference electrode, a platinum plate as the counter electrode, and the test steel with the dimension of 10 mm × 10 mm × 3 mm as the working electrode. The test specimens were sealed with epoxy resin except for the 1 cm^2 working area, and the working surface was ground with SiC paper sequentially to 3000 grits and washed with deionized water and alcohol, followed by blow drying. Prior to the test, the air-formed oxide film was removed through potentiostatic polarization at −1.2 V (vs. SCE) for 60 s, and then the open-circuit potential (OCP) was recorded for at least 30 min until reaching a steady state. The electrochemical impedance spectroscopy (EIS) measurements were performed at the AC excitation amplitude of 10 mV and the test frequency ranging from 100 kHz to 10 mHz. Potentiodynamic polarization measurements were conducted at the scan range of −1.2 V to −0.4 V (vs. SCE), and two scan rates were chosen: 0.5 mV/s to study the quasi-steady electrochemical process at the crack wall and 200 mV/s to study the non-steady electrochemical process at the crack tip [16]. All experiments were repeated more than three times in order to ensure repeatability and accuracy.

2.5. SSRT Tests

Slow strain rate tensile (SSRT) tests were performed to study the SCC susceptibility of the test steels. The dimension and the preparation of test specimens were the same as procedures described in Section 2.2. The SSRT tests were carried out at a strain rate of 10^{-6} s^{-1} with a WDML-30 kN material testing system. Each type of test steel was tested in air, under OCP and −1200 mV (vs. SCE) to study the SCC in the test solution without/with hydrogen charging. Prior to the test, specimens were preloaded with approximately 500 N for 12 h in the test solution in order to make sure that a uniform electrolyte film formed. For specimens tested at −1200 mV, potentiostatic polarization was imposed before and throughout the test. To ensure repeatability, all tests were performed at least three times. The elongation loss (I_σ) and the loss of the reduction in area (I_Ψ) were calculated with Equations (1) and (2) for characterizing the SCC susceptibility.

$$I_\sigma = \left(1 - \frac{\delta_s}{\delta_0}\right) \times 100\% \quad (1)$$

$$I_\psi = \left(1 - \frac{\psi_s}{\psi_0}\right) \times 100\% \quad (2)$$

where δ_s and δ_0 are the elongation rates of specimens tested in solution and air, respectively; Ψ_s and Ψ_0 are the reductions in area of the specimens tested in solution and air, respectively. However, the strengths of the four kinds of test steels are different, which makes the above equations fail to characterize the real SCC susceptibility differences among the different test steels. For example, under the same external stress, when the steel with a lower yield strength passes the yield point, the steel with a higher yield strength is below the yield point. This causes the steel with the higher yield strength to present lower I_σ and I_Ψ values than that with the lower yield strength. Therefore, we used the normalized elongation loss ($I_{\sigma x}{}'$) and the normalized loss of the reduction in area ($I_{\Psi x}{}'$) to evaluate SCC susceptibility, and the formulas for calculating them are as shown in (3) and (4):

$$I_{\sigma x}{}' = \frac{I_\sigma}{(\sigma_{sx}/\sigma_{s0})}\ (x = 1,2,3) \quad (3)$$

$$I_{\Psi x}{}' = \frac{I_\Psi}{(\sigma_{sx}/\sigma_{s0})}\ (x = 1,2,3) \quad (4)$$

where σ_{sx} (x = 1, 2, 3) is the yield strength of the 0.5NiCu, 1NiCu, and LNiCu anchor steel; σ_{s0} is the yield strength of the BM.

After the SSRT tests, the fracture surface was cut for the morphology observation, and then the corrosion products were removed by ultrasonic cleaning in the mixed solution of 500 mL HCl + 500 mL H_2O + 5 g hexamethylenetetramine for about 4~8 s. The fracture surface morphology and secondary cracks were observed with an FEI Quanta 250 SEM.

3. Results

3.1. Microstructural Characterization

The microstructure of the test steels is shown in Figure 3. The microstructure of the four kinds of anchor bolt steels is mainly composed of pearlite, which appears as a fine lamellar structure. A small amount proeutectoid phase also exists and is inhomogeneously distributed as a result of the proeutectoid phase transformation, which is common in hypoeutectoid steels [17,18]. The results indicated that microalloying and decarburization (as represented by LNiCu) had no obvious influence on the steel microstructure.

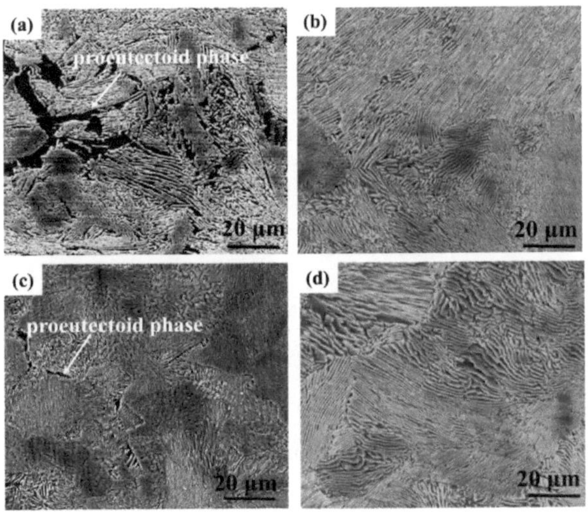

Figure 3. Microstructures of the prepared low-alloy, high-strength anchor steels: (**a**) BM; (**b**) 0.5NiCu; (**c**) 1NiCu; (**d**) LNiCu.

3.2. Mechanical Properties

Stress–strain curves for the four test steels are shown in Figure 4, and the mechanical properties are listed in Table 3. The tensile strength of 0.5NiCu is the highest (1046 MPa), while the elongation decreases slightly compared to the BM. Due to the addition of the alloying elements Cu, Nb, Ni, and Sb, the microalloyed anchor steels present a higher yield ratio. The tensile strength of the 1NiCu and LNiCu is slightly lower than that of the BM, and the tensile strength of the LNiCu is the lowest due to the loss of the solution-strengthening effect as a result of the decreased C content [19]. Therefore, decarburization causes a decrease in the tensile strength. The mechanical properties of the various designed microalloyed steels satisfy the basic requirements compared to the BM.

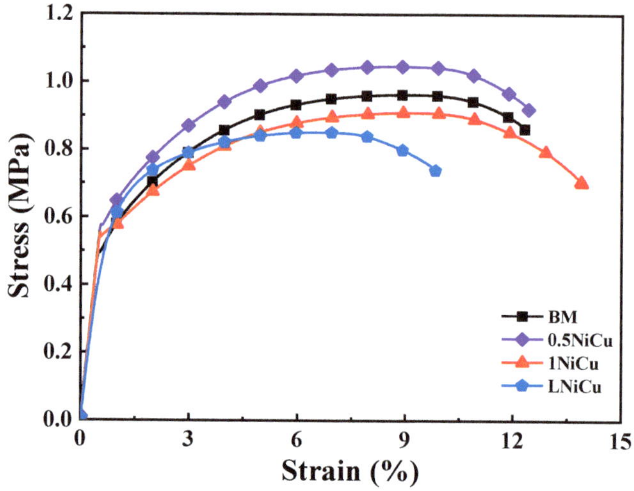

Figure 4. Stress–strain curves of the four kinds of anchor steels.

Table 3. The mechanical properties of the test steels.

Steels	Tensile Strength (MPa)	Yield Strength (MPa)	Elongation Rate (%)	Reduction in Area (%)
BM	962 ± 85	525 ± 57	9.14 ± 1.5	31.97 ± 2.9
0.5NiCu	1046 ± 99	604 ± 78	8.65 ± 1.2	32.73 ± 3.8
1NiCu	907 ± 83	557 ± 55	9.30 ± 2.0	45.65 ± 5.5
LNiCu	856 ± 89	551 ± 65	7.60 ± 1.5	37.01 ± 3.5

3.3. *Electrochemical Properties*

3.3.1. EIS Analysis

The EIS test results of the four test steels in the simulated mine environment are shown in Figure 5. All the Nyquist plots present a compressed semicircle, indicating a similar electrochemical mechanism. The semicircle radius of the 1NiCu steel is the biggest, which indicates that 1NiCu exerts the highest corrosion resistance. The Bode diagram indicates the existence of two time constants, and the Bode phase angle in Figure 5b is approximately 30° and far less than 90° on account of the porous corrosion product layer. Therefore, the equivalent electric circuit in the inset in Figure 5a with two time constants is used to fit the EIS curves, where R_s represents the solution resistance, and R_{ct} and R_{pore} are the charge transfer resistance and corrosion product film resistance. Q_{dl} and Q_f are the constant-phase elements (CPEs) of the electric double layer and the corrosion product film, and the corresponding n values describe the deviation from an ideal capacitor caused by the inhomogeneity of the electrodes [20,21].

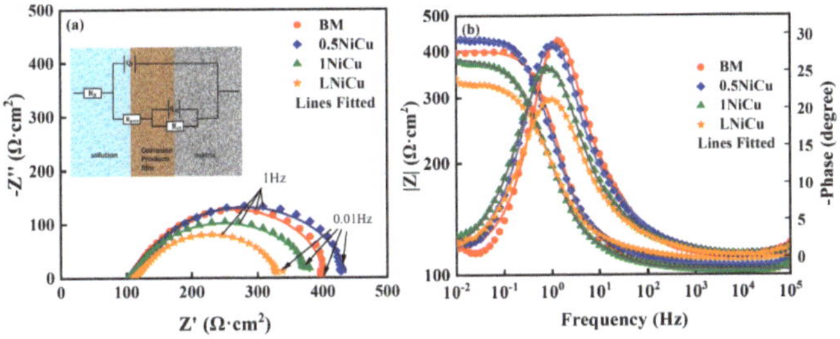

Figure 5. (**a**) Nyquist and (**b**) Bode plots of the prepared microalloyed anchor steels in a coal mine environment.

The fitted EIS values are listed in Table 4. The χ^2 values are all at the 10^{-4} level, which indicates a high fitting quality. The values of R_{ct} increase as a result of microalloying. In comparison to 1NiCu, R_{ct} of LNiCu decreases and is lower than the BM. Therefore, it can be deduced that microalloying can suppress the charge transfer process, while decarburization accelerates charge transfer. The polarization resistance R_p (R_{ct} + R_f) is used to characterize the corrosion resistance of the test steels, and the values are shown in Figure 6. It has been widely acknowledged that R_p is inversely proportional to the corrosion rate, so higher R_p values indicate a higher corrosion resistance [22]. By comparing R_p, we can conclude that 1NiCu presents the highest R_p value and corrosion resistance, and LNiCu exerts the lowest R_p and the worst corrosion resistance. In addition, 0.5NiCu exhibits a lower corrosion resistance than the BM, for Cu accelerates the generation of corrosion products and the corrosion process in the initial corrosion stage [23,24]. As a result, the corrosion resistance of Cr-bearing steels may decline. In addition, the beneficial effect of Cu on corrosion resistance makes a difference only when the content exceeds a certain value, so the corrosion resistance for 0.5NiCu is not markedly improved. The reason for the corrosion resistance differences is discussed thoroughly in the Discussion.

Table 4. Fitting results of the EIS curves of all test steels.

Steel	R_s/Ω·cm^2	$Q_f \times 10^{-4}$/Ω^{-1}cm^{-2}s^n	n_f	R_{pore}/Ω·cm^2	$Q_{dl} \times 10^{-4}$/Ω^{-1}cm^{-2}s^n	n_{dl}	R_{ct}/Ω·cm^2	$\chi^2 \times 10^{-4}$
BM	105.5	3.99	0.93	57.37	5.94	0.95	239.8	3.64
0.5NiCu	103.4	7.55	0.82	32.23	10.43	0.83	247.8	1.53
1NiCu	107.1	4.77	0.86	49.25	7.37	0.90	281.1	2.05
LNiCu	111.7	6.93	0.82	27.22	11.9	0.82	195.5	1.12

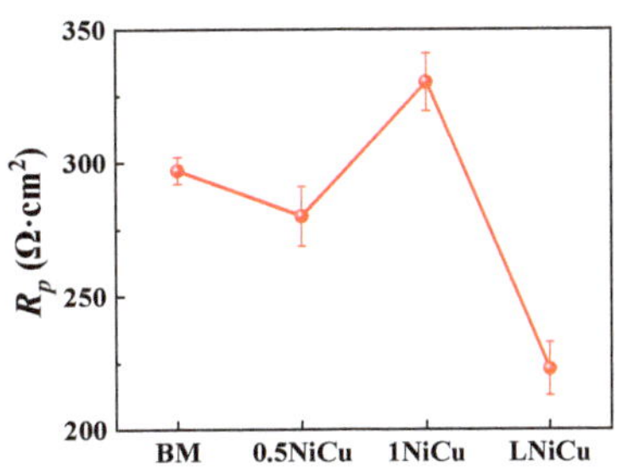

Figure 6. Statistical results of polarization resistance of the four kinds of anchor steels.

3.3.2. Polarization Curve Analysis

Potentiodynamic polarization curves for the anchor steels at the 0.5 mV/s scan rate are shown in Figure 7. All the curves present a similar shape, which is an indication of a similar corrosion mechanism. The anodic curves present the active dissolution reaction, and the cathodic branches exert a combination of the oxygen reduction reaction and hydrogen evolution reaction. The inset figures show that the anodic curve shifts to the left with the effect of alloy elements and decarburization in comparison to the BM, which indicates that both the microalloy and reducing the carbon content inhibit anodic dissolution. As for the cathodic branch, the cathodic polarization curves of 0.5NiCu and 1NiCu shift to the right, while the curve of LNiCu almost overlaps with that of the BM. Considering the opposite variation tendency of the cathodic branch and the anodic branch, we did not conduct Tafel fitting for corrosion rates as referred to in the literature [25]. The polarization curves of the three kinds of microalloyed steels vary slightly.

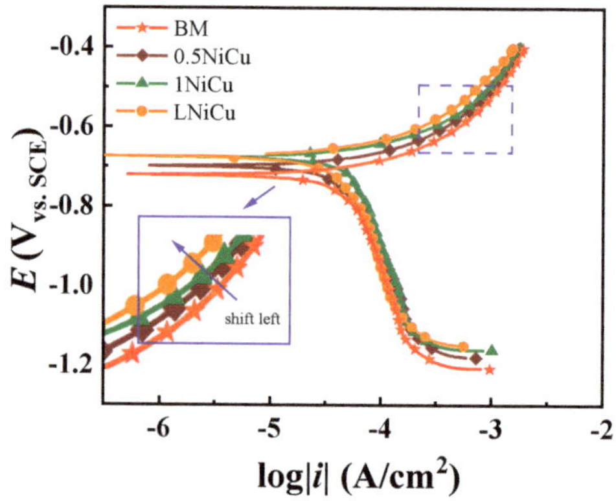

Figure 7. Polarization curves of the prepared microalloyed anchor steels in a coal mine environment.

3.4. *SSRT Results*

3.4.1. Stress–Strain Curves and SCC Susceptibility

Figure 8 shows the SSRT stress–strain curves of the test anchor steels. In addition, the elongation loss of 1NiCu is the lowest in the test solution, indicating the highest SCC resistance. It is worth noting that the elongation of 0.5NiCu at −1200 mV is significantly higher than that at OCP, which is probably due to the hydrogen-induced plasticity effect. This phenomenon has been observed in previous work, which proves that hydrogen can release stress concentration at crack initiation sites, decreases the stress intensity, and postpones SCC initiation at the applied cathodic polarization [26,27].

The normalized SCC susceptibility results of the test anchor steels are shown in Figure 9. When tested at −1200 mV, all microalloyed steels are less SCC-susceptible compared to the BM. Additionally, 1NiCu presents the highest SCC resistance at OCP and −1200 mV. Compared to the BM, the I_{σ}' of 1NiCu reduces by over 70% when tested at OCP, and reduces by 45% when tested at −1200 mV. LNiCu presents higher SCC susceptibility than 1NiCu. The reason is that the excessively low carbon content is harmful to the stability of austenite [28], which also causes microstructure degradation during phase transformation and deformation, leading to lower SCC susceptibility.

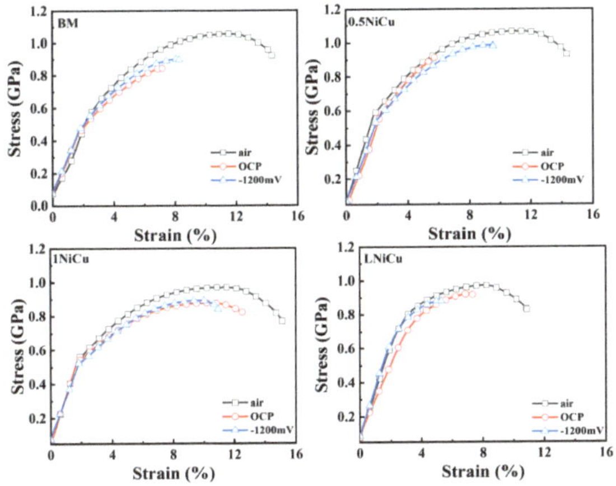

Figure 8. Stress–strain curves of the anchor steels after SSRT tests in air and simulated mine environment.

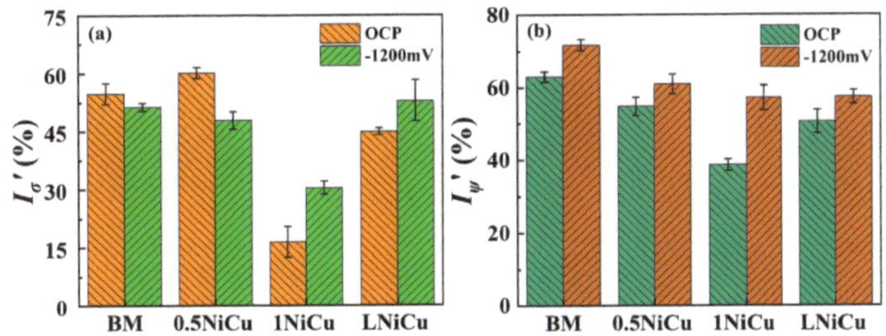

Figure 9. Normalized SCC susceptibility of the test anchor steels (**a**) $I_{\sigma X}'$ and (**b**) $I_{\Psi X}'$.

3.4.2. Fracture Analysis

After the SSRT tests, the fracture morphology of the anchor steels was carefully observed as shown in Figure 10. When the BM is tested in the air, there is obvious necking. In addition, dimples of different sizes are observed in the magnified view in Figure 10(a_1), which are typical ductile fracture features. When tested in solution at OCP, the size and number of dimples decrease sharply, and brittle characteristics appear, indicating high SCC susceptibility. At −1200 mV, the brittle characteristics become more obvious. The SCC initiation site at the edge of the specimen is characterized by river-like patterns with distinct tear ridges, which is typical of a quasi-cleavage fracture. In addition, the fracture morphology of the BM and 0.5NiCu under OCP exhibits a convergence of extended cracks in the center of the fracture. It is speculated that there are multiple crack initiation sites at the edge of the fracture, which propagate to the interior of the specimens and result in the final fracture at the center. In addition, at OCP, the fracture surface of the BM, 0.5NiCu, and LNiCu is flat, while the fracture morphology of 1NiCu still exhibits necking, demonstrating the highest SCC resistance, which is consistent with the results of SCC susceptibility in Figure 9.

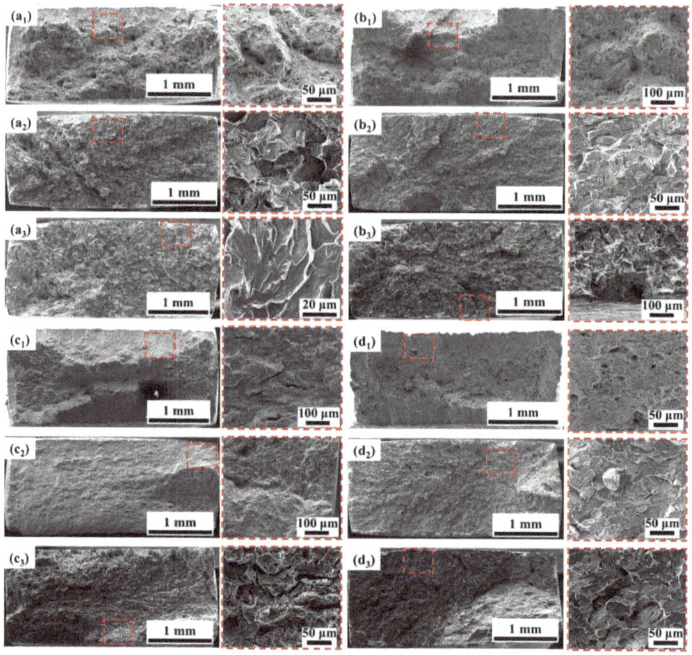

Figure 10. Fracture morphology of different test steels after SSRT tests in various test conditions: (**a$_1$**) BM in air; (**a$_2$**) BM at OCP; (**a$_3$**) BM at −1200 mV; (**b$_1$**) 0.5NiCu in air; (**b$_2$**) 0.5NiCu at OCP; (**b$_3$**) 0.5NiCu at −1200 mV; (**c$_1$**) 1NiCu in air; (**c$_2$**) 1NiCu at OCP; (**c$_3$**) 1NiCu at −1200 mV; (**d$_1$**) LNiCu in air; (**d$_2$**) LNiCu at OCP; (**d$_3$**) LNiCu at −1200 mV (The red squares show the crack initiation sites and the typical morphology).

Secondary cracks of the test anchor steels were also carefully analyzed after the SSRT tests, as seen in Figure 11, which shows the side surface of the specimens. When tested in air, secondary cracks are present on the side face of the tested BM specimen but are not apparent on the microalloyed steels. When tested in solution, secondary cracks can be observed in all test specimens. At OCP, cracks with a big, open mouth can be observed, which is a sign of SCC caused by localized AD [29]. When tested at −1200 mV, the cracks are thin and shallow, which are typical of hydrogen-induced cracking (HIC) [30]. In addition, at OCP, the BM and 0.5NiCu show obvious pitting corrosion, which is less evident in the 1NiCu and LNiCu specimens as more Ni and Cu inhibit localized AD. At −1200 mV, in comparison to 1NiCu, the secondary cracks of LNiCu are longer and deeper, which indicates that the decreased C content deteriorates HE resistance. In summary, 1NiCu possesses the highest SCC resistance for inhibiting both AD and HE.

Figure 11. Morphology of secondary cracks of different test steels after SSRT tests in various test conditions: (**a$_1$**) BM in air; (**a$_2$**) BM at OCP; (**a$_3$**) BM at −1200 mV; (**b$_1$**) 0.5NiCu in air; (**b$_2$**) 0.5NiCu at OCP; (**b$_3$**) 0.5NiCu at −1200 mV; (**c$_1$**) 1NiCu in air; (**c$_2$**) 1NiCu at OCP; (**c$_3$**) 1NiCu at −1200 mV; (**d$_1$**) LNiCu in air; (**d$_2$**) LNiCu at OCP; (**d$_3$**) LNiCu at −1200 mV (The red tangles are key observation areas and the morphology at magnified view).

4. Discussion

4.1. Effect of Alloy Elements on the Electrochemical Mechanism

The underground mine environment is a harsh corrosive medium with multiphase flow containing CO_2, SO_2, and H_2S. These gases are dissolved in water and form CO_3^{2-} and HSO_3^-. Therefore, the steel surface is covered with an acidic electrolyte film, which facilitates corrosion. In the coal mine environment, H_2S and SO_2 dissolve in the thin liquid film and form HSO_3^-, which acidifies the electrolyte film [11,31]. Additionally, HSO_3^- is unstable and can be easily oxidized to H_2SO_4, which further acidifies the electrolyte film, as shown in chemical reaction Equation (5) [32].

$$2HSO_3^- + O_2 \rightarrow SO_4^{2-} + 2H^+ \quad (5)$$

In consequence, anchor steels suffer from severe corrosion problems, and the electrochemical process is as follows. The anodic reaction is the dissolution of iron:

$$Fe + 2e^- \rightarrow Fe^{2+} \quad (6)$$

The cathodic reaction is the combination of oxygen reduction and hydrogen evolution:

$$O_2 + 4e^- + 4H^+ \rightarrow 2H_2O \quad (7)$$

$$2H^+ + 2e^- \rightarrow H_2 \quad (8)$$

The results in this work prove that microalloying improves corrosion resistance by making a difference in the electrochemical process, which inhibits anodic dissolution and accelerates the cathodic process as illustrated in Figure 7. Compared to the BM, the microalloyed steels Sb, Nb, Ni, and Cu were added, and the content of C was reduced in 1NiCu and LNiCu. The role of the discrete alloying elements is discussed as follows.

As for the effect of Sb, based on the Pourbaix diagram [33], Sb mainly exists in the form of Sb_2O_3 and Sb_2O_5 in the rust layer [34,35]. The reaction of Sb in the corrosion process is as follows:

$$2Sb + 3H_2O \rightarrow Sb_2O_3 + 6H^+ + 6e^- \quad (9)$$

$$Sb_2O_3 + 2H_2O \rightarrow Sb_2O_5 + 4H^+ + 4e^- \quad (10)$$

During the process, H^+ can be produced simultaneously, which can inhibit the anodic reactions and reduction of HSO_3^-. Thus, Sb inhibits anodic dissolution and enhances corrosion resistance. Additionally, Sb can interact with Cu and facilitate the formation of Cu-containing compounds in the product film, as reported in Le's work [36], inhibiting the anodic dissolution process. Nb can form NbC nanoprecipitates, which act as hydrogen traps and improve HE resistance.

As for Cu effects, Cu inhibits anodic dissolution by promoting the formation of protective corrosion product film, but the improvement is obvious only when the addition exceeds a critical value. In addition, at the early stage of corrosion before forming the protective rust layer, Cu accelerates the corrosion process. As a result, the polarization resistance of 0.5NiCu is lower than the BM. 1NiCu exhibits a high resistance to corrosion and SCC with greater Cu and Ni contents. Ni is present in the rust layer as protective divalent products, including $NiFe_2O_4$, which has a negative electrical effect [37]. $NiFe_2O_4$ can make the inner rust layer generate cation selectivity to effectively resist corrosive ions and significantly alleviate the corrosion behavior of low-alloy, high-strength steel [38].

As for the effect of C, studies have shown that many hydrogen defects appear in the microstructure of high-strength steel after fracture in a humid environment. Thus, hydrogen entering the matrix will accumulate in special microstructures, such as carbides, reduce the interface bonding force, increase the cracking tendency of the interface, and promote the initiation and expansion of microcracks under the action of tensile stress. When the carbides grow, the large matrix–carbide boundary becomes a strong hydrogen trap, so the susceptibility to hydrogen embrittlement increases [39,40]. Figure 11 proves that the BM steel has many deep corrosion pits, possibly due to the presence of carbides, and preferential corrosion occurs around the carbides. 0.5NiCu steel without decarburization treatment shows the same problem. After the decarburization treatment of 1NiCu and LNiCu, the number of corrosion pits on the side of the fracture significantly decreases. Therefore, the decarburization treatment helps to reduce the hydrogen collection in the matrix.

4.2. Effect of Alloy Elements on the SCC Mechanism

Potentiodynamic curves at the slow scan rate of 0.5 mV/s and fast scan rate of 200 mV/s were used to clarify the effect of microalloying on electrochemical factors. The potentiodynamic polarization curves are shown in Figure 12. The comparison of the steady state and steady-state stress corrosion mechanisms identifies three regions with different corrosion mechanisms: AD, AD + HE, and HE [41,42].

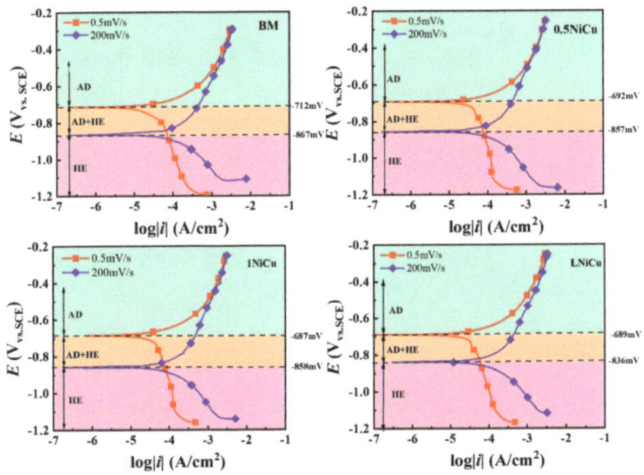

Figure 12. Potentiodynamic polarization curves of prepared steels at 0.5 mV/s and 200 mV/s sweep speeds.

Liu [42] proposed to quickly predict the potential range of each mechanism through the fast and slow sweep polarization curves and preliminarily determine the stress corrosion mechanism. According to Figure 12, the potential ranges that correspond to the stress corrosion mechanisms of the four low-alloy, high-strength steels can be determined as shown in Figure 12, and the specific data are shown in Table 5. Adding microalloying elements increases the potential range of the AD mechanism, indicating that Ni, Nb, and Sb have an inhibitory effect on anodic dissolution ability. In a similar study, Zhang [43] found that bainitic steels were more resistant to HE after adding 0.055% Nb. Both reversible and irreversible hydrogen traps are increased by Nb in steels [44,45]. It is speculated that the improvement of the hydrogen evolution corrosion potential largely depends on Nb.

Table 5. Potential range corresponding to three stress corrosion mechanisms of the anchor steels (V vs. SCE).

Steels	AD	AD + HE	HE
BM	>−0.712 V	−0.712~−0.867 V	<−0.867 V
0.5NiCu	>−0.692 V	−0.857~−0.692 V	<−0.857 V
1NiCu	>−0.687 V	−0.858~−0.687 V	<−0.858 V
LNiCu	>−0.689 V	−0.836~−0.688 V	<−0.836 V

With the decrease in carbon content, the anodic dissolution potential range of the 0.5NiCu and LNiCu low-alloy steels hardly changes, which indicates that the decarburization treatment does not significantly affect the anodic dissolution. With the addition of microalloying elements, the potential range of the HE mechanism slightly increases. With identical microalloying, the potential range of the HE mechanism significantly moves up in the decarburization treatment, which indicates that decarburization has a certain inhibitory effect on hydrogen embrittlement. Studies have shown that the formation of crack-sensitive paths in the process of stress corrosion may be closely related to C atoms from grain boundary defects. The presence of carbon or transition carbides at grain boundaries increases the stress corrosion sensitivity. It inhibits the dislocation movement away from the plastic deformation region associated with the crack, so it provides an adsorption site for anions on the crack surface, reduces the binding of adjacent iron atoms, and causes rapid failure.

5. Conclusions

Several conclusions can be drawn from the results and discussion:

(1) The four kinds of anchor steels have a similar microstructure, which is mainly composed of pearlite.
(2) The combined addition Ni, Cu, and Sb can improve the mechanical properties of the anchor steel, while reducing C content decreases tensile strength as a result of the loss of the solution-strengthening effect.
(3) The addition of Sb, Cu, Ni, and reducing the content of C enhances the resistance to corrosion and SCC by mitigating AD, while adding Nb improves SCC by inhibiting HE. The combined action of 1% Ni, 0.5% Cu, 0.05% Nb, 0.1% Sb, and 0.5% C presents the highest SCC resistance.

Author Contributions: Methodology, H.D.; Validation, Y.P.; Formal analysis, Y.L.; Investigation, H.D., X.W., Z.L., A.J. and Y.P.; Resources, X.W. and Y.L.; Data curation, X.W.; Writing—original draft, H.D. and N.A.; Writing—review & editing, Z.L. and Y.P.; Visualization, N.A., Y.L. and A.J.; Supervision, Z.L.; Project administration, R.Y. and X.L.; Funding acquisition, R.Y. and X.L. All authors have read and agreed to the published version of the manuscript.

Funding: This research was funded by the National Natural Science Foundation of China (No. 52071017) and National Science and Technology Plan Projects of State Administration for Market Regulation (No. 2021MK021).

Institutional Review Board Statement: Not applicable.

Informed Consent Statement: Not applicable.

Data Availability Statement: The raw/processed data required to reproduce these findings cannot be shared at this time as the data is related to an ongoing study.

Conflicts of Interest: The authors declare that they have no known competing financial interest or personal relationships that could have appeared to influence the work reported in this paper.

References

1. Li, X.; Zhang, D.; Liu, Z.; Li, Z.; Du, C.; Dong, C. Share Corrosion Data. *Nature* **2015**, *527*, 441–442. [CrossRef] [PubMed]
2. Wang, Y.; Sun, X.; Ren, A. Investigations of Rock Anchor Corrosion and Its Influence Factors by Exhumations in Four Typical Field Sites. *Eng. Fail. Anal.* **2019**, *101*, 357–382. [CrossRef]
3. Ananda Rao, M.; Sekhar Babu, R.; Pavan Kumar, M.V. Stress Corrosion Cracking Failure of a SS 316L High Pressure Heater Tube. *Eng. Fail. Anal.* **2018**, *90*, 14–22. [CrossRef]
4. Bestautte, J.; Kalácska, S.; Béchet, D.; Obadia, Z.; Christien, F. Investigation of Quasi-Cleavage in a Hydrogen Charged Maraging Stainless Steel. *Corros. Sci.* **2023**, *218*, 111163. [CrossRef]
5. Das, T.; Brahimi, S.V.; Song, J.; Yue, S. A Fast Fracture Approach to Assess Hydrogen Embrittlement (HE) Susceptibility and Mechanism(s) of High Strength Martensitic Steels. *Corros. Sci.* **2021**, *190*, 109701. [CrossRef]
6. Yang, L.; Liu, Z.; Yang, X.; Xu, X.; Cheng, X.; Li, X. Behavior and Evaluation of Stress Corrosion Cracking of Typical Anchor Bolt Steel in Simulated Crevice Environment. *J. Mater. Res. Technol.* **2023**, *25*, 7430–7443. [CrossRef]
7. Matos, T.S.; Portella, K.F.; Henke, S.L.; Bragança, M.O.G.P.; Galvão, M.P.; Dias, B.G.; Lagoeiro, L.; Almeida, L.M. Analysis of Anchor Rod Failure in a Guyed Transmission Tower: Influence of Microstructures and Corrosion Mechanisms. *Eng. Fail. Anal.* **2021**, *121*, 105166. [CrossRef]
8. Cano, H.; Díaz, I.; De La Fuente, D.; Chico, B.; Morcillo, M. Effect of Cu, Cr and Ni Alloying Elements on Mechanical Properties and Atmospheric Corrosion Resistance of Weathering Steels in Marine Atmospheres of Different Aggressivities. *Mater. Corros.* **2018**, *69*, 8–19. [CrossRef]

9. Zhou, Y.; Chen, J.; Xu, Y.; Liu, Z. Effects of Cr, Ni and Cu on the Corrosion Behavior of Low Carbon Microalloying Steel in a Cl− Containing Environment. *J. Mater. Sci. Technol.* **2013**, *29*, 168–174. [CrossRef]
10. Cano, H.; Neff, D.; Morcillo, M.; Dillmann, P.; Diaz, I.; De La Fuente, D. Characterization of Corrosion Products Formed on Ni 2.4 wt%–Cu 0.5 wt%–Cr 0.5 wt% Weathering Steel Exposed in Marine Atmospheres. *Corros. Sci.* **2014**, *87*, 438–451. [CrossRef]
11. Wu, W.; Liu, Z.; Wang, Q.; Li, X. Improving the Resistance of High-Strength Steel to SCC in a SO_2-Polluted Marine Atmosphere through Nb and Sb Microalloying. *Corros. Sci.* **2020**, *170*, 108693. [CrossRef]
12. Zhang, C.; Hui, W.; Zhao, X.; Zhang, Y.; Zhao, X. The Potential Significance of Microalloying with Nb in Enhancing the Resistance to Hydrogen-Induced Delayed Fracture of 1300-MPa-Grade High-Strength Bolt Steel. *Eng. Fail. Anal.* **2022**, *135*, 106144. [CrossRef]
13. Shi, R.; Ma, Y.; Wang, Z.; Gao, L.; Yang, X.-S.; Qiao, L.; Pang, X. Atomic-Scale Investigation of Deep Hydrogen Trapping in NbC/α-Fe Semi-Coherent Interfaces. *Acta Mater.* **2020**, *200*, 686–698. [CrossRef]
14. Chen, W.; Zhou, R.; Li, W.; Chen, Y.-H.; Chou, T.-H.; Wang, X.; Liu, Y.; Zhu, Y.; Huang, J.C. Effect of Interstitial Carbon and Nitrogen on Corrosion of FeCoCrNi Multi-Principal Element Alloys Made by Selective Laser Melting. *J. Mater. Sci. Technol.* **2023**, *148*, 52–63. [CrossRef]
15. *GB/T 228.1-2021*; Metallic materials—Tensile testing—Part 1: Method of test at room temperature. Standardization Administration of the People's Republic of China: Beijing, China, 2021.
16. Parkins, R.N. Predictive Approaches to Stress Corrosion Cracking Failure. *Corros. Sci.* **1980**, *20*, 147–166. [CrossRef]
17. Ding, Z.; Liang, B.; Xu, Z.; Tian, R.; Dong, L.; Zhao, J. Formation and Morphology of Grain Boundary Proeutectoid Pearlite in 100Mn13 Steel during Low Temperature Aging Treatment. *Mater. Today Commun.* **2022**, *33*, 104887. [CrossRef]
18. Jung, M.; Cho, W.; Park, J.; Jung, J.-G.; Lee, Y.-K. Variation of Carbon Concentration in Proeutectoid Ferrite during Austenitization in Hypoeutectoid Steel. *Mater. Charact.* **2014**, *94*, 161–169. [CrossRef]
19. Xiong, J.; Liu, E.; Zhang, C.; Kong, L.; Yang, H.; Zhang, X.; Wang, Y. Tuning Mechanical Behavior and Deformation Mechanisms in High-Manganese Steels via Carbon Content Modification. *Mater. Sci. Eng. A* **2023**, *881*, 145401. [CrossRef]
20. Qiao, C.; Wang, M.; Hao, L.; Jiang, X.; Liu, X.; Thee, C.; An, X. In-Situ EIS Study on the Initial Corrosion Evolution Behavior of SAC305 Solder Alloy Covered with NaCl Solution. *J. Alloys Compd.* **2021**, *852*, 156953. [CrossRef]
21. Brug, G.J.; van den Eeden, A.L.; Sluyters-Rehbach, M.; Sluyters, J.H. The Analysis of Electrode Impedances Complicated by the Presence of a Constant Phase Element. *J. Electroanal. Chem. Interfacial Electrochem.* **1984**, *176*, 275–295. [CrossRef]
22. Sun, B.; Pan, Y.; Yang, J.; Guo, J.; Zhao, B.; Liu, X.; Liu, Z.; Li, X. Microstructure Evolution and SSCC Behavior of Strain-Strengthened 304 SS Pre-Strained at Room Temperature and Cryogenic Temperature. *Corros. Sci.* **2023**, *210*, 110855. [CrossRef]
23. Zhang, T.; Liu, W.; Dong, B.; Yang, W.; Chen, L.; Sun, Y.; Li, H.; Zhang, B. Clarifying the Effect of Cu Element on the Corrosion Properties of Ni-Mo Low Alloy Steel in Marine Environment. *Corros. Sci.* **2023**, *216*, 111107. [CrossRef]
24. Zhang, T.; Liu, W.; Dong, B.; Mao, R.; Sun, Y.; Chen, L. Corrosion of Cu-Doped Ni–Mo Low-Alloy Steel in a Severe Marine Environment. *J. Phys. Chem. Solids* **2022**, *163*, 110584. [CrossRef]
25. Tian, H.; Cui, Z.; Ma, H.; Zhao, P.; Yan, M.; Wang, X.; Cui, H. Corrosion Evolution and Stress Corrosion Cracking Behavior of a Low Carbon Bainite Steel in the Marine Environments: Effect of the Marine Zones. *Corros. Sci.* **2022**, *206*, 110490. [CrossRef]
26. Liu, Z.Y.; Wang, X.Z.; Du, C.W.; Li, J.K.; Li, X.G. Effect of Hydrogen-Induced Plasticity on the Stress Corrosion Cracking of X70 Pipeline Steel in Simulated Soil Environments. *Mater. Sci. Eng. A* **2016**, *658*, 348–354. [CrossRef]
27. Pan, Y.; Liu, Z.; Zhang, Y.; Li, X.; Du, C. Effect of AC Current Density on the Stress Corrosion Cracking Behavior and Mechanism of E690 High-Strength Steel in Simulated Seawater. *J. Mater. Eng. Perform.* **2019**, *28*, 6931–6941. [CrossRef]
28. Seol, J.-B.; Jung, J.E.; Jang, Y.W.; Park, C.G. Influence of Carbon Content on the Microstructure, Martensitic Transformation and Mechanical Properties in Austenite/ε-Martensite Dual-Phase Fe–Mn–C Steels. *Acta Mater.* **2013**, *61*, 558–578. [CrossRef]
29. Liu, X.; Wu, M.; Wang, X.; Gong, K.; Du, J.; Huang, J.C.; Hu, M. Stress Corrosion Cracking Behavior of X80 Steel under the Combined Effects of Sulfide and Cathodic Potential. *J. Mater. Res. Technol.* **2023**, *24*, 3925–3936. [CrossRef]
30. Pan, Y.; Sun, B.; Liu, Z.; Wu, W.; Li, X. Hydrogen Effects on Passivation and SCC of 2205 DSS in Acidified Simulated Seawater. *Corros. Sci.* **2022**, *208*, 110640. [CrossRef]
31. Li, K.; Zeng, Y. Long-Term Corrosion and Stress Corrosion Cracking of X65 Steel in H_2O-Saturated Supercritical CO_2 with SO_2 and O_2 Impurities. *Constr. Build. Mater.* **2023**, *362*, 129746. [CrossRef]
32. Niu, J.; Tang, Z.; Huang, H.; Pei, J.; Zhang, H.; Yuan, G.; Ding, W. Research on a Zn-Cu Alloy as a Biodegradable Material for Potential Vascular Stents Application. *Mater. Sci. Eng. C* **2016**, *69*, 407–413. [CrossRef]
33. Gu, T.; Teng, W.; Liu, A.; Deng, Z.; Ling, L.; Zhang, W. Transformation of Nanoscale Zero-Valent Iron with Antimony: Effects of the Sb Spatial Configuration. *Chem. Eng. J.* **2021**, *416*, 129073. [CrossRef]
34. Yang, Y.; Cheng, X.; Zhao, J.; Fan, Y.; Li, X. A Study of Rust Layer of Low Alloy Structural Steel Containing 0.1% Sb in Atmospheric Environment of the Yellow Sea in China. *Corros. Sci.* **2021**, *188*, 109549. [CrossRef]
35. Li, J.; Chen, J.; Song, J.; Zhang, X.; Li, Z.; Luo, H.; Yu, W.; Dong, C.; Xiao, K. Research on Corrosion Behavior of Truck Body Steel in Chlorine-Containing Sulfuric Acid Environment. *J. Mater. Res. Technol.* **2022**, *21*, 1878–1889. [CrossRef]
36. Le, D.P.; Ji, W.S.; Kim, J.G.; Jeong, K.J.; Lee, S.H. Effect of Antimony on the Corrosion Behavior of Low-Alloy Steel for Flue Gas Desulfurization System. *Corros. Sci.* **2008**, *50*, 1195–1204. [CrossRef]
37. Wang, Y.; Li, J.; Zhang, L.; Zhang, L.; Wang, Q.; Wang, T. Structure of the Rust Layer of Weathering Steel in A High Chloride Environment: A Detailed Characterization via HRTEM, STEM-EDS, and FIB-SEM. *Corros. Sci.* **2020**, *177*, 108997. [CrossRef]

38. Yu, Q.; Yang, X.; Dong, W.; Wang, Q.; Zhang, F.; Gu, X. Layer-by-Layer Investigation of the Multilayer Corrosion Products for Different Ni Content Weathering Steel via a Novel Pull-off Testing. *Corros. Sci.* **2022**, *195*, 109988. [CrossRef]
39. Maetz, J.-Y.; Douillard, T.; Cazottes, S.; Verdu, C.; Kléber, X. $M_{23}C_6$ Carbides and Cr_2N Nitrides in Aged Duplex Stainless Steel: A SEM, TEM and FIB Tomography Investigation. *Micron* **2016**, *84*, 43–53. [CrossRef] [PubMed]
40. Wang, T.; Zhang, H.; Liang, W. Hydrogen Embrittlement Fracture Mechanism of 430 Ferritic Stainless Steel: The Significant Role of Carbides and Dislocations. *Mater. Sci. Eng. A* **2022**, *829*, 142043. [CrossRef]
41. Song, L.; Liu, Z.; Li, X.; Du, C. Characteristics of Hydrogen Embrittlement in High-PH Stress Corrosion Cracking of X100 Pipeline Steel in Carbonate/Bicarbonate Solution. *Constr. Build. Mater.* **2020**, *263*, 120124. [CrossRef]
42. Liu, Z.Y.; Lu, L.; Huang, Y.Z.; Du, C.W.; Li, X.G. Mechanistic Aspect of Non-Steady Electrochemical Characteristic During Stress Corrosion Cracking of an X70 Pipeline Steel in Simulated Underground Water. *Corrosion* **2014**, *70*, 678–685. [CrossRef] [PubMed]
43. Zhang, S.; Zhao, Q.; Liu, J.; Huang, F.; Huang, Y.; Li, X. Understanding the Effect of Niobium on Hydrogen-Induced Blistering in Pipeline Steel: A Combined Experimental and Theoretical Study. *Corros. Sci.* **2019**, *159*, 108142. [CrossRef]
44. Vercruysse, F.; Claeys, L.; Depover, T.; Verleysen, P.; Petrov, R.H.; Verbeken, K. The Effect of Nb on the Hydrogen Embrittlement Susceptibility of Q&P Steel under Static and Dynamic Loading. *Mater. Sci. Eng. A* **2022**, *852*, 143652. [CrossRef]
45. Samanta, S.; Gangavarapu, S.; Jayabalan, B.; Makineni, S.K.; Dutta, M.; Singh, S.B. Atomic-Scale Investigation of H-Trapping by Fine NbC Precipitates in a Low C Ferritic Steel. *Scr. Mater.* **2023**, *234*, 115537. [CrossRef]

materials

Review

Environment-Induced Degradation of Shape Memory Alloys: Role of Alloying and Nature of Environment

S. Santosh [1,*], W. B. Jefrin Harris [1] and T. S. Srivatsan [2]

[1] Department of Mechanical Engineering, Sri Sivasubramaniya Nadar College of Engineering, Chennai 603110, India
[2] Department of Mechanical Engineering, The University of Akron, Akron, OH 44325, USA; tss1@uakron.edu
* Correspondence: santoshs@ssn.edu.in

Abstract: Shape memory effects coupled with superelasticity are the distinctive characteristics of shape memory alloys (SMAs), a type of metal. When these alloys are subject to thermomechanical processing, they have the inherent ability to react to stimuli, such as heat. As a result, these alloys have established their usefulness in a variety of fields and have in recent years been chosen for use in stents, sensors, actuators, and several other forms of life-saving medical equipment. When it comes to the shape memory materials, nickel–titanium (Ni-Ti) alloys are in the forefront and have been chosen for use in a spectrum of demanding applications. As shape memory alloys (SMAs) are chosen for use in critical environments, such as blood streams (arteries and veins), orthodontic applications, orthopedic implants, and high temperature surroundings, such as actuators in aircraft engines, the phenomenon of environment-induced degradation is of both interest and concern. Hence, the environment-induced degradation behavior of the shape memory alloys (SMAs) needs to be studied to find viable ways to improve their resistance to an aggressive environment. The degradation that occurs upon exposure to an aggressive environment is often referred to as corrosion. Environment-induced degradation, or corrosion, being an unavoidable factor, certain techniques can be used for the purpose of enhancing the degradation resistance of shape memory alloys (SMAs). In this paper, we present and discuss the specific role of microstructure and contribution of environment to the degradation behavior of shape memory alloys (SMAs) while concurrently providing methods to resist both the development and growth of the degradation caused by the environment.

Keywords: shape memory alloy; smart materials; environment-induced degradation; corrosion behavior

Citation: Santosh, S.; Harris, W.B.J.; Srivatsan, T.S. Environment-Induced Degradation of Shape Memory Alloys: Role of Alloying and Nature of Environment. *Materials* **2023**, *16*, 5660. https://doi.org/10.3390/ma16165660

Academic Editors: Jingxiang Xu, Xingwei Zheng and Zhenhua Chu

Received: 22 June 2023
Revised: 9 August 2023
Accepted: 13 August 2023
Published: 17 August 2023

1. Introduction

A shape memory alloy (SMA) has few distinctive qualities to essentially include shape memory effect and superelasticity [1], which makes the alloy desirable for selection and use in biomedical and several other intricate applications. The shape memory alloys (SMAs), such as nickel–titanium (NiTi), have been chosen for use in both orthopedic applications and orthodontic applications for many years [2]. Through the years, these sectors have seen particularly noticeable improvements. The shape memory alloy is gradually replacing other materials as the preferred option for use as, e.g., filters, baskets, self-expansion stents, support system grafts, and a few other devices, which find use in minimally invasive interventional treatments. The possible impact on biocompatibility is still a source of concern. Also, the contradictory environment-induced degradation resistance, or corrosion, serves to complicate this issue. Nickel, which is necessary for both nourishment and bodily function [3], is found in human tissue in amounts of about 0.1 ppm. A higher nickel concentration that could leak out of the implant material could cause negative allergies, be toxic, and even promote cancerous reactions. Furthermore, NiTi and 316L stainless steel [4] MP35N (35 weight percent nickel) is another widely chosen and used high nickel-containing alloy that demonstrates strong biocompatibility and is often chosen for use as an implant for orthopedic applications spanning cardiovascular and orthodontic uses. Like the other non-noble implant metals, the environment-induced degradation resistance of a nickel–titanium (NiTi) alloy largely depends on the existence or presence of a passive coating on the surface. No general degradation upon exposure to aggressive environmental issues has to be anticipated so long as the passive film is in good condition. However, it should be kept in mind that the release of certain ions can also happen in the passive state. Additionally, the passive film could easily break down during both surgery and service life. Hence, the most concerning factor for an effective use of the shape memory alloys (SMAs) is the degradation that occurs upon exposure to an aggressive environment and there exists a need to work towards enabling an enhancement and/or improvement of the resistance to environment-induced degradation properties. In the following sections, a few ways to increase the resistance to environment-induced degradation of the different shape memory alloys (SMAs), along with adequate discussion of the specifics pertinent to the degradation response or corrosion behavior of a few shapes memory alloys (SMAs), are provided.

2. Environment-Induced Degradation and Its Mechanism

Environment-induced degradation, referred to as corrosion, is defined as the chemical–physical metal–environment interactions that often result in an observable modification to the properties of the chosen metal and the resultant deficit in the functioning of the metal, the medium, or the system, which essentially includes both the metal and the medium. When a substance is at its peak perfection, it is desirable to shield it from any adverse activity arising from being influenced and/or affected by an external action resulting from exposure to an aggressive environment. The best method to safeguard metal surfaces is to introduce coatings. Like mortality, the occurrence of environment-induced degradation or corrosion is unfortunate, and every effort is made to prevent it from occurring. Essentially, it is a phenomenon that we must learn to not only manage but to eventually conquer [4,5]. A system's decreased Gibbs free energy is the main contributing factor to degradation by corrosion. As a result, the specific metal has a significant inclination to revert to its original lower energy state. Environment-induced degradation or corrosion is the term given for this return to the original state and is shown in Figure 1. Although the occurrence of degradation by corrosion is inevitable, it can be slowed down to reach equilibrium. When considering a coating for the purpose of preventing environment-induced degradation, i.e., corrosion, it must provide a strong physical barrier that prevents the aggressive elements in the environment from reaching the metal surface. The degradation by corrosion of metals and alloys can safely be a complex issue that has plagued human beings for a considerable amount of time [5]. Due to their exceptional corrosion resistance, nickel-based alloys are frequently utilized in industry under challenging conditions. As a result, alloys based on nickel are commonly used in corrosive applications [6]. Hence, it is more important to study the behavior of these functional alloys under corrosive environment, as shown below.

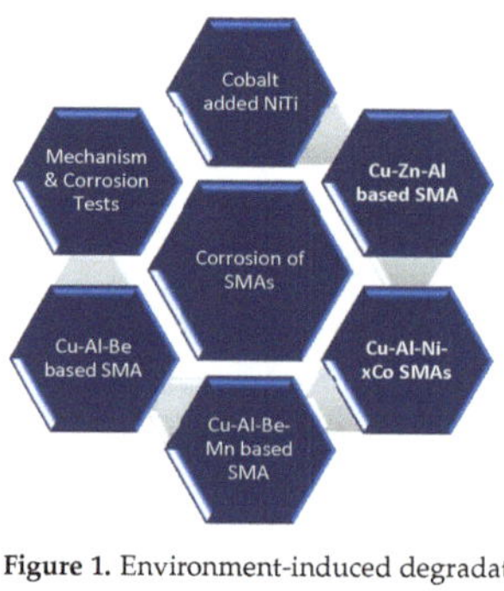

Figure 1. Environment-induced degradation or corrosion of various shape memory alloys (SMAs).

The mechanism illustrated above implies that mass loss with immersion time exhibits an erratic development at first, followed by a fast increase in the middle and, eventually, a tendency towards stability. This shows that the solid-state diffusion growth mechanism, which provides the protective behavior, is under control of the growth of corrosion products and that the corrosion products have a restricted rate of dissolution in molten salt.

3. Various Corrosion Resistance Tests and Their Procedures

3.1. Various Test Solutions Used

The solutions used in in vitro electrochemical testing are shown in Figure 2. The solution temperature in potentiostatic and potentiodynamic scratch tests were set to 4 °C. Modified ASTM tests were carried out at a temperature of 37 °C [7]. The two rising temperature tests were the following: (i) crevice temperature tests and (ii) potentiostatic critical pitting test. The corrosion test procedure and composition is presented in Figures 3 and 4, respectively.

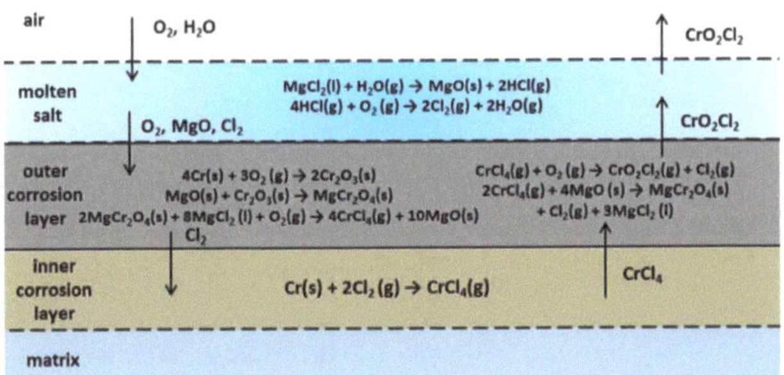

Figure 2. Corrosion mechanism of Ni-based alloy in NaCl-$CaCl_2$-$MgCl_2$. Environment [5] (reused with permission from Elsevier).

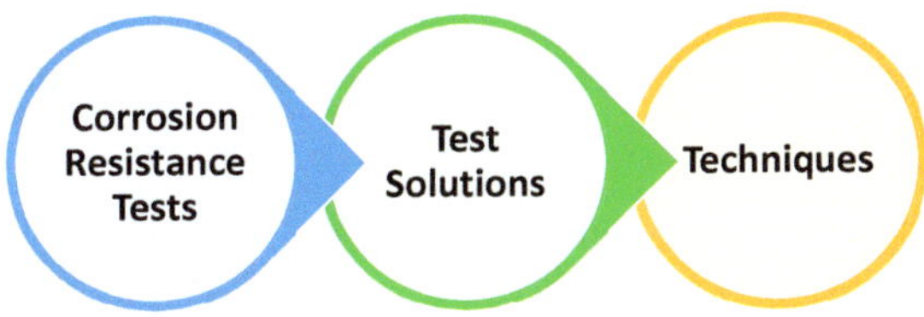

Figure 3. Depiction of the corrosion test and its procedure.

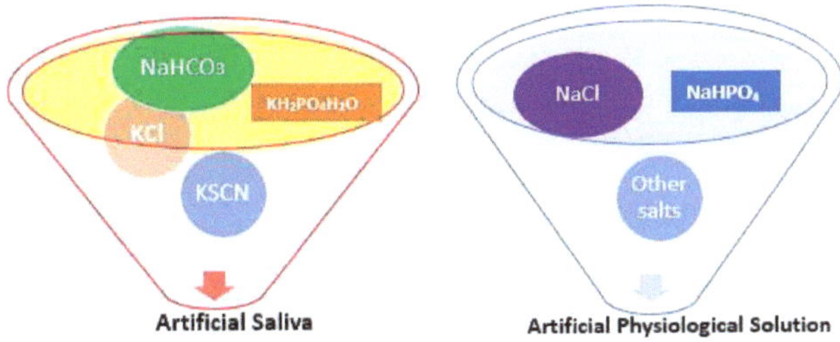

Figure 4. Schematic composition of the solutions used.

3.2. Techniques

The various types of tests used are shown in Figure 5. Polishing of the test specimens was performed using 1-micron diamond paste for both the potentiodynamic tests and the potentiostatic scratch tests. The rest of the test specimens were mechanically ground using a 600-mesh wheel or 600-mesh emery paper [8,9].

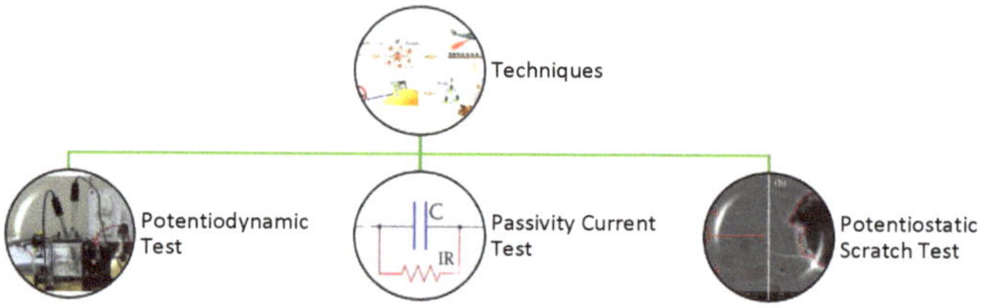

Figure 5. Types of tests to study environment-induced degradation.

3.3. The Potentiodynamic Test

The ASTM G5 practice-recommended experimental setup was used to conduct the cyclic potentiodynamic tests. The ASTM G61 practice was followed for flat specimens using a specimen holder and a scan rate of 600 mV/h and starting potential of 750 mV [10–12]. The direction of scanning was interchanged when the current density of the anode attained 100 $\mu A/cm^2$.

3.4. Passivity Current Test

Potentiostatic experiments were conducted at 100 mV in de-aerated settings versus standard calomel electrode (SCE). The current levels were periodically recorded for various times to both evaluate and establish the passivity current. Preparation of the cylindrical test specimens was very much in conformance with details specified in the standard ASTM G5.

3.5. Potentiostatic Scratch Test

Additionally, for the flat test specimen, the electrode assembly described in ASTM G61 standard practice was used. The scratch test approach, which was first presented by Pessal and Liu [13–16], enables an identification of the potential value at which repair of the injured passive film can be prevented. This technique involves delicately scratching the metal surface using a diamond tip while concurrently using a potentiostat to adjust the potential, with the initial potential being in the passive area. Further scratching is performed at a higher potential level should the scratched region re-passivate [17]. Therefore, it is possible to calculate the potential at which the chosen test specimen will not re-passivate.

4. The Copper-Aluminum-Beryllium (Cu-Al-Be)-Based Shape Memory Alloy

The copper-based shape memory alloys are less expensive, good to make, and, importantly, easy to work with. However, because they are both fragile and tedious to manufacture, the copper-based shape memory alloys (SMAs) cannot be readily chosen for use. Grain refining has often been used in an attempt to increase the resilience of multicrystalline copper-based shape memory alloys with varying degrees of success [18,19]. However, recent research studies have revealed that the copper-aluminium-manganese shape memory alloys (SMAs) have noticeably better ductility due to the presence of the progenitor phase, which essentially has an L21 morphology with less degree order. Shape memory alloys (SMAs) have been chosen for use in a variety of applications to esentially include the following: (i) valves, (ii) pipes, and (iii) fasteners. Shape memory alloys (SMAs) are often exposed to aggressive environments, spanning both aqueous and gaseous environments, during service in many applications. Degradation induced by the environment, referred to as corrosion, is primarily responsible for well over 75 percent of the gradual decay and/or deterioration experienced by the material, resulting in the occurrence of failure due to changes in properties while in use or service in a variety of applications [20,21]. Aschematic of the test is shown in Figure 6.

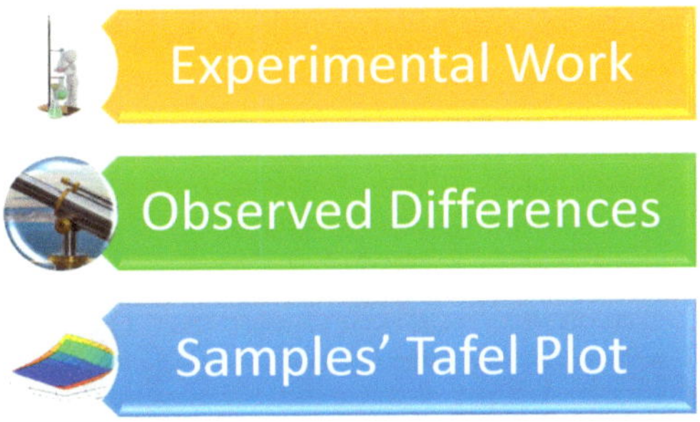

Figure 6. Layout of the copper-aluminum-beryllium (Cu-Al-Be)-based shape memory alloy (SMA).

4.1. Experimental Work

A composition of 11.5 weight percent aluminum, 0.44 weight percent beryllium (Be), and the remaining copper was chosen to complete the current research study as shown in Figure 7. Small bits of copper (Cu), aluminum (Al), and beryllium (Be), each weighing about 100 g for the chosen composition, were taken and then converted to the molten state in an induction furnace. The molten alloy was given time to harden in a permanent mold that measured 150 mm × 100 mm × 5 mm [22–24]. The ingots that were produced were subsequently homogenized. Subsequently, a spectrophotometer that measures plasma-optical emission was used to analyze composition of the alloy. Identical samples were rolled to a thickness of about 1 mm at 900 °C. The test samples were then subjected to the following treatment:

(i) Betatized for a full 30 min at 900 °C;
(ii) Step-cooled in water that was boiling (100 °C);
(iii) Finally cooled in a tub containing water at approximately 30 °C [25].

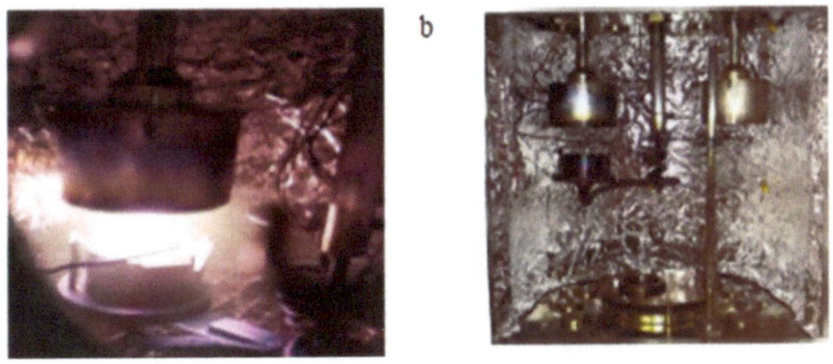

Figure 7. (**a**) Surface exposed to plasma arc and (**b**) apparatus used for sputtering [21]. (Reused with permission from Elsevier).

An optical microscope was used to examine the microstructure and morphology of the martensite that was produced. A bend test was used to check the samples produced for the shape memory effect. The coating for the reference sample was carried out utilizing the sputtering technique. The apparatus used for sputtering created a plasma phase while concurrently maintaining a deposition pressure of 6×10^{-3} millibar and a base pressure of

1×10^{-5} millibar. The sample essentially served as the anode, while titanium served as the cathode, with their separation being a distance of 45 mm. With a current of 120 amps and a DC bias voltage of 0.6 kV, the process was permitted to operate for a full 25 min [26–29].

The samples were chopped into rectangular shapes that were then utilized for conducting the corrosion test. Fresh water and Hank's solution were both chosen to be the aggressive aqueous media. A potentiostat was connected to the anode and opposite terminal after being placed in an appropriate holder [30]. During the test, a graphite rod served as the counter cathode and a wet calomel cathode as the reference anode. The product regulated both the environment-induced degradation process and its impact upon exposure to an aggressive environment. This helped in the following:

(a) Maintaining the desired voltage in the circuit;
(b) Estimating the characteristics of the current;
(c) Displaying the yield as a voltage (E) versus current (I) plot.

By using a voltage range between −2.4 V and 3.0 V and an output rate of 5.0 mV/s, anodic polarization curves of the chosen alloy were obtained [31].

4.2. Observed Differences

The shape memory effect (SME) of the chosen alloy that was identified using a bend test is summarized in Table 1. Without a coating, the shape memory alloys revealed an appreciable shape memory effect of 82 percent and 89 percent. However, with a coating on the shape memory alloy, the shape memory effect (SME) was 66 percent and 74 percent [32,33]. The amount of martensite that was changed and subsequently reverted to austenite determined how much strain could be recovered due to the shape memory effect (SME). The decrease in shape memory effect (SME) was made possible by the gradual growth of the intermetallic oxides, i.e., titanium dioxide (TiO_2). The stress caused by the presence of the titanium layer and its thermal expansion coefficient on the shape memory alloy (SMA) was a contributing factor to the observable decline of the shape memory effect (SME).

Table 1. Shape memory effect (%) exhibited after the bend test [21] (reused with permission from Elsevier).

Sample	Diameter	Thickness	Angle Recovered	SME%
CAB1	32 mm	1	72	81
CAB2	32 mm	1	80	88
CAB3	32 mm	1	60	65
CAB4	32 mm	1	66	75

CAB1 → Cu (88.01%) + Al (11.5%) + Be (0.44%) [Without coating]
CAB2 → Cu (88.05%) + Al (11.5%) + Be (0.45%) [Without coating]
CAB3 → Cu (88.01%) + Al (11.5%) + Be (0.44%) [With coating]
CAB4 → Cu (88.05%) + Al (11.5%) + Be (0.45%) [With coating]

For the combination tests considering two consuming media, the rate of degradation induced or caused by exposure to an aggressive environment was determined. $E_{(corr)}$ provided an indication of the starting point of the detachable layer's depletion, whereas E provided an indication of the point at which the detached layer had completely drained and thereby caused the formation and presence of pits in the chosen material as shown in Figures 8 and 9 [28–35]. Upon exposure to an aggressive aqueous environment, it was found that covering the copper-based alloys with titanium increased their corrosion resistance. The gradual decomposition of the admixture caused the release of cuprous ions that were harmful. The beryllium–copper alloy resisted marine fouling and was found to be suitable for use in marine applications [36–38].

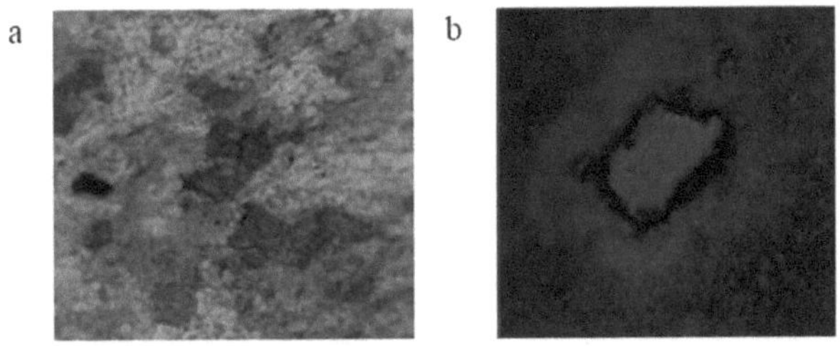

Figure 8. CAB1 sample optical micrograph at 20X resolution. (**a**) Evidence of degradation resulting from exposure to fresh water (H_2O) (50 nm). (**b**) Evidence of degradation resulting from exposure to Hank's solution (50 nm) and surface morphology of the chosen shape memory alloy (CAB1) [21] (reused with permission from Elsevier).

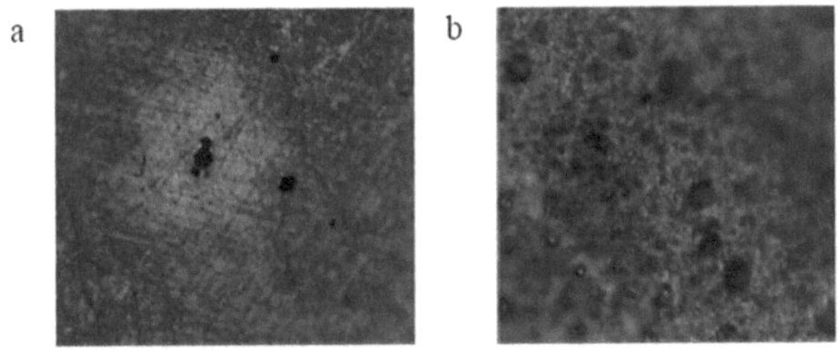

Figure 9. CAB3 alloy sample's optical micrograph at 20X. (**a**) Evidence of degradation resulting from exposure to pure H_2O (50 nm). (**b**) Evidence of degradation resulting from exposure to Hank's solution. (50 nm); surface morphology of the shape memory alloy (SMA: CAB3) [21] (reused with permission from Elsevier).

4.3. *The Samples' Tafel Plot*

The Tafel plot revealed both active and passive characterization of the test sample that was exposed to an aggressive aqueous environment as shown in Figure 10. In both situations, it was found that Hank's solution had a higher environment-induced degradation or corrosion potential than fresh water [39]. The current density of the coated specimen was found to decrease in the anodic zone and increase in the passive zone. The coated specimens had a larger pitting potential than the uncoated test specimens. The coating essentially shielded the metal from localized degradation caused by the environment [40].

In comparison with Hank's solution, fresh water showed a lower rate for environment-induced degradation of the chosen copper-aluminum-beryllium shape memory alloy (SMA). The resistance to environment-induced degradation, or corrosion, of the copper-aluminum-beryllium shape memory alloy was improved by the presence of a titanium coating [41]. It was assumed that the occurrence of environment-induced degradation, or corrosion, would tend to alter the innate shape memory characteristics of the chosen alloy since it tended to alter the composition of the alloy. Therefore, this feature needs to be carefully considered when selecting this alloy for a given application [42–44]. The titanium dioxide (TiO_2) layer formed as a passive layer on the surface of the alloy served as a barrier and slowed the process of degradation or corrosion induced by the aggressive environment. The covering tended to slow the occurrence of "pitting" corrosion. When compared with a shape memory alloy (SMA) that was not coated, current density was lower for the coated shape memory alloy (SMA) [45–47].

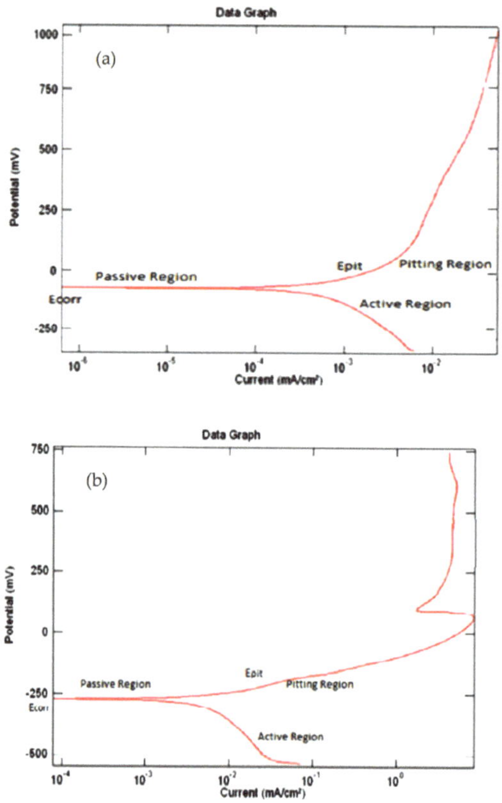

Figure 10. (**a**) Fresh water Tafel plot and (**b**) Hank's solution Tafel plot [21] (reused with permission from Elsevier).

5. Cu-Al-Be-Mn Tetrad Memory Alloys

The copper-aluminum-beryllium-manganese quaternary alloy has unique mechanical properties to include the following:

(a) Excellent shape remembrance;
(b) Appreciable mechanical strength;
(c) Enough immerge ability due to martensitic transfiguration and pseudo-elasticity;
(d) Speculative uniqueness to absorb sound, vibrations, and mechanical waves due to coarse grain.

The usage of shape memory alloys (SMAs) has been extremely successful in biomedical applications due essentially to the functionality properties of the alloys, which improve overall effectiveness, coupled with a potential for less invasive surgeries. Its biocompatibility is its primary biomedical application. Another significant area for the application of shape memory alloys (SMAs) is the domain of medicine, where pseudo-elasticity is effectively used to identify various components to include the following [48]:

(a) Filters for embolic protection;
(b) Tooth aligning wires;
(c) Cardiovascular stents;
(d) Microsurgical and endoscopic devices.

Let us consider four different copper-aluminum-beryllium–manganese (CABM) alloys having different chemical compositions as shown in Figure 11.

CABM1 ⟶ Cu (87.86%) + Al (11.5%) + Be (0.44%) + Mn (0.20%)

CABM2 ⟶ Cu (87.21%) + Al (12.00%) + Be (0.49%) + Mn (0.30%)

CABM3 ⟶ Cu (86.71%) + A l(12.50%) + Be(0.54%) + Mn(0.25%)

CABM4 ⟶ Cu (86.69%) + A l (12.50%) + Be (0.56%) + Mn (0.25%)

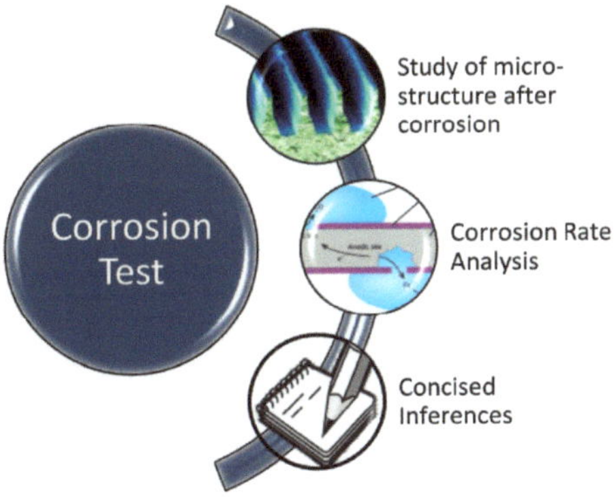

Figure 11. Skeletal structure of the Cu-Al-Be-Mn tetrad shape memory alloy.

5.1. Test Procedure for Evaluating the Influence of Environment

The created shape memory alloy (SMA) was pieced into a plane that was square in shape and measured 20.00 mm × 20.00 mm × 1.00 mm. The plane was then polished using emery paper of various sizes before being dusted with alumina (Al_2O_3) powder. The microstructure of the polished sample was examined before the test specimen was exposed to the three different aqueous environments, namely [49]:

(i) Ocean water (H_2O);
(ii) Fresh water (H_2O);
(iii) Hank's solution.

A pictorial representation of an electrochemical cell is shown in Figure 12a. As shown in the figure, 150 mL of the desired solution was taken for the standard test cell (Figure 12b). After positioning of the sample and electrodes in the proper locations, an area of 0.16 cm^2 was exposed to the aqueous solution. The standard calomel electrode, platinum electrode, and specimen electrode were preserved in their appropriate holders and linked to a potentiostat. Data specific to both voltage and current were collected using the Princeton applied research corrosion measurement program. The test data were displayed as the variation of voltage (E) versus current (I). A careful study and observation of the microstructure of the sample was made following exposure to an aqueous environment and the resultant environment-induced degradation, or corrosion, and the corrosion potentials E_{corr} and I_{corr} were established.

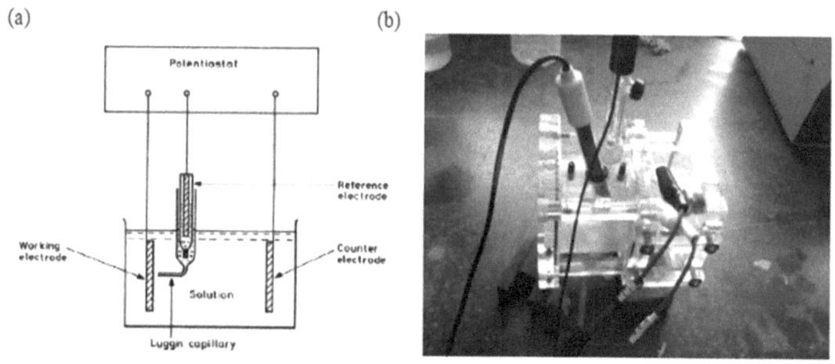

Figure 12. (**a**) Schematic representation of electrochemical cell. (**b**) Setup of electrochemical testing [50] (reused with permission from Elsevier).

5.2. Results

5.2.1. Study of Microstructure Following Exposure to Chosen Environment

The microstructure does exert a significant impact on both mechanical properties and physical properties, which has an influence on how these materials are chosen and used. Even though the alloys exhibit the austenitic parent phase upon casting, following step quenching, a complete change from austenite to lath martensite occurs. Step quenching demonstrates the entire transformation of austenite to martensite without a precipitate being formed [51]. Both shape memory effect and pseudo-elastic effect, as seen from the viewpoint of the microstructure, are outcomes of a solid phase microstructure transformation from austenite to martensite that could be triggered using a healthy synergism of mechanical pressure and thermal pressure. The chosen test piece was dried in laboratory air following exposure to an aggressive aqueous environment. Then, its microstructure was carefully examined. The micromorphology of the exposed and degraded test specimens of the copper-aluminum-beryllium–manganese sample (CABM4) in both sea water and Hank's solution are shown in Figure 13c,d. The micromorphology of the exposed and degraded test specimens of the CABM4 mixture following exposure to fresh water (H_2O) is shown in Figure 13b. It is clear from a comparison of the three microstructures that the environment of fresh water caused very little degradation by way of corrosion to the chosen sample. It was observed that the degradation that was experienced upon exposure to sea or marine water was much more complex than exposure of the test specimens to Hank's solution (Figure 13d). The pits resulting from environment-induced degradation, or corrosion, that were apparent in the microstructure are the dark patches [52].

5.2.2. Shape Memory Effect

The bend test, as shown in Figure 14, was used to assess the shape memory effect (SME) of the chosen alloy. The results are presented in Table 2. The formula used for the calculation of the shape memory effect is:

$$\textbf{Percentage of Shape Memory Effect} = \theta_m/(180 - \theta_e)$$

Table 2. Shape memory effect (SME) of the copper-aluminum-beryllium-manganese shape memory alloy (SMA). Ref. [50] (reused with permission from Elsevier).

Sample	Diameter	Thickness	Angle Recovered	SME%
CABM1	32 mm	1	155	82.39
CABM2	32 mm	1	148	77.93
CABM3	32 mm	1	157	83.57
CABM4	32 mm	1	163	88.78

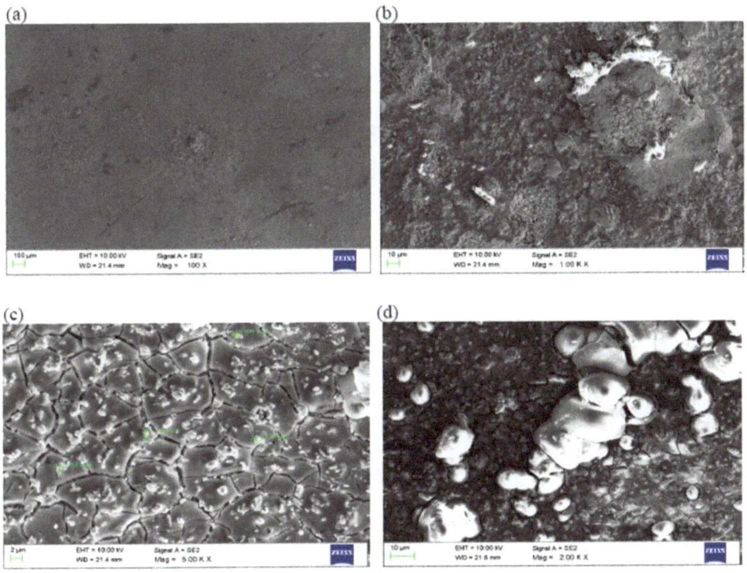

Figure 13. (**a**) CABM4 sample before corrosion. (**b**) CABM4 sample upon exposure to fresh water. (**c**) CABM4 sample after exposure to sea water. (**d**) CABM4 sample after exposure to Hank's solution [50] (reused with permission from Elsevier).

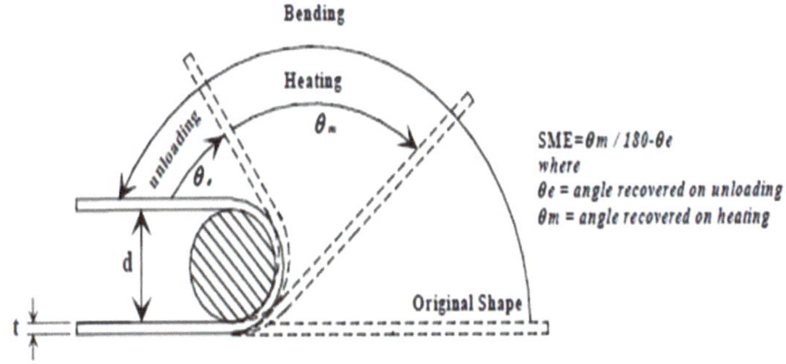

Figure 14. Bending test to find strain restoration [50] (reused with permission from Elsevier).

5.2.3. Analysis of Rate of Degradation Due to the Environment

A metallurgical corrosion analyzer system gill AC and an electrochemical corrosion cell were used to calculate both the rate of environment-induced degradation, or corrosive rate, and potential. E_{corr} was the actual location on the surface of the chosen material when the passive layer began to deplete [50]. It was seen that fresh water did not form pits. Additionally, it was noted that the rate of environment-induced degradation, i.e., the values of I_{corr}, and E_{corr}, were greater upon exposure to aqueous Hank's solution than the sea (marine) water solution. Results revealed the resistance to environment-induced degradation, or corrosion, to increase with an increase in beryllium content (weight percent) in the alloy.

The Tafel Plots

Tafel plots were able to provide a precise measurement of corrosion current that varied inversely with the rate of environment-induced degradation, or corrosion. When compared with the weight reduction methods, this process was quick. The Tafel plots were graphed with potential (E) along the Y-axis and current (I) along a logarithmic X-axis. Five regions made up the curve, as follows [53,54]:

(i) The passivation region.
(ii) The elementary passive zone.
(iii) The initiation of passivation.
(iv) The active region.
(v) The trans-passivation zone.

It took some trial and error to determine the initial potential. The potential once established was maintained for the entire family of chosen alloys. The starting potential and reverse potential both fell between −250 mV and +800 mV. The sweep rate was kept at 5 mV/s [55]. The potentiodynamic plot for the CABM4 alloy in pure water (H_2O), ocean (marine) water (H_2O), and Hank's solution is shown in Figure 15.

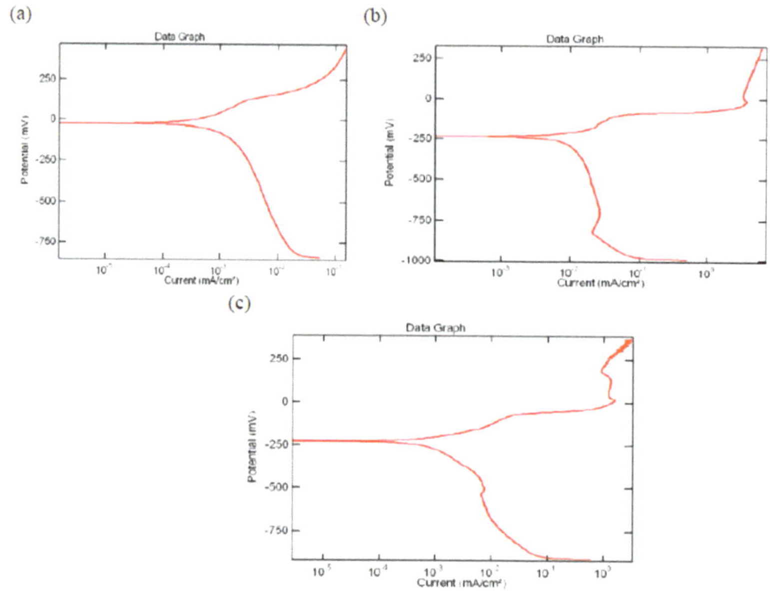

Figure 15. (**a**) Fresh water (H_2O) PDC for the CABM4 alloy. (**b**) Ocean water (H_2O) PDC for the CABM4 alloy. (**c**) Hank's solution PDC for the CABM4 alloy. Ref. [50]. (Reused with permission from Elsevier).

5.2.4. Observations

Using the ingot metallurgy technique, the copper-aluminum-beryllium–manganese shape memory alloy (SMA) mixtures were created. The shape memory effect (SME) of the alloys was good. The entire transition from the austenite phase to the martensite phase in these alloys resulted in a positive shape memory effect (SME) [56]. The results of our study revealed the following:

1. Exposure to freshwater resulted in a lower rate of corrosion for the copper-aluminum-beryllium–manganese shape memory alloy (SMA) when compared with Hank's solution and ocean water.
2. Hank's solution had a stronger resistance to environment-induced degradation on the copper-aluminum-beryllium–manganese shape memory alloys (SMAs) than ocean water.
3. By adding trace amounts of beryllium to the alloy, the copper-aluminum-beryllium–manganese quaternary shape memory alloys (SMAs) revealed an improved resistance to degradation induced by the aqueous environment.
4. The Cu-Al-Be-Mn alloy had a remarkable 88 percent shape memory effect (SME).

6. The Cu-Al-Ni-xCo Shape Memory Alloys Biformed with Low-Carbon Steel

This alloy has a high damping capability greater than nitinol; in recent years, it has become the most sought-after shape memory alloy (SMA). The Cu-Al-Ni alloy is appropriate for applications spanning fasteners, buildings, bridge-cushioning components, oil well extraction, microscopic elements, and structures. However, damage caused by the conjoint and mutually interactive influences of degradation resulting from exposure to an aggressive environment and erosion should be taken into consideration when put to use in applications that are often exposed to aggressive aqueous environments, such as in coastal areas and even oil rigs [57]. In addition, using the shape memory alloy mixture as an element but not as an entire system is both feasible and cost-effective. However, because the chosen shape memory alloy is often in touch with other metals, the tendency for galvanic corrosion to occur is favored. The Cu-Al-Ni alloy is one of the demanding shape memory alloy candidates for the purpose of industrial use, primarily because it offers good resistance to environment-induced degradation due in essence to the presence of an alumina layer and its contributing role as a passive film.

It is crucial to keep in mind that the Cu-Al-Ni alloy has the drawback in that it is susceptible to post-quench ageing, which does affect and/or influence its mechanical properties upon sustained exposure to high-temperature service conditions that tend to gradually worsen with time. Accordingly, extensive research has been carried out to enhance the characteristics of the copper-aluminum-nickel alloy in order to both enable and expand its use in space applications while concurrently satisfying industrial needs and requirements.

In recent years, numerous methods have been used to improve the properties of the shape memory alloys (SMAs). One such method is a refinement in the grain size. In this method, quaternary elements such as zirconium and titanium are added in small quantities or trace amounts to the shape memory alloy. During the additions of elements such as titanium (Ti) and manganese (Mn), the resistance to environment-induced degradation of the copper-aluminum-nickel shape memory alloy gradually increases when the grain size decreases. On the opposite end, it has been claimed that the addition of cobalt to a copper-aluminum-nickel alloy results in both an improvement in mechanical properties and the temperature at which the austenite phase transforms. This, thereby, allows for the use of this combination at higher temperatures [58–61]. The Cu-Al-Ni shape memory alloy is less appealing than the popular Ni-Ti alloy, although offering relatively good mechanical properties and acceptable resistance to environment-induced degradation or corrosion, which places a restriction on its utilization. By the addition of quaternary elements, such as cobalt (Co), titanium (Ti), and manganese (Mn), in different proportions and subjecting the alloy to heat treatment, researchers have been able to improve the mechanical properties of the copper-aluminum-nickel alloy. The capability of the alloy for selection and use at higher temperatures is made possible by the addition of cobalt, although this fact does not improve its applicability [62–66]. Additional research studies on environment influences on the behavior of the copper-aluminum-nickel shape memory alloys under these conditions is needed primarily because its applications may often require an exposure to aggressive aqueous environments.

6.1. Electrochemical Test on Sample Prepared

The Tafel electrochemical test was used to investigate both the corrosive behavior and the working ability of the copper-aluminum-nickel alloy with the addition of cobalt and without the addition of cobalt as the fourth alloying element, i.e., the alloys (Figure 16).

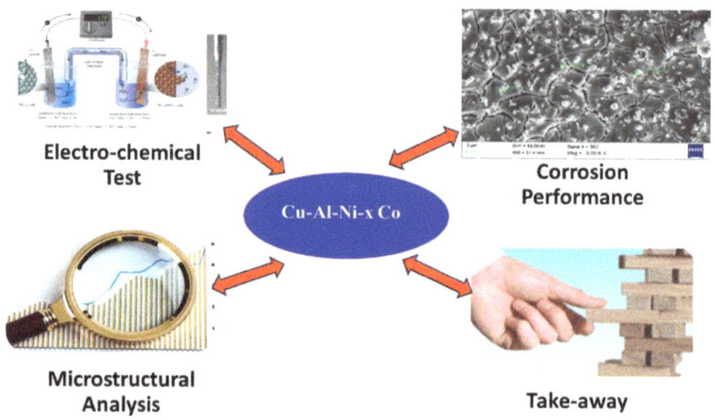

Figure 16. Corrosion study of the Cu-Al-Ni-xCo shape memory alloy.

(i) Copper-aluminum-nickel.
(ii) Copper-aluminum-nickel-1.0 wt% Co.
(iii) Copper-aluminum-nickel-0.4 wt% Co.

The test samples were coupled with low-carbon steel based on electrochemical measurements. The Cu-Al-Ni-(x) Co (x = 0 weight percent, 0.4 weight percent, and 1 weight percent) shape memory alloy (SMA) ingot was cut into tiny pieces that had dimensions of 25.00 mm (L) × 20.00 mm (W) × 2.00 mm (t) in order to make it both suitable and appropriate for the electrochemical test [67–71]. The samples were made from the as-homogenized ingots. The low-carbon steel bar was divided into pieces. Each piece measured 100.0 mm in length and 5.0 mm in width. Using epoxy, steel rebar and the biformed Cu-Al-Ni alloy were pieced or joined together. The Cu-Al-Ni-(x) Co shape memory alloys (SMAs) were also connected using copper wires. The samples were initially drilled and wires made of copper were then fastened to them using nuts and bolts. A fixed surface area of the rectangular test specimen was then subject to electrochemical tests. The tests were conducted using both a PARSTAT 2263 potentiostat and galvanostat at 20 °C in an amorphous glass cell that contained 325 milliliters of 3.50% sodium chloride solution. Potentiodynamic polarization studies were conducted using a three-electrode cell; a saturated calomel electrode (SCE) was used as the relating electrode [72]. The operating electrode chosen was the test specimen, while the counter electrode was a graphite rod. The start of each experiment was 225 mV below the open circuit voltage difference and the scan rate was maintained constant at 0.5 mV/s. Although the program permitted human control, fitting of the gathered test data was challenging. The Tafel plots were performed by choosing a section of the deterioration potential (φ_{corr}) versus current density (I_{corr}) curve and then estimating the potential value of the corrosion potential (φ_{corr}) [73]. Every potential pointed to the saturated calomel electrode (SCE); the test data reported were the mean values after considering the standard deviation. To ensure repeatability, each experiment was repeated three times.

6.2. Microstructural Analysis

Two separate phases were seen, namely (i) the plate-like γ and (ii) the needle-like β. One other independent study has also identified these two phases [74]. The plate-like γ phase essentially had a 2H structure. The grain size of the chosen and studied copper-aluminum-nickel shape memory alloys tended to grow larger as they aged. Furthermore, careful observations revealed that, with aging, the plate thickness of the γ phase increased while that of the needle-like β phase decreased. The peak became stronger with aging, which could be correlated with coarsening of the martensite phase. No other phases or peaks were found following the ageing treatment, based on X-ray diffraction data. At the opposite end, the electron micrographs clearly revealed that as the cobalt content in the alloy increased, the grain size of the copper-aluminum-nickel-(x) cobalt shape memory alloy decreased and a new phase, i.e., precipitate, developed in the matrix. The precipitate existed both at and along the grain boundary regions and gradually grew larger in size with an increase in cobalt content in the alloy. The energy dispersive X analysis (EDAX) revealed the precipitate to have a high cobalt concentration. Grain growth occurred upon aging of the –copper-aluminum-nickel-1.0 weight percent cobalt shape memory alloy, just as it did for the copper-aluminum-nickel alloy without the addition of cobalt. There was a noticeable difference in microstructure of the copper-aluminum-nickel-1.0 wt.% Co alloy both before the aging treatment and after the aging treatment. For the copper-aluminum-nickel-0.4 wt.% Co alloy and the copper-aluminum-nickel-1.0 wt.% Co alloy, the $Al_{75}Co_{22}Ni_3$ precipitates were both non-uniform in size and non-uniform in distribution prior to aging. Some of the precipitates had a diameter of 0.3 μm, while others had a diameter of only 0.1 micrometer. The smaller precipitates gradually disintegrated following the aging treatment, leaving behind only the larger precipitates, whose sizes had shrunk to about 0.25 μm [75]. The X-ray diffraction data were significantly impacted by both dissolution and size reduction of the $Al_{75}Co_{22}Ni_3$ precipitates. The peak intensity of the copper-aluminum-nickel-1.0 weight percent cobalt alloy revealed a noticeable decrease with ageing. However, after aging, the peak for the $Al_{75}Co_{22}Ni_3$ precipitate in the copper-aluminum-nickel-0.4 weight percent cobalt alloy increased [76]. This outcome was explained using the logic that the copper-aluminum-nickel-0.4 weight percent cobalt shape memory alloy (SMA) was still in the early stages of precipitate evolution.

6.3. Performance upon Exposure to an Aqueous Environment

After 40 min of exposure to an aqueous environment, the polarization curves for both the aged Cu-Al-Ni alloy and unaged copper-aluminum-nickel alloy biformed with low-carbon steel rebar and isolated with the low-carbon steel rebar were determined from electrochemical experiments using a 3.50 percent sodium chloride solution. The corrosion potential for the linked copper-aluminum-nickel/low-carbon steel was 313.78 mV, which was between the corrosion potentials of isolated copper-aluminum-nickel alloy (18.025 millivolt) and low-carbon steel (422.521 millivolt). However, the connected material's current density, which measured the rate of environment-induced degradation, gradually rose to 57.42 A/cm^2, thereby exceeding the current density of the two chosen materials, which was 19.231 A/cm^2 for the low-carbon steel and 28.592 A/cm^2 for the copper-aluminum-nickel alloy. Galvanic corrosion, which was favored to occur when the two distinct metals came into contact, was to blame for this event [77]. On the other hand, aging treatment had a substantial influence on environment-induced degradation behavior of the linked Cu-Al-Ni and low-carbon steel sample. Degradation-induced by the environment on the copper-aluminum-nickel alloy slowed down, following ageing treatment for both the coupled samples and the uncoupled samples. This outcome was explained by coarsening of the β phase during aging. The presence of aluminum in plate-like morphology may have made the passive film on the surface to be stable and thereby encouraged the formation and presence of layers of alumina (Al_2O_3) [78]. Additionally, the surface area diminished microscopically as the needle-like β phase in the alloy shrank. According to the test data, it was found that the rate of environment-induced degradation decreased as the level of cobalt increased. In a different way, with the addition of cobalt, the resistance offered to environment-induced degradation of the Cu-Al-Ni alloy improved. This feature could be attributed to the role and contribution of cobalt addition in refining the grain size. There have been few discoveries that have shown a smaller grain size and large volume percentage of the precipitates to improve both the compactness and stableness of the passive layer leading to an overall improvement in the shape memory alloys' (SMAs) resistance to environment-induced degradation or corrosion [79]. The trends shown by the connected samples mirrored those of the uncoupled samples. It was discovered that the copper-aluminum-nickel-0.4 wt percent cobalt alloy revealed an actual drop in the corrosion rate after ageing, which could be attributed to microstructural changes as a direct consequence of both ageing and the making of the alloy. For the case of the copper-aluminum-nickel 0.4 wt% cobalt alloy, the precipitate's volume fraction was at the initiation stage [80]. However, with the aging of the sample, the density of the fine precipitates gradually improved. The evolution of precipitates contributed to increasing the passive film's stability and overall compactness. Additionally, the size and dispersion of the precipitates had an impact on enhancing and/or improving the resistance to environment-induced degradation or corrosion. The corrosion resistance of the copper-aluminum-nickel-1.0 weight percent cobalt shape memory alloy (SMA), in which precipitation was numerous, was improved both by a noticeable reduction in size and an increase in homogeneity of the precipitates.

Additionally, a grain boundary, or other location, where the precipitates were separated often resulted in a change in the substrate. An equal anode–cathode dispersion would tend to decrease the localized galvanic corrosion between the matrix and the fine precipitates.

6.4. Interpretation on the Experimental Findings

1. Ageing treatment of the copper-aluminum-nickel alloy with the addition of cobalt as the fourth element promoted grain refinement and produced fine precipitates of the $Al_{75}Co_{22}Ni_3$ phase that contributed to increasing both the compactness and stability due to the formation and presence of a passive film and thereby improved the corrosive resistance.
2. After being adjusted with the addition of 1 weight percent of cobalt along with ageing treatment, the linked copper-aluminum-nickel and low-carbon steel shape memory alloy (SMA) revealed an optimum value of corrosion resistance. This resulted in reducing the rate of environment-induced degradation, or corrosion, by well over 50 percent when compared with the low-carbon steel sample (uncoupled).
3. After the addition of 1 weight percent of cobalt and ageing at 250 °C for 48 h, the copper-aluminum-nickel shape memory alloy revealed a microhardness of 340 Hv.

7. The Cu-Zn-Al Shape Memory Alloy in Monitored Ambience

Copper–zinc–aluminum alloys are often chosen for use in a broad scope of engineering fields, primarily because of a combination of unique properties that makes them simple to produce. The copper-based alloy has been extensively studied in a variety of environments, but additional research is still needed to fully understand how this alloy reacts upon exposure to an aggressive environment and the resultant degradation induced by the environment, i.e., corrosion [81]. Centrifugal casting was used to create the copper alloy, which contained 21% zinc and 5% aluminum.

Using a combination of induction melting and centrifugal casting, the CuZnAl shape memory alloy (SMA) was created. With the help of a ceramic crucible made of vitreous silica and zirconium dioxide (ZrO2), pure metals, such as electrolytic copper (99.99%), aluminum 1100 alloy in sheet form (99.95%), and zinc pieces (99.9%), were easily melted in an induction furnace in an environment of argon and at a pressure below the ambient pressure. In a graphite mold, the molten alloy was centrifugally cast. Since an observable loss of zinc occurred during the melting process, the cast specimens were carefully monitored to determine both their composition and shape memory characteristics. If required, the cast materials were remelted along with the addition of pure metals to get a composition that was close to the nominal composition [82–85]. The as-cast specimens were analyzed to ascertain both their composition and microstructure. Intricacies specific to the microstructure were studied using the techniques of

(a) X-ray diffraction;
(b) Electronic microscopy;
(c) Optical microscopy after chemically etching the polished surfaces using a chemical reagent. The reagent used was ferric chloride (FeCl3) in hydrochloric acid (HCl) solution.

The potentiodynamic anodic polarization measurements were used to systematically analyze the environment-induced degradation behavior or response of the chosen shape memory alloy (SMA). This was performed using a thin plate of the chosen shape memory alloy (SMA) that was taken out of a cast bar using a diamond blade. This was followed by polishing the thin plate using silicon carbide (SiC) -impregnated emery paper (up to 600 mesh) and subsequently cleaning the polished surface using acetone. Thereafter, the ASTM G5-compliant corrosion test was conducted. About 2 square cm of the surface was exposed to an aggressive environment that favored the occurrence of degradation, referred to as corrosion. The working electrode surface was close to the reference electrode. The working electrode and reference electrode were separated by roughly 1 square cm from an auxiliary electrode, which was essentially a wire mesh made of platinum.

To simulate the various conditions prevailing in reality, a variety of test solutions were used and essentially included the following:

(a) 3.5% weight sodium chloride (NaCl) aqueous solution that simulated a marine environment.
(b) Acid solutions of 1 M, 0.1 M, and 0.01 M nitric acid (HNO_3) that replicated acid rain in a metropolitan setting.
(c) Acid solutions of 1 M, 0.1 M, and 0.01 M sulfuric acid (H_2SO_4) that replicated acid rain in an industrial setting.

The Test Results

The surroundings utilized in the experiment exerted an influence on the behavior of the test specimens used that were subsequently analyzed. Martensitic lamellae were an easy target for the onset of environment-induced degradation in an environment of acid solution. The passivation phenomena were not easily visible and the environment-induced degradation rates tended to increase as a function of solution concentration [86].

The chosen shape memory alloys exhibited a decreasing trend when tested in concentrated sulfuric acid (H_2SO_4) and sodium chloride (NaCl) solutions. The current density also decreased and stabilized at high imposed potentials. Although these processes resembled passivation, the observed reduction in environment-induced degradation or corrosion was caused by a gradual buildup of layers of corrosive products. The pores present on the test sample surface permitted the degradation caused by the aqueous environment to continue by allowing or permitting a gradual diffusion of the chemical species from the interface of the deteriorating metal to the mixture. This layer eventually ruptured when a high potential (above 1.2 V) was reached, which resulted in the occurrence of pitting [87,88]. All the test specimens preserved the shape memory effect after the potentiodynamic test. The conditions suitable for oxidation resulted in the presence of flaws close to the grain boundary. The presence of these flaws caused the material to fail by the initiation of fine microscopic voids and their gradual growth and eventual coalescence to form one or more fine microscopic cracks that tended to propagate along the grain boundaries.

8. Nature of Degradation When Cobalt Is Added to the Nickel–Titanium Shape Memory Alloy (SMA) in Normal Saline Solution

Due to their strong biocompatibility, resistance to degradation-induced by the aqueous environment, and a high elastic modulus, the NiTi alloys are often chosen for use in medical devices. In addition, the NiTi alloys containing 1–2 percent cobalt offer the following attributes:

(a) A 30 percent higher modulus than the NiTi alloys.
(b) A distinct loading plateau and an unloading plateau.
(c) Non-reactivity in two tests, namely hemolysis and cytotoxicity.

An enhancement in yield strength was made possible for the NiTi alloys with the addition of 2–10 percent cobalt [89–91]. Since the martensitic transition temperature occurred by separation of the R phase with the addition of cobalt, the results could be better explained by a two-step transformation process.

With the use of wire electro discharge machining (EDM), the rate of material removal, microstructure, and hardness were also investigated for the alloys containing up to 10% cobalt. The effect of cobalt on overall resistance to environment-induced degradation of the chosen alloys must be investigated prior to their selection for use in a specific application. According to Huang and co-workers, upon investigation of the environment-induced degradation characteristics of the chosen alloy on exposure to an aggressive aqueous environment, they behaved like the NiTi alloys that did not contain cobalt and were not adversely influenced by the occurrence of galvanic corrosion.

The investigations were conducted in a mild climate where degradation due to both pitting and crevice deterioration were not typical. It was anticipated that the addition of enough cobalt (Co) to an alloy would boost its resistance to this type of deterioration, primarily because the cobalt-base alloys tended to resist the localized effects of environment-induced degradation, unless fretting and other mechanical deterioration occurred on the surface oxide [92]. Our recent study on investigating and understanding the corrosion and electrochemical behavior of the cobalt-based magnetic shape memory alloys (MSMAs) in 0.50 M sodium chloride solution revealed the vital corrosion resistance of the MSMAs. This was well supported by the following:

(i) Electrochemical tests;
(ii) XPS;
(iii) Scanning electron microscopy (SEM) observations;
(iv) Energy dispersive X-ray (EDX) analysis.

Several studies on both the role and contribution of the addition of cobalt (Co) on the environment-induced degradation behavior of NiTi shape memory alloys (SMAs) are limited. To accomplish its goal, a recent study used a variety of electrochemical techniques in addition to (i) scanning electron microscopy (SEM) observations, (ii) energy dispersive X-ray (EDX) analysis, and (iii) full XPS studies as presented in Figure 17. In both uniform degradation and pitting degradation experienced by the chosen NiTi(x)Co (x = 14.0%, 1.50%, and 4.0%) shape memory alloy (SMA) upon its exposure to a 0.90% sodium chloride solution at 38 °C, to know the influence on passivation of the alloy containing cobalt (Co), the impact of cobalt addition on passivation translated into a clear improvement in the overall resistance to environment-induced degradation of the chosen alloy [93–95].

8.1. Electrochemical Setup and Solutions

Calomel saturated electrode and platinum (Pt), which is a mesh, electrodes were used as the reference electrode and auxiliary electrode in a typical packed and covered three-electrode cell for the purpose of conducting electrochemical studies. A potentiostat connected to a PC along with a galvanostat were attached to an electrochemical cell to both run and concurrently observe the electrochemical procedures used in the investigation. Measurements were made using a freshly made 0.9% normal saline (NaCl) solution that was purified prior to use [96]. Analytical grade salt, which was acquired from Sigma-Aldrich (New York, NY, USA), was used for the experiment.

Using a temperature-monitored and controlled water bath, with water continuously circulating through the outer packed cell jacket, the temperature of the solution was kept constant at 37 °C.

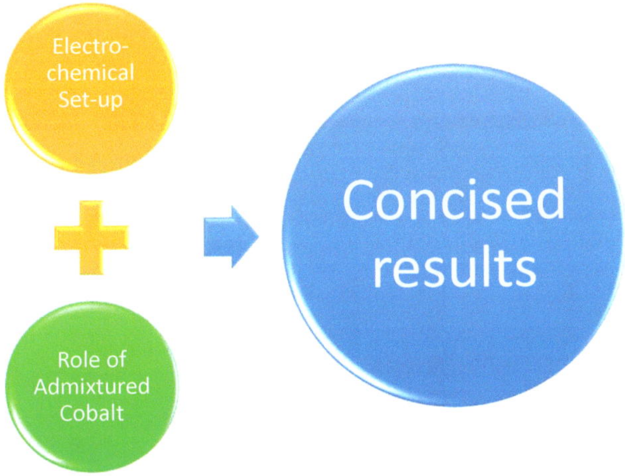

Figure 17. Study of the effect of cobalt added to nickel–titanium (NiTi).

8.2. Role of Addition of Cobalt

Study of Microstructure

In addition to precipitation and presence of the intermetallic phase (Ti_2Ni), it also included the parent phase (i.e., austenite) and the matrix phase (i.e., martensite). The addition of cobalt influenced the phase volume fraction (V_f) observed in the microstructure. When the percentage of cobalt was increased from 0 to 1.5 atomic percent, the volume fraction (V_f) of the parent phase (austenite) gradually decreased and even disappeared from the microstructure when the cobalt content was increased to 4 atomic percent. Additionally, density of the intermetallic phase (Ti_2Ni) in the microstructure of the titanium–nickel shape memory alloy (SMA) decreased with the addition of cobalt. In the microstructure of the NiTiCo (0 atomic percent Co) alloy, precipitates of the Ti_2Ni phase had a range of chemical composition. The nickel concentration was 34.15 atomic percent, while the titanium content in the Ti2Ni phase was 65.85 atomic percent. Additionally, as the level of cobalt content in the NiTi alloy increased, so did the cobalt content in the intermetallic phase (Ti_2Ni) [97].

In addition, for the TiNi (with 0 atomic percent cobalt) shape memory alloy (SMA) the matrix phase, i.e., martensite, had a chemical composition that included both nickel and titanium. Based on the amount of cobalt added to the TiNi shape memory alloy (SMA), different percentages of nickel (Ni), titanium (Ti), and cobalt (Co) were found to be present in the martensite phase.

8.3. Summary of the Results

The microstructure of the studied shape memory alloys (SMAs) was affected by the addition and/or presence of cobalt by decreasing both the average size of the Ti_2Ni intermetallic phase and the total area occupied by this phase. Cobalt affected overall resistance of the alloy to environment-induced degradation through a variety of mechanisms, including the following:

(a) Overall homogeneity of the surface electrochemical characteristics;
(b) Reduced activity of the microgalvanic cells.

With the addition of cobalt (Co), the number of chloride ions in the passive layer covering the surface of the chosen shape memory alloy (SMA) noticeably decreased. This demonstrated an overall stableness of the passive layer coupled with an improved resistance to the occurrence of both uniform environment-induced degradation or corrosion and pitting deterioration processes [98]. Despite this, the overall chemistry of the passive layer, which was made up of titanium dioxide (TiO_2), did not significantly change.

9. The Nickel-Titanium Shape Memory Alloy Sintered by Spark Plasma Sintering (SPS)

9.1. Development of the Sintered Nickel-Titanium Shape Memory Alloy (SMA)

To create 50-50 NiTi, nickel (average size 40 μm, 99.7% purity) and titanium (mean size 47 μm, 99.7% purity) powders were alloyed for 5 h in a planetary ball mill system at a high energy. The powder mixture was milled for a full 30 h in the same manner as before. The mixed powder could safely be classified to contain particles having an average size in the range 10 μm to 45 μm [98]. Both the milling parameters and process parameters chosen were essentially based on those used in an earlier study as shown in Figure 18. Since it exerted an influence on both size and uniformity of the powder particles, the duration of milling had a significant influence on alloying of the NiTi alloy.

Figure 18. Corrosion behavior of the sintered nickel–titanium alloy.

9.2. Characteristics upon Exposure to an Aggressive Environment

The NiTi alloys were sintered using the spark plasma sintering (SPS) method and in a vacuum environment, resulting thereby in high densification. Investigations were conducted into environment-related activity of the sintered alloy. Additionally, an evaluation of both the microstructure and dispersion of the elements in the degraded region was made. Using the method of spark plasma sintering (SPS), high density NiTi alloys having particles of size 10 μm were sintered. When the specimen was exposed to plastic deformation at a given pressure together with electrical discharge at ambient pressure, the goal was to achieve an improvement in the density coupled with a smaller grain size and a slender grain morphology. While the NiTi alloys were created using particles having a size of 10.0 μm and the spark plasma sintering (SPS) method at 910 °C, it was fully consolidated and had a dendritic structure. The presence of secondary phases in the microstructure, such as $NiTi_2$ and Ni_3Ti, prevented the occurrence of environment-induced degradation or corrosion. Low-corrosion current density and high-corrosion potential were easily attained for the nickel–titanium alloys that were created at a maximum binding temperature by using smaller size particles. Increased polarization current density did have an impact on the rate of degradation induced by the aqueous environment. The rate of degradation did affect the weight loss experienced by the NiTi alloys [98,99]. Thus, for the NiTi alloys that were produced using 10 μm particles and at a high SPS temperature, the loss in weight was essentially governed by the polarization current density. Due to a homogeneous distribution of the corrosion-resistant phases coupled with a decrease in the polarization current density, the rate of degradation caused by exposure to an aggressive aqueous environment was lowered by as much as 74 percent [100]. For the case of the nickel–titanium shape memory alloys (SMAs), a fine size of the starting particles coupled with a high sintering temperature essentially created a pore-free skin surface. This surface was beneficial for improving the overall resistance of the alloy to degradation caused by the aggressive aqueous environment.

10. Biocompatibility of Shape Memory Alloys and Its Progress

Understanding the corrosion process is the key to understanding metal biodegradation. Metals are biodegraded via electrochemical corrosion in the physiologically slightly alkaline environment (pH 7.4), which is often made up of anodic and cathodic processes [100,101]. It has been discovered that anodic processes include both metal breakdown and electron release:

$$M - ne^{-} \rightarrow M^{n+} \text{ (anodic reaction)} \quad (1)$$

The subsequent cathodic processes that take place in the electrolyte to capture these electrons result in the production of hydroxide ions.

$$M^{n+} + n\text{OH}^{-} \rightarrow M(\text{OH})_{n} \text{ (corrosion product)} \quad (2)$$

In order to increase the overall corrosion rate of the alloy during the manipulation of biodegradation performance, it is essential to enhance the electrochemical corrosion effects happening between the SMA matrix and electrolytes or other phases. Structure, implantation duration, and SMA phase transition temperatures are additional variables that may impact SMA biodegradation efficacy [102]. Both macro- and microstructures have a significant impact on biodegradations. For instance, NiTi SMA is shown to have reduced corrosion resistance as porosity increases due to bigger pore openings, but considerably higher corrosion rates [103]. Biocompatibility of Ni-based shape memory alloys prove that they are more suitable for biomedical implants and pursue their highest degree of functionality in the applications they are involved in.

10.1. Methods to Improve Biocompatibility

10.1.1. Grain Refinement

It has been generally documented that grain refining works well to slow the rate of corrosion. The SME and SE/PE of SMAs are not anticipated to be impacted by grain refining, since phase transitions are often not linked to this process. In addition, grain refining may eliminate any biocompatibility issues brought on by the inclusion of additional components and further increase the mechanical characteristics of SMAs based on the Hall–Petch connection [104]. The formation of a more cohesive and compact protective corrosion layer, the prevention of preferred crystallographic pitting restrained by increased grain boundaries, and the improvement of basal plane intensity can all be used to explain the reduction in corrosion rates caused by grain refinement.

10.1.2. Surface Coatings

Additionally, Ni-based SMA's medicinal uses may be expanded by surface coatings with additional bioactive or therapeutic properties, including antibacterial and drug-loading properties. It maintains the mechanical characteristics and shape memory behaviors of the underlying metal substrate without changing its microstructure [105]. To avoid abnormally low biodegradation rates, the coating layer should preferably be degraded or detached after the creation of the passivation layers. In situ conversion coating and ex situ deposition coating are two options for achieving the needed transitory corrosion protection. For the conversion coating, in situ interfacial interactions between the metal substrate and coating solution produce the noble coating layer. The native surface of the metal substrate might change into oxide or salts depending on the coating solution. Due to the in situ growth of the coating layer from the metal substrate, a suitable interfacial adhesion may be achieved to avoid premature coating layer detachment or potential penetration of corrosive media into the metal–coating layer interface [106].

10.2. Challenges in Employing Shape Memory Alloys at Bioapplications

(1) To fully comprehend the causes and mechanisms underlying temperature- or stress-induced phase changes between austenite and martensite, further microstructural insights into the mechanisms underlying SMA properties are needed. Understanding these mechanisms can help design SMA and manipulate corrosion with beneficial recommendations. Through in situ studies of SMA deformation and mechanical reactions during the phase transition, advanced microscopy is anticipated to play

a significant role in this respect [106]. At the austenite–martensite contacts, crucial information may include the lattice resistance, steps, and dislocation arrays.

(2) These materials must be modified to be suited for more nuanced biological applications, which calls for precise control of the SMA transition temperature and stress. There is evidence that alloy composition and thermomechanical treatments can adjust SMA transition temperatures [107]. The difficulty in designing a functional device stem from the fact that such modulation is not precise enough to achieve the precise needed values. In order to overcome this issue, computational intelligence may be useful, as topological models, artificial neural networks, and Gaussian process regression may all be used to anticipate transformation temperature and stress.

(3) More research on passive films and film–metal interactions may provide new insights into the behavior of SMA corrosion. According to reports, TiO_2 is crucial in preventing NiTi corrosion and the passive coatings on NiTi SMA display n-type semiconductor characteristics. While efficient TiO_2 dissolution may be achieved by lowering the pH of the corrosion environment, doping levels can be enhanced by donor production at the metal–film interface to reduce the corrosion resistance of passive films [108]. Future work may focus on altering the chemical makeup of passive films to control their corrosion resistances or adjusting the doping levels of semiconducting passive films to produce metastable or stable pits and voids at the metal–film interface to speed up corrosion.

11. Conclusions

Based on a study aimed at understanding the effects of alloying and environment on the degradation response or corrosion behavior of shape memory alloys, the key findings are as follows:

1. The trend towards the use of less invasive techniques and microscopic applications will continue.
2. Processing capabilities are being noticeably improved and the shape memory alloys (SMAs) are gradually gaining for themselves a dominant place for due consideration by all engineers for selection and use in medical design-related applications and even technologies specific to emerging smart materials.
3. An increase in the selection and use of the shape memory alloys (SMAs) both in medicine and sensor technology can be expected. The shape memory alloys are currently being chosen for use in critical environments, such as high temperature vital fluids, i.e., the blood stream.
4. The nature of environment-induced degradation, or corrosion, of the shape memory alloys (SMAs) is presented and examined in this paper based on results obtained from tests conducted in aggressive aqueous environments.
5. Improving the resistance to environment-induced degradation of the shape memory alloys (SMAs) will pave the way for their selection and use in a sizeable number of applications. Further, discovering ways to resist degradation induced by the environment, through the development of passive films and coatings is both essential and desirable.

Author Contributions: Conceptualization, S.S. and W.B.J.H.; methodology, S.S., T.S.S. and W.B.J.H.; formal analysis, S.S. and W.B.J.H.; investigation, S.S.; resources, S.S. and T.S.S.; writing—original draft preparation, S.S. and W.B.J.H.; writing—review and editing, T S.S.; visualization, S.S.; supervision, S.S.; project administration, S.S.; funding acquisition, S.S. All authors have read and agreed to the published version of the manuscript.

Funding: One of the authors, Santosh Sampath gratefully acknowledges the funding from Science and Engineering Research Board (SERB), India and Sri Sivasubramaniya Nadar Trust through file numbers, SUR/2022/001060 and SSN/IFFP/DECEMBER 2021/1-21/08, respectively.

Institutional Review Board Statement: Not applicable.

Informed Consent Statement: Not applicable.

Data Availability Statement: Data sharing not applicable. No new data were created or analyzed in this study. Data sharing is not applicable to this article.

Conflicts of Interest: The authors declare no conflict of interest.

References

1. Otsuka, C.M.K. *Wayman: Shape Memory Materials*; Cambridge University Press: Cambridge, UK, 1998; pp. 1–90.
2. Santosh, S.; Praveen, R.; Sampath, V. Influence of Cobalt on the Hot Deformation Characteristics of an NiTi Shape Memory Alloy. *Trans. Indian. Inst. Met.* **2019**, *72*, 1465–1468. [CrossRef]
3. Santosh, S.; Nithyanandh, G.; Ashwath, J.; Kishore, K.L. Comparison of Internal friction measurements on Ni-Ti reinforced smart composites prepared by Additive Manufacturing. *J. Alloys Compd.* **2022**, *924*, 166027. [CrossRef]
4. Evirgen, A.; Ma, J.; Karaman, I.; Luo, Z.P.; Chumlyakov, Y.I. Effect of aging on the super elastic response of a single crystalline FeNiCoAlTa shape memory alloy. *Scr. Mater.* **2012**, *67*, 475–478. [CrossRef]
5. Liu, B.; Wei, X.; Wang, W.; Lu, J.; Ding, J. Corrosion behavior of Ni-based alloys in molten NaCl-$CaCl_2$-$MgCl_2$ eutectic salt for concentrating solar power. *Sol. Energy Mater. Sol. Cells* **2017**, *170*, 77–86. [CrossRef]
6. Fadlallah, S.A.; El-Bagoury, N.; El-Rab, S.M.F.G.; Ahmed, R.A.; El-Ousamii, G. An ov view of NiTi shape memory alloy: Corrosion resistance and antibacterial inhibition for dental application. *J. Alloys Compd.* **2014**, *583*, 445–464. [CrossRef]
7. Mallik, U.S.; Sampath, V. Influence of Aluminum and Manganese Concentration on the Shape Memory Characteristics of Cu-Al-Mn Shape Memory Alloys. *J. Alloys Compd.* **2008**, *459*, 142–147. [CrossRef]
8. Castleman, L.S.; Motzki, S.M.; Williams, D.F. Biocompatibility of nitinol. In *Biocompatibility of Clinical Implant Material*; CRC Press: Boca Raton, FL, USA, 1981; Volume 1, pp. 129–154.
9. Buehler, J.; Wang, F.E. Summary research in the nitinol alloys and their potential applications in ocean engineering. *Ocean. Eng.* **1968**, *1*, 105–108. [CrossRef]
10. Spruit, D.; Bongaarts, P.J.M.; Malden, K.E. Dermatological effects of nickel. In *Nickel in the Environment*; Nriagu, J.O., Ed.; Wiley: New York, NY, USA, 1980; pp. 601–609.
11. Revie, R.W.; Green, N.D. Corrosion behaviour of surgical implant materials: I Effect of sterilization. *Corms Sci.* **1969**, *10*, 755–762. [CrossRef]
12. ASTM International. *Annual Book of ASTM Standard*; American Society for Testing and Materials: Philadelphia, PA, USA, 1986; Volume 03.02, pp. 560–575.
13. Pessal, N.; Liu, C. Determination of critical potentials of stainless steels in aqueous chloride environments. *Electrochem. Acta.* **1971**, *16*, 1987–2003. [CrossRef]
14. Malki, B.; Berthomé, G.; Souier, T.; Boissy, C.; Guillotte, I.; Baroux, B. A Combined Experimental and Computational Approach to Study Crevice Corrosion of Stainless Steels. *J. Electrochem. Soc.* **2021**, *168*, 101504. [CrossRef]
15. Sohmura, T.; Kimura, H. Shape recovery in Ti-V-Fe- Al alloy and its application to dental implant. In Proceedings of the International Conference on Martensitic Transformations, Nara, Japan, 26–30 August 1986; The Japan Institute of Metals: Nara, Japan, 1986; pp. 1065–1070.
16. Edie, J.W.; Andreasen, G.F.; Zaytoun, M.P. Surface corrosion of nitinol and stainless steel under clinical conditions. *Angle Orthod.* **1981**, *51*, 319–324.
17. Sarkar, N.K.; Redmond, W.; Schwaninger, W. The chloride corrosion behaviour of four orthodontic wires. *J. Dent. Res.* **1979**, *10*, 53–98. [CrossRef] [PubMed]
18. Nakayama, Y.; Yamamuro, T.; Kotoura, Y.; Oka, M. Electrochemical corrosion study of surgical implant metals in rabbit back muscle with comparison to Ringer's solution. In Proceedings of the Third World Biomaterials Congress, Kyoto, Japan, 21–25 April 1988; p. 202.
19. Nakayama, Y.; Yamamuro, T.; Kotoura, Y.; Oka, M. In vivo measurement of anodic polarization of orthopaedic implant alloys: Comparative study of in viva and in vitro experiments. *Biomaterials* **1989**, *10*, 420–424. [CrossRef] [PubMed]
20. Weinstein, A.M.; Spires, W.P.; Klawitter, J.J.; Clemow, A.J.T.; Edmunds, J.O. Orthopaedic implant retrieval and analysis study. In *Corrosion and Degradation of Implant Materials*; Syrett, B.C., Acharya, A., Eds.; ASTM International: West Conshohocken, PA, USA, 1979; pp. 212–228.
21. Prashanth, S.; Shivasiddaramaiah, A.G.; Mallikarjun, U.S. Corrosion Behaviour of Cu-Al-Be Based Shape Memory Alloy With and Without Coating. *Mater. Today Proc.* **2019**, *17*, 147–154. [CrossRef]
22. Shivasiddaramaiah, A.G.; Manjunath, S.Y.; Singh, P.; Mallikarjun, U.S. Synthesis and Evaluation of Mechanical Properties of Cu-Al-Be- Mn Quaternary Shape Memory Alloys. *Int. J. Appl. Eng. Res.* **2015**, *55*, 356–390.
23. Priyanga, G.S.; Mattur, M.N. Mattur, Structural and mechanical properties of NiTiAg shape memory alloys: ab-initio study. *Modelling Simul. Mater. Sci. Eng.* **2023**, *31*, 065004. [CrossRef]
24. Chen, B.; Liang, C.; Fu, D. Pitting Corrosion of Cu-Zn-Al Shape Memory Alloy in Simulated Uterine Fluid. *J. Mater. Sci. Technol.* **2005**, *21*, 226–230.
25. Shivasiddaramaiah, A.G.; Das, B.R.; Singh, P. Study on Corrosion Behaviour of Cu-Al-Be-Mn Quaternary Shape Memory Alloy At Room Temperature. *Int. J. Appl. Eng. Res.* **2015**, *55*, 1115–1130.

26. Shivasiddaramaiah, A.G.; Singh, P.; Manjunath, S.Y.; Mallikarjun, U.S. Microstructure and Shape Memory Effect of Cu-Al-Be-Mn Quaternary Shape Memory Alloys. *Appl. Mech. Mater.* **2015**, *813–814*, 213–217. [CrossRef]
27. Wu, S.; Lin, H. Recent development of TiNi-based shape memory alloys in Taiwan. *Mater. Chem. Phys.* **2011**, *64*, 81–92. [CrossRef]
28. Duerig, T.W.; Melton, K.N.; Stockel, D.; Wayman, C.M. *Engineering Aspects of Shape Memory Alloys, Books on Demand*; Butterworth Heinemann Publishing: London, UK, 1990.
29. Lopez, G.A.; Barrado, M.; Bocanegra, E.H.; San Juan, J.M.; No, M.L. Cu–Al–Ni shape memory alloy composites with very high damping capacity. In *International conference on martensitic transformations (ICOMAT)*; John Wiley & Sons, Inc.: Hoboken, NJ, USA, 2008; Volume 75, pp. 231–238.
30. Song, G.; Patil, D.; Kocurek, C.; Bartos, J. *Earth and Space 2010: Engineering, Science, Construction, and Operations in Challenging Environments*; ASCE Publications: League City, TX, USA, 2010; pp. 1551–1567.
31. Prashanth, S.; Shivasiddaramaiah, A.G. Shape memory effect in Cu-Sn-Mn ternary shape memory alloy processed by ingot metallurgy. *Int. J. Metall. Mater. Sci.* **2012**, *75*, 850–895.
32. Wei, Z.G.; Sandstrom, R. Shape-memory materials and hybrid composites for smart systems. *J. Mater. Sci.* **1998**, *33*, 3743–3762. [CrossRef]
33. Davis, J.R. Copper and Copper Alloys. *ASM Spec. Handb.* **2001**, *2*, 759–780.
34. Abid, K.A.R.A.; Al-Tai, T.Z. The Effect of Iron Addition on the Dry Sliding Wear and Corrosion Behavior of Cu Al Ni Shape Memory Alloy. *Eng. Technol. J.* **2010**, *28*, 6888–6902.
35. Christian, L. *Shape-Memory Alloys Handbook*; John Wiley & Sons: Hoboken, NJ, USA, 2013; Volume 8, pp. 61–108.
36. Chenxu, Z. Shape Memory Stainless Steel. In *Advanced Materials & Processing*; Xi'an Jiao tong University: Xi'an, China, 2011; pp. 2015–2045.
37. Sevost'yanov, M.A.; Nasakina, E.O.; Baikin, A.S.; Sergienko, K.V.; Konushkin, S.V.; Kaplan, M.A.; Seregin, A.V.; Leonov, A.V.; Kozlov, V.A.; Shkirin, A.V.; et al. 2018 Biocompatibility of new materials based on nano-structured nitinol with titanium and tantalum composite surface layers: Experimental analysis in vitro and in vivo. *J. Mater. Sci. Mater. Med.* **2012**, *29*, 33–60. [CrossRef]
38. Xu, Y.; Wang, J.; Zhang, X.; Wang, P.; Shi, J.; Huo, F. Corrosion Behaviour of Ti–6Al–4V Alloy as Dental Implant Containing Fluoride Ions. *Int. J. Electrochemical. Sci.* **2017**, *12*, 10308–10316. [CrossRef]
39. Sevostyanov, M.A.; Baikin, A.S.; Shatova, L.A.; Nasakina, E.O.; Berezhnov, A.V.; Gudkov, S.V.; Sergienko, K.V.; Konushkin, S.V.; Baskakova, M.I. Biocompatibility of the Ti81Nb13Ta3Zr3 Alloy Doklady Chemistry. *J. Biomed. Mater. Res.* **2018**, *52*, 395–403482.
40. Kim, J.I.; Kim, H.Y.; Inamura, T.; Hosoda, H.; Miyazaki, S. Shape memory characteristics of Ti– 22Nb–(2–8)Zr(at.%) biomedical alloys. *Mater. Sci. Eng. A* **2005**, *403*, 334–339. [CrossRef]
41. Dubinsky, S.M.; Proboski, S.D.; Brailovski, V.; Inaekyan, K.; Korotitskiy, A.V.; Filonov, M.R.; Petrzhik, M.I. Structure Formation During Thermomechanical Processing of Ti-Nb-(Zr, Ta) Alloys and the Manifestation of the Shape-Memory Effect. *Phys. Met. Metallogr.* **2011**, *112*, 529–542. [CrossRef]
42. de Salazar, J.M.G.; Soria, A.; Barrena, M.I. Corrosion behavior of Cu-based shape memory alloys diffusion bonded. *J. Alloys Compd.* **2005**, *387*, 109–114. [CrossRef]
43. Heusler, K.E.; Landolt, D.; Trasaiti, S. Electrochemical Corrosion Nomenclature. *Pure Appl. Chem.* **1989**, *61*, 19–22. [CrossRef]
44. Trethewey, K.R.; Chamberlain, J. *Corrosion for Students of Science and Engineering*, 3rd ed.; John Wiley & Sons Inc: Hoboken, NJ, USA, 1995; Volume 18, pp. 2000–2045.
45. Shaw, B.A.; Kelly, R.G. What is corrosion? *Electrochem. Soc. Interface* **2006**, *15*, 24–26. [CrossRef]
46. Zhang, Q.; Xu, P.; Pang, C.; Yub, K.C.C.; Huang, L. A superhydrophobic surface with a synergistic abrasion–corrosion resistance effect prepared by femtosecond laser treatment on an FeMnSiCrNiNb shape memory alloy coating. *New J. Chem.* **2022**, *46*, 19188. [CrossRef]
47. Velmurugan, C.; Kumar, V.S.; Kamala, P.S. Microstructure and corrosion behavior of NiTi shape memory alloys sintered in the SPS process. *Int. J. Miner Metall. Mater.* **2019**, *42*, 1311. [CrossRef]
48. Zhong, Y.; Zhu, T. Phase-field modelling of martensitic microstructure in NiTi shape memory alloys. *Acta Mater.* **2014**, *75*, 337. [CrossRef]
49. Elayaperumal, K.; Raja, V.S.; Failures, C.; Studies, C. *Corrosion Failures: Theory, Case Studies, and Solutions*, 1st ed.; John Wiley & Sons: Hoboken, NJ, USA, 2015; Volume 24, p. 26.
50. Mallik, A.G.S.U.S.; Mahato, R.; Shekar, C.S. Evaluation of corrosion behaviour of Cu-Al-Be-Mn quaternary shape memory alloys. *Mater. Today Proc.* **2017**, *4*, 10971–10977.
51. Stergioudi, F.; Vogiatzis, C.A.; Pavlidou, E.; Skolianos, S.; Michailidis, N. Corrosion resistance of porous NiTi biomedical alloy in simulated body fluids. *Smart Mater. Struct.* **2016**, *25*, 322–345. [CrossRef]
52. Jani, J.M.; Leary, M.; Subic, A.; Gibson, M.A. A review of shape memory alloy research, applications and opportunities. *Mater. Des.* **2014**, *56*, 1078. [CrossRef]
53. Novák, P.; Pokorný, P.; Vojtěch, V.; Knaislová, A.; Školáková, A.; Čapek, J.; Karlík, M.; Kopecek, J. Formation of Ni–Ti intermetallic during reactive sintering at 500–650 °C. *Mater. Chem. Phys.* **2015**, *155*, 113. [CrossRef]
54. Elahinia, M.H.; Hashemi, M.; Tabesh, M.; Bhaduri, S.B. Manufacturing, and processing of NiTi implants: A review. *Prog. Mater. Sci.* **2012**, *57*, 911. [CrossRef]

55. Elahinia, M.; Moghaddam, N.S.; Andani, M.T.; Amerinatanzi, A.; Bamber, B.A.; Hamilton, R.F. Fabrication of NiTi through additive manufacturing: A review. *Prog. Mater. Sci.* **2016**, *83*, 630. [CrossRef]
56. Morgan, N.B.; Broadley, M. Taking the art out of smart-Forming processes and durability issues for the application of NiTi shape memory alloys in medical devices. In Proceedings of the Materials and Processes for Medical Devices Conference, St. Paul, MN, USA, 25–27 August 2004; Volume 6, p. 247.
57. Zhang, L.; Ren, D.; Ji, H.; Ma, A.; Daniel, E.F.; Li, S.; Jin, W.; Zheng, Y. Study on the corrosion behavior of NiTi shape memory alloys fabricated by electron beam melting. *NPJ Mater Degrad* **2022**, *6*, 79. [CrossRef]
58. Yang, C.; Liu, L.; Cheng, Q.; You, D.; Li, Y. Equiaxed grained structure: A structure in titanium alloys with higher compressive mechanical properties. *Mater. Sci. Eng. A* **2013**, *580*, 397–405. [CrossRef]
59. Hang, R.Q.; Ma, S.L.; Ji, V.; Chu, P.K. Corrosion behavior of NiTi alloy in foetal bovine serum. *Electrochim. Acta* **2010**, *55*, 5551.
60. Dong, H.; Ju, X.; Yang, H.; Qian, L.; Zhou, Z. Effect of ceramic conversion treatments on the surface damage and nickel ion release of NiTi alloys under fretting corrosion conditions. *J. Mater. Sci. Mater. Med.* **2008**, *19*, 937. [CrossRef] [PubMed]
61. Chembath, M.; Balaraju, J.N.; Sujata, M. In vitro corrosion studies of surface modified NiTi alloy for biomedical applications. *Adv. Biomater.* **2014**, *20*, 229–263.
62. Kim, J.; Park, J.K.; Kim, H.K.; Unnithan, A.R.; Kim, C.S.; Park, C.H. Optimization of electropolishing on NiTi alloy stents and its influence on corrosion behaviour. *J. Nanosci. Nanotechnol.* **2017**, *17*, 460–470.
63. Hosseini, S.A.; Akbarinia, S.; Mohammadyani, D.; Sadrnezhaad, S.K. Enhanced corrosion resistance of porous NiTi with plasma sprayed alumina coating. *Corros. Eng. Sci. Technol.* **2015**, *50*, 595.
64. Velmurugan, C.; Kumar, V.S. The effect of Cu addition on the morphological, structural and mechanical characteristics of nanocrystalline NiTi shape memory alloys. *J. Alloys Compd.* **2018**, *767*, 944. [CrossRef]
65. Fu, Y.Q.; Gu, Y.W.; Sherwood, C.; Luo, J.K.; Flewitt, A.J.; Milne, W.I. Spark plasma sintering of TiNi nano-powders for biological application. *Nanotechnology* **2006**, *17*, 5293. [CrossRef]
66. Verdian, M.M.; Raeissi, K.; Salehi, M.; Sabooni, S. Characterization and corrosion behavior of NiTi–Ti2Ni–Ni3Ti multiphase intermetallic produced by vacuum sintering. *Vacuum* **2011**, *86*, 91. [CrossRef]
67. Chen, J.; Qiao, Y.; Meng, F.; Wang, Y. Corrosion and Degradation of Materials. *Coatings* **2022**, *2005*, 12–969.
68. Chen, J.; Rasmussen, R.M.; Zagidulin, D.; Noel, J.J.; Shoesmith, D.W. Electrochemical and corrosion behavior of a 304 stainless-steel-based metal alloy waste form in dilute aqueous environments. *Corros. Sci.* **2013**, *66*, 142–152. [CrossRef]
69. Qiao, Y.; Wang, X.; Yang, L.; Wang, X.; Chen, J.; Wang, Z.; Zhou, H.; Zou, J.; Wang, F. Effect of aging treatment on microstructure and corrosion behavior of a Fe-18Cr-15Mn-0.66N stainless steel. *J. Mater. Sci. Technol.* **2022**, *107*, 197–206.
70. Hu, T.; Chu, C.L.; Xin, Y.; Paul, K.C. Corrosion products and mechanism on NiTi shape memory alloy in physiological environment. *J. Mater. Res.* **2019**, *25*, 350–358. [CrossRef]
71. Widu, F.; Drescher, D.; Junker, R.; Bourauel, C. Corrosion and biocompatibility of orthodontic wires. *J. Mater. Sci.-Mater. Med.* **1999**, *10*, 275. [CrossRef]
72. Cruz, J.C.; Nascimentoa, M.A.; Amaral, H.A.V.; Lima, D.S.D.; Teixeira, A.P.C.; Renata, P. Lopes Synthesis and characterization of cobalt nanoparticles for application in the removal of textile dye. *J. Environ. Manag.* **2019**, *242*, 220–228. [CrossRef]
73. Huang, H.H.; Chiu, Y.H.; Lee, T.H.; Wu, S.C.; Yang, H.W.; Su, K.H.; Hsu, C.C. Ion release from NiTi orthodontic wires in artificial saliva with various acidities. *Biomaterials* **2003**, *24*, 3585. [CrossRef]
74. Mediha, K.; Zardari, H.S.A.; Qader, I.N.; Kanca, M.S. The effects of cobalt elements addition on Ti2Ni phases, thermodynamics parameters, crystal structure and transformation temperature of NiTi shape memory alloys. *Eur. Phys. J. Plus* **2019**, *134*, 197.
75. Santosh, S.; Sampath, V. Effect of Ternary Addition of Cobalt on Shape Memory Characteristics of Ni–Ti Alloys. *Trans. Indian Inst. Met.* **2019**, *72*, 1481–1484. [CrossRef]
76. Najib, A.S.M.; Saud, S.N.; Hamzah, E. Corrosion Behavior of Cu–Al–Ni–*x*Co Shape Memory Alloys Coupled with Low-Carbon Steel for Civil Engineering Applications. *J. Bio-Tribo-Corros.* **2019**, *5*, 47. [CrossRef]
77. Tang, C.Y.; Zhang, L.N.; Wong, C.T.; Chan, K.C.; Yue, T.M. Fabrication and characteristics of porous NiTi shape memory alloy synthesized by microwave sintering. *Mater. Sci. Eng. A* **2011**, *528*, 6006–6011. [CrossRef]
78. Kapanen, A.; Ryann, J.; Danilov, A.; Tuukkanen, J. Effect of nickel–titanium shape memory metal alloy on bone formation. *Biomaterials* **2001**, *22*, 2475–2480. [CrossRef] [PubMed]
79. Mentz, J.; Frenzel, J.; Wagner, M.F.-X.; Neuking, K.; Eggeler, G.; Buchkremer, H.P.; Stover, D. Powder metallurgical processing of NiTi shape memory alloys with elevated transformation temperatures. *Mater. Sci. Eng. A* **2008**, *491*, 270–278. [CrossRef]
80. Velmurugan, C.; Kumar, V.S.; Biswas, K.; Yadav, S. Densification and microstructural evolution of spark plasma sintered NiTi shape memory alloy. *Adv. Powder Technol.* **2018**, *45*, 2456–2462. [CrossRef]
81. Diouf, S.; Molinari, A. Densification mechanisms in spark plasma sintering: Effect of particle size and pressure. *Powder Technol* **2012**, *221*, 220–227. [CrossRef]
82. Cheng, Y.; Cui, Z.; Cheng, L.; Gong, D.; Wang, W. Effect of particle size on densification of pure magnesium during spark plasma sintering. *Adv. Powder Technol.* **2017**, *28*, 1129–1135. [CrossRef]
83. Falodun, O.E.; Obadele, B.A.; Oke, S.R.; Maja, M.E.; Olubambi, P.A. Effect of sintering parameters on densification and microstructural evolution of nano- sized titanium nitride reinforced titanium alloys. *J. Alloys Comp.* **2018**, *736*, 202–210. [CrossRef]
84. Lee, G.; McKittrick, J.; Ivanov, E.; Olevsky, E.A. Densification mechanism and mechanical properties of tungsten powder consolidated by spark plasma sintering. *Int. J. Refract. Met. Hard Mater.* **2016**, *61*, 22–29. [CrossRef]

85. Zhang, L.; Zhang, Y.Q.; Jiang, Y.H.; Zhou, R. Super elastic behaviours of biomedical porous NiTi alloy with high porosity and large pore size prepared by spark plasma sintering. *J. Alloys Comp.* **2015**, *644*, 513–522. [CrossRef]
86. Chiyasak, P.; Kalnaowakul, P.; Rodchanarowan, A. The investigation of magnetic property and corrosion resistance of the assisted Sol-Gel synthesis of Ti/Ni/Co nanocomposites. *Surf. Coat. Technol.* **2020**, *393*, 2125–2150. [CrossRef]
87. Nam, T.H.; Lee, J.H.; Jung, D.W.; Yu, C.A.; Liu, Y.; Kim, Y.W. Transformation behaviour of Ti-Ni and Ti-Ni-Cu alloy ribbons with nano Ti2Ni particles. *Mater. Sci. Eng. A* **2007**, *449–451*, 1041–1044. [CrossRef]
88. Alqarni, N.D.; Wysocki, J.; El-Bagoury, N.; Reel, J.; Mohammed, A. Amin Rabah Boukherroub Effect of cobalt addition on the corrosion behavior of near equiatomic NiTi shape memory alloy in normal saline solution: Electrochemical and XPS studies. *RSC Adv.* **2018**, *8*, 19289. [CrossRef]
89. Han, Y.; Kong, F.L.; Han, F.F.; Inoue, A.; Zhu, S.L.; Shalaan, E.; Al-Marzouki, F. New Fe-based soft magnetic amorphous alloys with high saturation magnetization and good corrosion resistance for dust core application. *Intermetallic* **2016**, *76*, 18–25. [CrossRef]
90. de Araújo, C.J.; da Silva, N.J.; da Silva, M.M.; Gonzalez, C.H. A comparative study of Ni–Ti and Ni–Ti–Cu shape memory alloy processed by plasma melting and injection molding. *Mater. Des.* **2011**, *32*, 4925–4930. [CrossRef]
91. da Silva, N.J.; Grassi, E.N.D.; de Araújo, C.J. Dynamic properties of NiTi shape memory alloy and classic structural materials: A comparative analysis. *Mater. Sci. Forum* **2010**, *643*, 37–41. [CrossRef]
92. Machado, L.G.; Savi, M.A. Medical applications of shape memory alloys. *Braz. J. Med. Biol. Res.* **2002**, *36*, 683–691. [CrossRef]
93. Gonzalez, C.H.; Oliveira, C.A.N.; Pina, E.A.C.; Urtiga Filho, S.L.; Araújo Filho, O.O.; de Araújo, C.J. Heat treatments and thermomechanical cycling influences on the R-phase in Ti–Ni shape memory alloys. *Mater. Res.* **2010**, *13*, 325–331. [CrossRef]
94. Cai, W.; Men, X.L.; Zhao, L.C. Recent development of TiNi-based shape memory alloys. *Solid. State Mater. Sci.* **2005**, *9*, 296–302. [CrossRef]
95. Pasebani, S.; Dutt, A.K.; Burns, J.; Charit, I.; Rajiv, S. Mishra Oxide dispersion strengthened nickel-based Alloys via spark plasma sintering. *Mater. Sci. Eng. A* **2015**, *45*, 1506–1530.
96. Makena, I.M.; Shongwe, M.B.; Ramakokovhu, M.M.; Moipone, L. Letha bane A Review on Sintered Nickel based Alloys. In Proceedings of the World Congress on Engineering, London, UK, 5–7 July 2017.
97. Songwe, M.B.; Diouf, S.; Durowoju, M.O.; Olubambi, P.A. Effect of sintering temperature on the microstructure and mechanical properties of Fe-30%Ni alloys produced by spark plasma sintering. *J. Alloys Compd.* **2015**, *649*, 824–832. [CrossRef]
98. Mignanelli, P. Processing nickel-base superalloys. *Naked Sci.* **2012**, *51*, 4015–4020.
99. Metikos-Hukovic, M.; Babic, R. Passivation and corrosion behaviours of cobalt and cobalt–chromium–molybdenum alloy. *Corros. Sci.* **2007**, *49*, 3570–3579. [CrossRef]
100. Shao, X.; Guo, X.; Han, Y.; Lin, Z.; Qin, J.; Lu, W.; Zhang, D. Preparation of TiNi films by diffusion technology and the study of the formation sequence of the intermetallic in Ti-Ni systems. *J. Mater. Res.* **2014**, *29*, 2707–2716. [CrossRef]
101. Nazeer, A.A.; Madkour, M. Potential use of smart coatings for corrosion protection of metals and alloys: A review. *J. Mol. Liq.* **2018**, *253*, 11–22. [CrossRef]
102. Kawahara, Y. High temperature corrosion mechanisms and effect of alloying elements for materials used in waste incineration environment. *Corros. Sci.* **2002**, *44*, 223–245. [CrossRef]
103. Liu, M.; Zheng, J.Y.; Lu, Y.L.; Li, Z.J.; Zou, Y.; Yu, X.H.; Zhou, X.T. Investigation on corrosion behavior of Ni-based alloys in molten fluoride salt using synchrotron radiation techniques. *J. Nucl. Mater.* **2013**, *440*, 124–128. [CrossRef]
104. Walther, G.; Kl, B.; Büttner, T.; Weißg, T.; Kieback, B.; Böhm, A.; Naumann, D.; Saberi, S.; Timberg, L. A new class of high temperature and corrosion resistant nickel-based open-cell foams. *Adv. Eng. Mater.* **2008**, *10*, 803–811. [CrossRef]
105. Fu, Y.; Wu, X.; Han, E.; Ke, W.; Yang, K.; Jiang, Z. Influence of cold work on pitting corrosion behavior of a high nitrogen stainless steel. *J. Electrochem. Soc.* **2008**, *155*, 8. [CrossRef]
106. Brook, G.B. Applications of titanium-nickel shape memory alloys. *Mater. Des.* **1983**, *4*, 835–840. [CrossRef]
107. Kujala, S.; Ryh, J.; Jämsä, T.; Danilov, A.; Saaranen, J.; Pramila, A.; Tuukkanen, J. Bone modeling controlled by a nickel–titanium shape memory alloy intramedullary nail. *Biomaterials* **2002**, *23*, 2535–2543. [CrossRef] [PubMed]
108. Kasano, F.; Morimitsu, T. Utilization of nickel-titanium shape memory alloy for stapes prosthesis. *Auris Nasus Larynx* **1997**, *24*, 137–142. [CrossRef] [PubMed]

metals

MDPI

Article

The Influence of the Tantalum Content on the Main Properties of the TixTa9Nb8Zr2Ag Alloy

Gabriel Dobri [1,*], Alexandra Banu [1,*], Cristina Donath [2] and Maria Marcu [2]

[1] Faculty of Industrial Engineering and Robotics, Politehnica University of Bucharest, Splaiul Independentei 313, 060042 Bucharest, Romania

[2] Institute of Physical Chemistry "Ilie Murgulescu", Splaiul Independentei 202, 060021 Bucharest, Romania; cristidonath@yahoo.com (C.D.); m_marcu@icf.ro (M.M.)

* Correspondence: gabriel.dobri@upb.ro (G.D.); alexandra.banu@upb.ro (A.B.)

Abstract: This study presents the influence of different contents of tantalum alloying elements on the mechanical and electrochemical properties of TixTa9Nb8Zr2Ag alloys and their corrosion resistance in a 3% NaCl solution. These alloys exhibit a structure with more than 80% of the beta phase, a Young's modulus between 82 and 55 GPa close to human bone, and good corrosion resistance, with a corrosion rate between 5 and 47 $\mu m\ y^{-1}$. Furthermore, the excellent corrosion behavior of the TixTa9Nb8Zr2Ag alloy with 10 and 15% tantalum content is highlighted, revealed by a nobler corrosion potential, low corrosion rate, and a high passivation tendency in a 3% NaCl solution. The results reported in this work allow us to consider that titanium alloys TixTa9Nb8Zr2Ag with 10–20% Ta could be a valid alternative for use in orthopedic surgery, and the level of tantalum can be customized depending on the nature of the treated bone and the complexity and difficulty of the implant machining, i.e., of the required optimum hardness.

Keywords: titanium alloy; biomaterials; corrosion; mechanical properties

Citation: Dobri, G.; Banu, A.; Donath, C.; Marcu, M. The Influence of the Tantalum Content on the Main Properties of the TixTa9Nb8Zr2Ag Alloy. *Metals* **2023**, *13*, 1294. https://doi.org/10.3390/met13071294

Academic Editors: Jingxiang Xu, Zhenhua Chu and Xingwei Zheng

Received: 24 May 2023
Revised: 7 July 2023
Accepted: 13 July 2023
Published: 19 July 2023

1. Introduction

The use of metallic materials for surgical devices demands the fulfillment of biocompatibility requirements [1] next to that of corrosion resistance and bioadaptability. The corrosion resistance of metallic materials depends on their thermo-mechanical history but also on the aggressiveness of the contact environment. In the last years, a series of stainless steels [2], titanium-based alloys [3,4], and cobalt-based alloys [5] have been developed with very good corrosion resistance in biological environments. However, a good material for implantology applications must meet all requirements, and bioadaptability is an essential criterion with significant implications on the healing time and functionality of a surgical implant. Thus, the reduction in post-operative stress is an intense concern for doctors and a requirement addressed to researchers in the field of materials, to find formulas and combinations of elements to obtain new alloys, which best meet the criteria of materials used in implantology. It is well known that post-operative stress is generated by the different responses to taking over the mechanical loads at the interface between the implantable device and the bone, which leads to the appearance of the phenomenon of bone resorption. The high density of metallic materials, but also the incompatibility of mechanical properties, compared to those of human bone, are responsible for this unwanted effect. The high value of the Young modulus of various materials such as 190–210 GPa for stainless steels [6], 220 GPa for Co-Cr alloys [6], and 90–110 for titanium and its alloys [6,7], compared to that of human bone (10–40 GPa depending on the nature of the bone and the state of health [8,9]) has the effect of taking over the loads by the implant, protecting the bone and causing bone resorption [10,11]. It is recognized that titanium and its alloys have high biocompatibility for medical applications, and the Ti6Al4V alloy with a biphasic structure (α + β) and an elastic modulus value of about 110 GPa are widely used [12], although Al and V have proven to be toxic to the human body. This aspect determines their replacement with non-toxic beta-stabilizer elements such as Mo, Nb, and Ta [13,14], aiming to obtain a structure with a low content of the alpha phase for a low Young modulus simultaneously. Thus, special attention is given to a new range of alloys in the titanium–niobium–tantalum–zirconium alloy system (TNTZ), with good mechanical properties, a low modulus of elasticity, high corrosion resistance, and good biocompatibility [14–17]. Moreover, the extensive mechanical properties of these alloys are still to be enhanced in clinical practice. Accordingly, the design concept of these alloys can be used to obtain biomedical alloys with excellent appropriate properties to meet demand.

In addition, another property of tantalum that decides the presence of this element in the titanium–niobium–tantalum–zirconium alloy system is its low magnetic susceptibility, leading to a high quality of images and fewer distortion in post-operative magnetic resonance imaging [10]. In this regard, special attention is paid to a new range of alloys from the titanium–niobium–tantalum–zirconium (TNTZ) system with good mechanical properties, a low modulus of elasticity, high corrosion resistance, and good biocompatibility. In this vein, four alloys with a 10Nb 8Zr 2Ag Ti base were designed with a tantalum content between 10 and 20%, and after the morpho structural characterization, the influence of the Ta content on the electrochemical properties was investigated in the very severe conditions of 3% NaCl, the temperature of 37 °C, and 168 h of immersion in order to identify the alloy with the modulus of elasticity as close as possible to that of human bone (40 GPa) and maximum resistance to corrosion. Many research studies focused on a series of synthetic biofluids used in the field [18–20], but this paper presents the results in the most aggressive environment to test the limits of the material from this point of view.

2. Materials and Methods

The alloys were obtained in a vacuum arc melting furnace (VAR) ABD MRF 900 in the form of ingots with a diameter of 10 mm and a length of 150 mm with chemical compositions according to Table 1. The chemical composition of the experimental alloys was determined by energy-dispersive X-ray spectroscopy (EDS, FEI, Eindhoven, The Netherlands). It should be noted that during the EDS analysis, the only elements considered were those of interest, mainly Ti, Nb, Zr, Ta, and Ag. EDS analysis was performed on 5 random regions for each sample, and the average values and standard deviations are presented in Table 1.

Table 1. The chemical composition (% wt. and standard error) of experimental ingots.

Specimen	% Nb	% Zr	% Ag	% Ta	% Ti
A0	9.22 ± 0.41	8.28 ± 0.19	1.86 ± 0.16	0.0	Balance
A10	10.26 ± 0.68	7.95 ± 0.28	1.97 ± 0.09	8.94 ± 0.36	Balance
A15	9.23 ± 0.25	7.80 ± 0.12	1.87 ± 0.09	15.57 ± 0.60	Balance
A20	9.01 ± 0.16	7.88 ± 0.17	1.92 ± 0.08	20.02 ± 0.64	Balance

2.1. Morphology and Structural Analysis

Morphology and structural analysis was carried out by scanning electron microscopy (SEM) using the FEI Inspect F50 electron microscope (FEI Company, Eindhoven, The Netherlands), and phase analysis was performed with X-ray diffraction (XRD) using an X'PERT-PRO PANalytic diffractometer (Malvern Panalytical Ltd., Malvern, UK) ($K\alpha 1$ = 1.5405980 Å) with Cu-$K\alpha$ radiation at 45 kV, 4 mA, and a scan range starting from 10° to 90°.

2.2. The Young's Modulus

To measure the Young modulus, tensile tests were performed using a Walter + Bai LFV300 universal testing machine, according to [21], and the Vickers hardness was determined with a Shimadzu HMV 2TE hardness tester (Shimadzu Corporation, Kyoto, Japan) using a load of 1.961 N with a holding time of 15 s, according to the [22].

2.3. The Electrochemical Characterization

The electrochemical characterization was carried out by the method of immersion in a 3% NaCl solution at a temperature of 37.5 °C through the following techniques: monitoring the stationary potential for 168 h, the Tafel slopes method for determining corrosion rates, and electrochemical impedance spectroscopy (EIS). The stationary potential, corresponding to the steady state established at the interface between the alloy surface and the NaCl solution was monitored for 168 h, and, after that, the Tafel plots were recorded in the potential range from −0.25 V to + 0.25 V vs. OCP with a scan rate of 2.5 mV s^{-1}. EIS measurements were carried out potentiostatically, at the OCP, with an AC amplitude of 10 mV over the frequency range of 100 kHz to 0.01 Hz, after 168 h of exposure in 3% NaCl solution. EChemAnalyst (Gamry Instruments Inc., Warminster, PA, USA) and ZView specialized software (ZView 3.3, Scribner Associates, Inc., Souther Pines, NC, USA) were used to estimate the electrochemical parameters (E_{cor}, i_{cor}, r_{cor}) and to analyze EIS data.

A Reference 600 Gamry potentiostat/galvanostat (Gamry Instruments Inc., Warminster, PA, USA) with a classic electrochemical cell with three electrodes—a working electrode with a surface of 2 cm^2, a platinum counter electrode with a surface of 6 cm^2, and as a reference electrode, a Ag/AgCl 3M electrode—was used for all electrochemical measurements. Before the electrochemical tests, the active surfaces of the samples (2 cm^2) were processed by mechanical grinding, in stages, until metallographic quality; then, they were cleaned in an ultrasonic field with a degreasing solution, followed by ethanol and double-distilled water washing. The electrical contacts were made of titanium wire and insulated with insulating varnish and Teflon tape like the inactive surfaces. In order to prove reproducibility of the measurements, three specimens for each type of alloy were tested.

3. Results and Discussions

3.1. Microstructure and Morphology Analysis

Figures 1 and 2 show the SEM aspects of the microstructures of the analyzed samples. All the samples have specific structures for titanium alloys in the cast state; from a morphological point of view, they are lamellar biphasic, α + β, the proportion of the two phases changing depending on the tantalum content in the alloy. It can be observed that the reference alloy, without tantalum (A0), consists of a mixture of relatively balanced α and β solid solutions (Figure 1a) with α lamellae of considerable thickness. When adding a 9% amount of tantalum, the morphology does not change significantly (Figure 1b sample A10) with the mention that, as the concentration of Ta increases, the thickness of the α lamellae tends to decrease. At a content of 15% Ta, (A15), the morphology changes drastically, the microstructure consists mainly of a beta-solid solution with colonies of thin alpha lamellae mostly grouped at the grain boundaries, as seen in Figure 2a. The structure of the A20 alloy, with 20% Ta, is like A15, the proportion of the α phase decreases, the alloy has a predominantly β structure, over 90% [23], and the α phase is organized mainly in the form of rosettes, which induces changes in the mechanical and electrochemical behavior of alloys (Figure 2b).

The evolution of the alloy structure from biphasic α + β (A0) to the majority β structure (A20) is also highlighted by the XRD analysis (Figure 4), by the decrease in the intensity of signals specific to the α phase until the disappearance of some of them.

The XRD patterns were compared to reference cards JCPDS No: 44-1294 for α Titanium (P63/mmc) and 44-1288 for β Titanium (Im^3 m). The experimental peaks were slightly shifted to the left with respect to the ones from the reference cards, suggesting that the lattice parameter has changed by the dissolution of the alloying elements in titanium.

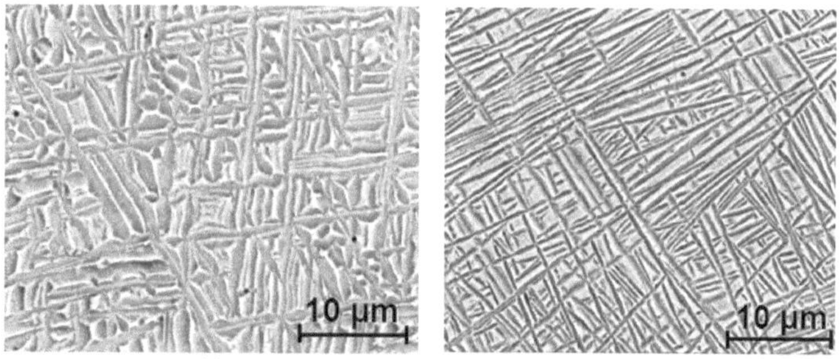

(**a**) (**b**)

Figure 1. Scanning electron micrographs of the experimental alloys: (**a**) A0; (**b**) A10.

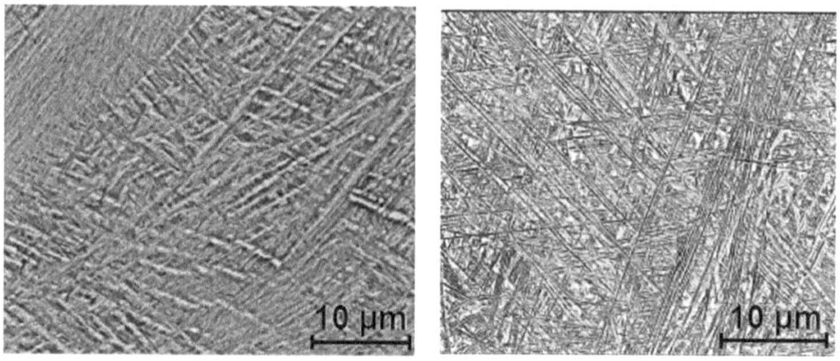

(**a**) (**b**)

Figure 2. Scanning electron micrographs of the experimental alloys: (**a**) A15; (**b**) A20.

It should be noted that in all cases, the distribution of the phases is relatively disordered, and random, with an insular organization largely like Widmannstätten-type structures similar to those reported by [23,24] for cast titanium alloys (Figure 3 low-magnification overview of sample A10).

Figure 3. Microstructure of sample A10 at small magnification.

The specific peaks for the β phase are difficult to identify because of the overlapping of α-(002)/β-(110), α-(102)/β-(200), α-(103)/β-(211), and α-(004)/β-(220). In the current configuration of the experimental setup, a quantitative determination is impossible, and because of this aspect, the Rietveld analysis failed to converge or provided erroneous results.

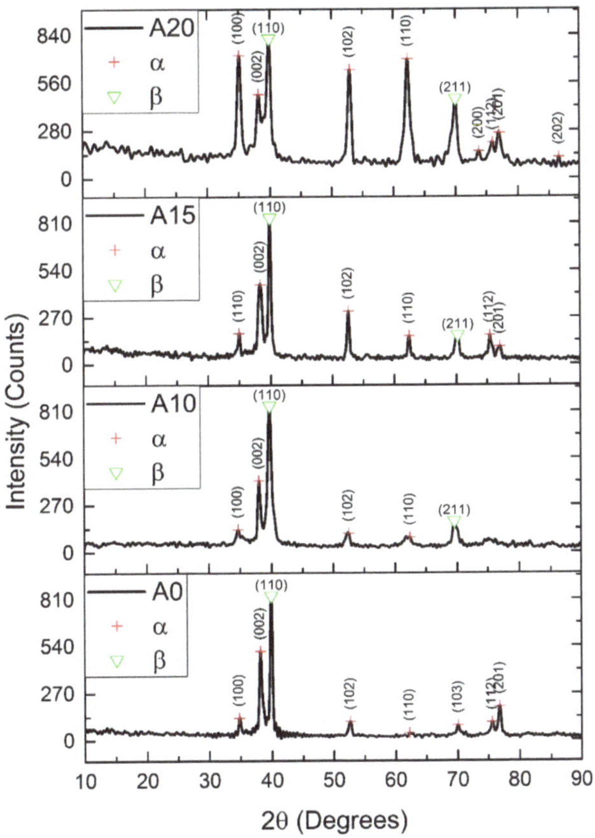

Figure 4. XRD patterns of experimental titanium alloys.

An interesting aspect that could indicate the increase in the β phase proportion with tantalum content resides in the intensity increase for the peak for β-(211) positioned at around 69.95°: as tantalum content increases, so does its intensity, as can be observed in Figure 4. A similar behavior is observed for the peaks present at around 53.03° that correspond to α-(102), they are in the proximity of those for β-(200), and a combined effect can be noticed as well. Regarding the mechanical behavior of the beta titanium alloys studied, the Vickers hardness and the modulus of elasticity were of interest for this study. To measure the Vickers hardness, cylindrical samples were used, according to ISO 6507-1 [22]. The measurements were performed at five points on each sample, and the obtained values were averaged. Figure 5 shows the dependence of the mean Vickers hardness on the tantalum concentration in the alloy.

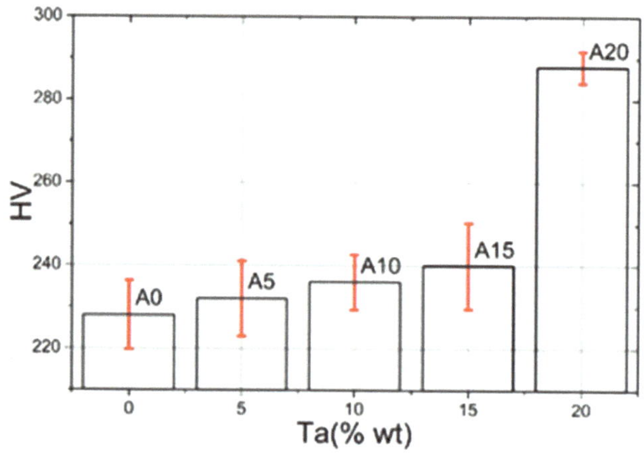

Figure 5. Vickers hardness dependence on tantalum concentration in alloys.

As the tantalum concentration increases from zero to 15% Ta, a slight increase in Vickers hardness can be observed, in the range of 230–240 Vickers units, followed by a sudden increase in hardness up to 290 units when concentration is 20% by the weight of Ta.

The evaluation of the elasticity moduli of the alloys was based on the results obtained during the tensile tests and presented in Supplementary Materials (Figure S1 and Table S1). The results highlighted that the modulus of elasticity decreases with increasing tantalum content from 100 GPa for alloy A0 to 82 GPa for A10 and 55 GPa for the A20 alloy, which is in full correlation with the evolution of the microstructure from a balanced biphasic beta to a monophasic around 90% β phase [13]. The obtained values are in good agreement with those reported in the specialized literature for the alloys with a much higher content of niobium (35%) [25,26]. These findings are a plus of this research, the elaboration of an alloy with a modulus of elasticity close to that of human bone and with a lower content of niobium, considering the scientific controversies surrounding this alloying element.

3.2. *Electrochemical Behavior of Alloys*

The electrochemical behavior of the experimentally developed alloys, in a 3% NaCl solution, was studied by monitoring the stationary potential (OCP) for 168 h at a temperature of 37.5 °C, i.e., in a steady state without external polarization, completed with electrochemical impedance measurements (EIS) at the OCP. Corrosion rates were determined by the Tafel slope method in a potential range of ± 250 mV around the corrosion potential. The time evolution of the OCP in the 3% NaCl solution is presented in Figure 6, from which it can be seen that all alloys initially have negative values of the OCP that move slowly, with less than 2 mV/h, toward positive values, which indicates the tendency of passivation in the presence of water. All alloys with tantalum content (A10, A15, A20 samples) have more negative values (with about 150 mV) compared to the alloy without tantalum A0, and this tendency is maintained throughout immersion.

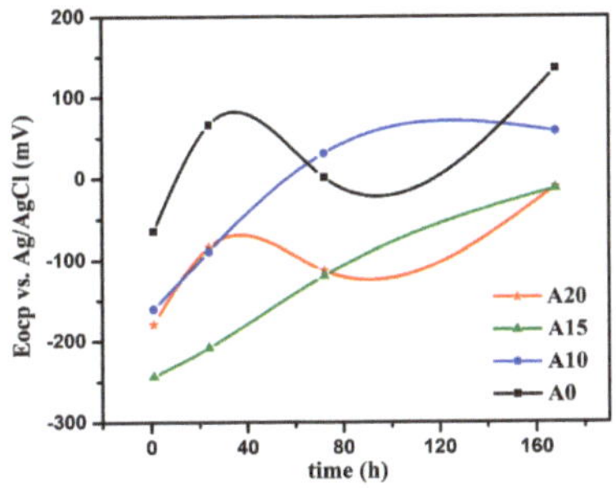

Figure 6. Time evolution of the OCP of the experimental alloys immersed in 3% NaCl solution at a temperature of 37.5 °C for 168 h.

As Figure 6 illustrates, a similar behavior is observed for the A10 and A15 samples, with a tendency to reach a steady state after 168 h, while the A0 and A20 samples show a different response. After 70 h of immersion, for the alloys with 15% and 20% Ta, the stationary potential values are practically identical in spite of dissimilar behavior suggesting that on the A20 sample, the surface of the passive film is an active state. According to literature report [8], this action is due to the preferential adsorption of Cl^- ions on the passive film grown on the surface during immersion. At the end of the period of immersion, only the alloy with 10% Ta (A10) seems to have reached the steady state; all the others show a relatively monotonously increasing variation, which may mean a consolidation of the passive state of the surface.

Microscopic analysis of the surface after 168 h of immersion in the 3% NaCl solution did not reveal any localized corrosion, unlike other types of titanium alloys, for example, TiNi which, according to [27], in a 0.9% NaCl solution at 25 °C showed the tendency of depassivation and pitting corrosion, one of the most common forms of corrosion that appear on orthodontic materials [28]. On the other hand, all the potentials have more electropositive values compared to those of other alloys based on titanium TiNi, Ti6Al4V, or even titanium CP [29]. This behavior can be attributed to the spontaneous passivation in the air of the alloys but also to the contribution of silver, the potential of the standard Ag electrode being known to be very electropositive (E_{Ag0} = +0.799 V) according to [30]. Regarding the electrochemical parameters extracted from the Tafel curves (Figure 7) systematized in Table 2, it is observed that after 168 h of immersion in the 3% NaCl solution, the equilibrium potentials of all the alloys are much more electropositive than the corrosion potentials, which means that the alloys are spontaneously in a state of passivity.

Table 2. Electrochemical parameters after 168 h of immersion in 3% NaCl solution and artificial saliva at temperature of 37.5 °C (Tafel curves).

Specimen	3% NaCl				Artificial Saliva			
	E_{corr}, mV	i_{cor}, μA cm^{-2}	r_{cor}, μm y^{-1}	E_{oc}, mV	E_{corr}, mV	i_{cor}, μA cm^{-2}	r_{cor}, μm y^{-1}	E_{oc}, mV
A0	14.3	0.027	0.9	135	11.5	0.011	0.41	−10.5
A10	−69.8	0.13	5	57	0.002	0.09	4.6	0.002
A15	−90.3	0.5	18	−12	0.008	0.15	17.2	0.008
A20	−72.8	1.1	47	−10	0.011	42.25	34	0.011

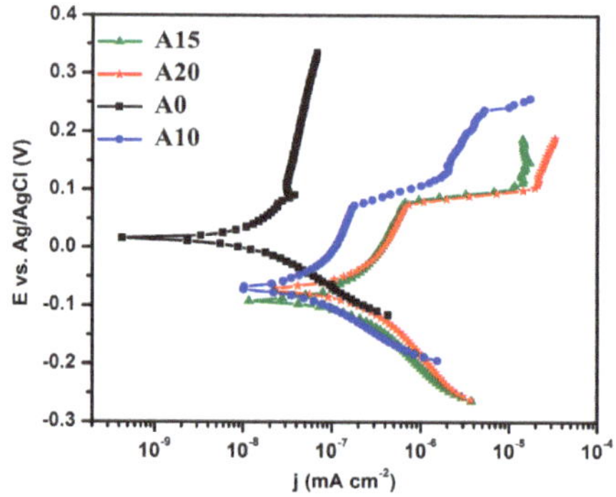

Figure 7. Tafel curves of tested alloys after 168 h of exposure in 3% NaCl solution at temperature of 37.5 °C.

Also, it can be observed that the addition of Ta moves the corrosion potential toward more electronegative values, a fact reported by other authors [31]. Regarding the corrosion rates in the NaCl environment with increased aggressiveness and due to the temperature of 37.5 °C, they increase as the tantalum content in the alloy increases but remain at lower values than those reported by [31] for the alloys commonly used in orthopedic surgery, Ti6Al4V. Thus, SK Yen [32] reported corrosion current densities greater than 45.93 μA cm^{-2} in a 0.6 M NaCl solution at room temperature, while the A 20 alloy with the smallest modulus of elasticity showed i_{cor} = 1.1 μA cm^{-2} at a temperature of 37.5 °C. The superior corrosion behavior of these alloys can also be attributed to the beneficial influence of silver, explained in the literature also by catalyzing the oxygen reduction reaction and favoring passivation [33]. Concerning the corrosion rates, although they increase with the increase in the concentration of tantalum in the alloy, they remain at relatively low values between 5 μm y^{-1} and 47 μm y^{-1}. Moreover, the preliminary research carried out on these materials in artificial saliva highlights a relatively similar influence of tantalum on the electrochemical parameters (Table 2).

The electrochemical behavior of the alloys discussed above was also confirmed by the EIS tests whose response is presented in the form of Nyquist and Bode curves (Figure 8). Figure 8a corresponding to Nyquist plots discloses the decrease in polarization resistance with the increase in tantalum content in the alloy, simultaneously with the increase in resistive behavior. Samples A0, A10, and A15 showed similar behavior, while for the A20 sample, a significant difference is evidenced (Figure 8a inset). However, it is important to notice that the A0, A10, and A15 samples present higher impedance related to good corrosion resistance, while for the A20 sample with a high content of tantalum, more susceptibility to corrosion in a NaCl solution was observed. From Bode plots (Figure 8b), the presence of two-time constants can be seen, one at high frequencies due to the non-compensation of the ohmic drop in the electrolyte according to [34], and in the low-frequency range, the second time constant appears because of the reaction or diffusion of the species, especially of the chlorine ions, through the oxide film formed. This behavior is much more obvious at the A15 and A20 alloys, while the A10 alloy, in which the phase angle at low frequencies (Figure 8b inset) remains relatively constant at a value close to 80°, showing a pseudocapacitive character of surface passivation [35,36]. Instead, the lower phase angle of the A20 sample reveals a defective capacitor, a less protective layer that inherently decreases the corrosion resistance of the alloy with a high content of tantalum.

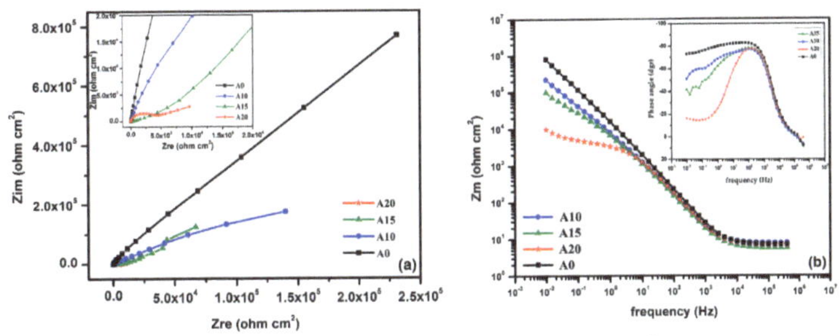

(a) (b)

Figure 8. EIS plots obtained after 168 h of immersion in 3% NaCl solution at 37.5 °C on all samples: (**a**) Nyquist plots (inset: magnification of plots at high frequency); (**b**) Bode plots (inset: phase angle).

In order to go deeper, the EIS spectra of all samples were fitted using an electric equivalent circuit model (EEC) illustrated in Figure 9. The equivalent circuit herein used consists of R_s, R_{ox}, R_{ct}, Q_1, and Q_2 where R_s represents the resistance of the electrolyte, the system R_{ox} and Q_1 simulates the processes that take place through the oxide film, and R_{ct} simulates the charge transfer resistance, respectively, a constant-phase element Q_2 instead of a capacitance that describes the capacitive characteristic of the oxide layer. Since the oxide layers formed on Ti-based alloys regularly exhibit a duplex structure that resides in a barrier-type thin compact inner one and a porous external one [37], the Q elements, which describe the deviation of the electrochemical reaction from the ideal capacitive behavior, are more suited for better fitting the EIS spectra.

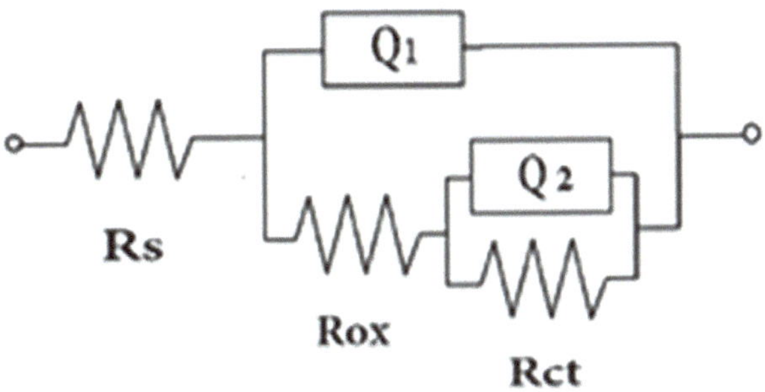

Figure 9. Equivalent electrical circuit model.

The results of fitting the experimental data with the proposed EEC are shown in Table 3. As a general remark, the higher R_{ox} or R_{ct} values are an indicator of better surface protection [38]. From the parameters presented in Table 3, one may estimate that the R_{ox} and R_{ct} of the A10 sample (i.e., 9.5 KΩ cm^2 and 258.2 KΩ cm^2) are superior to those of the A15 (i.e., 8.1 KΩ cm^2 and 82.1 KΩ cm^2) and A20 samples (i.e., 2.2 KΩ cm^2 and 5.04 KΩ cm^2), respectively. On the other hand, the alpha-exponent of the constant-phase element for all samples has values of about 0.9, which indicates a behavior close to an ideal capacity of the film formed on titanium alloys [34] and confirms the tendency of titanium alloys to passivate [38].

Table 3. The obtained EIS parameters from fitting experimental data on the proposed equivalent electrical circuit (Figure 9).

Specimen	R_s (Ω cm^2)	R_{ox} (KΩ cm^2)	Q_1 (Ω^{-1} cm^{-2} s^n)	n_1	R_{ct} (KΩ cm^2)	Q_2 (Ω^{-1} cm^{-2} s^n)	n_2	Chi
A0	3.09	37	6.6×10^{-6}	0.62	1022	11.5×10^{-6}	0.9	0.4
A10	3.4	9.5	2.68×10^{-6}	0.72	258.2	12.68×10^{-6}	0.88	0.335
A15	2.8	8.1	13.37×10^{-6}	0.69	82.1	97.03×10^{-6}	0.89	0.538
A20	4.6	2.2	861×10^{-6}	0.60	5.04	821.9×10^{-6}	0.92	0.439

These results suggest a better corrosion resistance of the sample with a lower content of tantalum in chemical compositions which is in line with literature reports [4,18].

It is known that an aggressive environment with a high concentration of chloride ions (Cl^-) furthers the corrosion processes, leading to a decrease in the corrosion resistance of the alloys. The corrosion parameters estimated for the alloys presented in this study highlighted good resistance corrosion after 168 h of immersion in a 3% NaCl solution, placing these alloys in the very stable (A10 sample) and stable (A15 and A20 samples) corrosion resistance classes according to [39]. This good resistance corrosion of TixTa9Zr8Nb2Ag alloys might also occur due to a large content of the β phase in the alloy structure evidenced by the XRD results, which are in agreement with literature report [40]. Moreover, from EIS results, which evidenced higher R_{ct} and R_{ox} for the A10 sample and a lower corrosion rate (<10 µm y^{-1}), one may state that the Ti10Ta9Zr8Nb2Ag alloy provides better corrosion resistance in a 3% NaCl solution. Furthermore, a low value of the elasticity modulus obtained for the Ti10Ta9Zr8Nb2Ag alloy (A10 sample) close to that of human bone makes it suitable for surgical application.

In summary, the results presented in this study highlighted the influence of the tantalum content in the chemical composition of the alloy on the mechanical properties, as well as on corrosion resistance, and point out that the Ti10Ta9Zr8Nb2Ag alloy might be successfully used for medical implants.

4. Conclusions

The alloying of the Ti9Zr8Nb2Ag alloy with tantalum significantly changes its structure in the cast state, in the sense of increasing the proportion of the beta phase.

The alloy with 20% tantalum contains around 90% of the β phase and can be considered monophasic.

Structure changing produces a change in the mechanical and electrochemical properties of alloys.

After adding 10% tantalum to the alloy, the modulus of elasticity decreased by 18% from 100 GPa to 83 GPa, and at a content of 20% Ta, it reached the value of 55 GPa, closer to the modulus of human bone compared with most used alloys such as CoCr or Ti-6Al-4V.

Vickers hardness increases with the tantalum concentration increasing in alloys.

Tantalum shifts the stationary potential in a 3% NaCl solution at 37.5 °C toward more electronegative values by approximately 150–180 mV.

Alloys with a Ta content between 10 and 20% show, in a 3% NaCl solution, stationary potentials that are more electropositive than corrosion potentials, which highlights the passivation tendency.

The corrosion rates increase with the increase in the concentration of tantalum in the alloy, but they remain at relatively low values between 5 µm y^{-1} and 47 µm y^{-1}.

Supplementary Materials: The following supporting information can be downloaded at https://www.mdpi.com/article/10.3390/met13071294/s1, Figure S1: Stress—strain curves for a selection of test samples. On the chart plots showing the linear—elastic region of elastic moduli of 100 GPa, 50 GPa and 20 GPa are presented; Table S1: Strength parameters for the experimental alloys.

Author Contributions: G.D.—Methodology, Investigation, and Formal Analysis; A.B.—Writing Original Draft, Writing–Review and Editing, Supervision; C.D.—Investigation and Formal Analysis; M.M.—Conceptualization, Writing–Review. All authors have read and agreed to the published version of the manuscript.

Funding: This research was funded by the European Social Fund from the Sectoral Operational Program Human Capital 2014–2020, through the Financial Agreement with the title "Training of PhD students and postdoctoral researchers in order to acquire applied research skills—SMART" grant number 13530/16.06.2022—SMIS code: 153734.

Data Availability Statement: No new data were created or analyzed in this study. Data sharing does not apply to this article.

Conflicts of Interest: The authors declare that they have no known competing financial interests or personal relationships that could have appeared to influence the work reported in this paper.

References

1. Davis, J.R. *Handbook of Materials for Medical Devices, Overview of Biomaterials and Their Use in Medical Devices*; Davis, J.R., Ed.; ASM International: Detroit, MI, USA, 2003.
2. Banu, A.; Radovici, O.; Marcu, M. The Alloying Influence on Corrosion Behaviour of Chromium Surgical Alloys. *Rev. Roum. Chim.* **2008**, *53*, 947–953.
3. Ijaz, M.F.; Vasilescu, C.; Drob, S.I.; Osiceanu, P.; Marcu, M.; Kim, H.Y.; Miyazaki, S. Electrochemical characterization of the superelastic (Ti-Zr)-Mo-Sn biomedical alloy displaying a large recovery strain. *Mater. Corros.* **2017**, *68*, 1220–1227. [CrossRef]
4. Li, Z.; Lai, W.; Wang, B.; Tong, X.; You, D.; Li, W.; Wang, X. A novel $Ti_{42.5}Zr_{42.5}Nb_5Ta_{10}$ multi-principal element alloy with excellent properties for biomedical application. *Intermetallics* **2022**, *151*, 107731. [CrossRef]
5. Banu, A.; Marcu, M.; Juganaru, C.; Osiceanu, P.; Anastasescu, M.; Capra, L. Corrosion behavior of CoCrMoW cast alloy in lactic environment. *Arab. J. Chem.* **2019**, *12*, 2007–2016. [CrossRef]
6. Kaur, M.; Singh, K. Review on titanium and titanium-based alloys as biomaterials for orthopaedic applications. *Mater. Sci. Eng. C* **2019**, *102*, 844–862. [CrossRef]
7. Yang, W.; Pang, S.; Liu, Y.; Wang, Q.; Liaw, P.K.; Zhang, T. Design and properties of novel Ti–Zr–Hf–Nb–Ta high-entropy alloys for biomedical applications. *Intermetallics* **2022**, *141*, 107421. [CrossRef]
8. Cui, Y.-W.; Chen, L.-Y.; Chu, Y.-H.; Zhang, L.; Li, R.; Lu, S.; Wang, L.; Zhang, L.-C. Metastable pitting corrosion behavior and characteristics of passive film of laser powder bed fusion produced Ti–6Al–4V in NaCl solutions with different concentrations. *Corros. Sci.* **2023**, *215*, 111017. [CrossRef]
9. Guo, L.; Naghavi, S.A.; Wang, Z.; Varma, S.N.; Han, Z.; Yao, Z.; Wang, L.; Wang, L.; Liu, C. On the design evolution of hip implants: A review. *Mater. Des.* **2022**, *216*, 110552. [CrossRef]
10. Thampi, A.; Ramanathan, S. Corrosion behavior of anodized Ti-Ta binary surface alloys in various physiological fluids for implant applications. *Corros. Sci.* **2023**, *219*, 111233. [CrossRef]
11. Hua, N.; Wang, W.; Wang, Q.; Ye, Y.; Lin, S.; Zhang, L.; Guo, Q.; Brechtl, J.; Liaw, P.K. Mechanical, corrosion, and wear properties of biomedical Ti-Zr-Nb-Ta-Mo high entropy alloys. *J. Alloys Compd.* **2021**, *861*, 157997. [CrossRef]
12. Banu, A.; Preda, L.; Marcu, M.; Dinca, L.L.; Maxim, M.E.; Dobri, G. Electrochemical behavior of SLM Ti-6Al-4V alloy after long time of immersion in lactic acid environment. *Met. Mater. Trans. A* **2022**, *53*, 2060–2070. [CrossRef]
13. Biesiekierski, A.; Wang, J.; Abdel-Hady Gepreel, M.; Wen, C. A new look at biomedical Ti-based shape memory alloys. *Acta Biomater.* **2012**, *8*, 1661–1669. [CrossRef] [PubMed]
14. Zareidoost, A.; Yousefpour, M. A study on the mechanical properties and corrosion behavior of the new ascast TZNT alloys for biomedical applications. *Mater. Sci. Eng. C* **2020**, *110*, 110725. [CrossRef] [PubMed]
15. Wong, K.-K.; Hsu, H.-C.; Wu, S.-C.; Hung, T.-L.; Ho, W.-F. Structure, Properties, and Corrosion Behavior of Ti-Rich TiZrNbTa Medium-Entropy Alloys with $\beta+\alpha''+\alpha'$ for Biomedical Application. *Materials* **2022**, *15*, 7953. [CrossRef] [PubMed]
16. Rossi, M.C.; Ventura, B.N.; Milián, L.; Escuder, A.V.; Borrás, V.A. Study of Electrochemical and Biological Characteristics of As-Cast Ti-Nb-Zr-Ta System Based on Its Microstructure. *Metals* **2022**, *12*, 476. [CrossRef]
17. Feng, J.; Wei, D.; Zhang, P.; Yu, Z.; Lui, C.; Lu, W.; Wang, K.; Yan, H.; Zhang, L.; Wang, L. Preparation of TiNbTaZrMo high-entropy alloy with tunable Young's modulus by selective laser melting. *J. Manuf. Process.* **2023**, *85*, 160–165. [CrossRef]
18. Vasilescu, C.; Drob, S.I.; Osiceanu, P.; Calderon Moreno, J.M.; Prodana, M.; Ionita, D.; Demetrescu, I.; Marcu, M.; Popovici, I.A.; Vasilescu, E. Microstructure, Surface characterization and electrochemical behavior of new Ti-Zr-Ta-Ag alloy in simulated human electrolyte. *Metall. Mater. Trans A* **2017**, *48*, 513–523. [CrossRef]
19. Gurel, S.; Yagci, M.B.; Bal, B.; Canadinc, D. Corrosion behavior of novel Titanium-based high entropy alloys designed for medical implants. *Mater. Chem. Phys.* **2020**, *254*, 123377. [CrossRef]
20. Chen, Y.H.; Chuang, W.S.; Huanga, J.C.; Wang, X.; Chou, H.S.; Lai, Y.J.; Lin, P.H. On the bio-corrosion and biocompatibility of TiTaNb medium entropy alloy films. *Appl. Surf. Sci.* **2020**, *508*, 145307. [CrossRef]
21. *ISO 6892-1*; Metallic materials—Tensile Testing—Part 1: Method of Test at Room Temperature. ISO: Geneva, Switzerland, 2019.

22. *ISO 6507-1*; Metallic Materials—Vickers Hardness Test—Part 1: Test Method. ISO: Geneva, Switzerland, 2018.
23. Dobri, G.; Berbecaru, A.; Paraschiv, A.; Geanta, V.; Ciuca, S.; Banu, A. The effect of tantalum content on microstructure and vickers hardness of TiNbZrTaAg alloy. *U.P.B. Sci. Bull. Ser. B* **2023**, *85*, 167–178.
24. Preisler, D.; Strasky, J.; Harcuba, P.; Kosutova, T.; Dopita, M.; Janovska, M.; Hajek, M.; Janecek, M. Phase transformations in Ti-Nb-Zr-Ta-O β titanium alloys with high oxygen and reduced Nb and Ta content. *Acta Crystallogr. A* **2021**, *77*, C1251. [CrossRef]
25. Yang, Y.; Songxiao, H.; Wenjun, Y.; Baiking, X. Mechanical properties and microstructure of an α + β titanium alloy with high strength and fracture toughness. *Rare Met.* **2009**, *28*, 346. [CrossRef]
26. Zhong, C.; Liu, J.; Zhao, T.; Schopphoven, T.; Fu, J.; Gasser, A.; Schleifenbaum, J.H. Laser Metal Deposition of Ti6Al4V—A Brief Review. *Appl. Sci.* **2020**, *10*, 764. [CrossRef]
27. Yang, W.; Pang, S.-J.; Wang, G.; Liu, Y.; Liaw, P.K.; Zhang, T. Ti–Zr–Hf–Nb–Ta–Sn high-entropy alloys with good properties as potential biomaterials. *Rare Met.* **2022**, *41*, 2305–2315. [CrossRef]
28. Hansen, A.W.; Fuhr, L.T.; Antonini, L.M.; Villarinho, D.J.; Marino, C.E.B.; Malfatti, C.D.F. The Electrochemical Behavior of the NiTi Alloy in Different Simulated Body Fluids. *Mat. Res.* **2015**, *18*, 184–190. [CrossRef]
29. Vasilescu, C.; Osiceanu, P.; Calderon Moreno, J.M.; Drob, S.I.; Preda, S.; Popa, M.; Dan, I.; Marcu, M.; Prodana, M.; Popovici, I.A.; et al. Microstructure, surface characterization and long-term stability of new quaternary Ti-Zr-Ta-Ag alloy for implant use. *Mater. Sci. Eng. C* **2017**, *71*, 322–334. [CrossRef]
30. Sargeant, A.; Goswami, T. Hip implants: Paper V. Physiological effects. *Mater. Des.* **2006**, *27*, 287–307. [CrossRef]
31. Sherif, E.-S.M.; Bahri, Y.A.; Alharbi, H.F.; Ijaz, M.F.; Alnaser, I.A. Influence of Tantalum Addition on the Corrosion Passivation of Titanium-Zirconium Alloy in Simulated Body Fluid. *Materials* **2022**, *15*, 8812. [CrossRef]
32. Yen, S.K.; Huang, T.Y. Characterization of the electrolytic ZrO_2 coating on Ti-6A1-4V. *Mater. Chem. Phys.* **1998**, *56*, 214–221. [CrossRef]
33. Oh, K.-T.; Joo, U.-H.; Park, G.-H.; Hwang, C.-J.; Kim, K.-N. Effect of Silver Addition on the Properties of Nickel–Titanium Alloys for Dental Application. *J. Biomed. Mater. Res. B Appl. Biomater.* **2006**, *76*, 306–314. [CrossRef]
34. Tamilselvi, S.; Murugaraj, S.R.; Rajendran, N. Electrochemical impedance spectroscopic studies of titanium and its alloys in saline medium. *Mater. Corros.* **2007**, *58*, 113–120. [CrossRef]
35. Preda, L.; Negrila, C.; Lazarescu, M.F.; Enache, M.; Anastasescu, M.; Toader, A.M.; Ionescu, S.; Lazarescu, V. Ga and As Competition for Thiolate Formation at p-GaAs(111) Surfaces. *Electrochim. Acta* **2013**, *104*, 1–11. [CrossRef]
36. Chelariu, R.; Bolat, G.; Izquierdo, J.; Mareci, D.; Gordin, D.M.; Gloriant, T.; Souto, R.M. Metastable beta Ti-Nb-Mo alloys with improved corrosion resistance in saline solution. *Electrochim. Acta* **2014**, *137*, 280–289. [CrossRef]
37. Xie, F.X.; He, X.B.; Cao, S.L.; Lu, X.; Qu, X.H. Structural characterization and electrochemical behavior of a laser-sintered porous Ti-10Mo alloy. *Corros. Sci.* **2013**, *67*, 217–224. [CrossRef]
38. Assis, S.I.; Wolynek, S.; Costa, I. Corrosion characterization of titanium alloys by electrochemical techniques. *Electrochim. Acta* **2006**, *51*, 1815–1819. [CrossRef]
39. *ISO 8044*; Corrosion of Metals and Alloys—Basic Terms and Definitions. ISO: Geneva, Switzerland, 2020.
40. Yang, J.; Yang, H.; Yu, H.; Wang, Z.; Zeng, X.A. Corrosion Behavior of Additive Manufactured Ti6Al4V Alloy in NaCl Solution. *Metall. Mater. Trans. A* **2017**, *48*, 3583–3593. [CrossRef]

coatings

MDPI

Article

Structural and Thermal Stability of CrZrON Coatings Synthesized via Reactive Magnetron Sputtering

Sung-Min Kim [1] and Sang-Yul Lee [2,*]

[1] Heat and Surface Technology R&D Department, Korea Institute of Industrial Technology (KITECH), Incheon 21999, Republic of Korea; sungminkim@kitech.re.kr

[2] Department of Materials Engineering, Korea Aerospace University, Goyang 10540, Republic of Korea

* Correspondence: sylee@kau.ac.kr

Abstract: This research manuscript investigates the structural and thermal stability of CrZrON coatings synthesized through reactive magnetron sputtering. The coatings were deposited at different temperatures with 120 °C and 400 °C, and with varying oxygen-to-reactive gas ratios in the range of 8.3% to 25.7%. The average chemical composition, crystallographic orientation, microstructure, lattice parameter, crystallite size, and hardness of the coatings were evaluated. The results revealed that the coatings deposited at a lower temperature of 120 °C exhibited a columnar structure, while those deposited at a higher temperature of 400 °C showed a transition towards a featureless or amorphous structure. The lattice parameter and crystallite size were influenced by the deposition temperature and oxygen ratio, indicating the incorporation of oxygen into the coatings. Hardness measurements demonstrated that the coatings' hardness decreased from 33.7 GPa to 28.6 GPa for a process temperature of 120 °C and from 32.1 GPa to 25.7 GPa for 400 °C with an increase in the oxygen ratio, primarily due to the formation of oxygen-rich compounds or oxides. Additionally, annealing experiments indicated that the coatings with featureless or amorphous structures exhibited improved thermal stability, as they maintained their structural integrity without delamination even at high annealing temperatures.

Keywords: CrZrON; oxynitride; hardness; thermal stability

Citation: Kim, S.-M.; Lee, S.-Y. Structural and Thermal Stability of CrZrON Coatings Synthesized via Reactive Magnetron Sputtering. *Coatings* **2023**, *13*, 1254. https://doi.org/10.3390/coatings13071254

Academic Editor: Sergey N. Grigoriev

Received: 19 June 2023
Revised: 11 July 2023
Accepted: 13 July 2023
Published: 16 July 2023

1. Introduction

CrN coatings have been widely used in the tool and forming industries due to their excellent mechanical properties and wear behaviors [1–3]. However, further research has been devoted to the development of ternary systems to meet the harsh criteria of a fast-changing industry. Especially in an attempt to improve the hardness, friction resistance, and oxidation-resistance of CrN, the incorporation of elements such as Zr, Al, or Si into a Cr-N coating has been actively introduced [4–14].

Among the Cr-based ternary transition-metal nitrides, CrZrN coatings provide quite unique characteristics relative to CrN [4,6,8,9]. Kim et al. [4] reported that the maximum hardness of CrZrN with an increase of up to 12.1 at.% of Zr content reached 34 GPa, as well as the tribological properties from the ball-on-disc wear test, which revealed no film failure compared to CrN. Feng et al. [8] also found that the content ratio of Zr to Cr played an important role in enhancing mechanical and tribological properties. They proposed that the increased hardness of Cr-Zr-N originated from solid solution hardening and the lattice distortion in the film. However, as reported in our previous studies, the thermal stability of the CrZrN coating significantly deteriorates at elevated temperatures up to 500 °C, resulting in a hardness and friction coefficient almost identical to those of Cr-N [6]. This phenomenon iss mostly due to the fact that the CrZrN coating was oxidized primarily by the inward transport of oxygen, along with the outward diffusion of Cr and Zr to a small extent [15]. To overcome the inadequate high-temperature applications of CrZrN coatings, structural modification is necessary to ensure the blockage of the oxygen pathway in CrZrN coatings. One of the strategic candidates for enhancing thermal stability is modulating structure by adding an active oxygen element, i.e., the formation of oxynitride. It has been well established that oxynitride coatings have a beneficial effect on thermal stability since the formation of a protective oxide layer on the surface impedes the inward diffusion of oxygen [16,17]. Nevertheless, the oxidation resistance and thermal stability of CrZr oxynitride coatings have never been investigated in depth until now.

As the thermal stability of the CrZrN coatings is limited to less than 500 °C, aim of this study is to investigate the structural evolution and thermal stability of CrZrON coatings as a function of the oxygen-to-nitrogen flow ratio and deposition temperature during the sputtering process. The development of thermally stable CrZrON coatings by the incorporation of oxygen into CrZrN would become valuable insights for the advancement of scientific knowledge and practical applications in industries such as tools and forming.

2. Materials and Experimental Procedures

2.1. Deposition Conditions

CrZrON coatings were deposited on two types of substrates, Si (100) wafers, using a reactive magnetron sputtering system. A segment target composed of Cr and Zr (volume fraction 1:1) was used as a source material [18]. Prior to deposition, the base pressure of the sputtering chamber was pumped down to less than 2.6×10^{-3} Pa, and pre-sputtering was carried out to clean the substrate surface at the Ar pressure of 0.4 Pa with a pulsed DC power of 0.5 kW. Subsequently, the CrZrON coatings were synthesized to be ca. 2 µm thick using a pulsed DC power of 0.7 kW (frequency: 25 kHz, duty ratio: 70%). The working pressure with the mixture of Ar of 0.4 Pa and $N_2 + O_2$ of 0.16 Pa gas was fixed at a total pressure of 0.56 Pa. To modulate the structure of coatings, the ratio of $O_2/(O_2 + N_2)$ gas pressure was varied to be 8.3%, 16.7%, and 25%, and the deposition temperature was controlled to be 120 °C and 400 °C, respectively. During deposition, the bias voltage and rotation speed of the substrate were maintained at −100 V and 10 rpm, respectively.

2.2. Characterization of Coatings

Chemical compositions of the coatings were acquired using energy-dispersive X-ray spectroscopy (EDX) installed on scanning electron microscopy (SEM, Tescan/VEGA II LMU, Brno, Czech Republic). The crystalline structure and phase were characterized via X-ray diffraction (XRD, Rigaku/Ultima IV, Tokyo, Japan) with Cu Kα radiation (λ = 0.15418 nm). Based on XRD results, Scherrer equation was used to calculate the crystallite size as follows [19]:

$$D = \frac{0.9\,\lambda}{\mathrm{d}\,\cos\theta} \tag{1}$$

where D is crystallite size, λ is wavelength (λ = 0.15418 nm), d is FWHM (full width at half maximum intensity of the peak in radians), and θ is Bragg's diffraction angle. In addition, the lattice parameters were calculated from the interplanar spacing using the Miller indices of the crystal planes.

Hardness was measured via a microhardness-testing system (Fischerscope H100C, Sindelfingen, Germany) with a Poisson ratio of 0.3, a load of 25 mN, and 10 s of loading time by measuring a depth less than 0.2 µm to avoid a possible substrate effect. For the reliability of the hardness value, the hardness of films was measured at least 10 times.

For evaluating the thermal stability of the coatings, the annealing was performed at a temperature of 500 °C and 600 °C in air for an hour, and then they were cooled down in a furnace.

3. Results and Discussion

The CrZrON coatings were synthesized via the reactive magnetron sputtering process, which employed the simultaneous sputter of a segment target composed of Cr and Zr (vol. ratio of 1:1) in the presence of O_2 and N_2 reactive gases. Using this strategy, the average chemical composition of CrZrON coatings was measured via a SEM-equipped EDX, as summarized in Table 1 and Figure 1. As shown in Figure 1a, the ratio of metallic to non-metallic elements was calculated as a function of $O_2/(O_2 + N_2)$ gas flow. The (Cr + Zr)/(O + N) ratio was calculated to be <1 for all quaternary CrZrON coatings, which suggests that the metal is under stoichiometry. With increasing the O_2 gas with a deposition temperature of 120 °C, the (Cr + Zr)/(O + N) ratio tended to decrease in the range of 0.75 to 0.62, which is lower relative to the ratio range of 0.9 to 0.7 with the deposition temperature of 400 °C. Typically, the ratio of metallic to non-metallic elements for pure oxide or high oxygen content is approximately 0.66, implying the stoichiometric formation of $(Cr_xZr_{1-x})_2(O_{1-y}N_y)_3$ (see dashed line in Figure 1a) [20,21]. Although the CrZrON coatings synthesized with a deposition temperature of 120 °C seem to have an oxygen-rich composition, as evidenced by a (Cr + Zr)/(O + N) ratio similar to 0.66 with an increase in the O_2 gas ratio, it was found that the oxygen content and the ratio of oxygen to nitrogen in coatings are much lower than those at a deposition temperature of 400 °C (see Table 1 and Figure 1b). Furthermore, the Cr/Zr content ratio of the coatings deposited at 400 °C (*ca.* 1.1) decreased compared to deposition at 120 °C (*ca.* 2.1). This result can be explained by the fact that the reactivity of O_2 gas with metallic species, especially Zr, among possible reactions in the chamber, was accelerated with increasing deposition temperature. Therefore, the CrZrON coatings synthesized at low deposition temperatures consist of ternary CrZr nitride rather than CrZr oxynitride or CrZr oxide, whereas high deposition temperatures give rise to the formation of oxygen-rich compounds. Such phase formation in CrZrON coatings is comprehensively discussed in the XRD analysis.

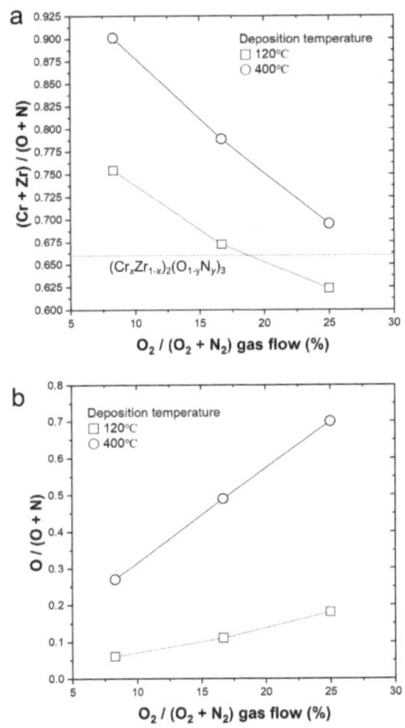

Figure 1. (**a**) (Cr + Zr)/(O + N) ratio and (**b**) O/(O + N) of CrZrON coatings corresponding to (**a**) (Cr + Zr)/(O + N) ratio and (**b**) O/(O + N) ratio as a function of $O_2/(O_2 + N_2)$ gas flow.

Table 1. Chemical composition of CrZrON coatings measured via EDS.

Deposition Temperature (°C)	$O_2/(O_2 + N_2)$ Gas Flow Ratio (%)	Chemical Composition (at.%)
120	8.3	$Cr_{29.0}\,Zr_{14.0}\,O_{3.4}\,N_{53.6}$
	16.7	$Cr_{27.4}\,Zr_{12.8}\,O_{6.7}\,N_{53.1}$
	25.0	$Cr_{26.5}\,Zr_{11.9}\,O_{11.1}\,N_{50.5}$
400	8.3	$Cr_{24.9}\,Zr_{22.5}\,O_{14.1}\,N_{38.5}$
	16.7	$Cr_{23.3}\,Zr_{20.8}\,O_{27.2}\,N_{28.7}$
	25.0	$Cr_{21.8}\,Zr_{19.3}\,O_{41.2}\,N_{17.9}$

The XRD analysis was performed on CrZrON coatings deposited under different conditions, specifically at deposition temperatures of 120 and 400 °C, as shown in Figure 2. The influence of the $O_2/(O_2 + N_2)$ gas flow ratio on the crystallographic orientation and structure of the films was investigated. At a deposition temperature of 120 °C, the XRD results revealed interesting observations. When the $O_2/(O_2 + N_2)$ ratio was initially set at 8.3%, a strong CrZrN (200) peak was observed, indicating a preferential growth of the CrZrN phase. CrN and ZrN both crystallize in a rock-salt (NaCl) structure and have similar lattice parameters. This similarity allows for the formation of a solid solution between the CrN (200) and ZrN (200) planes, resulting in the CrZrN (200) phase observed in the XRD analysis. However, as the $O_2/(O_2 + N_2)$ ratio increased to 25%, the intensity of the CrZrN (200) peak gradually decreased. Concurrently, the appearance of a strong (104) Cr_2O_3 peak in the XRD pattern suggested the formation of Cr_2O_3 alongside the CrZrN phase. Moreover, the (111) and (200) peaks of CrZrN were accompanied by the emergence of a strong Cr_2O_3 peak, indicating the coexistence of these phases in the deposited coating. In the case of a deposition temperature of 400 °C, the XRD results demonstrated distinct characteristics. Initially, at an $O_2/(O_2 + N_2)$ ratio of 8.3%, a dominant peak corresponding to the CrZrN (200) plane was observed, indicating a preferred crystallographic orientation of the film. However, as the $O_2/(O_2 + N_2)$ ratio increased to 25%, the intensity of the CrZrN (200) peak gradually decreased. Notably, at the 25% $O_2/(O_2 + N_2)$ ratio, the CrZrN peaks were no longer discernible. Additionally, the sharp peaks observed for Cr_2O_3 at lower $O_2/(O_2 + N_2)$ ratios were nearly diminished, and broad peaks were observed instead. These broad peaks suggest a potential transformation towards an amorphous or featureless structure in the coating. The formation of an amorphous structure at relatively high temperatures and high O_2 ratios could be attributed to the thermodynamic and kinetic processes occurring during the deposition of the CrZrON coatings. At high temperatures, atoms or ions possess higher kinetic energy, promoting their mobility and facilitating diffusion. This enhanced diffusion could lead to atomic rearrangements and the breakdown of long-range order, contributing to the formation of an amorphous structure [22–24]. In addition, in the presence of high oxygen content, chemical reactions between oxygen and the material's constituents can occur. These reactions may involve the formation of oxides, oxygen-rich compounds, or chemical reactions with existing crystalline phases. These chemical transformations disrupt the crystalline lattice and favor the formation of an amorphous structure by making the energy required to maintain a crystalline structure unfavorable [23,25].

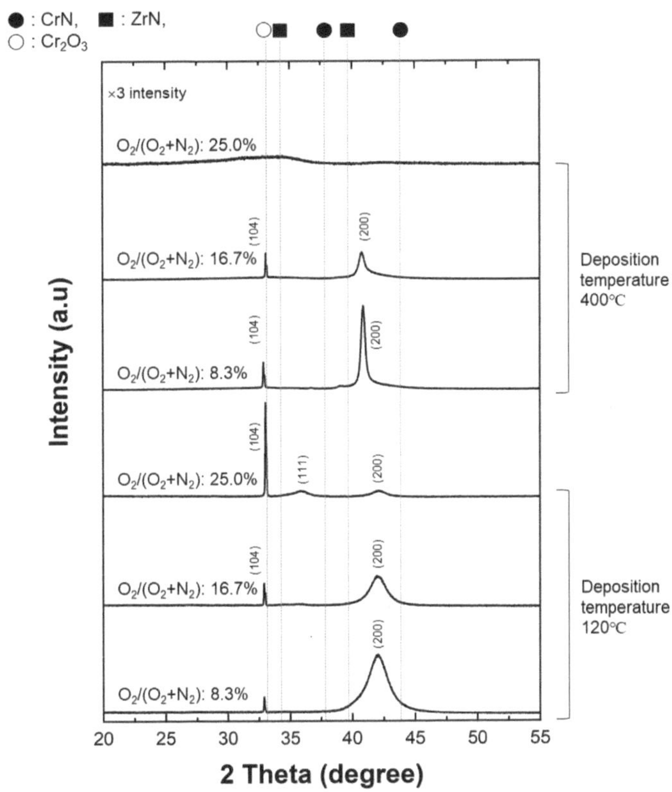

Figure 2. XRD of as-prepared CrZrON coatings as a function of $O_2/(O_2 + N_2)$ gas flow and deposition temperature.

The lattice parameter, crystallite size, and hardness of the CrZrON coatings were evaluated to understand the influence of the $O_2/(O_2 + N_2)$ gas flow ratio and deposition temperature on their structural and mechanical properties, as shown in Figure 3. In terms of the lattice parameter, at a deposition temperature of 120 °C, an increase in the $O_2/(O_2 + N_2)$ ratio from 8.3% to 25% led to a gradual decrease in the lattice parameter from 0.435 nm to 0.429 nm. This reduction can be attributed to the incorporation of oxygen into the coatings, resulting in the formation of oxygen-rich compounds. On the other hand, at a higher deposition temperature of 400 °C, the lattice parameter slightly decreased from 0.442 nm to 0.44 nm as the $O_2/(O_2 + N_2)$ ratio increased from 8.3% to 16.7%. However, at the highest $O_2/(O_2 + N_2)$ ratio of 25.0% and 400 °C, no clear lattice parameter was observed, suggesting the presence of a potentially amorphous or featureless structure. The crystallite size of the coatings exhibited different behaviors with respect to deposition temperature. At 120 °C, an increase in the $O_2/(O_2 + N_2)$ ratio from 8.3% to 25% resulted in a decrease in the crystallite size from 5.9 nm to 5.3 nm. This reduction can be attributed to enhanced nucleation and grain growth inhibition due to the presence of oxygen during deposition. An increase in the $O_2/(O_2 + N_2)$ ratio from 8.3% to 16.7% resulted in a decrease in the crystallite size from 26.2 nm to 15 nm. This decrease can be attributed to the inhibition of grain growth and coalescence due to the higher concentration of oxygen in the coating. However, at the highest $O_2/(O_2 + N_2)$ ratio of 25.0% and 400 °C, no clear crystallite size was observed, indicating the potential transformation towards an amorphous or featureless structure. Regarding hardness, at 120 °C, the hardness values decreased from 33.7 GPa to 28.6 GPa as the $O_2/(O_2 + N_2)$ ratio increased from 8.3% to 25%. This decrease can be attributed to the incorporation of oxygen into the coatings, which is known to reduce the hardness of nitride films. At 400 °C, the hardness values decreased from 32.1 GPa to 25.7 GPa with an increasing $O_2/(O_2 + N_2)$ ratio from 8.3% to 25.0%. This reduction in hardness can be attributed to the combined effects of increased oxygen content. The introduction of oxygen can lead to the formation of oxygen-rich compounds, or oxides, within the coating. These compounds often have a lower hardness compared to the corresponding nitride phases. The presence of softer phases or compounds within the coating matrix can contribute to a decrease in hardness [26,27]. When oxygen is introduced into the crystal lattice, it can disrupt the arrangement of atoms and cause structural changes, leading to the formation of softer phases or compounds. The presence of these softer phases within the coating matrix can contribute to a decrease in hardness [28].

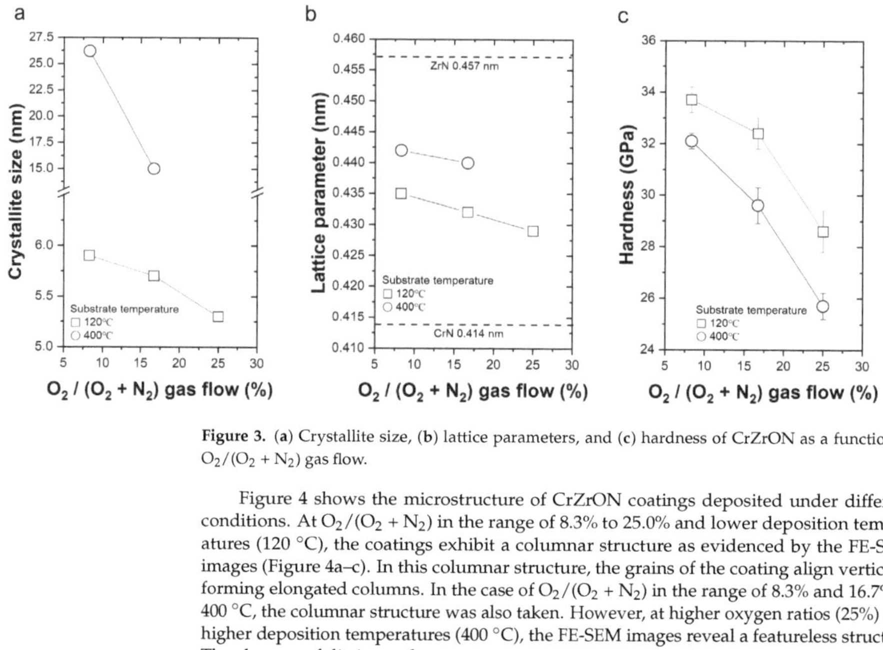

Figure 3. (**a**) Crystallite size, (**b**) lattice parameters, and (**c**) hardness of CrZrON as a function of $O_2/(O_2 + N_2)$ gas flow.

Figure 4 shows the microstructure of CrZrON coatings deposited under different conditions. At $O_2/(O_2 + N_2)$ in the range of 8.3% to 25.0% and lower deposition temperatures (120 °C), the coatings exhibit a columnar structure as evidenced by the FE-SEM images (Figure 4a–c). In this columnar structure, the grains of the coating align vertically, forming elongated columns. In the case of $O_2/(O_2 + N_2)$ in the range of 8.3% and 16.7% at 400 °C, the columnar structure was also taken. However, at higher oxygen ratios (25%) and higher deposition temperatures (400 °C), the FE-SEM images reveal a featureless structure. The absence of distinct columnar structure suggests a more homogeneous and possibly amorphous or nanocrystalline microstructure. The observed transition from a columnar structure to a featureless structure with an increase in oxygen ratio and deposition temperature may be attributed to the influence of oxygen on the growth kinetics and microstructure evolution during deposition. The presence of oxygen can affect the nucleation and growth processes, leading to different crystallographic orientations and grain structures.

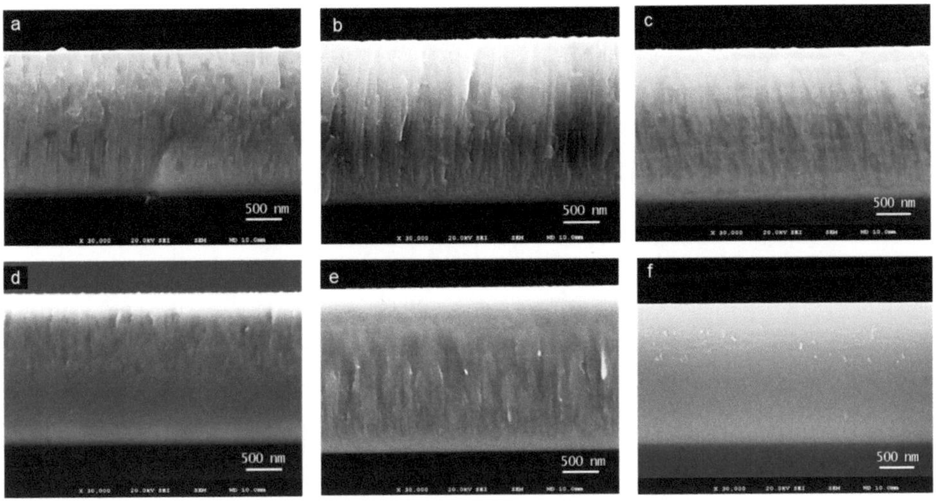

Figure 4. Cross-sectional FE-SEM images of CrZrON coatings corresponding to (**a**) 8.3% of $O_2/(O_2 + N_2)$ at 120 °C, (**b**) 16.7% of $O_2/(O_2 + N_2)$ at 120 °C, (**c**) 25.0% of $O_2/(O_2 + N_2)$ at 120 °C, (**d**) 8.3% of $O_2/(O_2 + N_2)$ at 400 °C, (**e**) 16.7% of $O_2/(O_2 + N_2)$ at 400 °C, and (**f**) 25.0% of $O_2/(O_2 + N_2)$ at 400 °C.

The thermal stability of the CrZrON coatings was assessed by examining their hardness under different annealing temperatures, as shown in Figure 5. For the coatings synthesized at an oxygen ratio of 8.3% and a deposition temperature of 120 °C, a high hardness of 33.7 GPa was observed at room temperature (RT). However, upon annealing at 500 °C, the hardness decreased to 9.8 GPa, indicating a reduction in the coating's strength as a result of thermal relaxation and potential grain growth. At 600 °C, delamination occurred, indicating a loss of structural integrity within the coating. Similarly, for the coatings synthesized at an $O_2/(O_2 + N_2)$ of 25% and a 120 °C deposition temperature, the hardness was 28.6 GPa at RT. After annealing at 500 °C, the hardness further decreased to 12.9 GPa, and subsequently, delamination was observed at 600 °C, indicating a critical temperature beyond which the coating's structure became unstable. In contrast, the coatings deposited at a higher temperature of 400 °C exhibited higher hardness values after annealing compared to those at 120 °C. For the 8.3% oxygen ratio, the hardness was 32.1 GPa at RT. Upon annealing at 500 °C, the hardness decreased to 11.5 GPa, and delamination also occurred at 600 °C. For the 25% oxygen ratio at 400 °C, the coating exhibited a hardness of 25.7 GPa at RT, which was relatively lower compared to the 8.3% oxygen ratio. Upon annealing at 500 °C, the hardness of the coatings with a 25% oxygen ratio decreased to 20.5 GPa. Subsequently, at 600 °C, the hardness further dropped to 18.9 GPa. Notably, while a decrease in hardness was observed, no delamination of the coatings was observed under these annealing conditions. This was due to the absence of grain boundaries and defects, which are typically present in crystalline materials and can act as initiation sites for thermal degradation processes [29,30]. In featureless or amorphous structures, the absence of long-range order and the presence of a disordered atomic arrangement result in improved resistance to grain growth, diffusion, and structural transformations at elevated temperatures [31,32]. Consequently, coatings with featureless or amorphous structures have the potential to maintain their integrity and mechanical properties even under high-temperature conditions, making them desirable for applications requiring superior thermal stability.

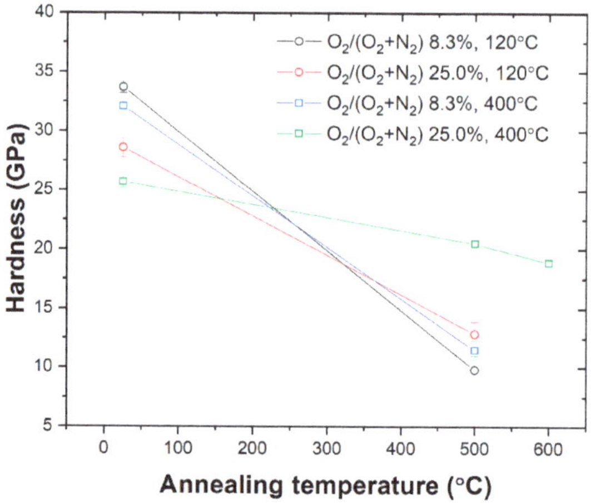

Figure 5. Hardness of CrZrON coatings after annealing in the range of room temperature to 600 °C.

4. Conclusions

In this study, the structural and thermal stability of CrZrON coatings synthesized via reactive magnetron sputtering were investigated in terms of the deposition temperature and oxygen-to-reactive gas ratio. The aim was to explore the relationship between the modified structure and the thermal stability of the coatings and to provide insights into the development of thermally stable CrZrON coatings. The main conclusions are as follows:

(1) Coatings deposited at lower temperatures exhibited a columnar structure, while those deposited at higher temperatures showed a transition towards a featureless or amorphous structure.
(2) Increasing the $O_2/(O_2 + N_2)$ ratio led to a decrease in lattice parameter, suggesting the incorporation of oxygen and the formation of oxygen-rich compounds in the coatings.
(3) Crystallite size exhibited different behaviors with respect to deposition temperature and $O_2/(O_2 + N_2)$ ratio, showing reductions due to enhanced nucleation, grain growth inhibition, and the presence of oxygen during deposition.
(4) Hardness values decreased with increasing $O_2/(O_2 + N_2)$ ratio, attributed to oxygen incorporation and the formation of oxygen-rich compounds with lower hardness compared to nitride phases.
(5) Annealing experiments demonstrated that coatings with featureless or amorphous structures exhibited superior thermal stability, maintaining their structural integrity even at high temperatures.
(6) The thermal stability of CrZrN coatings limited to 500 °C can be enhanced by incorporating oxygen, allowing CrZrON coatings to be utilized up to 600 °C, thus opening new avenues for their application in various industries.

Author Contributions: Conceptualization, S.-M.K.; methodology, S.-M.K.; validation, S.-M.K. and S.-Y.L.; formal analysis, S.-M.K.; investigation, S.-M.K.; resources, S.-M.K.; data curation, S.-M.K.; writing—original draft preparation, S.-M.K.; writing—review and editing, S.-Y.L.; visualization, S.-M.K.; supervision, S.-Y.L.; project administration, S.-Y.L.; funding acquisition, S.-Y.L. All authors have read and agreed to the published version of the manuscript.

Funding: This study was supported by the National Research Foundation of Korea (NRF) grant funded by the Korea government (MSIT) (No. 2021R1A2C1010058).

Institutional Review Board Statement: Not applicable.

Informed Consent Statement: Not applicable.

Data Availability Statement: Not applicable.

Conflicts of Interest: The authors declare no conflict of interest.

References

1. Mayrhofer, P.H.; Mitterer, C.; Hultman, L.; Clemens, H. Microstructural design of hard coatings. *Prog. Mater. Sci.* **2006**, *51*, 1032–1114. [CrossRef]
2. Polcar, T.; Parreira, N.M.G.; Novak, R. Friction and wear behaviour of CrN coating at temperatures up to 500 degrees C. *Surf. Coat. Technol.* **2007**, *201*, 5228–5235. [CrossRef]
3. Lorenzo-Martin, C.; Ajayi, O.; Erdemir, A.; Fenske, G.R.; Wei, R. Effect of microstructure and thickness on the friction and wear behavior of CrN coatings. *Wear* **2013**, *302*, 963–971. [CrossRef]
4. Kim, G.; Kim, B.; Lee, S.; Hahn, J. Structure and mechanical properties of Cr-Zr-N films synthesized by closed field unbalanced magnetron sputtering with vertical magnetron sources. *Surf. Coat. Technol.* **2005**, *200*, 1669–1675. [CrossRef]
5. Li, W.Z.; Evaristo, M.; Cavaleiro, A. Influence of Al on the microstructure and mechanical properties of Cr-Zr-(Al-)N coatings with low and high Zr content. *Surf. Coat. Technol.* **2012**, *206*, 3764–3771. [CrossRef]
6. Kim, S.M.; Kim, B.S.; Kim, G.S.; Lee, S.Y.; Lee, B.Y. Evaluation of the high temperature characteristics of the CrZrN coatings. *Surf. Coat. Technol.* **2008**, *202*, 5521–5525. [CrossRef]
7. Kim, H.K.; Kim, S.M.; Lee, S.Y. Mechanical Properties and Thermal Stability of CrZrN/CrZrSiN Multilayer Coatings with Different Bilayer Periods. *Coatings* **2022**, *12*, 1025. [CrossRef]
8. Peng, X.G.; Zhang, K.F.; Zheng, Y.G.; Zhou, H.; Wan, Z.H. Structure, morphologies and mechanical properties study of Cr-Zr-N films. *Nucl. Instrum. Meth B* **2018**, *436*, 112–118. [CrossRef]
9. Khamseh, S.; Araghi, H. A study of the oxidation behavior of CrN and CrZrN ceramic thin films prepared in a magnetron sputtering system. *Ceram. Int.* **2016**, *42*, 9988–9994. [CrossRef]
10. Kim, S.M.; Kim, G.S.; Lee, S.Y.; Lee, J.W.; Lee, J.Y.; Lee, B.Y. Mechanical Properties and Thermal Stability of CrSiN/AlN Multilayer Coatings Using a Closed-Field Unbalanced Magnetron Sputtering System. *J. Korean Phys. Soc.* **2009**, *54*, 1109–1114. [CrossRef]
11. Kim, G.S.; Kim, S.M.; Lee, S.Y.; Lee, B.Y. Comparative studies on the thermal stability and corrosion resistance of CrN, CrSiN, and CrSiN/AlN coatings. *J. Vac. Sci. Technol. A* **2009**, *27*, 873–879. [CrossRef]
12. Kim, D.J.; La, J.H.; Kim, K.S.; Kim, S.M.; Lee, S.Y. Tribological properties of CrZr-Si-N films synthesized using Cr-Zr-Si segment targets. *Surf. Coat. Technol.* **2014**, *259*, 71–76. [CrossRef]
13. Kim, H.K.; Kim, S.M.; Lee, S.Y. Influence of Interlayer Materials on the Mechanical Properties and Thermal Stability of a CrAlN Coating on a Tungsten Carbide Substrate. *Coatings* **2022**, *12*, 1134. [CrossRef]
14. Kim, D.; La, J.; Kim, S.M.; Lee, S.Y. The tribological performance of laser surface treated CrZrSiN thin films. *Mater. Res. Bull.* **2014**, *58*, 39–43. [CrossRef]
15. Gan, M.D.; Chong, X.Y.; Yu, W.; Xiao, B.; Feng, J. Understanding the ultralow lattice thermal conductivity of monoclinic RETaO4 from acoustic-optical phonon anti-crossing property and a comparison with ZrO2. *J. Am. Ceram. Soc.* **2023**, *106*, 3103–3115. [CrossRef]
16. Castaldi, L.; Kurapov, D.; Reiter, A.; Shkover, V.; Schwaller, P.; Patscheider, J. Effect of the oxygen content on the structure, morphology and oxidation resistance of Cr-O-N coatings. *Surf. Coat. Technol.* **2008**, *203*, 545–549. [CrossRef]
17. Du, J.W.; Chen, L.; Chen, J.; Hu, C. Influence of oxygen addition on the structure, mechanical and thermal properties of CrN coating. *Surf. Coat. Technol.* **2021**, *411*, 126992. [CrossRef]
18. Kim, D.J.; Kim, S.M.; La, J.H.; Lee, S.Y.; Hong, Y.S.; Lee, M.H. Synthesis and characterization of CrZrAlN films using unbalanced magnetron sputtering with segment targets. *Met. Mater. Int.* **2013**, *19*, 1295–1299. [CrossRef]
19. Arreguin-Campos, M.; Gutierrez, Z.B.K.; Quinones-Galvan, J.G.; Santos-Cruz, J.; Mayen-Hernandez, S.A.; Zelaya-Angel, O.; Olvera, M.D.; Contreras-Puente, G.; de Moure-Flores, F. Fabrication of CdS/CdTe Heterostructures by Chemical Synthesis Using a p-Type CdTe Film Grown by Electrodeposition Employing EDTA as Strong Complexing Agent. *J. Electron. Mater.* **2019**, *48*, 3595–3602. [CrossRef]
20. Khatibi, A.; Sjolen, J.; Greczynski, G.; Jensen, J.; Eklund, P.; Hultman, L. Structural and mechanical properties of Cr-Al-O-N thin films grown by cathodic arc deposition. *Acta Mater.* **2012**, *60*, 6494–6507. [CrossRef]
21. Landalv, L.; Lu, J.; Spitz, S.; Leiste, H.; Ulrich, S.; Johansson-Joesaar, M.P.; Ahlgren, M.; Gothelid, E.; Alling, B.; Hultman, L.; et al. Structural evolution in reactive RF magnetron sputtered (Cr,Zr)(2)O-3 coatings during annealing. *Acta Mater.* **2017**, *131*, 543–552. [CrossRef]
22. Kolobov, A.V.; Krbal, M.; Fons, P.; Tominaga, J.; Uruga, T. Distortion-triggered loss of long-range order in solids with bonding energy hierarchy. *Nat. Chem.* **2011**, *3*, 311–316. [CrossRef]
23. Zhu, Y.H.; Zhou, Y.K.; Wang, A.M.; Li, H.; Fu, H.M.; Zhang, H.W.; Zhang, H.F.; Zhu, Z.W. Atomic-scale icosahedral short-range ordering in a rejuvenated Zr-based bulk metallic glass upon deep cryogenic treatment. *Mat. Sci. Eng. Struct.* **2022**, *850*, 143565. [CrossRef]

24. Qiao, J.C.; Wang, Q.; Pelletier, J.M.; Kato, H.; Casalini, R.; Crespo, D.; Pineda, E.; Yao, Y.; Yang, Y. Structural heterogeneities and mechanical behavior of amorphous alloys. *Prog. Mater. Sci.* **2019**, *104*, 250–329. [CrossRef]
25. Hu, Q.; Guo, S.; Wang, J.M.; Yan, Y.H.; Chen, S.S.; Lu, D.P.; Liu, K.M.; Zou, J.Z.; Zeng, X.R. Parametric Study of Amorphous High-Entropy Alloys formation from two New Perspectives: Atomic Radius Modification and Crystalline Structure of Alloying Elements. *Sci. Rep.* **2017**, *7*, 39917. [CrossRef] [PubMed]
26. Suzuki, T.; Inoue, J.; Saito, H.; Hirai, M.; Suematsu, H.; Jiang, W.H.; Yatsui, K. Influence of oxygen content on structure and hardness of Cr-N-O thin films prepared by pulsed laser deposition. *Thin Solid Films* **2006**, *515*, 2161–2166. [CrossRef]
27. Bottger, P.H.M.; Lewin, E.; Patscheider, J.; Shklover, V.; Cahill, D.G.; Ghisleni, R.; Sobiech, M. Thermal conductivity of hard oxynitride coatings. *Thin Solid Films* **2013**, *549*, 232–238. [CrossRef]
28. Smirniotis, P.G.; Boningari, T.; Damma, D.; Inturi, S.N.R. Single-step rapid aerosol synthesis of N-doped TiO_2 for enhanced visible light photocatalytic activity. *Catal. Commun.* **2018**, *113*, 1–5. [CrossRef]
29. Garg, P.; Rupert, T.J. Grain incompatibility determines the local structure of amorphous grain boundary complexions. *Acta Mater.* **2023**, *244*, 118599. [CrossRef]
30. Lortaraprasert, C.; Shiomi, J. Robust combined modeling of crystalline and amorphous silicon grain boundary conductance by machine learning. *NPJ Comput. Mater.* **2022**, *8*, 219. [CrossRef]
31. Yang, J.Q.; Shang, L.L.; Sun, J.Y.; Bai, S.Y.; Wang, S.Z.; Liu, J.K.; Yun, D.; Ma, D.Y. Restraining the Cr-Zr interdiffusion of Cr-coated Zr alloys in high temperature environment: A Cr/CrN/Cr coating approach. *Corros. Sci.* **2023**, *214*, 111015. [CrossRef]
32. Yang, J.Q.; Bai, S.Y.; Sun, J.Y.; Wu, H.; Sun, S.H.; Wang, S.Z.; Li, Y.H.; Ma, W.H.; Tang, X.Y.; Xu, D.H. Microstructural understanding of the oxidation and inter-diffusion behavior of Cr-coated Alloy 800H in supercritical water. *Corros. Sci.* **2023**, *211*, 110910. [CrossRef]

coatings

Article

Effect of Heat Treatment on the Passive Film and Depassivation Behavior of Cr-Bearing Steel Reinforcement in an Alkaline Environment

Yuwan Tian, Cheng Wen *, Xiaohui Xi, Deyue Yang and Peichang Deng

Guangdong Provincial Ocean Equipment and Manufacturing Engineering Technology Research Center, School of Mechanical and Power Engineering, Guangdong Ocean University, Zhanjiang 524088, China
* Correspondence: wcheng.3jia@163.com

Abstract: Using Cr-bearing low-alloy steel is an effective preventive measure for marine structures, as it offers superior corrosion resistance when compared to plain carbon steel. However, it remains unclear how quenching and tempering heat treatment, which is commonly applied to steel reinforcement in some specific environments to improve its mechanical properties, affects its corrosion resistance. In the present work, the impact of heat treatment on the passive film and depassivation behavior of the 0.2C-1.4Mn-0.6Si-5Cr steel are studied. The results reveal that quenching and tempering result in grain refinement of the Cr-bearing steel, which increases its hardness. However, this refinement causes significant degradation in its corrosion resistance. The critical $[Cl^-]/[OH^-]$ ratio after quenching and tempering is determined to be approximately 6.6 times lower than that after normalization, and the corrosion rate is 1.6 times higher. After quenching and tempering, the passive film predominantly comprises iron oxides and hydroxides, with relatively high water content and defect density. Additionally, the Fe^{II}/Fe^{III} ratio and film resistance are relatively low. In comparison, after normalization, the steel exhibits high corrosion resistance, with the passive film formed offering the highest level of protection.

Keywords: chromium-bearing steel; corrosion; grain size; heat treatment; passive film

Citation: Tian, Y.; Wen, C.; Xi, X.; Yang, D.; Deng, P. Effect of Heat Treatment on the Passive Film and Depassivation Behavior of Cr-Bearing Steel Reinforcement in an Alkaline Environment. *Coatings* **2023**, *13*, 964. https://doi.org/10.3390/coatings13050964

Academic Editor: Jose Luis Viesca

Received: 29 March 2023
Revised: 6 May 2023
Accepted: 8 May 2023
Published: 22 May 2023

1. Introduction

Chloride-induced corrosion of steel reinforcement precisely limits the service life of marine structures. The use of low-alloy steel reinforcement with higher corrosion resistance than plain carbon steel is an efficient preventative method in highly aggressive marine environments [1]. In recent years, Cr-bearing low-alloy steel has attracted increasing attention due to its superior corrosion resistance, mechanical properties, weldability, and low cost compared to traditional carbon steel and stainless steel.

A thin, protective passive film can be formed on steel reinforcement under the alkaline conditions of ordinary concrete. This film has a strong influence on the durability of the structure because in the chloride-contaminated concrete, the service life is usually assumed to be equal to the initiation timeframe when the breakdown of passive film occurs (Figure 1) [2]. Several studies indicate that the alloy element Cr is involved in the formation of passive films on Cr-bearing steel, as Cr facilitates the formation of Cr_2O_3 and $Cr(OH)_3$ and thus inhibits the further oxidation and hydration of passive films [3,4]. The modified passive film of Cr-bearing steel improves the corrosion resistance of steel: numerous studies have proven that the critical chloride content and resistance to pitting corrosion initiation are higher after Cr addition to carbon steel [5–9]. As a result, the passive film and its depassivation behavior for Cr-bearing steel play important roles in increasing corrosion resistance.

Generally, hot-rolled steel bars are used directly in practical applications. In some specific environments, such as prestressed concrete sleepers, heat-treated steel reinforcement is applied to improve the mechanical properties of the materials used in terms of strength, ductility, yield ratio, and cold bending performance [10]. Heat treatment can lead to a change in the microstructure of steels in terms of the carbides, grain size, and so on. Numerous works have observed a direct relationship between localized corrosion susceptibility and the presence of carbide phases in steel [11,12], since carbides with a noble electrochemical equilibrium potential accelerate galvanic corrosion and ferrite dissolution [13,14]. Even worse, carbides such as cementite can continuously aggravate galvanic corrosion during corrosion evolution processes due to the sustained accumulation of cementite on the steel surface, which can lead to an increase in the surface area ratio between the cathode and anode [12,15]. However, a recent study reported that an increase in the carbide content could slow the corrosion rate for steel, as it can improve the formation kinetics for passive films [16]. The grain size also has a considerable effect on the corrosion process [17–20]. Wang et al. noted that the relationship between the grain size and corrosion rate depends on the corrosive medium [21]. In an active corrosion system, grain refinement can accelerate corrosion due to the high electrochemical activity of grain boundaries, while in a passive solution, grain refinement can lead to an improvement in corrosion resistance [21–23]. In addition, some researchers have suggested the importance of structural heterogeneity (simultaneous presence of small and large grains) in decreasing the corrosion resistance of steel, and they observed that the corrosion rate is more dependent on grain size and the corresponding dispersion of grains with different grades [24].

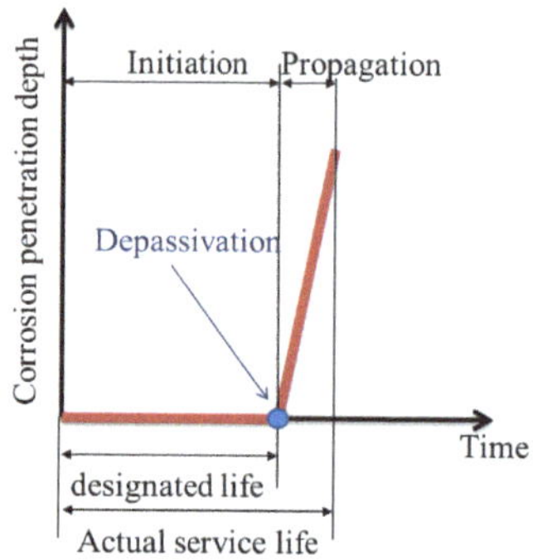

Figure 1. The service life including the initiation and propagation periods for corrosion in a reinforced concrete structure (Tuutti's model) [2].

However, the effect of heat treatment on the passivation and depassivation behavior of Cr-bearing low-alloy steel remains unclear. Does heat treatment influence the protective ability of passive film and thus affect the structural durability? How does heat treatment influence the property of passive film, that is, the Cr distribution, chemical composition, structure, and electrical characteristics in the passive film? In fact, previous studies have established the mechanism of effect of heat treatment on corrosion behavior of plain steel, but few studies are available regarding the use of different heat treatment processes for Cr-bearing steel reinforcement.

In this study, three types of heat treatment, namely, annealing, normalizing, and quenching and tempering, were applied to the as-received 5 wt.% Cr steel (0.2C-1.4Mn-0.6Si-5Cr). Then, electrochemical measurements, immersion tests, and physical characterization techniques were employed to investigate the effect of heat treatment on the microstructure evolution, passivation, and corrosion behavior of Cr-bearing steel in a simulated marine concrete environment.

2. Experimental Procedures

In order to evaluate the effect of heat treatment on the corrosion resistance of Cr-bearing steel, different heat treatments were carried out on the steel specimens, followed by physical characterization of the surface passive film, and finally the corrosion rate was detected. The specific workflow is shown in Figure 2.

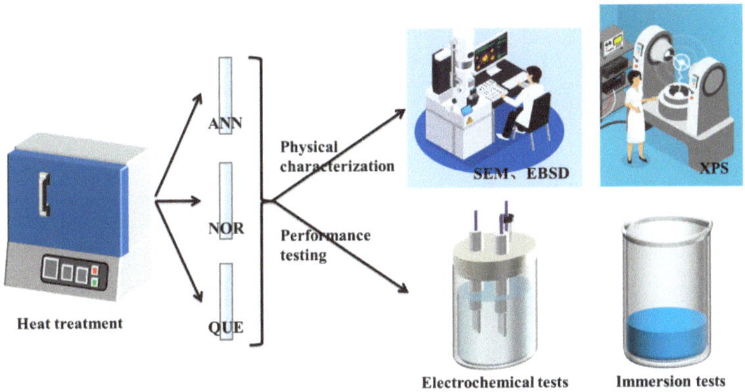

Figure 2. The workflow of the present work.

2.1. Material and Heat Treatment

The Cr-bearing steel reinforcement used in the present study consisted of a 400 MPa-grade hot-rolled ribbed bar, and its chemical composition (wt.%) was 0.19 C, 0.60 Si, 1.39 Mn, 0.02 P, 0.09 S, 5.20 Cr, and Fe for balance. The Cr-bearing steel rebar was machined into sheets with a size of 20 mm × 20 mm × 10 mm; ground with SiC emery paper; cleaned with saturated calcium hydroxide, water, and alcohol; and dried with nitrogen.

Three different heat treatment routes without a controlled atmosphere were adopted. The steel rebar specimens were kept at 900 °C for 40 min, and then cooled to room temperature in a furnace, air, and water and labelled ANN, NOR, and QUE, respectively. The QUE specimens were then tempered at 480 °C for 20 min and air-cooled to room temperature.

2.2. Microstructure

The heat-treated samples were then mechanically ground with SiC paper, polished with Ar ions, and etched with 4 vol.% nitric alcohol. Optical microscopy (Leica Microsystems, DVM6, Würzburg, Germany, and accuracy of 1 μm) and field emission scanning electron microscopy (TESCAN SEM, CLARA-GMH, Brno, Czech Republic, and accuracy of 0.8 nm) were utilized to observe the microstructures of Cr-bearing steels. The grain distribution and orientation were observed by electron backscattering diffraction analysis (Oxford Instruments EBSD, Symmetry, Oxford, UK, and accuracy of 0.1°). The crystal phase composition was examined in the angular range of 5–90° with X-ray diffraction (Bruker XRD equipment, D8 DISCOVER, Canton, USA, and accuracy of 0.001°) equipment with a Cu target.

2.3. Passive Film

X-ray photoelectron spectroscopy (Thermo Fisher XPS equipment, Thermo ESCALAB 250Xi, Wilmington, USA, and energy resolution of 15 meV) was used to characterize the Cr-bearing steel passive film after immersion in saturated calcium hydroxide, and all XPS spectra were calibrated by assuming a C_{1s} peak at 284.8 eV.

Electrochemical tests (Autolab electrochemical workstation, 302 N, Utrecht, Netherlands, voltage accuracy of 0.1 mV, and current accuracy of 10 nA) including impedance spectroscopy (EIS) results and Mott-Schottky (MS) curves for the Cr-bearing steel were analyzed to characterize the properties of the passive film. A classic three-electrode cell in simulated concrete pore solution was used, with a Pt foil counter electrode, a Cr-bearing steel working electrode, and an SCE reference electrode. The steel rebar was machined into dimensions of 10 mm × 10 mm × 5 mm, welded with copper wire, embedded in epoxy resin with a working surface of 1 cm^2, and ground with #2000 emery papers. EIS was measured from 100 kHz to 10 mHz at the open-circuit potential with a sinusoidal potential perturbation of 10 mV. MS was tested from 1 to -1 V_{SCE} with a step rate of 50 mV/s using a sinusoidal signal with a frequency of 1 kHz and an amplitude of 10 mV. All electrochemical measurements were repeated at least three times to ensure the repeatability of the data.

2.4. *Depassivation*

To determine the critical concentration of the chloride threshold (C_{crit}), the open-circuit potential (OCP) was continuously recorded for Cr-bearing steel in saturated calcium hydroxide with the gradual addition of 0.05 M chloride ion per 24 h. When the OCP displayed a sharp shift to a negative value, the corresponding value for the chloride concentration was defined as C_{crit}.

The corrosion behavior of Cr-bearing steel was also studied using potentiodynamic polarization measurements from -300 mV_{OCP} to 700 mV_{SCE} at a sweep rate of 0.1667 mV/s. The working electrode was immersed for 2 h to satisfy the stability requirements for an open-circuit potential (OCP) fluctuation lower than 5 mV/10 min. The electrolyte was composed of saturated $Ca(OH)_2$ + $NaHCO_3$ + Na_2CO_3 + 1.25 M NaCl (pH 11) to simulate aggressive concrete due to carbonation or penetration of salts.

3. Results

3.1. *Microstructure and Phase Distribution*

Figure 3 shows the metallographic structure for Cr-bearing steels with different heat treatments. The microstructure of the ANN sample is mainly ferrite and contains a minor amount of pearlite. Normalization increases the pearlite content and further refines the grain size of Cr-bearing steel. The Vickers hardness of ANN and NOR is low, with values of 110 HV and 161 HV, respectively. For the QUE sample, the microstructure presents less clear characteristics for lamellar pearlite, and the steel hardness increases to 249 HV.

Figure 4 shows the EBSD results obtained for Cr-bearing steels with different heat treatments. All the samples show a uniform microstructure. The major differences in grain size and grain boundary density can be observed, while the misorientation angle and its distribution are similar for all the samples.

3.2. *Passive Film*

Figure 5 shows the XPS spectra measured for the passive film on the Cr-bearing steels in saturated $Ca(OH)_2$ solution without any aggressive ions. The major peaks of O, Cr, Fe, C, and Ca can be observed in the survey spectra measured for the different surface films. The C and Ca elements are not considered because their presence may be attributed to the contamination of the immersion solution and the surrounding environment. Thus, the passive films are mainly composed of oxides and hydroxides of Fe and Cr. These conclusions are consistent with those in the literature [25–27], in which a bilayer structure with an outer Fe^{III}-enriched layer and an inner protective layer composed of anhydrous mixed Fe-Cr oxide have been reported. Since the XPS element sensitivity factors for Cr and Fe are very similar, the ratio of Cr oxides/Fe oxides in the passive film can be roughly assessed based on the peak intensity observed in Figure 5a. Clearly, the content of chromium oxides in the passive film is higher than that of the iron species for ANN and NOR, whereas iron oxides are predominant in the QUE passive film. The chromium in the oxide film has been confirmed to play a positive role in corrosion resistance since Cr can act as a barrier to the transfer of electrons and holes in the space-charge layer [28,29], and Cr oxides exhibit a compact spinel structure. This indicates that the passive films on ANN and NOR are more stable and protective than those on QUE, which will influence the electrochemical properties and corrosion resistance of steel in corrosive media.

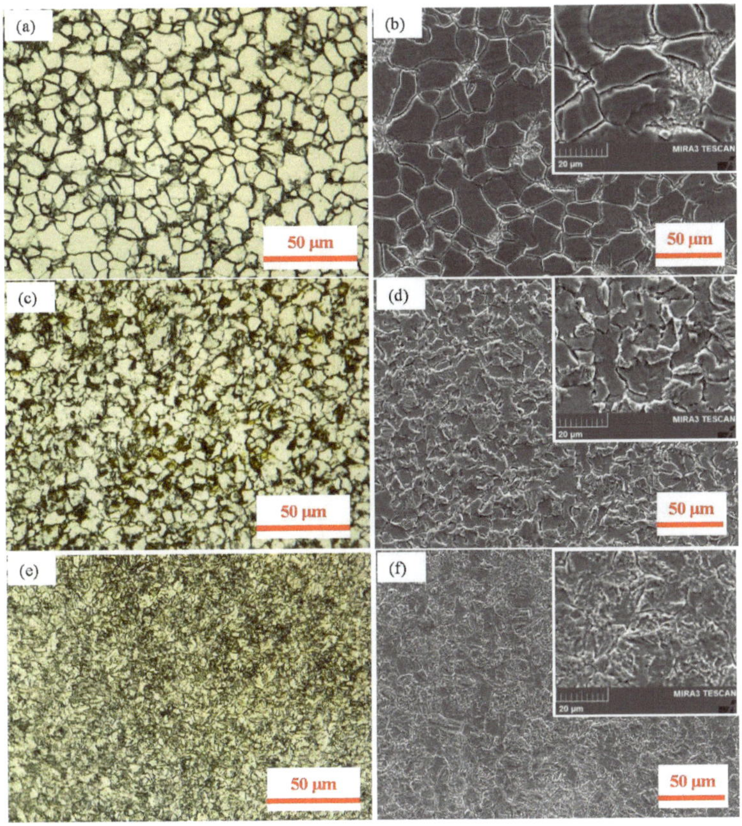

Figure 3. Microstructures for Cr-bearing steel with different heat treatments using optical and scanning electron microscopy: (**a**,**b**) ANN, (**c**,**d**) NOR, (**e**,**f**) QUE.

The typical Fe $2p_{3/2}$ and O 1s high-resolution XPS spectra shown in Figure 5c,d consist of different components, for which the corresponding binding energies and full widths at half maxima (FWHM) are listed in Table 1 [30–34]. All peaks were calibrated to the hydrocarbon (C 1s) signal set at 284.8 eV. The Shirley background correction algorithm was used to determine the optimal spectral baseline and peaks. The Fe 2p spectrum consists of a doublet structure of Fe $2p_{3/2}$ and Fe $2p_{1/2}$, and only the Fe $2p_{3/2}$ with a higher intensity was used in the present analysis. Quantitative composition information was determined using the corresponding area for each peak. From the Fe $2p_{3/2}$ high resolution spectra measured for ANN, NOR, and QUE, the proportion of Fe metal is determined to be 39%, 44%, and 25%, and the relative water content is 23%, 7%, and 23%, respectively. Since the experimental environment and parameters are consistent for all tests, the effective volume for the XPS detection should likewise remain relatively constant. Thus, the passive film on NOR is the thinnest but the most dehydrated, while QUE displays the thickest passive film with a higher content of bound water. This is reasonable since there are two stages in the formation process of passive films [35]: the formation of a protective anhydrous Fe^{II} oxide layer directly on the top of the metal substrate and the formation of a nonprotective hydrous Fe^{III} layer due to the thickening and hydration of the Fe^{II} layer. Thus, the high water content in the passive film of the QUE increases the electrochemical activity and accelerates the growth of the outer film, which results in a thicker but less protective oxide film.

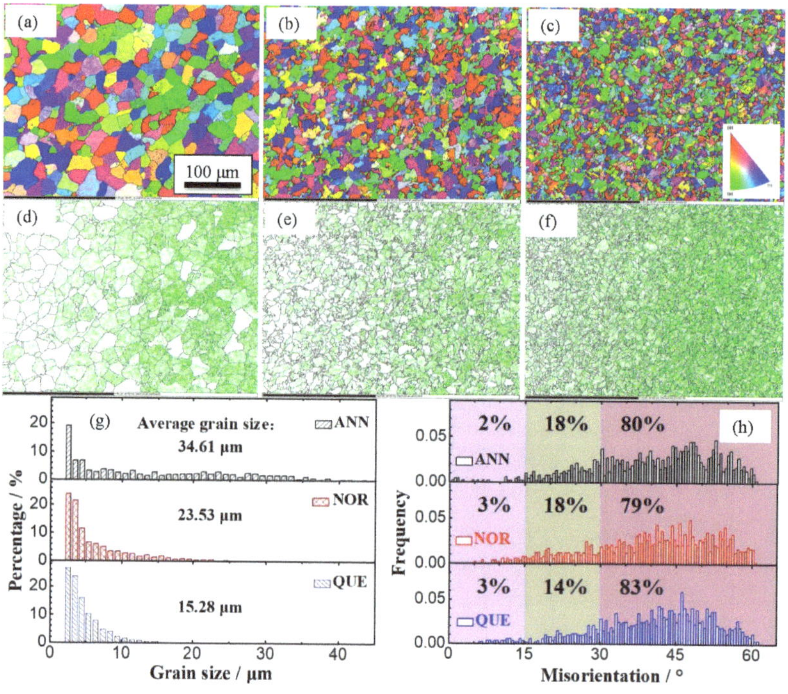

Figure 4. EBSD results for Cr-bearing steel with different heat treatments using scanning electron microscopy equipped with electron backscattered diffraction: the inverse pole figure (IPF) maps and corresponding grain boundary maps of ANN (**a**,**d**), NOR (**b**,**e**), and QUE (**c**,**f**); the black lines stand for the high-angle grain boundaries (HAGBs) and the green lines for the low-angle grain boundaries (LAGBs); (**g**) the distribution of grain sizes; (**h**) the distribution of misorientation angles.

The percentage contents for different iron oxides and hydroxides in the passive films are illustrated in Figure 5e. The Fe^{II} oxide fraction is approximately 41% for ANN and 49% for NOR, while it is only 27% for QUE, which implies a much thicker inner Fe^{II} layer for the passive film on NOR. These results are consistent with the presence of the thinnest passive film on NOR, as the thick Fe^{II} layer inhibits the growth of the outer Fe^{III} layer and thus decreases the total thickness of the passive film.

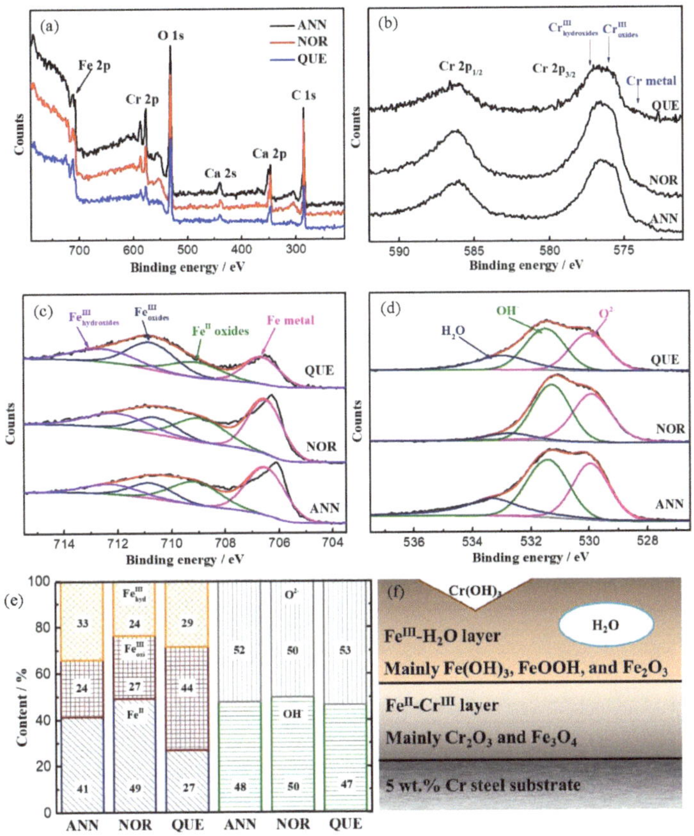

Figure 5. (**a**) XPS survey spectra of the passive film on Cr-bearing steel in the saturated $Ca(OH)_2$ solution. (**b**) High-resolution XPS spectra of Cr 2p. (**c**) High-resolution XPS spectra of Fe $2p_{3/2}$. (**d**) High-resolution XPS spectra of O 1 s. (**e**) Fractions of the iron oxides and hydroxides, (**f**) A model of the passive film on Cr-bearing steel.

Table 1. The peak parameters of Fe 2p and O 2p XPS spectrum used in the present work.

	Assignment	Fe Metal	Fe^{II}	Fe^{III} in Oxides	Fe^{III} in Hydroxides
Fe 2p	Binding energy/eV	706.9 ± 0.3	709.2 ± 0.3	711.0 ± 0.3	712.7 ± 0.3
	FWHM/eV	1.5 ± 0.2	2 ± 0.2	2 ± 0.2	2.5 ± 0.2
O 1s	Assignment	O^{2-}	OH^-	H_2O	-
	Binding energy/eV	530.2 ± 0.3	531.4 ± 0.3	533 ± 0.3	-
	FWHM/eV	1.5 ± 0.2	1.5 ± 0.2	2 ± 0.2	-

In addition, all three types of samples display an O^{2-}/OH^- ratio that is almost equal to 1, which conflicts with the analyses for the Fe^{II} content and Cr/Fe ratio in the passive film.

Based on the literature, hydroxides are supposed to be mainly enriched in the outer layer and represented in the forms of Fe(OH)3, FeOOH, and Cr(OH)3, as XPS depth profiling analyses show that OH- exhibits a declining tendency with increasing distance from the free surface of the passive film and becomes more difficult to detect at depths greater than several nanometers [3,36]. As a result, at least theoretically, the O^{2-}/OH^{-} ratio should follow the order of NOR > ANN > QUE. One possible reason for this phenomenon is the contamination of Ca(OH)2.

A model for the distribution of different oxidation states in the passive film is illustrated in Figure 5f. In terms of the overall performance, the passive film on NOR can be considered the most protective compared to that for ANN and QUE. Even though the total thickness of the passive film on NOR is very thin, it exhibits the best resistance to aggressive ions, due to the barrier effect of the thickest inner Fe^{II} layer, electrochemical inactivation due to the low water content, and modification due to the high Cr/Fe ratio.

Figure 6 shows the Mott-Schottky curves for Cr-bearing steels in a saturated $Ca(OH)_2$ solution without any aggressive ions. The semiconductive behavior of the passive film on the steel can be assessed according to the Mott-Schottky relationship [37]:

$$\frac{1}{C^2} = \frac{1}{C_H^2} + \frac{2}{\varepsilon\varepsilon_0 e N_q}\left(E - E_{FB} - \frac{kT}{e}\right)$$

where C_H is the Helmholtz capacitance (22 μFcm^2), ε is the dielectric constant (12), ε_0 is the vacuum permittivity (8.854×10^{-12} F/m), e is the elementary charge (1.602×10^{-19} C), N_q is the carrier concentration, E is the applied potential, E_{FB} is the flatband potential, k is the Boltzmann constant (1.381×10^{-23} J/K), and T is the absolute temperature. Table 2 and Figure 6b show the calculated parameters for the electronic properties of the passive films, and detailed information for the formulas is provided in the literature [37].

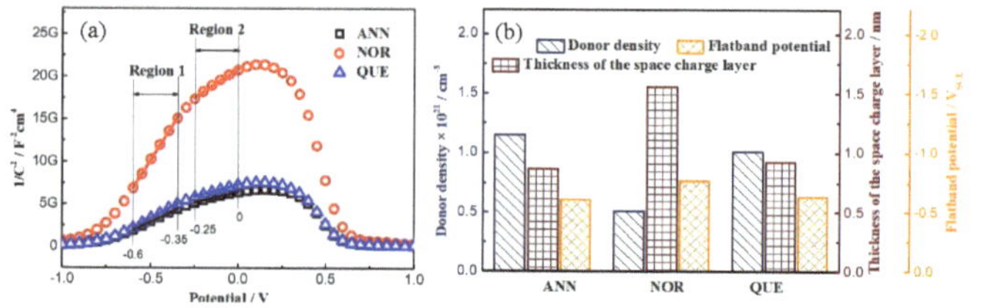

Figure 6. (**a**) Mott-Schottky curves for Cr-bearing steel in the saturated $Ca(OH)_2$ solution. (**b**) Electronic properties of the passive films.

Table 2. Fitting results for the Mott-Schottky plots.

Sample	$Slope_1$	$Slope_2$	Donor Density, cm^{-3}	Flatband Potential, mV_{SCE}	Thickness, nm
ANN	$5.02 \pm 0.09 \times 10^9$	$9.87 \pm 0.04 \times 10^9$	$1.15 \pm 0.01 \times 10^{21}$	-608 ± 3	0.869 ± 0.08
NOR	$13.5 \pm 0.03 \times 10^9$	$33.0 \pm 0.04 \times 10^9$	$0.51 \pm 0.01 \times 10^{21}$	-773 ± 8	1.553 ± 0.03
QUE	$5.74 \pm 0.10 \times 10^9$	$11.3 \pm 0.11 \times 10^9$	$1.00 \pm 0.06 \times 10^{21}$	-638 ± 17	0.926 ± 0.09

The passive films on all three samples exhibit n-type semiconductive behavior, and the charge carriers are mainly in the donor state, as evidenced by the positive slopes observed in the swept potential range. The two separate linear regions indicate two different donor states, with Fe^{II} donors being oxidized from tetrahedral and octahedral sites in the oxide crystal lattice [37]. The Fe^{II} donors at the tetrahedral sites are more important in the passivation/depassivation processes because they are more readily excited at low potential and room temperature. The Fe^{II} donor density at tetrahedral sites of NOR is significantly lower, down to approximately half that of ANN and QUE. It is slightly higher for ANN than for QUE but remains on the order of 10^{21} cm^{-3}. The opposite trend for the space-charge layer thicknesses can be observed, i.e., the donor density displays an inverse relation with the thickness of the space-charge layer. The thickness of the space-charge layer for ANN, NOR, and QUE is 0.87, 1.55, and 0.93 nm, respectively, which is within the thickness range of the protective Fe^{II} layer of the passive film [38]. This is a reasonable result since the oxidation and hydration processes for the inner Fe^{II} layer to the outer Fe^{III} layer will lead to the injection of numerous impurities into the film together with a thinning of the inner protective layer. Moreover, the thickness of the space-charge layer is in accordance with the XPS results for the Fe^{II} content. Evidently, the passive film on NOR is less disordered, and the inner protective layer is thicker when compared with the layers on ANN and QUE. The flatband potential is much more negative for NOR than for ANN and QUE, and the difference can reach as high as 0.15 V, revealing the different states of the electrode and the electrolyte.

Figure 7 shows the electrochemical impedance spectroscopy for Cr-bearing steels in the saturated $Ca(OH)_2$ solution without any aggressive ions. The equivalent circuit containing two time constants in parallel [39] is used to describe the interfaces between the working electrode and electrolyte, as shown in Figure 7c, in which R_s represents the solution resistance, Q_{dl} is the electric double-layer capacitance, R_{dl} is the charge transfer resistance, Q_f is the film capacitance, R_f is the film resistance, and R_p is the polarization resistance. The most widely used equivalent circuit with two hierarchical parallel RC loops in series [40] is not used here due to the fine protective efficiency of the passive film in highly alkaline media and also due to the fact that this equivalent circuit gives less fitting parameters in the present work. The impedance function of the $R(QR)(QR)$ model [41] is:

$$Z = R_s + \frac{R_{dl}}{1 + (j\omega)^{\alpha} R_{dl} Q_{dl}} + \frac{R_f}{1 + (j\omega)^{\alpha} R_f Q_f}$$

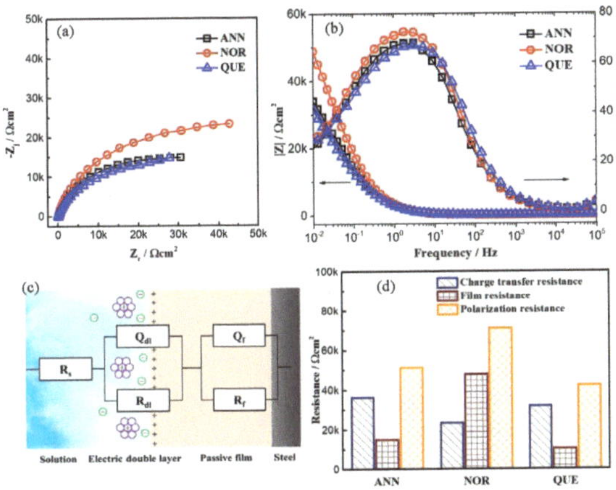

Figure 7. The electrochemical impedance spectroscopy for Cr-bearing steel in the saturated $Ca(OH)_2$ solution: (**a**) Nyquist plots, (**b**) Bode plots, (**c**) equivalent circuit, (**d**) fitted parameters.

The time constants for the medium and low frequencies are assigned to the charge transfer process in the double layer (R_{dl}-Q_{dl} couple) and the redox reaction in the passive film (R_f-Q_f couple), respectively. As shown in Table 3 and Figure 7d, the polarization resistance and film resistance can be established to follow the same order in terms of the passivation ability, with NOR > ANN > QUE, indicating that the dissolution rate of the passive film on NOR is the slowest. In particular, the transport process for redox species (OH^-/O_2 or Fe_3O_4/FeOOH) at the pores on the film/solution interface is greatly suppressed for NOR, as the NOR film resistance is increased by 3.2 and 4.5 times compared to that for ANN and QUE, respectively. The results of electric resistances are consistent with the conclusions made from the XPS and MS measurements, demonstrating the formation of a compact and protective passive film on the surface of NOR. However, the charge transfer resistance for NOR is slightly decreased by 1.3–1.5 times compared to that for ANN and QUE.

Table 3. Fitting results for the impedance spectra.

Sample	R_s Ωcm^2	Q_{dl} $Fcm^{-2}s^{n-1}$	n_{dl} s^n	R_{dl} Ωcm^2	Q_f $Fcm^{-2}s^{n-1}$	n_f s^n	R_f Ωcm^2
ANN	118 ± 23	$2.50 \pm 0.08 \times 10^{-4}$	0.77	$3.64 \pm 0.07 \times 10^{6}$	$1.12 \pm 0.02 \times 10^{-4}$	0.85	$1.50 \pm 0.09 \times 10^{4}$
NOR	112 ± 10	$7.97 \pm 0.01 \times 10^{-4}$	0.87	$2.35 \pm 0.01 \times 10^{6}$	$2.54 \pm 0.05 \times 10^{-4}$	0.88	$4.78 \pm 0.03 \times 10^{4}$
QUE	90 ± 17	$2.42 \pm 0.14 \times 10^{-4}$	0.83	$3.19 \pm 0.10 \times 10^{6}$	$1.07 \pm 0.02 \times 10^{-4}$	0.83	$1.05 \pm 0.11 \times 10^{4}$

3.3. *Depassivation and Corrosion*

Figure 8 shows the open-circuit potential evolution for Cr-bearing steels with respect to the incremental addition of chloride ions. At the initial stage of immersion, the OCPs range between −360 and −300 mV_{SCE}, which is lower than the empirical value of −200 mV_{SCE} obtained in an actual concrete structure. These rather low potentials may be attributed to the low oxygen availability under the full immersion condition. Upon the gradual addition of chloride to the saturated $Ca(OH)_2$ solution, the OCPs change little initially when the chloride contents are relatively low. In fact, the passive film on steel transitions slowly from an ordered state to an amorphous-like oxide structure with increasing chloride content before breakdown [37], even though the corrosion potential appears to remain unchanged. The sharp shift in the OCP in the negative direction indicates a passivity breakdown for the steel, and the corresponding chloride content is defined as the chloride threshold [42]. The chloride threshold is sequenced as NOR (critical $[Cl^-]/[OH^-]$ of 43.94, critical Cl^- content of 3.37 wt.%) > ANN (critical $[Cl^-]/[OH^-]$ of 20.81, critical Cl^- content of 1.60 wt.%) > QUE (critical $[Cl^-]/[OH^-]$ of 6.94, critical Cl^- content of 0.53 wt.%), suggesting that the breakdown of the passive film on NOR requires significantly higher chloride levels, while QUE is much more susceptible to chloride-induced corrosion.

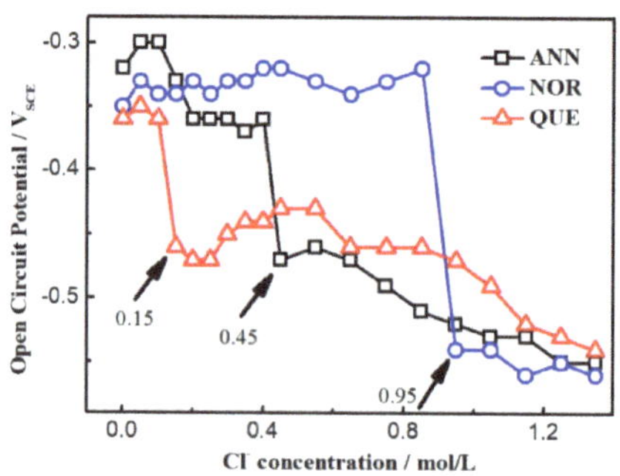

Figure 8. OCPs' evolution for Cr-bearing steels in the saturated $Ca(OH)_2$ solution with incremental addition of chloride ions.

Figure 9 shows the potentiodynamic polarization curves for Cr-bearing steels in a saturated $Ca(OH)_2$ solution with 1.25 M chloride ions. The corrosion processes include the cathodic reaction of oxygen reduction and the anodic reaction of iron dissolution. The differences in anodic current density corresponding to a particular overpotential (η) among the three samples are much more notable than those for the cathodic current density, indicating that the heat treatment mainly influences the anodic process. The Tafel fitting results for the polarization curves are listed in Table 4. The significantly higher pitting potential (E_{pit}) for NOR suggests the formation of a more protective passive film against chloride penetration, which is consistent with the chloride threshold results. Table 4 also shows that NOR exhibits the lowest corrosion current density and the highest anodic Tafel slopes. The corrosion rate for NOR is approximately 1.61 times lower than that obtained for QUE. However, the difference in the corrosion rate is less considerable, indicating that the effect of heat treatment on corrosion propagation is less evident than on corrosion initiation.

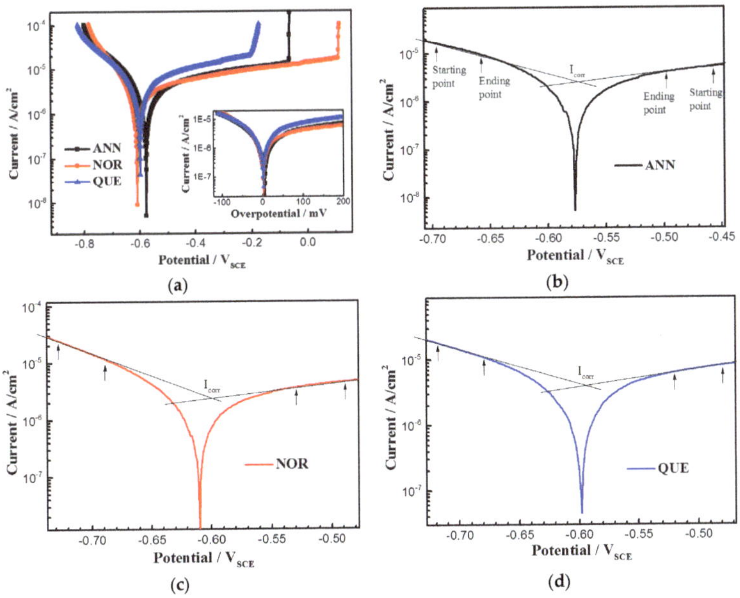

Figure 9. (**a**) Potentiodynamic polarization curves for Cr-bearing steels in the saturated $Ca(OH)_2$ solution with 1.25 M NaCl. (**b**) The fitting curves for potentiodynamic polarization curves of ANN (**b**), NOR (**c**), and QUE (**d**).

Table 4. Tafel fitting results of the potentiodynamic polarization curves.

Sample	E_{corr} mV_{SCE}	E_{pit} mV_{SCE}	β_c mV/dec	β_a mV/dec	I_{corr} μAcm^{-2}
ANN	-580 ± 5	-70 ± 21	161.85 ± 6.38	386.86 ± 1.02	2.72 ± 0.05
NOR	-610 ± 10	110 ± 6	120.77 ± 9.51	410.9 ± 7.55	2.44 ± 0.10
QUE	-600 ± 8	-200 ± 37	176.46 ± 9.77	376.19 ± 6.90	3.92 ± 0.07

Figure 10 shows the corrosion weight loss and corrosion morphology of steel samples after 10 days immersion in the saturated $Ca(OH)_2$ with 1.25 M NaCl. The dry-wet cycle experiment was chosen to simulate the extremely aggressive environment. The corrosion attacks are localized, and strong local corrosion occurred on the QUE sample. NOR exhibits the lowest corrosion rate, which is approximately 56% of QUE. The weight loss is in line with the results from potentiodynamic polarization curves.

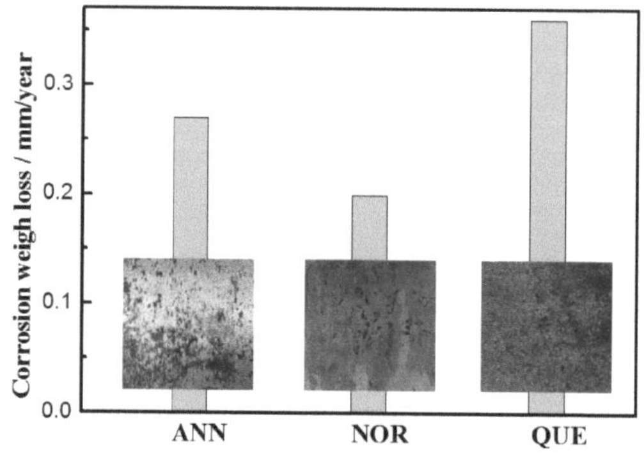

Figure 10. Corrosion weight loss and corrosion morphology for Cr-bearing steels in the saturated $Ca(OH)_2$ solution with 1.25 M NaCl.

4. Discussion

Heat treatment of quenching and tempering is commonly applied for steel reinforcement in order to optimize its mechanical properties [43]. However, according to the findings presented in Section 3, quenching and tempering can have adverse effects on the corrosion resistance of Cr-bearing steel in alkaline environments with chlorides. Specifically, after undergoing quenching and tempering heat treatment, the Cr-bearing steel reinforcement displays the poorest corrosion resistance, with the lowest critical $[Cl^-]/[OH^-]$ and the highest corrosion rate. In contrast, normalization resulted in the best corrosion resistance for the samples.

From Figures 3 and 4, the main distinguishing factor in the microstructure for the three types of steel samples is the grain size. Generally, grain refinement is beneficial in terms of encouraging passivation and the formation of passive films. This is due to the high energy state and fast diffusion coefficient of the intercrystalline region. Ralston et al. [17] proposed an equation to model the correlation between corrosion rate and grain size for passive metals (corrosion rate < 10 μAcm^{-2}):

$$I_{corr} = A + B \times gs^{-0.5}$$

where A and B are constants, I_{corr} is the corrosion rate (μAcm^{-2}), and gs is the grain size (μm). In the present study, it is observed that the corrosion resistance does not always increase with decreasing grain size under alkaline conditions, which contradicts the commonly accepted notion. For ANN, NOR, and QUE, the grain sizes are 19, 7, and 4 μm, respectively, their corresponding critical $[Cl^-]/[OH^-]$ values are 20.81, 43.94, and 6.94, respectively, and the corrosion rates are 2.72, 2.44, and 3.92 μAcm^{-2}, respectively. That is, even though the specimen after quenching and tempering shows a fine-grained microstructure, its corrosion resistance remains suboptimal.

Theoretically, this phenomenon can likely be attributed to the redistribution of substitutional alloying elements in the alloy steel after heat treatment. The Cr-bearing steel, under thermodynamic equilibrium conditions, experiences significant enrichment of Cr at the cementite/matrix interface in the cementite, with a corresponding depletion in the matrix surrounding the carbide. However, when both a high degree of supercooling (quenching) and a low degree of tempering are present, the Cr enrichment process in the cementite is hindered during carbide precipitation from the supersaturated ferrite [44]. The different partitions of Cr in ferrite and cementite on the phase and grain scales can lead to remarkable differences in the characteristics and protective properties of passive films.

However, the present work does not address experimental studies at the atomic scale, and, therefore, it is uncertain how heat treatment specifically affects the Cr element segregation and thus the protective properties of passive film. Relevant experiments, such as atomic force microscopy, Auger electron spectroscopy, and simulation calculations, will be further explored in subsequent research.

5. Conclusions

Low-alloy steels containing Cr are often utilized instead of ordinary carbon steel in coastal concrete structures to improve the durability of reinforced concrete. In some complex, high-stress concrete environments, the Cr-bearing steel undergoes heat treatment to enhance its strength, with quenching and tempering methods being the most widely employed due to their ability to refine grain and improve mechanical properties. The present research indicates that the Cr-bearing steel reinforcement exhibits microstructures with pearlite and ferrite after undergoing various heat treatments, including annealing, normalizing, and quenching and tempering. The grain sizes differ significantly among the three sample types, with values of 19 μm for ANN, 7 μm for NOR, and 4 μm for QUE. Furthermore, the Vickers hardness results indicate considerable variation, with respective values of 110 HV for ANN, 161 HV for NOR, and 249 HV for QUE.

Based on this study, although quenching and tempering improve the mechanical properties of Cr-bearing steel, their corrosion resistance significantly decreases in an alkaline environment containing chloride. On one hand, after quenching and tempering, the passive film offers the least protection for Cr-bearing steel, while it offers the most protection after normalization. The Cr/Fe and Fe^{II}/Fe^{III} ratios in the passive film formed on QUE are much lower than those in NOR. Additionally, the space-charge layer of the passive film in QUE is much thinner, with a much higher defect density, and the film resistance is almost five times lower than that of NOR. On the other hand, the resistance to chloride-induced depassivation is lower for Cr-bearing steel after quenching and tempering than after normalization. The critical $[Cl^-]/[OH^-]$ value for QUE is approximately 6.6 times lower than that for NOR, and the corrosion rate is 1.6 times higher. Therefore, when designing or selecting the processes of heat treatment, the comprehensive effects of heat treatment on the mechanical and corrosion properties of low-alloy steel should be considered.

Author Contributions: Conceptualization, Y.T. and C.W.; methodology, Y.T. and D.Y.; validation, C.W., X.X. and P.D.; formal analysis, Y.T. and X.X.; data curation, D.Y.; writing—original draft preparation, Y.T.; writing—review and editing, C.W.; funding acquisition, Y.T. All authors have read and agreed to the published version of the manuscript.

Funding: This research was funded by [Natural Science Foundation of Guangdong Province] grant number [2021A1515012129] and [Science and technology program of Zhanjiang] grant number [2021E05005].

Institutional Review Board Statement: Not applicable.

Informed Consent Statement: Not applicable.

Data Availability Statement: The datasets generated and/or analyzed during the current study are available from the corresponding author on reasonable request.

Conflicts of Interest: The authors declare no conflict of interest.

References

1. Winslow, D.N. High-strength low-alloy, weathering, steel as reinforcement in the presence of chloride ions. *Cem. Concr. Res.* **1986**, *16*, 491. [CrossRef]
2. Bertolini, L.; Elsener, B.; Pedeferri, P.; Redaelli, E.; Polder, R.B. *Corrosion of Steel in Concrete*; WILEY-VCH Verlag GmbH and Co. KgaA: Weinheim, Germany, 2013.

3. Ming, J.; Wu, M.; Shi, J. Passive film modification by concrete carbonation: Re-visiting a corrosion-resistant steel with Cr and Mo. *Cem. Concr. Compos.* **2021**, *123*, 104178. [CrossRef]
4. Shi, J.; Ming, J.; Wu, M. Electrochemical behavior and corrosion products of Cr-modified reinforcing steels in saturated Ca $(OH)_2$ solution with chlorides. *Cem. Concr. Compos.* **2020**, *110*, 103587. [CrossRef]
5. Liu, M.; Cheng, X.; Li, X.; Pan, Y.; Li, J. Effect of Cr on the passive film formation mechanism of steel rebar in saturated calcium hydroxide solution. *Appl. Surf. Sci.* **2016**, *389*, 1182. [CrossRef]
6. Callaghan, B. The performance of a 12% chromium steel in concrete in severe marine environments. *Corros. Sci.* **1993**, *35*, 1535. [CrossRef]
7. Callaghan, B. The use of 3CR12 as reinforcing in concrete. *Constr. Build. Mater.* **1993**, *7*, 131. [CrossRef]
8. Shi, J.; Ming, J.; Wang, D.; Wu, M. Improved corrosion resistance of a new 6% Cr steel in simulated concrete pore solution contaminated by chlorides. *Corros. Sci.* **2020**, *174*, 108851. [CrossRef]
9. Shi, J.J.; Zou, Y.; Ming, J.; Wu, M. Effect of DC stray current on electrochemical behavior of low-carbon steel and 10%Cr steel in saturated $Ca(OH)_2$ solution. *Corros. Sci.* **2020**, *169*, 108610. [CrossRef]
10. Ammarullah, M.I.; Santoso, G.; Sugiharto, S.; Supriyono, T.; Kurdi, O.; Tauviqirrahman, M.; Winarni, T.; Jamari, J. Tresca stress study of CoCrMo-on-CoCrMo bearings based on body mass index using 2D computational model. *J. Tribol.* **2022**, *33*, 31.
11. Clover, D.; Kinsella, B.; Pejcic, B.; De Marco, R. The influence of microstructure on the corrosion rate of various carbon steels. *J. Appl. Electrochem.* **2005**, *35*, 139. [CrossRef]
12. Wei, J.; Zhou, Y.; Dong, J.; He, X.; Ke, W. Effect of cementite spheroidization on improving corrosion resistance of pearlitic steel under simulated bottom plate environment of cargo oil tank. *Materialia* **2019**, *6*, 100316. [CrossRef]
13. Sun, J.B.; Zhang, G.A.; Liu, W.; Lu, M.X. The formation mechanism of corrosion scale and electrochemical characteristic of low alloy steel in carbon dioxide-saturated solution. *Corros. Sci.* **2012**, *57*, 131. [CrossRef]
14. Lu, B.; Li, W.; Luo, J.; Chiovelli, S. Pitting susceptibility of induction-quenched pipeline with microstructural heterogeneity. *Mater. Sci.* **2012**, *47*, 6823. [CrossRef]
15. Hao, X.H.; Dong, J.H.; Etim, I.N.; Wei, J.; Ke, W. Sustained effect of remaining cementite on the corrosion behavior of ferrite-pearlite steel under the simulated bottom plate environment of cargo oil tank. *Corros. Sci.* **2016**, *110*, 296. [CrossRef]
16. Zhao, Y.G.; Liu, W.; Fan, Y.M.; Fan, E.D.; Dong, B.J.; Zhang, T.Y.; Li, X.G. Effect of Cr content on the passivation behavior of Cr alloy steel in a CO_2 aqueous environment containing silty sand. *Corros. Sci.* **2020**, *168*, 108591. [CrossRef]
17. Ralston, K.D.; Birbilis, N.; Davies, C.H.J. Revealing the relationship between grain size and corrosion rate of metals. *Scr. Mater.* **2010**, *63*, 1201. [CrossRef]
18. Argade, G.R.; Panigrahi, S.K.; Mishra, R.S. Effects of grain size on the corrosion resistance of wrought magnesium alloys containing neodymium. *Corros. Sci.* **2012**, *58*, 145. [CrossRef]
19. Hadzima, B.; Janeček, M.; Estrin, Y.; Kim, H.S. Microstructure and corrosion properties of ultrafine-grained interstitial free steel. *Mater. Sci. Eng. A* **2007**, *462*, 243. [CrossRef]
20. Wang, S.G.; Shen, C.B.; Long, K.; Zhang, T.; Wang, F.H.; Zhang, Z.D. The electrochemical corrosion of bulk nanocrystalline ingot iron in acidic sulfate solution. *Phys. Chem. B* **2006**, *110*, 377. [CrossRef]
21. Wang, P.; Ma, L.; Cheng, X.; Li, X. Effect of grain size and crystallographic orientation on the corrosion behaviors of low alloy steel. *Alloys Compd.* **2021**, *857*, 158258. [CrossRef]
22. Rai, P.K.; Shekhar, S.; Mondal, K. Development of gradient microstructure in mild steel and grain size dependence of its electrochemical response. *Corros. Sci.* **2018**, *138*, 85. [CrossRef]
23. Zhang, L.; Ma, A.; Jiang, J.; Jie, X. Effect of processing methods on microhardness and acid corrosion behavior of low-carbon steel. *Mater. Des.* **2015**, *65*, 115. [CrossRef]
24. Novikov, V.F.; Sokolov, R.A.; Muratov, K.R.; Venediktov, A.N. Determination of influence of grain size factor on the corrosion speed of structural steel. *Mater. Today Proc.* **2021**, *38*, 1749. [CrossRef]
25. Carmezim, M.J.; Simoes, A.M.; Montemor, M.F.; Belo, M.D.C. Capacitance behaviour of passive films on ferritic and austenitic stainless steel. *Corros. Sci.* **2005**, *47*, 581. [CrossRef]
26. Hakiki, N.E.; Montemor, M.F.; Ferreira MG, S.; da Cunha Belo, M. Semiconducting properties of thermally grown oxide films on AISI 304 stainless steel. *Corros. Sci.* **2000**, *42*, 687. [CrossRef]
27. Ghods, P.; Isgor, O.B.; Bensebaa, F.; Kingston, D. Angle-resolved XPS study of carbon steel passivity and chloride-induced depassivation in simulated concrete pore solution. *Corros. Sci.* **2012**, *58*, 159–167. [CrossRef]
28. Xuan, Q.L.; Rondot, B.; Belo, M.D.C. Relations entre la structure electronique des films de passivation formes sur les aciers inoxydables et la susceptibilite de ces derniers a la corrosion par piqures. *Chim. Sci. Mater.* **1998**, *23*, 607.
29. Betova, I.; Bojinov, M.; Karastoyanov, V.; Kinnunen, P.; Saario, T. Estimation of kinetic and transport parameters by quantitative evaluation of EIS and XPS data. *Electrochim. Acta* **2010**, *55*, 6163. [CrossRef]
30. Zanna, S.; Seyeux, A.; Allion-Maurer, A.; Marcus, P. Escherichia coli siderophore-induced modification of passive films on stainless steel. *Corros. Sci.* **2020**, *175*, 108872. [CrossRef]
31. Wang, Z.; Feng, Z.; Zhang, L. Effect of high temperature on the corrosion behavior and passive film composition of 316 L stainless steel in high H2S-containing environments. *Corros. Sci.* **2020**, *174*, 108844. [CrossRef]
32. Chan-Rosado, G.; Pech-Canul, M.A. Influence of native oxide film age on the passivation of carbon steel in neutral aqueous solutions with a dicarboxylic acid. *Corros. Sci.* **2019**, *153*, 19–31. [CrossRef]

33. Chen, R.; Hu, J.; Ma, Y.W.; Guo, W.H.; Huang, H.L.; Wei, J.X.; Yin, S.H.; Yu, Q.J. Characterization of the passive film formed on the reinforcement surface in alkali activated fly ash: Surface analysis and electrochemical evaluation. *Corros. Sci.* **2020**, *163*, 108393. [CrossRef]
34. Lynch, B.; Neupane, S.; Wiame, F.; Seyeux, A.; Maurice, V.; Marcus, P. An XPS and ToF-SIMS study of the passive film formed on a model FeCrNiMo stainless steel surface in aqueous media after thermal pre-oxidation at ultra-low oxygen pressure. *Appl. Surf. Sci.* **2021**, *554*, 149435. [CrossRef]
35. Sánchez, M.; Gregori, J.; Alonso, C.; García-Jareño, J.J.; Takenouti, H.; Vicente, F. Electrochemical impedance spectroscopy for studying passive layers on steel rebars immersed in alkaline solutions simulating concrete pores. *Electrochim. Acta* **2007**, *52*, 7634.
36. Wang, Z.; Zhang, L.; Zhang, Z.R.; Lu, M.X. Combined effect of pH and H_2S on the structure of passive film formed on type 316L stainless steel. *Appl. Surf. Sci.* **2018**, *458*, 686. [CrossRef]
37. Williamson, J.; Isgor, O.B. The effect of simulated concrete pore solution composition and chlorides on the electronic properties of passive films on carbon steel rebar. *Corros. Sci.* **2016**, *106*, 82. [CrossRef]
38. Ghods, P.; Isgor, O.B.; Carpenter GJ, C.; Li, J.; McRae, G.A.; Gu, G.P. Nano-scale study of passive films and chloride-induced depassivation of carbon steel rebar in simulated concrete pore solutions using FIB/TEM. *Cem. Concr. Res.* **2013**, *47*, 55. [CrossRef]
39. Blanco, G.; Bautista, A.; Takenouti, H. EIS study of passivation of austenitic and duplex stainless steels reinforcements in simulated pore solutions. *Cem. Concr. Compos.* **2006**, *28*, 212. [CrossRef]
40. Freire, L.; Carmezim, M.J.; Ferreira, M.A.; Montemor, M.F. The passive behaviour of AISI 316 in alkaline media and the effect of pH: A combined electrochemical and analytical study. *Electrochim. Acta* **2010**, *55*, 6174. [CrossRef]
41. Joiret, S.; Keddam, M.; Nóvoa, X.R.; Pérez, M.C.; Rangel, C.; Takenouti, H. Use of EIS, ring-disk electrode, EQCM and Raman spectroscopy to study the film of oxides formed on iron in 1 M NaOH. *Cem. Concr. Compos.* **2002**, *24*, 7. [CrossRef]
42. Li, C.Z.; Jiang, L.H.; Li, S.S. Effect of limestone powder addition on threshold chloride concentration for steel corrosion in reinforced concrete. *Cem. Concr. Res.* **2020**, *131*, 106018. [CrossRef]
43. Ammarullah, M.I.; Afif, I.Y.; Maula, M.I.; Winarni, T.I.; Tauviqirrahman, M.; Akbar, I.; Basri, H.; van der Heide, E.; Jamari, J. Tresca stress simulation of metal-on-metal total hip arthroplasty during normal walking activity. *Materials* **2021**, *14*, 7554. [CrossRef] [PubMed]
44. Thomson, R.C.; Miller, M.K. Carbide precipitation in martensite during the early stages of tempering Cr- andMo-containing low alloy steels. *Acta Mater.* **1998**, *46*, 2203. [CrossRef]

metals

Article

Determination of Corrosion Resistance of High-Silicon Ductile Iron Alloyed with Nb

Carlos Rodrigo Muñiz Valdez [1], Daniel García Navarro [2], Jesús Salvador Galindo Valdés [1], Félix Alan Montes González [2], Efrain Almanza Casas [2] and Nelly Abigail Rodríguez Rosales [2,*]

[1] Facultad de Ingeniería, Universidad Autónoma de Coahuila, Fundadores km. 13, Las Glorias, Ciudad Universitaria, Arteaga 25350, Coahuila, Mexico; rodrigo.muniz@uadec.edu.mx (C.R.M.V.); s_galindo@uadec.edu.mx (J.S.G.V.)

[2] Tecnológico Nacional de México/I.T. Saltillo, Blvd. Venustiano Carranza 2400, Col. Tecnológico, Saltillo 25280, Coahuila, Mexico; d15051930@saltillo.tecnm.mx (D.G.N.); d13050533@saltillo.tecnm.mx (F.A.M.G.); efrain.ac@saltillo.tecnm.mx (E.A.C.)

* Correspondence: nelly.rr@saltillo.tecnm.mx

Abstract: In this study, the effects of Nb on the microstructural characteristics, hardness, and corrosion resistance of high-silicon ductile cast iron (HSDI)-3.6 wt.% Si were investigated. Samples from different castings with 0–0.9 wt.% Nb were obtained and compared to a commercial ductile iron. Microstructures showed that the amount of ferrite in the matrix increased with increasing Nb content, from 34% for unalloyed HSDI to 88% for HSDI-0.9 wt.% Nb. The presence of randomly distributed NbC carbides was identified by EDX for all the samples alloyed with Nb, and the hardness of the HSDI increased with the Nb content. To evaluate the influence of the Nb content on the corrosion resistance of HSDI, potentiodynamic tests were carried out in a solution of H_2SO_4. The highest corrosion rate on HSDI was obtained for the HSDI-0.3 wt.% Nb sample, with 2802 mills per year, due to the amount of pearlite present and the lowest presence of NbC carbides, compared to the HSDI-0.9 wt.% Nb, with 986 mills per year. This behavior was attributed to the ferrite matrix obtained because of a high Si content in the DI, which delayed the anodic dissolution of the alloy and suppressed the pearlitizing effect of Nb for contents greater than 0.3 wt.%, as well as to the effect of NbC carbides, which acted as inhibitors.

Keywords: high-silicon ductile iron; corrosion; NbC carbides

Citation: Muñiz Valdez, C.R.; García Navarro, D.; Galindo Valdés, J.S.; Montes González, F.A.; Almanza Casas, E.; Rodríguez Rosales, N.A. Determination of Corrosion Resistance of High-Silicon Ductile Iron Alloyed with Nb. *Metals* **2023**, *13*, 917. https://doi.org/10.3390/met13050917

Academic Editor: Nong Gao

Received: 18 April 2023
Revised: 3 May 2023
Accepted: 6 May 2023
Published: 9 May 2023

1. Introduction

Ductile iron (DI) is a type of cast iron with a metallic matrix of ferrite, pearlite, or a mixture of the two phases with graphite nodules embedded. Due to this morphology, DI has high strength, ductility, and good wear and fatigue resistance; therefore, DI is used in the automotive, energy, and agricultural industries, among others [1,2]. In addition, the microstructure and mechanical properties can be modified by adding different alloying elements or by heat treatment.

Silicon is an important element in DI, as it promotes ferrite formation in as-cast conditions and allows graphite the ability to precipitate during solidification by sulfide formation, surrounded by complex oxides of silicon and magnesium, which serve as nucleation sites for graphite [3]. Furthermore, Si reduces the carbon content of the eutectic composition, increases the amount of precipitated graphite, and prevents the formation of eutectic carbides [4–7]. In terms of mechanical properties, Si has a significant effect on the toughness and tensile strength of ferritic DI. As the Si content increases, the yield strength and tensile strength also increase [8–11]. A new DI classification called High-Silicon Ductile Iron (HSDI) has recently been introduced for DI with Si content ranging from 3.2 to 4.3 wt.%. It has been reported that with increasing Si content, hardness, tensile strength, and yield strength properties are promoted, while elongation to failure decreases [8,12–14] and machinability and corrosion resistance improve [15–18]. These results suggest a potential advantage in the production process and performance over DI with lower Si content [10,11].

With the purpose of modifying the structure and properties of HSDI, the addition of different carbide-forming elements, such as V, Cr, Mo, and Mn has been investigated by several authors [18–20]. Riebisch et al. [18] have studied the effect of carbide-forming elements on HSDI with 3.8 wt.% Si. They have found that V has a significant negative influence on graphite morphology, as it promotes the formation of chunky graphite due to a long solidification time and its effect of slowing down the diffusion of C into austenite; on the other hand, Cr and Mn have the greatest influence on pearlite area fraction, due to their segregation. Similarly, Górny et al. [19] studied the influence of 0.65 wt.% Mo on HSDI with 4.36 wt.% Si and reported a microstructure of graphite nodules in a ferrite matrix with a small amount of pearlite, as well as the formation of Fe–Mo carbides, which were related to segregation during solidification rather than to cooling rates to the eutectoid temperature. Although some authors have studied HSDI with carbide-forming elements, the role of Si in such castings must be better understood due to the wide range of HSDI applications.

Niobium has a significant impact on DI due to its high affinity for carbon, thus contributing to the formation of NbC carbides. Additionally, Nb has low solubility, especially in ferrous alloys with relatively high carbon content, such as cast iron. The presence of these particles in DI has been reported to improve corrosion resistance and provide good wear resistance with little change in austenite stability and graphite refinement [21–26]. However, the effects of a combination of high Si and Nb content on DI have not been reported in the literature.

Sckudlarek et al. [27] reported the formation of dispersed NbC carbides in the microstructure, refinement of graphite nodules with nodularity higher than 90%, and refinement and increase of pearlite volume fraction from 49 to 62% in DI with 2.46 wt.% Si and 0.35 wt.% Nb. On the other hand, Chen et al. [28] found that the addition of Nb up to 0.11 wt.% in the as-cast condition DI had a limited effect on the nodularity and roundness of the graphite nodule. Among the Nb containing DI, the optimum Nb addition in the as-cast condition was about 0.08 wt.% Nb, which gave the highest tensile strength and an elongation of 746 MPa and 8.0%, respectively. This was attributed to the significant microstructural improvement of the pearlite compared to the deterioration of the graphite morphology in DI due to the Nb addition. Furthermore, Ahmed et al. [29] reported that increasing to 0.1 wt.% Nb content in the DI increased the graphite nodule count and eutectic cells and refined the graphite structure, but at the same time, a slight decrement in nodularity was observed. Microstructural analysis showed that by increasing the Nb content, a finer pearlite structure was formed in the matrix.

In contrast, Alias et al. [30] reported an increase in tensile and impact strengths from 26 to 34% and 30 to 80%, respectively, in DI with 1.9 wt.% Si and Nb content in the range of 0.5–2 wt.% Nb compared to unalloyed DI. The formation of NbC carbides and the effect of finer pearlite in the microstructure have been attributed to such increases. In some studies, Nb addition was found to increase cast iron hardness due to the refinement of pearlitic interlamellar spacing, austenitic grain size, and the presence of NbC carbides [31–33].

Most studies on Nb addition to DI focus on mechanical properties, leaving aside the study of corrosion resistance, which is an important feature to consider since it can be used in the agriculture, automotive, and aerospace industries, etc. Alloying elements such as Si, Ni, Cr, and Cu, as well as carbide-forming elements (V, Ti, Mo) are commonly used to improve the corrosion resistance of DI [34,35]. Gutiérrez et al. [35] reported that DI alloyed with V and Mo and exposed to three different solutions (H_2SO_4, NaCl, and NaOH) showed corrosion damage due to the formation of galvanic couples between the graphite nodule and the ferrite. However, a high concentration of V and Mo carbides improves the corrosion resistance in H_2SO_4, and Si improves the scale resistance of cast iron by forming a light-resistant surface oxide in oxidizing atmospheres, such as H_2SO_4, HNO_3, HCl, and CH_3COOH [36]. Recently, Çelik et al. [37] have designed a novel ductile cast iron alloyed with a composition of 3.5 wt.% C, 4 wt.% Si, 1 wt.% Nb and 0–4 wt.% Al. They found, for all alloys, the room temperature phases were graphite, NbC carbides, and ferrite. In addition, they reported that the A_1 temperature of the alloy with the 4 wt.% Al addition reached 960 °C, indicating that this alloy can be used at a higher temperature, compared to the commercial SiMo cast iron; and with the increase in Al content, the thermal expansivity of ferrite increased, the graphite content decreased, and the graphite nodularity changed from spheroidal to vermicular. These results are similar to those obtained by Adebayo [38] and Sandikoglu [14] in DI alloyed with Al.

There are some studies about the fabrication of composite coatings on DI deposited by means of the diffusion thermo-reactive treatment with the objective to improve wear resistance [39,40] or corrosion resistance [41], and it has been reported that carbide formation on the substrate surface protects the ductile iron from corrosion in a NaCl environment.

Although the addition of some carbide-forming elements in HSDI has been studied, specific corrosion studies on the effects of Nb are not yet available. Therefore, the main objective of this work is to determine the influence of the high Si content together with the variation in the Nb contents alloyed in HSDI on its microstructure, and the effect on the corrosion rate when exposed to an acid environment.

2. Materials and Methods

The HSDI was produced in a medium-frequency induction furnace. Steel scrap was used as raw material, e.g., crankshafts and foundry waste. Graphite and copper C11000 were used to adjust the composition of the DI. Nodularization and inoculation processes were carried out with FeSi (75 wt.% Si) and FeSiMg (7 wt.% Mg) alloys by sandwich and ladle methods, respectively. In order to obtain different Nb contents in the HSDI, an FeNb alloy (65 wt.% Nb) in powder form was added to the ladle to achieve better dissolution during casting. After maintaining a temperature of 1400 ± 20 °C, the molten iron was poured into Y-shaped molds (175 × 150 × 25 mm^3) according to ASTM A-536-84 [42]. Molds were allowed to cool for 24 h before the samples were cut out of the useful area of the Y-block to obtain smaller samples of 25 × 25 × 15 mm^3, as shown in Figure 1.

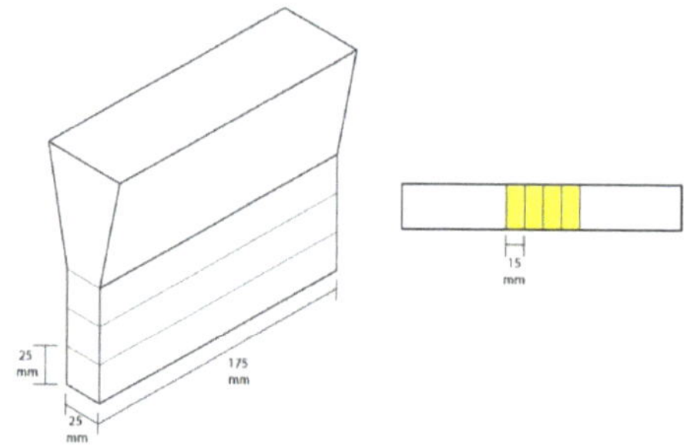

Figure 1. Y-block schematic representation and samples extraction region.

Chemical composition was determined using a Thermo ARL 3460 Optical Emission Spectrometer (Thermo Fisher Scientific, Waltham, MA, USA), taking the average of three measurements of each casting alloy. Samples were polished to a mirror finish and examined using an Olympus BX-51 optical microscope (OM, Olympus, Tokyo, Japan). Phase identification, nodule count, and nodularity were measured and quantified using Image Pro-Plus software (version 6.0, Rockville, MD, USA).

The surface of each sample was etched with Nital 2% for 20 s to reveal the phases and evaluate the microstructural evolution. In addition, the samples were analyzed in a JSM-6610LV scanning electron microscope (SEM, JEOL USA, Inc., Pleasanton, CA, USA), along with Energy-dispersive X-ray Spectroscopy (EDS, JEOL USA, Inc., Pleasanton, CA, USA) to determine the presence of Nb in the samples and the corroded surfaces.

Three corrosion tests were performed for each HSDI composition. Each sample was exposed to a 1N H_2SO_4 solution at room temperature in a flat cell using a three-electrode Gamry 1000 A potentiostat/galvanostat with an Ag/AgCl reference electrode and a scan rate of 1.0 mV/s ranging from −1000 to 1000 mV vs. E_{corr}. Corrosion data (E_{corr} and i_{corr}) were graphically extrapolated using the Tafel method from potentiodynamic curves.

In addition, Rockwell C hardness measurements were performed in an MACROMET®5100 Rockwell Type Hardness Tester, (Rockwell, Lake Bluff, IL, USA) with a test load of 150 kN to determine the influence of Nb content on HSDI.

3. Results and Discussion

3.1. Chemical Composition

The chemical composition of commercial DI and the different HSDIs produced in this study are shown in Table 1. It can be seen that the Si content in all alloys was higher than the conventional DI (>2.6 wt.% Si); according to [43], they correspond to HSDI. As Nb content varied from 0.3–0.9 wt.%, the samples were designated as U-HSDI for unalloyed HSDI, and depending on the Nb content, as HSDI-0.3% Nb, HSDI-0.6% Nb, and HSDI-0.9% Nb.

Table 1. Chemical composition for HSDI-Nb alloys studied (wt.%).

	C	Si	Mn	S	Cu	P	Nb	Mg	Ce	Cr	V	Mo	Fe
Commercial DI	3.25	2.60	0.85	0.006	0.71	0.018	-	0.053	0.031	-	-	-	bal.
U-HSDI	3.40	3.76	0.42	0.006	1.12	0.03	0.004	0.026	0.033	0.07	0.013	0.011	bal.
HSDI-0.3% Nb	3.42	3.76	0.42	0.005	1.12	0.03	0.296	0.024	0.033	0.08	0.011	0.012	bal.
HSDI-0.6% Nb	3.39	3.76	0.42	0.006	1.12	0.03	0.592	0.021	0.031	0.06	0.015	0.011	bal.
HSDI-0.9% Nb	3.40	3.75	0.42	0.007	1.12	0.03	0.921	0.020	0.031	0.06	0.012	0.014	bal.

3.2. Microstructural Analysis

Figure 2 shows the different microstructures identified by SEM for all samples. Graphite nodules embedded in a full pearlitic matrix can be seen for commercial DI in Figure 2a, whereas a pearlitic-ferritic matrix for U-HSDI is observed in Figure 2b. For all HSDI samples with different Nb content, a change in the phase fraction in the metallic matrix was observed, as well as the presence of small irregular particles randomly distributed and corresponding to polygonal primary NbC carbides. These particles form before the casting due to the high affinity between Nb and C atoms, so they are called primary carbides [18,21,22,24–31]. Unlike other carbide-forming elements (Mo, Cr, and V), precipitation of NbC carbides occurs randomly throughout the material rather than by positive segregation. Regarding the changes in the matrix, it is observed that for the HSDI-0.3% Nb sample, there is a refinement in the pearlite in comparison with the U-HSDI and the commercial DI samples, as reported by [27–29]; however, for the HSDI-0.6% Nb and HSDI-0.9% Nb samples, this refinement is not observed. On the contrary, a mostly ferritic matrix was obtained, which is attributable to the high silicon content in the alloy.

To quantify the size and amount of NbC carbides in each sample and the Nb distribution in the matrix using EDX mapping, five different fields were taken from each HSDI-Nb sample at 1000×. It can be seen that the distribution of NbC carbides is random both within the matrix and the graphite nodules, and the carbide size varies with Nb content, whereas the distribution of Nb is shown homogeneously in the matrix. In the HSDI-0.9% Nb sample, the size and amount of NbC carbides varies more than in the other two HSDI-Nb samples, as shown in Figure 3; i.e., the presence of NbC carbides increases with Nb content in HSDI. The average number of NbC carbides over a 1200 μm^2 area for each HSDI with 0.3, 0.6, and 0.9 wt.% Nb was 7, 12, and 16 particles, respectively. The NbC carbides exhibit a polygonal shape, as shown in Figure 2c–e, with average sizes of 0.5 and 4.5 µm for all samples, and a 4.49 µm size for the HSDI-0.3% Nb sample, showing a slight decrease in size as the Nb content increases to 3.48 µm for the HSDI-0.9% Nb sample. These variations occurred because FeNb was added in powder form during melting to achieve better dissolution efficiency into the liquid, resulting in fine carbides, similar to those reported in [21,24,28].

Summarizing, as the Nb content of HSDI increases, smaller precipitates are obtained, but in greater proportion. Figure 4 shows the average size (μm) and the number of NbC carbides in the HSDI-Nb.

Figure 5a shows the EDX analysis performed on the polygonal particles and the graphite nodule center of the HSDI-0.9% Nb to determine the presence of Nb, as shown in Figure 5b,c, respectively. This indicates that for this composition, the precipitation of primary carbides produces heterogeneous nucleation for graphite. Thus, in addition to the inoculant, some of the NbC carbides act as nucleation sites for graphite nodule growth.

Similarly, mapping was obtained by EDX at 1000× to determine the general distribution of the alloying elements HSDI-Nb, for C, Si, and Cu, which are shown to be homogeneously distributed, as shown Figure 6, with no segregation visible.

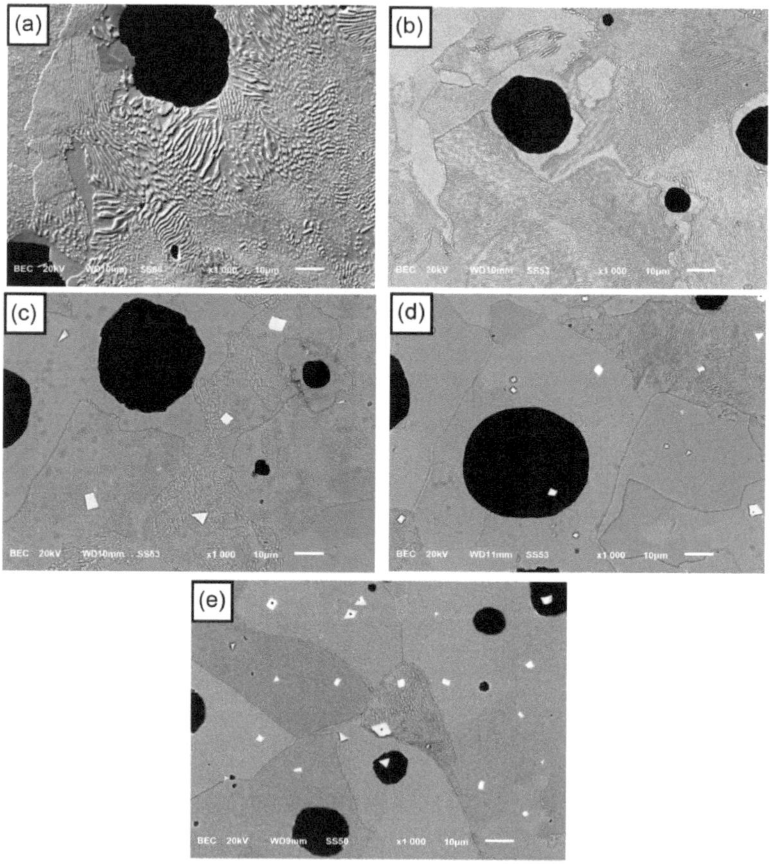

Figure 2. SEM micrographs of (**a**) commercial DI, (**b**) U-HSDI, (**c**) HSDI-0.3% Nb, (**d**) HSDI-0.6% Nb, (**e**) HSDI-0.9% Nb.

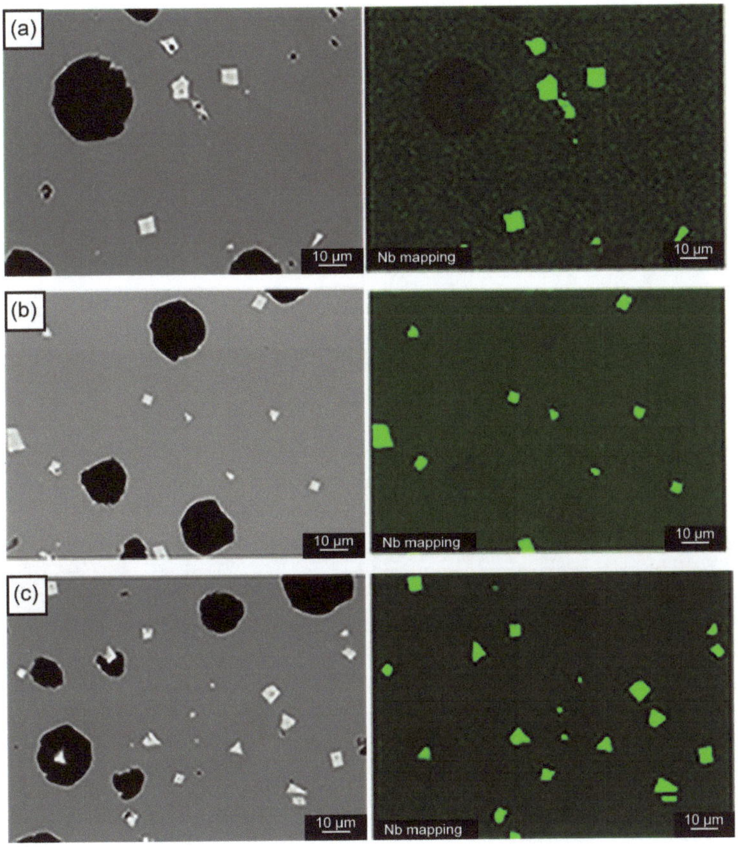

Figure 3. Distributions of Nb and NbC carbides in samples: (**a**) HSDI-0.3% Nb, (**b**) HSDI-0.6% Nb, (**c**) HSDI-0.9% Nb, obtained by EDX at 1000×.

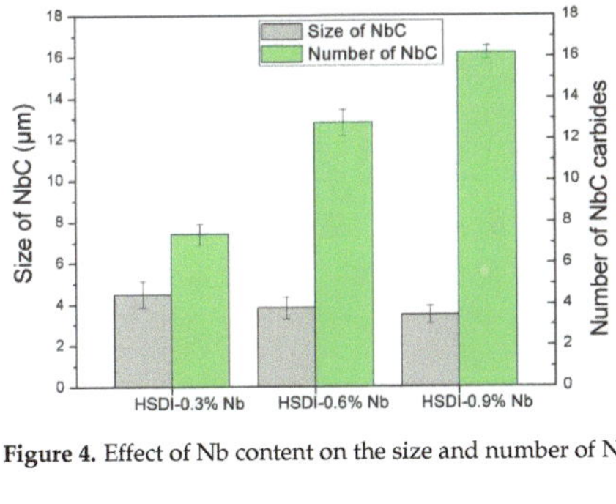

Figure 4. Effect of Nb content on the size and number of NbC carbides on HSDI.

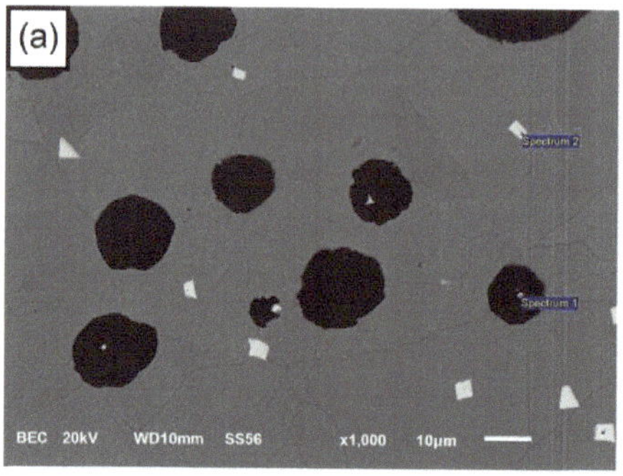

Element	Weight% Spectrum 1	Weight% Spectrum 2
C K	34.33	36.21
Nb L	65.67	63.79
Totals	100.00	100.00

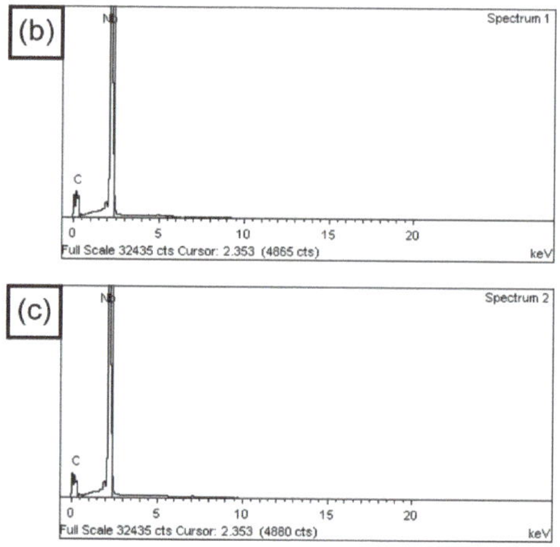

Figure 5. (**a**) SEM micrograph of HSDI-0.9% Nb sample showing the NbC carbides distribution, (**b**) EDX of NbC carbide inside the graphite nodule, (**c**) EDX of NbC carbide.

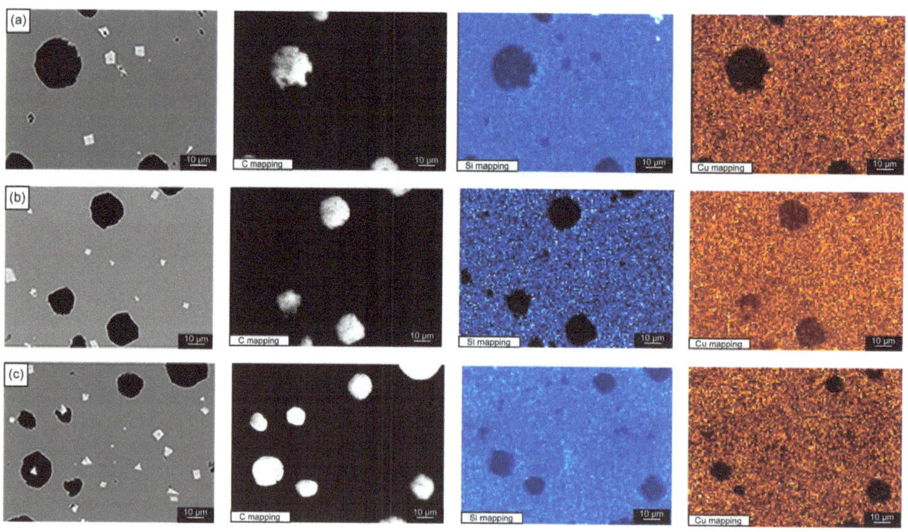

Figure 6. Distributions of C, Si, and Cu in samples: (**a**) HSDI-0.3% Nb, (**b**) HSDI-0.6% Nb, (**c**) HSDI-0.9% Nb, obtained by EDX at 1000×.

Nodularity showed a linear trend between 86 and 90% for the four different compositions. This can be explained by the effects of the corresponding amount of cerium oxide and the residual amount of Mg in the alloy [44–47]. In contrast, nodule count decreased with the presence of Nb in the HSDI, from 182 n/mm^2 for U-HSDI to a range of 147–154 n/mm^2 for HSDI-Nb, as shown in Figure 7. The solidification of conventional DI starts with the nucleation of the graphite nodules, whereas for HSDI-Nb, the primary NbC carbides were simultaneously formed in addition to the nucleation of the graphite nodules. These results indicate that HSDI-Nb has a lower amount of free carbon due to the formation of NbC carbides, reducing the nodule count. Some authors [25,27,48,49] report a similar pattern in which the nodule count is reduced by the addition of various carbide-forming elements such as V, Cr, and Nb. Moreover, the addition of more Nb had little effect on the nodule count. This can be confirmed with microstructures shown in Figure 2, where the HSDI-0.9% Nb sample has smaller NbC carbides in the center of the graphite nodules compared to the HSDI-0.3% Nb sample. This is due to the heterogeneous nucleation that occurs with a higher Nb content and keeps the nodules in the microstructure.

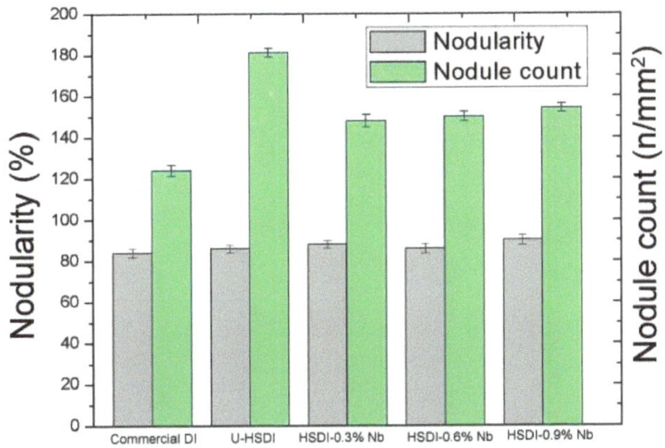

Figure 7. Effect of Nb on the nodularity and nodule count of HSDI samples.

The changes in the HSDI matrix due to the variation in Nb content are shown in OM microstructures of Figure 8. Whereas an increase in pearlite (dark phase) is observed in the HSDI-0.3% Nb sample compared to the U-HSDI sample, ferrite (light phase) increased in the HSDI-0.6% Nb and HSDI-0.9% Nb samples, until the ferritic matrix is almost complete. Although Nb is an element that refines pearlite [21,33,50,51], the results show that when the Nb content exceeds 0.3 wt.% in HSDI, this effect is attenuated by the formation of larger amounts of ferrite, which is mainly due to the fact that Si is a ferrite-stabilizing primary element, and its content in HSDI helps to suppress the pearlitizing effect of Nb in the matrix [52,53]. In addition, the strong affinity between Nb and C to form carbides reduces the amount of carbon available to form pearlite when the Nb content in HSDI increases.

The eutectoid transformation was limited by the reduction of free carbon in the liquid, and as a result of the precipitation of graphite and the NbC primary carbides, the ferrite content in the microstructure varies from 34% for U-HSDI to 88% for HSDI-0.9% Nb. Nevertheless, when only the metallic matrix is considered, the ferrite content increased from 39 to 98% with the addition of Nb, as shown in Figure 9. According to some studies [21,24,26,50], the addition of Nb as a microalloying element (0.09–0.11 wt.%) increases hardenability and the equilibrium carbon content in austenite, and causes further pearlitic transformation to a cast structure of DI. In the present study, the resulting microstructures of different HSDI-Nbs show that the pearlite is refined only to HSDI-0.3% Nb sample in comparison to U-HSDI; nevertheless, an increase of more than 0.3 wt.% Nb ended up decreasing the pearlite volume fraction.

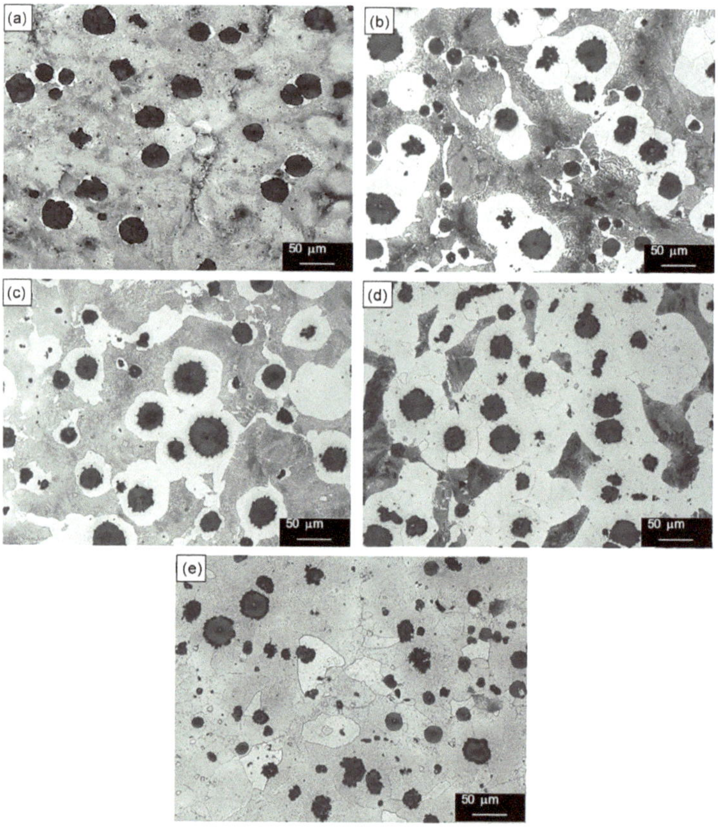

Figure 8. OM microstructures of samples: (**a**) commercial DI, (**b**) U-HSDI, (**c**) HSDI-0.3% Nb, (**d**) HSDI-0.6% Nb, (**e**) HSDI-0.9% Nb.

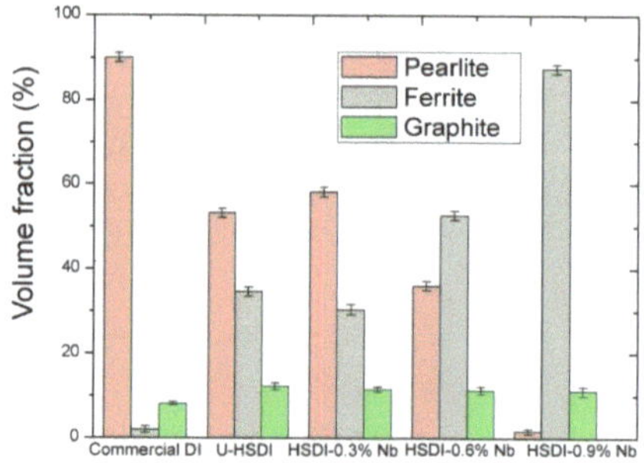

Figure 9. Variation of volume fraction of phases in commercial DI and HSDI alloyed with Nb.

3.3. *Hardness*

Hardness measurements were made for each composition to determine the effects of high Si and NbC carbide precipitation, as shown in Figure 10. The commercial DI presented with 29.9 HRC due to the pearlitic matrix, whereas in the Nb-HSDI samples, hardness incremented from 30.44 HRC for U-HSDI to 32.87, 35.93, and 38.38 HRC as Nb content increased. These values were attributed to the higher NbC precipitation with the Nb addition. These carbides increase hardness in HSDI; however, it is important to highlight that this property is a function of the overall microstructure present [24,25,29,30].

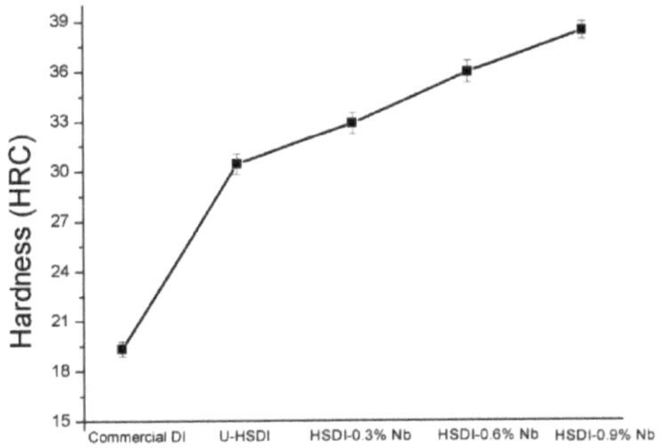

Figure 10. Hardness profile of commercial DI and HSDI with different Nb content.

Compared to U-HSDI, HSDI-0.3% Nb increased hardness by about 8%, and as would be expected with increasing Nb content in HSDI, HSDI-0.9% Nb increased hardness by about 26%. Si is one of the most common elements always present in DI chemical composition and exerts a strong effect on the mechanical properties of ferritic DI, as mentioned before [54,55]. Due to Si promoting ferrite formation in as-cast conditions, the presence of this element in the present HSDI-Nb conditions resulted in a fully ferritic microstructure, especially in the HSDI-0.9% Nb sample, as shown in Figure 8. Because NbC carbides could act as metallic matrix reinforcement, it would be expected that with higher Nb content, hardness and mechanical resistance would increase due to carbide precipitation. However, as the solubility limit of Nb is exceeded, the morphology, distribution, and segregation of carbides reduce the toughness of DI [44,45]. The results obtained indicate that the increase in hardness is mainly due to the greater presence of NbC carbides, but that the high Si content also favors the formation of ferrite, resulting in a moderate increase in hardness in HSDI alloyed up to 0.3 wt.% Nb.

3.4. *Corrosion*

The corrosion resistance of DI and HSDI samples was evaluated by potentiodynamic tests in an acidic medium (H_2SO_4). Figure 11 shows the polarization curves obtained for the different samples with similar behavior between them. The DI curve presents a lower potential than those of Nb-HSDI, so there are no significant differences in corrosion potentials. However, as the Nb content of HSDI increased, the current density shifted to lower values, indicating a decrease in the corrosion rate in HSDI-Nb.

Nevertheless, the HSDI-0.3 wt.% Nb sample has a higher current density value than the other samples, resulting in lower corrosion resistance in sulfuric acid. This behavior can be explained by a higher percentage of pearlite in the matrix of HSDI, as presented in Figure 9. Since the pearlite matrix is a laminar mixture of ferrite and cementite, there is a larger reaction surface between these phases, which exhibits electrochemical properties similar to those of grain boundaries and produces more high-energy and chemically active sites [55].

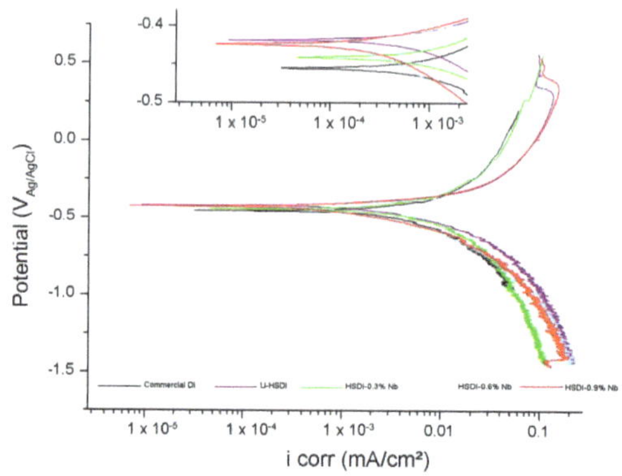

Figure 11. Potentiodynamic curves of commercial DI and HSDI with different Nb contents.

Figure 12 shows the corrosion rate of commercial DI and HSDI with different Nb contents. For DI with a conventional Si content, a corrosion rate of 4174 mpy (mills per year) was obtained, which is 3.5 times higher than that of the U-HSDI (1187 mpy). Thus, it was determined that the addition of Si to the HSDI significantly reduced the corrosion rate. In addition, it was found that the corrosion rate decreases as ferrite content increases, which is explained by the distribution of carbon between graphite and NbC carbides during the solidification of HSDI. Since the corrosion resistance of the phases present in Fe–C alloys depends on the amount of interstitial carbon present, ferrite is highly susceptible to pitting corrosion in acidic media. However, cast iron contains Si in addition to being an Fe–C alloy, which leads to the formation of a layer that acts as a protective barrier [56,57]. In an acidic medium, the corrosion of iron is controlled by the cathodic reaction, and the reaction rate increases when the pH decreases, which means that the corrosion rate is determined by the rate of hydrogen evolution [35]. As a result of the higher proportion of Si-saturated ferrite in HSDI, the surface develops a more compact and complete passive layer when interacting with the environment, which reduces the corrosion rate. In addition, Si reduces the potential difference between ferrite and graphite, increases the ferrite potential, and protects the substrate [57].

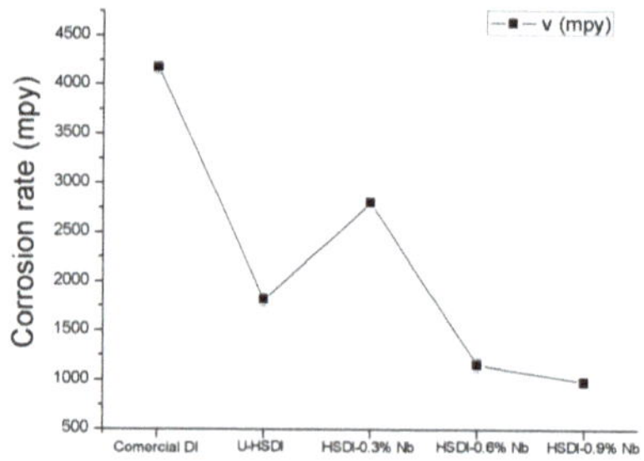

Figure 12. Corrosion rates of commercial DI and HSDI with different Nb contents.

It is obvious that the addition of Nb at concentrations higher than 0.3 wt.% to HSDI decreases the corrosion rate when comparing the U-HSDI sample with the HSDI-% Nb samples. The amount of fine NbC carbides precipitated in the metal matrix increased the corrosion resistance; as a result, the corrosion rate was reduced by 59% (1156 mpy) and

65% (986 mpy) for the HSDI-0.6 wt.% Nb and HSDI-0.9 wt.% Nb samples, respectively, compared to the DI-0.3 wt.% Nb sample (2802 mpy).

The corrosion resistance of cast iron alloys is primarily determined by their chemical composition, which causes changes in the microstructure. Figure 13 shows the SEM microstructures of all corroded surfaces where a preferential attack of pearlite in the matrix is evident upon exposure to acid solution; it can be assumed that a higher pearlite volume fraction reduced the corrosion resistance of HSDI in this environment.

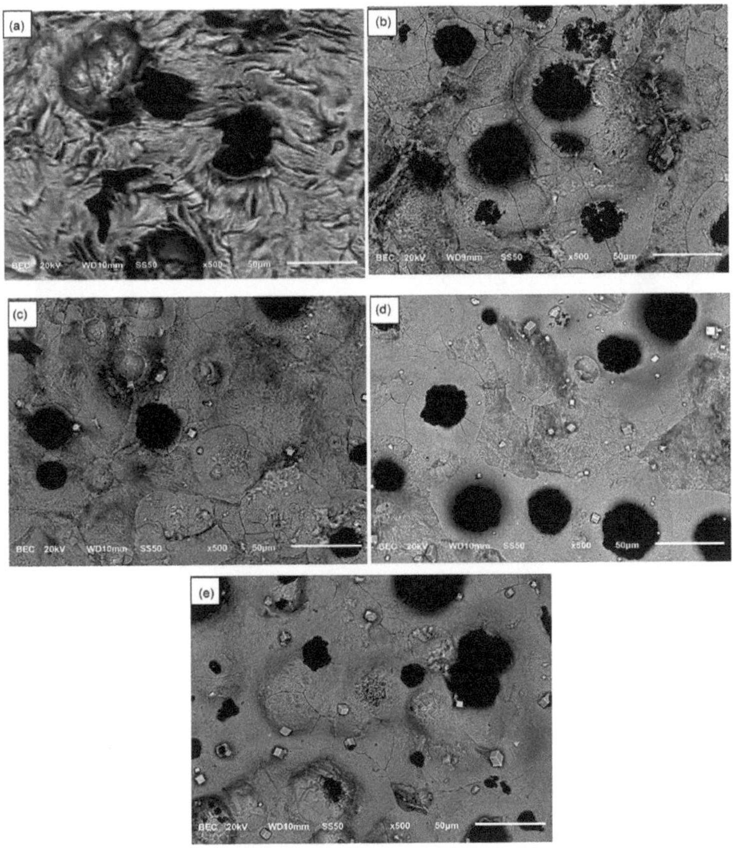

Figure 13. SEM microstructures of all corroded surfaces of HSDI with different Nb contents: (**a**) commercial DI, (**b**) U-HSDI, (**c**) HSDI-0.3% Nb, (**d**) HSDI-0.6% Nb, (**e**) HSDI-0.9% Nb.

Figure 14 shows the scheme of the corrosion mechanism of Nb-HSDI exposed to H_2SO_4, as a result of potentiodynamic curves and microstructures analysis. In Figure 14a the HSDI is presented with a mixture of ferritic–pearlitic matrix, with embedded graphite nodules and NbC dispersed within the microstructure. At this stage, it was not yet in contact with the electrolyte; that is, it did not start the corrosion process. At the beginning of the corrosion process, as can be seen in Figure 14b, upon contact with the electrolyte, galvanic pairs were formed between the graphite and matrix (Figure 14c), resulting in a corroded surface; however, the graphite nodules were unaffected, as shown in Figure 14d.

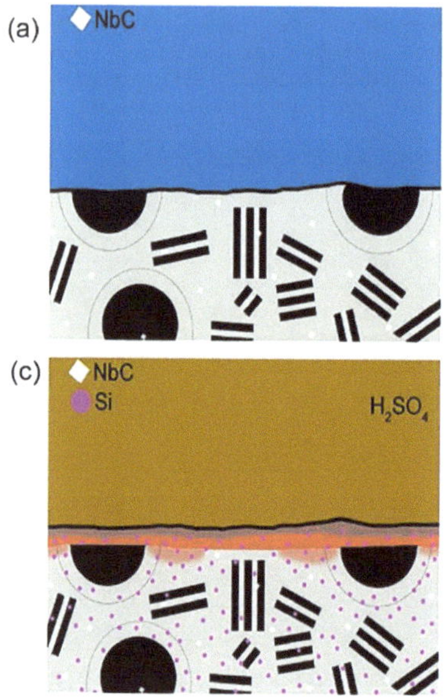

Figure 14. The schematic mechanism on HSDI corrosion exposed to H_2SO_4. (**a**) unaffected HSDI; (**b**) HSDI is initially exposed at H_2SO_4 solution; (**c**) galvanic pairs were formed between the graphite and HSDI matrix; (**d**) formation of a protective layer of silicon oxide (SiO_2) on HSDI surface.

Si atoms play an important role in current foundries because adding Si in cast iron produces cathodic protection [58,59]; as they are present in large numbers, it is possible to find them throughout the matrix. However, in the ferritic matrix, these atoms are more readily available to associate with oxygen, unlike pearlite. Figure 14c is an illustration of an arrangement of Si atoms, while Figure 14d shows that the residual Si atoms are oxidized to form a protective layer of silicon oxide (SiO_2). Iron with a high content of Si can easily form a SiO_2 passivation layer to protect the material [58,59] because the enthalpy of Si–O chemical bond formation is significantly more negative than that of Fe–O.

In the present study it was found that the electrochemical corrosion occurs due to the formation of a galvanic couple between the graphite and the metallic matrix. Therefore, it can be concluded that the corrosion resistance of HSDI in an acidic environment is significantly influenced by the high Si content and Nb content, which contribute to the reduction of the corrosion rate. Nb promotes carbide formation, while Si promotes a ferritic matrix, which in combination results in HSDI being passive. Therefore, Nb content higher than 0.3 wt.% and high Si content in DI can be used as protective systems against cathodic corrosion.

4. Conclusions

The current study shows the effects of Nb on the microstructural characteristics, hardness, and corrosion resistance of high-silicon ductile cast iron (HSDI)-3.6 wt.% Si. It was found that:

- The formation of polygonal NbC carbides was observed in HSDI upon the Nb addition in a range of 0.3–0.9 wt.%. These carbides have a size between 0.5 and 4.5 μm, which decreases with increasing Nb content in HSDI.
- The presence of NbC carbides increases with Nb content, resulting in heterogeneous graphite nodule nucleation.
- Nodularity maintained a linear trend between 86 and 90%, and the nodule count decreased with the presence of Nb in the HSDI, from 182 n/mm^2 for U-HSDI to 147–154 n/mm^2 for HSDI-% Nb samples.
- An increment in hardness is due to the greater presence of NbC carbides in combination with high Si, resulting in an increase from 30.44 to 38.38 HRC for U-HSDI and HSDI-0.9 wt.% Nb, respectively.
- Electrochemical corrosion is achieved by the formation of a galvanic couple between the graphite and the metal matrix which preferentially attacks the pearlite, with the greatest corrosion occurring at HSDI-0.3 wt.% Nb.
- The combination of Nb with Si in HSDI helps to reduce the corrosion rate, since Nb promotes the development of NbC carbides and Si promotes the ferritic matrix. If the Nb content exceeds 0.3 wt.% and the Si content in the DI is more than 2.8 wt.%, protective systems against cathodic corrosion can be applied to the HSDI and it can be used in agriculture applications.

For future work, HSDI austempering is one of the main interests of the authors as a highly developed alternative for applications in the agricultural industry, since with this research it has been shown that Si and Nb in ductile iron play an important role in the resulting microstructures as well as in the distribution of NbC, affecting the mechanical properties and the corrosion resistance of such castings.

Author Contributions: Conceptualization, N.A.R.R., C.R.M.V. and D.G.N.; Methodology, N.A.R.R. and C.R.M.V.; Validation, J.S.G.V. and E.A.C.; Formal Analysis, N.A.R.R., D.G.N. and F.A.M.G.; Investigation, D.G.N. and N.A.R.R.; Resources, C.R.M.V.; Data Curation, E.A.C. and D.G.N.; Writing—Original Draft Preparation, N.A.R.R. and D.G.N., Software, D.G.N., E.A.C. and F.A.M.G.; Writing—Review and Editing, N.A.R.R. and D.G.N.; Supervision, J.S.G.V. All authors have read and agreed to the published version of the manuscript.

Funding: This research received no external funding.

Data Availability Statement: The research data and methods used in the investigation are presented in sufficient detail in the paper to allow other researchers to reproduce the work.

Acknowledgments: The authors are grateful to the TECNM (project: 11548.21-P) and the Universidad Autónoma de Coahuila for the facilities granted.

Conflicts of Interest: The authors declare no conflict of interest.

References

1. Chikali, P.; Shinde, V. Analysis of machinability in ductile iron casting. *Mater. Today Proc.* **2020**, *27*, 584–588. [CrossRef]
2. Scorza, D.; Ronchei, C.; Vantadori, S.; Zanichelli, A.; Carpinteri, A. A novel methodology for fatigue assessment of Ductile Cast Iron (DCI) with solidification defects. *Procedia Struct. Integr.* **2022**, *41*, 500–504. [CrossRef]
3. Skaland, T.; Grong, O.; Grong, T. A model for the graphite formation in ductile cast iron: Part I. Inoculation Mechanisms. *Metall. Mater. Trans. A* **1993**, *24*, 2321–2345. [CrossRef]
4. González Martínez, R.; Sertucha, J.; Lacaze, J. Effects of cobalt on mechanical properties of high silicon ductile irons. *Mater. Sci. Technol.* **2020**, *36*, 1292–1300. [CrossRef]
5. Khalil-Allafi, J.; Amin-Ahmadi, B. Influence of Mold Preheating and Silicon Content on Microstructure and Casting Properties of Ductile Iron in Permanent Mold. *J. Iron Steel Res. Int.* **2011**, *18*, 34–39. [CrossRef]
6. Riebisch, M.; Seiler, C.; Pustal, B.; Bührig-Polaczek, A. Microstructure of As-Cast High-Silicon Ductile Iron Produced via Permanent Mold Casting. *Int. J. Met.* **2018**, *13*, 112–120. [CrossRef]
7. Almanza, A.; Dewald, D.; Licavoli, J. Effect of Cobalt Additions on the Microstructure and Mechanical Properties of As-Cast Thin-Wall Ductile Iron. *Int. Met.* **2021**, *15*, 417–432. [CrossRef]
8. Franzen, D.; Weiß, P.; Pustal, B.; Bührig-Polaczek, A. Modification of silicon microsegregation in solid-solution-strengthened ductile iron by alloying with aluminum. *Int. J. Met.* **2020**, *14*, 1105–1120. [CrossRef]

9. Almanza, A.; Dewald, D.; Licavoli, J.; Sanders, P.G. Influence of cobalt in the tensile properties of inch ductile iron Y-blocks. *Int. Met.* **2020**, *15*, 433–446. [CrossRef]
10. Mikoleizik, P.; Geier, G. SiWind—Development of materials for offshore wind power plants of the multi megawatt range. *Cast. Plant Technol. Int. (CP+T)* **2014**, *9*, 64–69.
11. Werner, H.; Lappat, I.; Aurich, B. Mischkristallverfestigte EN-GJS-Werkstoffe für Groβ-und Schwergussteile. *Giesserei* **2016**, *103*, 38–42.
12. De la Torre, U.; Loizaga, A.; Lacaze, J.; Sertucha, J. As cast high silicon ductile irons with optimised mechanical properties and remarkable fatigue properties. *Mater. Sci. Technol.* **2014**, *30*, 1425–1431. [CrossRef]
13. Ikeda, T.; Noda, N.; Sano, Y. Conditions for notch strength to be higher than static tensile strength in high–strength ductile cast iron. *Eng. Fract. Mech.* **2019**, *206*, 75–88. [CrossRef]
14. Sandikoglu, A.; Gecu, R. Microstructural, mechanical and tribological characterization of aluminum-alloyed ductile cast irons based on aluminum content. *J. Alloys Compd.* **2021**, *879*, 160428. [CrossRef]
15. Kasvayee, K.A.; Ghassemali, E.; Svensson, I.L.; Olofsson, J.; Jarfors Anders, E.W. Characterization and modeling of the mechanical behavior of high silicon ductile iron. *Mater. Sci. Eng. A* **2017**, *708*, 159–170. [CrossRef]
16. Bjorkgen, L.E.; Hamberg, K. Silicon alloyed ductile iron with excellent ductility and machinability. *Foundryman* **2001**, *94*, 42–51.
17. Foglio, E.; Gelfi, M.; Pola, A.; Goffelli, S.; Lusuardi, D. Fatigue Characterization and Optimization of the Production Process of Heavy Section Ductile Iron Castings. *Int. J. Met.* **2016**, *11*, 33–43. [CrossRef]
18. Riebisch, M.; Pustal, B.; Bührig-Polaczek, A. Influence of carbide-promoting elements on the microstructure of high-silicon ductile iron. *Int. J. Met.* **2020**, *14*, 1152–1161. [CrossRef]
19. Górny, M.; Kawalec, M.; Gracz, B.; Tupaj, M. Influence of Cooling Rate on Microstructure Formation of Si–Mo Ductile Iron Castings. *Metals* **2021**, *11*, 1634. [CrossRef]
20. Stets, W.; Löblich, H.; Gassner, G. Solution Strengthened Ferritic Ductile Cast Iron Properties, Production and Application. *Int. Met.* **2014**, *8*, 35–40. [CrossRef]
21. Chen, X.; Xu, J.; Hu, H.; Mohrbacher, H.; Kang, M.; Zhang, W.; Guo, A.; Zhai, Q. Effects of niobium addition on microstructure and tensile behavior of as-cast ductile iron. *Mater. Sci. Eng. A* **2017**, *688*, 416–428. [CrossRef]
22. Souza, T.N.F.; Nogueira, R.A.P.S.; Franco, F.J.S.; Aguilar, M.T.P.; Cetlin, P.R. Mechanical and microstructural characterization of Nodular Cast Iron (NCI) with Niobium Additions. *Mater. Res.* **2014**, *17*, 1167–1172. [CrossRef]
23. Zhang, H.; Wu, Y.; Li, Q.; Hong, X. Mechanical properties and rolling-sliding wear performance of dual phase austempered ductile iron as potential metro wheel material. *Wear* **2018**, *406–407*, 156–165. [CrossRef]
24. Bedolla-Jacuinde, A.; Solis, E.; Hernandez, B. Effect of niobium in medium alloyed ductile cast irons. *Int. J. Cast Met. Res.* **2003**, *16*, 481–486. [CrossRef]
25. Sun, X.; Yin, Z.; Chang, L.; Hua, Q.; Zhai, Q. Existence of Niobium in Ductile Iron and Its Effect on the Morphology of Graphite Ball. *Suppl. Proc.* **2012**, *2*, 603–611.
26. Li, H.; Yang, C.; Wei, Z.; Chen, X.; Zhai, Q. Morphology and Distribution of Nb-Rich Phase and Graphite in Nb Microalloyed Ductile Iron. *Charact. Miner. Met. Mater.* **2015**, *2015*, 579–585.
27. Sckudlarek, W.; Krmasha, M.N.; Al-Rubaie, K.S.; Preti, O.; Milan, J.C.; da Costa, C.E. Effect of austempering temperature on microstructure and mechanical properties of ductile cast iron modified by niobium. *J. Mater. Res. Technol.* **2021**, *12*, 2414–2425. [CrossRef]
28. Chen, X.; Zhao, L.; Zhang, W.; Mohrbacher, H.; Wang, W.; Guo, A.; Zhai, Q. Effects of niobium alloying on microstructure, toughness, and wear resistance of austempered ductile iron. *Mater. Sci. Eng. A* **2019**, *760*, 186–194. [CrossRef]
29. Ahmed, M.; Soliman, M.; Youssef, M.; Bähr, R.; Nofal, A. Effect of niobium on the microstructure and mechanical properties of alloyed ductile irons and austempered ductile irons. *Metals* **2021**, *11*, 703. [CrossRef]
30. Alias, S.K.; Abdullah, B.; Jaffar, A.; Abdullah, A.H.; Jenal, N. Development of high strength ductile iron with niobium addition. *Adv. Mater. Res.* **2012**, *576*, 366–369. [CrossRef]
31. Souza Oliveira, P.A.; Luiz, G.W.; Rodrigues Custódio da Silva, W.J.; Dolabella Portella, P.; Woydt, M.; Burbank, J. Abrasive wear behavior of austempered ductile iron with niobium additions. *Wear* **2019**, *440–441*, 203065. [CrossRef]
32. Zhou, W.; Zhu, H.; Zheng, D.; Zheng, H.; Hua, Q.; Zhai, Q. Niobium alloying effect in high carbon equivalent grey cast iron. *China Foundry* **2011**, *08*, 36–40.
33. Mohrbacher, H.; Zhai, Q. Niobium alloying in grey cast iron for vehicle brake discs. In Proceedings of the Materials Science & Technology Conference, Columbus, OH, USA, 16–20 October 2011; ASM International: Detroit, MI, USA, 2011; pp. 434–445.
34. Sun, Y.; Hu, S.; Xiao, Z.; You, S.; Zhao, J.; Lv, Y. Effects of nickel on low-temperature impact toughness and corrosion resistance of high-ductility ductile iron. *Mater. Des.* **2012**, *41*, 37–42. [CrossRef]
35. Gutiérrez Pérez, V.H.; Cruz Ramírez, A.; Olvera Vázquez, S.L.; Colin García, E.; Sánchez Alvarado, R.G.; Delgado Pámanes, M.F.; Rivera Salinas, J.E. Study of the effects of vanadium and molybdenum on the microstructure of ductile iron (DI) and austempered ductile iron (ADI) and their corrosion resistance. *Acta Univ.* **2022**, *32*, 1–20. [CrossRef]
36. Elliott, R. *Cast Iron Technology*; Butterworth-Heinemann: Oxford, UK, 1988.
37. Celik Çelik, G.A.; Tzini, M.-I.T.; Polat, S.; Hakan Atapek, Ş.H.; Haidemenopoulos, G.N. Thermal and microstructural characterization of a novel ductile cast iron modified by aluminum addition. *Int. J. Miner. Met. Mater.* **2020**, *27*, 190–199. [CrossRef]

38. Adebayo, A.O.; Alaneme, K.K.; Oyetunji, A. Corrosion evaluation of austempered aluminium-alloyed ductile irons in well water and 0.5M NaCl solution. *J. Chem. Technol. Metall.* **2021**, *56*, 180–193.
39. Günen, A.; Kalkandelen, M.; Gök, M.S.; Kanca, E.; Kurt, B.; Karakaş, M.S.; Karahan, I.H.; Çetin, M. Characteristics and high temperature wear behavior of chrome vanadium carbide composite coatings produced by thermo-reactive diffusion. *Surf. Coat. Technol.* **2020**, *402*, 126402. [CrossRef]
40. Günen, A.; Soylu, B.; Karakaş, O. Titanium carbide coating to improve surface characteristic, wear and corrosion resistance of spheroidal graphite cast irons. *Surf. Coat. Technol.* **2022**, *437*, 128280. [CrossRef]
41. Mariani, F.E.; Takeya, G.S.; Lombardi, A.N.; Picone, C.A.; Casteletti, L.C. Wear and corrosion resistance of Nb-V carbide layers produced in vermicular cast iron using TRD treatments. *Surf. Coat. Technol.* **2020**, *397*, 126050. [CrossRef]
42. *ASTM A536-84*; Standard Specification for Ductile Iron Castings. ASTM International: West Conshohocken, PA, USA, 2019.
43. *SIST EN 1563:2012*; Founding—Spheroidal Graphite Cast Irons. European Standard: Brussels, Belgium, 2012.
44. Chen, X.R.; Zhai, Q.J.; Dong, H.; Dai, B.H.; Mohrbacher, H. Molybdenum alloying in cast iron and steel. *Adv. Manuf.* **2019**, *8*, 3–14. [CrossRef]
45. Wang, B.; Barber, G.C.; Qiu, F.; Zou, Q.; Yang, H. A review: Phase transformation and wear mechanisms of single-step and dual-step austempered ductile irons. *J. Mater. Res. Technol.* **2020**, *9*, 1054–1069. [CrossRef]
46. Vicente, A.D.A.; Sartori Moreno, J.R.; Santos, T.F.D.A.; Espinosa, D.C.R.; Tenório, J.A.S. Nucleation and growth of graphite particles in ductile cast iron. *J. Alloys Compd.* **2019**, *775*, 1230–1234. [CrossRef]
47. Lacaze, J.; Sertucha, J.; Åberg, L.M. Microstructure of As-cast Ferritic-pearlitic Nodular Cast Irons. *ISIJ Int.* **2016**, *56*, 1606–1615. [CrossRef]
48. Vdovin, K.N.; Gorlenko, D.A.; Feoktistov, N.A.; Kuryaev, D.V. Study of the effect of isothermal holding on parameters of graphite phase in indefinite chromium-nickel cast iron alloyed by nitrogen and vanadium. *CIS Iron Steel Rev.* **2019**, *17*, 30–33. [CrossRef]
49. Rezvani, M.; Harding, R.A.; Campbell, J. The effect of vanadium in as-cast ductile iron. *Int. J. Cast Met. Res.* **1997**, *10*, 1–15. [CrossRef]
50. Yang, C.; Zhang, W.; Li, H.; Chen, X.; Zhai, Q. Effects of Niobium Microalloying on the Pearlite of Ductile Iron. *Charact. Miner. Met. Mater.* **2015**, *2015*, 601–606.
51. Alhussein, A.; Risbet, M.; Bastien, A.; Chobaut, J.; Balloy, D.; Favergeon, J. Influence of silicon and addition elements on the mechanical behavior of ferritic ductile cast iron. *Mater. Sci. Eng. A* **2014**, *605*, 222–228. [CrossRef]
52. Basso, A.; Caldera, M.; Rivera, G.; Sikora, J. High Silicon Ductile Iron: Possible Uses in the Production of Parts with Dual Phase ADI; Microstructure. *ISIJ Int.* **2012**, *52*, 1130–1134. [CrossRef]
53. González, F.; Houbaert, Y. A review of ordering phenomena in iron-silicon alloys. *Rev. Metal.* **2013**, *49*, 178–199. [CrossRef]
54. Weiß, P.; Tekavčič, A.; Bührig-Polaczek, A. Mechanistic approach to new design concepts for high silicon ductile iron. *Mater. Sci. Eng. A* **2018**, *713*, 67–74. [CrossRef]
55. Ahmad, Z. *Principles of Corrosion Engineering and Corrosion Control*, 1st ed.; Butterworth-Heinemann: Oxford, UK, 2006.
56. You, Z.; Lai, Y.; Zeng, H.; Yang, Y. Influence of water and sodium chloride content on corrosion behavior of cast iron in silty clay. *Constr. Build. Mater.* **2020**, *238*, 117762. [CrossRef]
57. Ding, D.; Zhang, Y.; Yu, X.; Fang, B.; Guo, J.; Li, J.; Liu, L.; Du, C. Effects of environmental factors on corrosion behavior of high-silicon cast iron in Shanxi soil medium. *Anti-Corros. Methods Mater.* **2018**, *65*, 538–546. [CrossRef]
58. Krawiec, H.; Stypuła, B.; Stoch, J.; Mikołajczyk, M. Corrosion behaviour and structure of the surface layer formed on austempered ductile iron in concentrated sulphuric acid. *Corros. Sci.* **2006**, *48*, 595–607. [CrossRef]
59. Zhang, Y.; Xiao, J.; Zhang, Y.; Liu, W.; Pei, W.; Zhao, A.; Zhang, W.; Zeng, L. The study on corrosion behavior and corrosion resistance of ultralow carbon high silicon iron-based alloy. *Mater. Res. Express* **2021**, *8*, 026504. [CrossRef]

coatings

MDPI

Article

Effect of Energy Density on the Microstructure and Wear Resistance of Nickel-Based WC Coatings by Laser Cladding of Preset Zr702 Alloy Plates

Guangsheng Zhang [1,2,3], Aixin Feng [1,2,3,*], Pu Zhao [1], Xiaoming Pan [1] and Huibin Feng [4]

1 College of Mechanical and Electrical Engineering, Wenzhou University, Wenzhou 325035, China; 20461439004@stu.wzu.edu.cn (G.Z.)
2 Rui'an Graduate College, Wenzhou University, Wenzhou 325206, China
3 Zhejiang Provincial Key Laboratory of Laser Processing Robots, Machinery Industry Key Laboratory of Laser Processing and Testing, Wenzhou 325035, China
4 Zhejiang Linuo Fluid Control Technology Co., Ltd., Wenzhou 325200, China
* Correspondence: aixfeng@wzu.edu.cn; Tel.: +86-0577-8659-6099

Abstract: This study aimed to evaluate the microstructure and wear resistance of laser cladding coatings with different energy densities in the case of a preset 0.5 mm thick Zr702 alloy plate to determine the specific present form of Zr elements and the optimal laser energy density. Thereby, microscopic characterization and performance tests were carried out by the microhardness tester, X-ray diffraction analysis (XRD), scanning electron microscopy (SEM), and tribometer. The results showed that, at different energy densities, the Zr elements in the coating were mainly in the form of ZrC and (Zr,W)C, which are hard particles with high wear resistance, and diffusely distributed in the coating to have second-phase-strengthening effects. Moreover, when the energy density was 28.3 J/mm^2, the coating was well fused and had the highest microhardness of 936.4 HV$_{0.2}$. The wear rate of the coating was the lowest at 90.8 μm^3/(m·N). The wear was characterized by hard particle spalling and abrasive wear.

Keywords: laser cladding; energy density; Zr702 alloy; (Zr,W)C; wear resistance; microstructure

Citation: Zhang, G.; Feng, A.; Zhao, P.; Pan, X.; Feng, H. Effect of Energy Density on the Microstructure and Wear Resistance of Nickel-Based WC Coatings by Laser Cladding of Preset Zr702 Alloy Plates. *Coatings* **2023**, *13*, 826. https://doi.org/10.3390/coatings13050826

Academic Editor: Alina Vladescu

Received: 13 March 2023
Revised: 17 April 2023
Accepted: 19 April 2023
Published: 25 April 2023

1. Introduction

Solar energy has the advantages of being environmentally friendly, sustainable, and green [1] and is one of the important development directions for building a world of clean energy. Silicon-based solar batteries occupy more than 90% of the market share in the solar batteries field due to their excellent stability [2]. Polysilicon is the most-common raw material for the production of silicon-based solar batteries. However, the sealing surfaces of control valves used in the polysilicon preparation process are susceptible to failure because of wear under the harsh conditions of ultra-hard silicon powder particles (Mohs hardness of 7.5), resulting in long downtimes of the polysilicon production line for maintenance, which greatly increases the costs of production and time. Therefore, the development of ultra-high wear-resistant coatings for control valve sealing surfaces in the field of polysilicon preparation has become a key problem to be solved.

Strengthening treatment on traditional control valve sealing surfaces mainly uses arc welding, plasma arc welding, and other overlay welding technology. However, it has the disadvantageous problems of cracking, an uneven hardness distribution, and coarse crystals. As a laser cladding technology based on overlay welding technology, it has received extensive attention and research from experts and scholars at home and abroad because of its dense microstructure [3], fine grain size [4], and good mechanical properties [5]. From the perspective of the addition and in situ generation of hard phases, a large number of scholars have carried out research on reinforcing phases such as WC, SiC,

NbC, Al_2O_3, and BN [6–10]. However, it is worth noting that compounds of Zr such as ZrB_2 (microhardness of 22 GPa), ZrN (19.6 GPa), ZrC (29.2 GPa), and ZrO_2 (12 GPa) have excellent mechanical properties with high hardness. Yang, C. and X. Li [11,12] investigated the effect of different ZrN additions on the microstructure and properties of TC11 titanium alloy by laser cladding, and the results showed that the addition of ZrN can greatly improve the wear resistance of the coating. Y.-W. Yong [13] showed that a nickel-based alloy coating with ZrO_2 + C was melted on AISI 1045, and the results showed that fine ZrC particles were generated in situ in the coating and the wear resistance was 2.7 times that of Ni25. Guo, J. [14] showed that a 35 wt.% ZrB_2 + Ni60A coating was produced on pure titanium substrates, and he found that the coating had higher hardness than conventional NiCrBSi composite coatings. The addition of ZrB_2 effectively reduced the adhesive wear of the NiCrBSi coatings. Liu Kun [15] prepared NiCrBSi coating on a Zr702 substrate, and the results showed that the coating microstructure consisted of NiZr solid solution and bulk reinforcement including $NiZr_2$, $Zr_5(SixNi_{1-x})_4$, and ZrB_2. The average microhardness reached 1200–1300 $HV_{0.2}$, nearly seven times that of the Zr702, overturning the usual thinking that lower dilution is better in laser cladding.

Currently, no studies have been found to improve the properties with Zr702 in nickel-based coatings. Therefore, this paper creatively adopted a preset Zr702 alloy plate to investigate the effect of different energy densities on the microstructure and wear resistance of the laser cladding of Ni60A + 20 wt.% WC composite coatings, providing a reference for the research and application in the field of laser cladding of preset high-performance alloy plates. Figure 1 shows the schematic diagram and the expected Microstructure of the laser melting of the preset Zr702 plates.

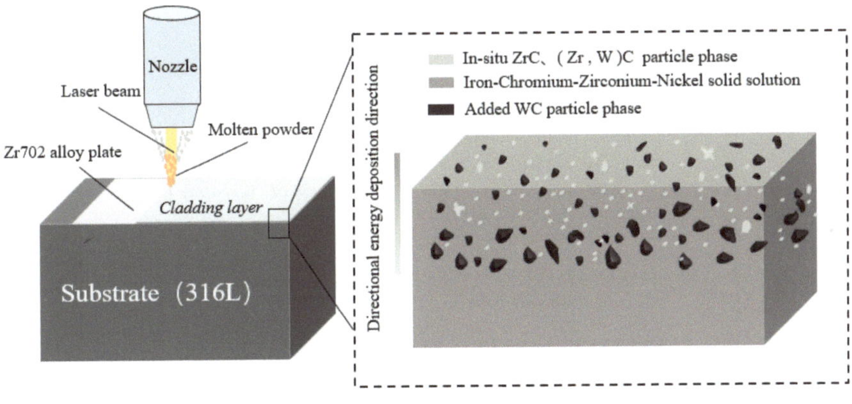

Figure 1. Schematic diagram and expected microstructure for laser cladding of preset Zr702 alloy plates.

2. Materials and Methods

The bond phase powder used in this experiment is the 45–106 μm Ni60A alloy powder(Xindian Materials Co., Ltd., Shanghai, China). The hard phase is the 150–325 mesh powders of WC (Nangong Xindun Alloy Welding Material Spraying Co., Ltd., Xingtai, China). Zr702 (Shanxi TOP Metal Co., Ltd., Shanxi, China) is a 0.5 mm-thick zirconium hafnium alloy plate with dimensions of 90 mm × 60 mm. 316 L was selected as the substrate with dimensions of 90 mm × 60 mm × 9 mm. After presetting the Zr702 alloy, the coating was prepared by Model RFL-C4000 fiber-coupled semiconductor laser cladding equipment with a powder mixture of Ni60A + 20 wt.% WC under an argon gas-protected environment. The laser specifications are as follows: a wavelength of about 1064 nm and a peak power of 4000 W. The friction wear test of the coating was performed by a tribometer (UMT-5, Bruker, Billerica, MA, USA) with a wear time of 30 min under the parameters of a normal load of 15 N at room temperature, a reciprocating stroke of 2 mm, a frequency of 5 Hz, and a Si_3N_4 friction pair of 4 mm in diameter. After taking the 3D morphology of wear surface of the coating with the PCB 3D optical profiler, the wear loss volume was calculated by the Vision software. The specific composition of the experimental material is shown in Table 1. The microhardness from the top of the cladding layer to the substrate was measured by a microhardness tester (MVA-402TS), and the load and duration used for each hardness measurement were 200 g and 20 s, respectively. Microhardness experiments were performed three times. XRD (FRINGE EV, LANScientific, Suzhou, China) and SEM (Nova 200 Nano SEM, FEI, Hillsboro, United States) with energy dispersive spectrometry (EDS, FEI, Hillsboro, OR, USA) were used to characterize the microstructure. The laser parameters of the cladding coating were designed as shown in Table 2.

Table 1. Elemental composition (wt.%) of Ni60A and Zr702.

Element	Zr + Hf	Hf	Si	Cr	Fe	Ni	C	B	N	H	O
Zr702 alloy plates	≥99.2	≤4.5			0.2		0.05		0.025	0.005	0.16
Ni60A power			4.3	13.7	4.43	Bal.	0.6	3.18			

Table 2. Experimental laser parameters design for composite coatings.

Sample	Off-Focus Volume (mm)	Laser Power (W)	Scanning Speed (mm/s)	Feed Rate (r/min)	Beam Diameter (mm)	Energy Density (J/mm^2)
S1	2.5	800	6	16	3	22.6
S2	2.5	1000	6	16	3	28.3
S3	2.5	1300	6	16	3	36.8
S4	2.5	1500	6	16	3	42.5

3. Results and Discussion

3.1. Section Morphology and Microhardness

Scholars have opined that a large dilution of the coating will weaken the mechanical properties [16], generally not exceeding 20%. However, for this experiment, the dilution rate had to be increased to melt through the Zr702 alloy plate and melt a small portion of the 316 L substrate to achieve good metallurgical bonding of the dissimilar metals under the condition of preset 0.5 mm-thick Zr702 alloy plates. As can be seen from Figure 2, there was no obvious cracking in the overall cladding section and only a few pores. When the laser energy density was 22.6 J/mm^2, the S1 coating just melted through the Zr702 alloy plate and melted the 316 L substrate with less dilution into the melt pool. In Figure 2c, the article defines the melt height of the cladding layer as L_1 and the melt depth as L_2. The melt height and width were measured by the LEXT software (OLS4100 3.1.7) attached to the confocal microscope, and the dilution rates of the coatings in Figure 2a–d were 33.4%, 38.6%, 43.9%, and 48.8% respectively. The dilution rate increases linearly with increasing laser energy density.

Figure 3 presents microhardness diagrams of the top of the laser cladding coating of preset Zr702 to the 316 L substrate at different laser energy densities. The average microhardness value of the S1 to S4 coatings are 885.8 $HV_{0.2}$, 936.4 $HV_{0.2}$, 778.9 $HV_{0.2}$, and 736.2 $HV_{0.2}$, respectively. With increasing energy density, the microhardness showed a trend of increasing and then decreasing, which is consistent with the hardness trend in literature studies [17]. When the energy density was 22.6 J/mm^2, the S1 coating was not sufficiently melted and recrystallized with less eutectic phase precipitation and a more γ-(Fe,Ni) solid solution. The insufficiently melted particles slowed down the Marangoni convection inside the melt pool; dislocations at the grain boundaries were hindered and stresses were more concentrated, resulting in poor coating quality and a lower hardness of the S1 coating. When the energy density was 28.3 J/mm^2, the S2 coating microhardness reached a maximum value of 936.4 HV$_{0.2}$. With the increasing energy density, the microhardness of the S3 and S4 coatings gradually decreased, which was mainly attributed to the following two factors. On the one hand, the increase in energy density inevitably led to the dilution of 316 L base material elements into the coating, which weakened the mechanical properties of the coating [18]. On the other hand, due to the increasing energy input, the grain size was coarse, and the hardness decreased according to the Hall–Petch equation [19].

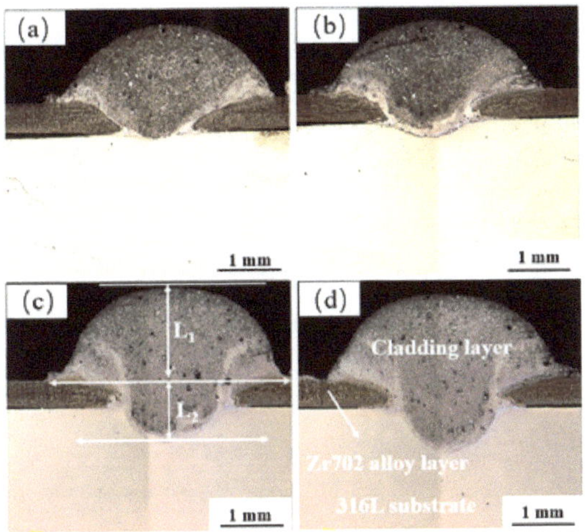

Figure 2. Single-pass cross-sectional morphology of the cladding layer at different energy densities: (**a**) S1; (**b**) S2; (**c**) S3; (**d**) S4.

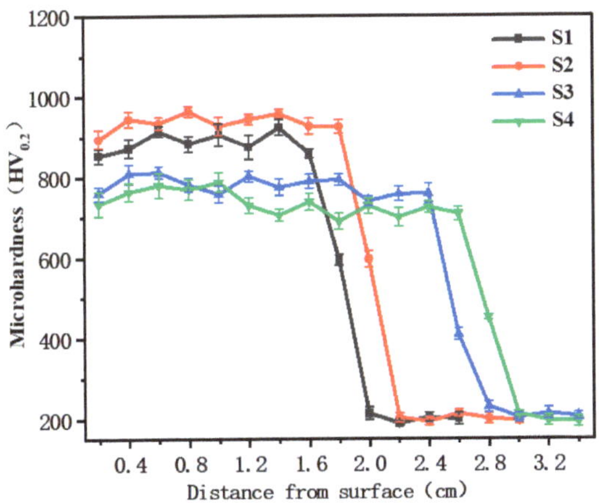

Figure 3. Microhardness diagram of preset Zr702 alloy coating at different energy densities.

3.2. *Microstructure Analysis*

3.2.1. Composition of the Material Phase

To reveal the composition of the coating phase, the XRD diffraction analysis was carried out. By comparing the PDF standard cards, the physical phases are shown in Figure 4. Although coatings with different laser energy densities were obtained, the main phases of the individual coatings differed little and were γ-(Fe, Ni), Ni_3Fe, WC, $Cr_{23}C_6$, Fe_3N, (Zr,W)C, $Ni_{10}Zr_7$, ZrC, and $M_{23}C_6$. As typical hard phases, $Cr_{23}C_6$ and ZrC play a precipitation-strengthening role in the coating. At high temperatures, the WC in the coating decomposed to produce free C. As the Zr and Hf elements have similar chemical properties, the free C and the Hf and Zr atoms in the Zr702 alloy were bonded to produce HfC and ZrC at specific temperatures. When the energy density was 28.3 J/mm^2, the new ZrN and ZrO_2 phases in the S2 coating had high hardness and chemical stability, and the coating wear resistance was enhanced to a certain extent. Unlike the literature [20], due to the low introduction of Zr, there were no multiple intermetallic compounds of NiZr, Ni_2Zr, and $Ni_{10}Zr_7$ as mentioned in the literature, and only $Ni_{10}Zr_7$ was detected by XRD in this experimental coating. The average grain sizes calculated from S1 to S4 by the Debye–Scherrer formula [21] were 27 nm, 27.9 nm, 29.6 nm, and 31.5 nm, respectively.

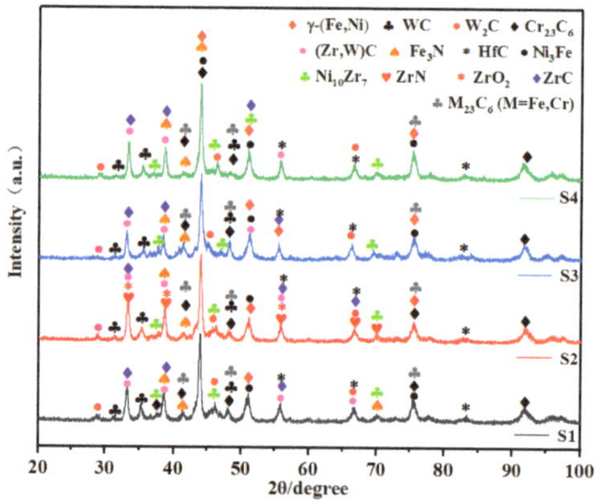

Figure 4. XRD diffraction results of S1 to S4 cladding coatings at different energy densities.

3.2.2. Microstructure and (Zr,W)C Phase Evolution Pattern

In Figure 5, the microstructures of S2 and S4 cladding coatings all contain the γ-(Fe,Ni) solid solution, eutectic organization ($M_{23}C_6$, $Cr_{23}C_6$, $Ni_{10}Zr_7$), incompletely decomposed WC, and diffusely distributed ZrC, (W,Zr)C. Table 3 shows the EDS point scan results of each marker in Figure 5. The flower-like Marker 1 and small particle Marker 2 in Figure 5 contain a large amount of W and Zr elements, and about 2.4 wt.% of the Hf elements were detected for both of them. Based on the facts that the HfC phase was detected in XRD and the literature on the generation of (W,Zr)C at high temperatures [22], Markers 1 and 2 can be further analyzed to determine (W,Zr)C, as well as trace amounts of HfC. As can be seen in Figure 5(a1), Markers 3 and 4 showed serrated and needle-like morphologies. The results of the point scan showed that the content of each element in the two markers was similar, mainly containing W, Cr, and a small amount of C. Therefore, Markers 3 and 4 were judged as the incompletely decomposed WC and the carbide of Cr formed around the WC. Since free C is the first to form a ceramic phase with Zr and Hf, then, the C concentration decreases and inhibits the growth of M_7C_3 [23], while fine $M_{23}C_6$ is formed in the eutectic microstructure. Marker 5 contained 76.88 wt.% of Ni elements and a small amount of

Cr, Zr, and Fe elements, so the analysis determined that it was a γ-(Fe, Ni) solid solution containing Cr and Zr. It is noteworthy that the eutectic microstructure (Marker 6) in the region of Figure 5(a2) contains a fence-like tissue, while it was also found in Figure 5(b2). Compared with Marker 5, Marker 6 showed 3.98 wt% of Zr elements and increased Ni elements, and $Ni_{10}Zr_7$ was detected with XRD, so it was judged as the γ-(Fe, Ni) solid solution and surface precipitation of reticulated $Ni_{10}Zr_7$ intermetallic compound.

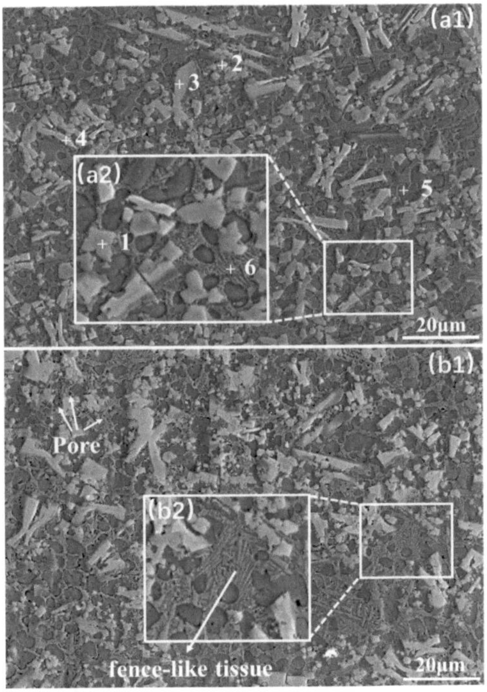

Figure 5. SEM microstructure of the central part of the cladding coating of preset Zr702 alloy at different energy densities: (**a1**,**a2**): S2; (**b1**,**b2**): S4.

Table 3. EDS point scan results at different locations in Figure 5 (wt.%).

Element	1	2	3	4	5	6
C	6.30	5.75	6.61	8.23	0.71	1.07
O	0.52	0.48				
Si					1.13	1.82
Cr	2.25	1.73	18.43	17.80	11.31	3.48
Fe	1.48	1.32	1.61	2.65	7.46	3.07
Ni	17.93	14.45	5.55	7.29	76.88	86.58
Zr	40.82	42.64			0.57	3.98
Hf	2.54	2.40				
W	28.17	31.24	67.8	66.49	1.93	

After presetting a 0.5 mm-thick Zr702 alloy plate, to investigate the distribution of the Zr element in the laser cladding Ni60A-20wt.%WC coating and the chemistry morphology of the Zr element, EDS surface scans were performed on the central microzone of the S2 cladding coating. In Figure 6, it can be seen that the Zr elements are more uniformly distributed, showing irregular shapes such as flowers and squares. Zr (0.167 nm) and W (0.141 nm) have similar atomic radii with a radius difference of only 0.026 nm, implying a certain degree of intercalation. Research work has also reported [24] that the solubility of W in ZrC is approximately 19.5%. Given that the solubility of Zr in WC is almost negligible [25], it can be determined that Markers 2 and 5 in Figure 5(a1) were formed (Zr,W)C around ZrC. The fact that the Zr, W elemental maps did not completely overlap and that ZrC was detected in XRD indicated that ZrC was not completely dissolved to form (Zr,W)C. Therefore, the Zr element ended up in the form of ZrC, (Zr,W)C hard particles as the reinforcing phase of the coating. Since ZrC and (Zr,W)C hard phase particles are highly thermally stable and almost insoluble in the γ-(Fe, Ni) solid solution, most of them were diffusely distributed in the eutectic microstructure between the equiaxed crystals.

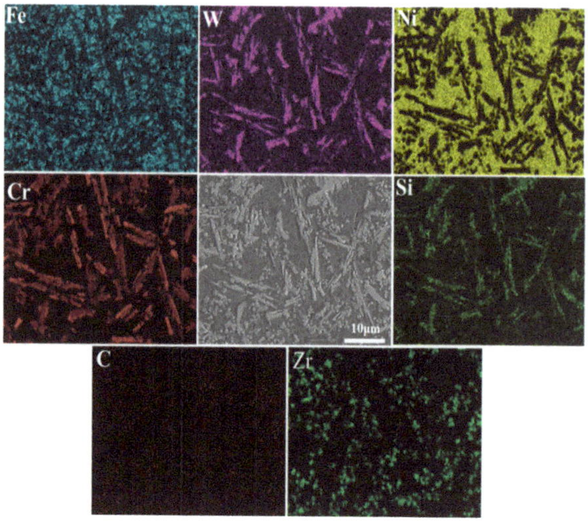

Figure 6. Distribution of elements in the middle area of the S2 cladding coating.

3.3. *Wear Properties and Wear Mechanisms*

The wear resistance of laser cladding coatings can be evaluated by the coefficient of friction and the wear rate [26]. The distribution of the coefficient of friction with time and the variation of the average coefficient of friction of the laser cladding coating at different energy densities are shown in Figure 7. As can be seen in Figure 7a, the curve is divided into two main stages: the initial ascent stage and the stable stage. The coefficient of friction first underwent a period of oscillatory rise for about 10 min. This was because the surfaces of the coating and the friction pair were not completely smooth, and the tiny convexity on them contacts, resulting in a larger sliding resistance, so the coefficient of friction oscillated up. As the friction proceeded, the surface of the friction pair and the coating was smoother, the wear behavior entered a dynamic and steady stage, and the friction coefficient was more stable.

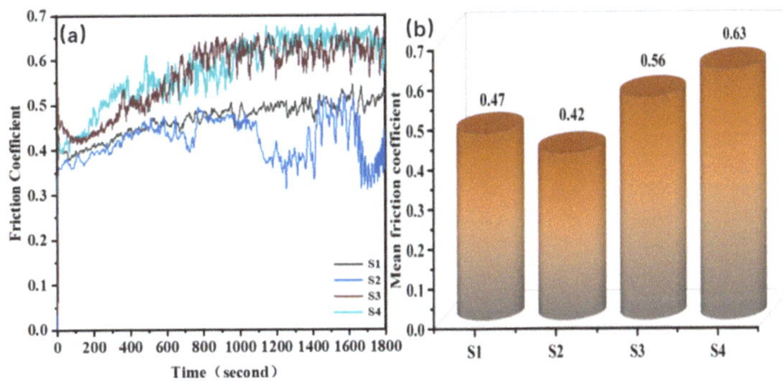

Figure 7. Tribological properties of the S1 to S4 cladding coatings at different energy densities: (**a**) distribution of friction coefficients; (**b**) the average coefficient of friction.

The average coefficients of friction of the coatings S1 to S4 were, respectively, 0.47, 0.42, 0.56, and 0.63, as shown in Figure 7b. The wear rates of the coatings S1 to S4 in Figure 8 were 131.3 $\mu m^3/(m\cdot N)$, 90.8 $\mu m^3/(m\cdot N)$, 145 $\mu m^3/(m\cdot N)$, and 178 $\mu m^3/(m\cdot N)$. Both of them showed a trend of decreasing and then increasing with the increase in the energy density. In Figure 8, it can be seen that the S2 coating had the smallest wear rate of 90.8 $\mu m^3/(m\cdot N)$. It is worth mentioning that the S2 coating appeared to have two suddenly lower coefficients of friction, which was mainly due to the detection of the ZrO_2 phase by XRD and the fact that ZrO_2 is a good self-lubricating ceramic material [27], reducing the coefficient of friction. The wear rate of the S1 and S2 coatings was lower than that of the S3 and S4 coatings. This was mainly because the S1 and S2 coatings contained a higher volume fraction of hard-phase ZrC and HfC particles. When the laser energy density was greater than the critical value of the energy density for melting through Zr702, more 316 L matrix elements diluted into the melt pool, decreasing the mechanical properties of the coating, and there were some ZrC particles in the melt depth part, decreasing the volume fraction of ZrC; therefore, the wear resistance of the S3 and S4 coatings was poorer.

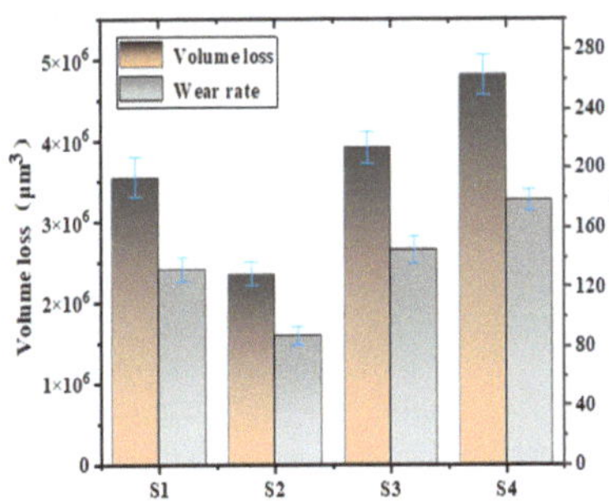

Figure 8. Volume loss and wear rate of the S1 to S4 cladding coatings at different energy densities.

Figure 9(a1,a2) show that the S1 coating surface had large lamellar adhesions and obvious plow grooves. This was because the plow chips of the bonded phase were taken out of the coating under the action of abrasive plowing and crushed by Si_3N_4 against the abrasive part to form the laminate, and the hard particles were shed and acted as abrasive particles, forming "three-body wear" [28]. Figure 9(b1,b2) show that, when the energy density was 28.3 J/mm^2, the S2 coating had only slight scratches and small hard particle spalling pits. The slight wear behavior above was mainly attributed to the following three aspects: On the one hand, the lattice distortion caused by Zr atoms solidly dissolved in the Ni60A alloy, having solid solution strengthening effects. Therefore, the resistance of the nickel-based solid solution to plastic deformation increased. On the other hand, due to the moderate energy input, the Zr atoms saturated in the γ-(Fe,Ni) solid solution and then segregated at the grain boundaries, forming a large number of Ni-Cr-B-Si eutectics and inhibiting grain growth, having fine crystal strengthening effects; furthermore, a large number of hard phases such as in situ ZrC, (Zr,W)C, and WC particles that are not completely decomposed played an important role in the diffusion-strengthening and second-phase-strengthening effects. From the XRD grain size calculation results, it can be seen that, with the further increase of the energy density, the grain size of the S3 and S4 coatings became larger, decreasing the grain boundary length; the ability to prevent dislocation movement was weakened [29], and the resistance to plastic deformation was reduced. Therefore, as shown in Figure 9(c1,c2,d1,d2), there were obvious plow grooves, spalling pits, and a small number of abrasive chips on the surface of S3. The wear surface morphology of the S4 coating showed wear debris accumulation, mutual parallel abrasion marks, and fatigue cracks formed by the laminae due to cyclic loading, and its wear mechanism was abrasive wear.

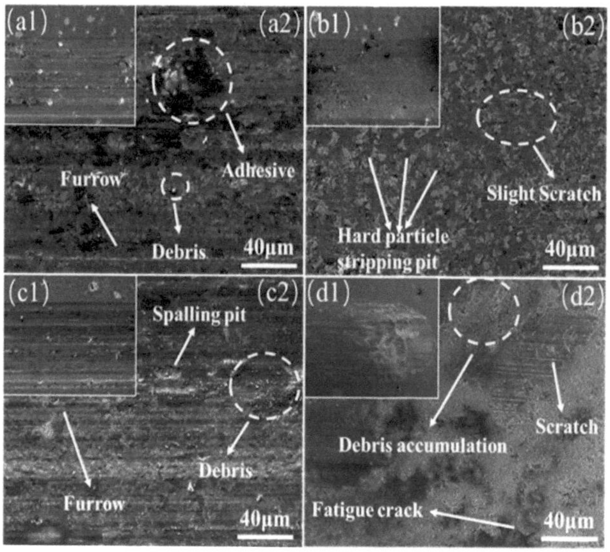

Figure 9. Wear morphology of S1 to S4 cladding coatings at different energy densities: (**a1**,**a2**) S1; (**b1**,**b2**) S2; (**c1**,**c2**) S3; (**d1**,**d2**) S4.

4. Conclusions

In this paper, the effects of different laser energy densities on the microstructure and mechanical properties of Ni60A-20wt.%WC cladding layers were investigated by preplacing a 0.5 mm-thick Zr702 alloy plate on a 316 L substrate, and the main conclusions were as follows:

(1) When the energy density was 28.3 J/mm^2, the coating microhardness was the highest among the specimens of four groups, reaching 936.4 HV$_{0.2}$. With the increase in

energy density, the microhardness of the coating showed a trend of increasing and then decreasing.

(2) After presetting the Zr702 alloy, the coatings with different energy densities were mainly the γ-(Fe,Ni), $M_{23}C_6$, $Cr_{23}C_6$, Ni_3Fe, undecomposed WC, Fe_3N, ZrC, (W,Zr)C, HfC, and $Ni_{10}Zr_7$ phases. The fence-like tissue found in the microstructure was determined to be $Ni_{10}Zr_7$ after EDS point scanning with the XRD results. The Zr element ended up in the form of ZrC, (Zr,W)C hard particles as the reinforcing phase of the coating.

(3) With the increase in the energy density, the wear rate and coefficient of friction showed a trend of decreasing and then increasing. When the energy density was 28.3 J/mm^2, the wear rate of the coating was the lowest at 90.8 $\mu m^3/(m \cdot N)$. The wear was characterized by hard particle spalling and abrasive wear. The enhancement mechanism was mainly attributed to the low dilution of 316 L into the melt pool, the refinement of the grain, and the formation of a large number of ZrC, (Zr,W)C, which were hard particles with high wear resistance and distributed at the grain boundaries to have second-phase-strengthening effects.

Author Contributions: Conceptualization, G.Z. and P.Z.; methodology, A.F., G.Z. and P.Z.; investigation, A.F. and G.Z.; data curation, X.P. and H.F.; writing—original draft preparation, A.F. and G.Z.; writing—review and editing, A.F., G.Z. and P.Z. All authors have read and agreed to the published version of the manuscript.

Funding: This study was supported by the Natural Science Foundation of Zhejiang Province (LY20E050027), the Major Special Project of Wenzhou Science and Technology Innovation and Tackling Industry (ZG2020029), and the Innovation Fund of Wenzhou University Rui'an Graduate College (YC202212011).

Institutional Review Board Statement: Not applicable.

Informed Consent Statement: Not applicable.

Data Availability Statement: Not applicable.

Conflicts of Interest: The authors declare no conflict of interest. The funders had no role in the design of the study; in the collection, analyses, or interpretation of the data; in the writing of the manuscript; nor in the decision to publish the results.

References

1. Li, P.; Dong, L.; Hu, Z.; Ren, S.; Tan, Y.; Jiang, D.; You, X. Recycling of silicon scraps by electron beam melting coupled with directional solidification and its mechanism of impurity removal. *J. Clean. Prod.* **2022**, *355*, 131716. [CrossRef]
2. Vähänissi, V.; Laine, H.S.; Liu, Z.; Yli-Koski, M.; Haarahiltunen, A.; Savin, H. Full recovery of red zone in p-type high-performance multicrystalline silicon. *Sol. Energy Mater. Sol. Cells* **2017**, *173*, 120–127. [CrossRef]
3. Li, Q.; Li, Y.; Bai, Q.; Chen, C.; Zhao, C.; Liu, Y. Effect of power spinning combined with heat treatment on the organization and wear resistance of high-speed laser cladding coatings. *Mater. Lett.* **2023**, *333*, 133594. [CrossRef]
4. Lian, G.; Liu, Z.; Zhang, Y.; Que, L.; Chen, C.; Yue, K. Grey relational analysis–based composite coating property optimization fabricated by laser cladding. *Int. J. Adv. Manuf. Technol.* **2022**, *120*, 7985–7996. [CrossRef]
5. Shi, C.; Lei, J.; Zhou, S.; Dai, X.; Zhang, L.-C. Microstructure and mechanical properties of carbon fibers strengthened Ni-based coatings by laser cladding: The effect of carbon fiber contents. *J. Alloys Compd.* **2018**, *744*, 146–155. [CrossRef]
6. Sun, S.T.; Fu, H.G.; Ping, X.L.; Guo, X.Y.; Lin, J.; Lei, Y.P.; Wu, W.B.; Zhou, J.X. Formation mechanism and mechanical properties of titanium-doped NbC reinforced Ni-based composite coatings. *Appl. Surf. Sci.* **2019**, *476*, 914–927. [CrossRef]
7. Tan, C.; Zou, J.; Wang, D.; Ma, W.; Zhou, K. Duplex strengthening via SiC addition and in-situ precipitation in additively manufactured composite materials. *Compos. Part B Eng.* **2022**, *236*, 109820. [CrossRef]
8. Li, W.; Yang, X.; Xiao, J.; Hou, Q. Effect of WC mass fraction on the microstructure and friction properties of WC/Ni60 laser cladding layer of brake discs. *Ceram. Int.* **2021**, *47*, 28754–28763. [CrossRef]
9. Zhao, Y.; Feng, K.; Yao, C.W.; Li, Z.G. Effect of MoO_3 on the microstructure and tribological properties of laser-clad Ni_{60}/nanoCu/h-BN/MoO_3 composite coatings over wide temperature range. *Surf. Coat. Technol.* **2020**, *387*, 125477. [CrossRef]
10. Zhu, Z.; Li, J.; Peng, Y.; Shen, G. In-situ synthesized novel eyeball-like Al_2O_3/TiC composite ceramics reinforced Fe-based alloy coating produced by laser cladding. *Surf. Coat. Technol.* **2020**, *391*, 125671. [CrossRef]

11. Yang, C.; Cheng, X.; Tang, H.; Tian, X.; Liu, D. Influence of microstructures and wear behaviors of the microalloyed coatings on TC11 alloy surface using laser cladding technique. *Surf. Coat. Technol.* **2018**, *337*, 97–103. [CrossRef]
12. Li, X.; Liu, S.; Wang, J.; Yu, M.; Tang, H. Effect of Different ZrN Addition on Microstructure and Wear Properties of Titanium Based Coatings by Laser Cladding Technique. *Coatings* **2019**, *9*, 261. [CrossRef]
13. Yong, Y.-W.; Fu, W.; Deng, Q.-L.; Yang, J.-G. Mechanism of Zr in in situ-synthesized particle reinforced composite coatings by laser cladding. *Rare Met.* **2017**, *36*, 934–941. [CrossRef]
14. Guo, C.; Zhou, J.; Zhao, J.; Chen, J. Effect of ZrB_2 on the Microstructure and Wear Resistance of Ni-Based Composite Coating Produced on Pure Ti by Laser Cladding. *Tribol. Trans.* **2010**, *54*, 80–86. [CrossRef]
15. Liu, K.; Li, Y.; Wang, J.; Ma, Q. In-situ synthesized Ni–Zr intermetallic/ceramic reinforced composite coatings on zirconium substrate by high power diode laser. *J. Alloys Compd.* **2015**, *624*, 234–240. [CrossRef]
16. Zhu, S.; Chen, W.; Zhan, X.; Ding, L.; Wang, E. Optimization of dilution rate of laser cladding repair based on deep learning. *Int. J. Adv. Manuf. Technol.* **2020**, *110*, 1471–1484. [CrossRef]
17. Rui, H.; Meiping, W.; Chen, C.; Dadong, J.; Yuling, G.; Xiaojin, M. Microstructure evolution, mechanical properties of FeCrNiMnAl high entropy alloy coatings fabricated by laser cladding. *Surf. Coat. Technol.* **2022**, *447*, 128851. [CrossRef]
18. Cui, C.; Wu, M.; Miao, X.; Gong, Y.; Zhao, Z. The effect of laser energy density on the geometric characteristics, microstructure and corrosion resistance of Co-based coatings by laser cladding. *J. Mater. Res. Technol.* **2021**, *15*, 2405–2418. [CrossRef]
19. Ghasempour-Mouziraji, M.; Limouei, M.B.; Najafizadeh, M.; Hosseinzadeh, M.; Cavaliere, P. The effect of simple shear extrusion on the mechanical properties and microstructure of copper. *Mater. Lett.* **2023**, *335*, 133815. [CrossRef]
20. Liu, K.; Li, Y.; Wang, J.; Ma, Q. Effect of high dilution on the in situ synthesis of Ni–Zr/Zr–Si(B, C) reinforced composite coating on zirconium alloy substrate by laser cladding. *Mater. Des.* **2015**, *87*, 66–74. [CrossRef]
21. Berretta, S.; Wang, Y.; Davies, R.; Ghita, O.R. Polymer viscosity, particle coalescence and mechanical performance in high-temperature laser sintering. *J. Mater. Sci.* **2016**, *51*, 4778–4794. [CrossRef]
22. Song, G.-M.; Wang, Y.-J.; Zhou, Y. The mechanical and thermophysical properties of ZrC/W composites at elevated temperature. *Mater. Sci. Eng. A* **2002**, *334*, 223–232. [CrossRef]
23. Yu, T.; Deng, Q.; Dong, G.; Yang, J. Effects of Ta on microstructure and microhardness of Ni based laser clad coating. *Appl. Surf. Sci.* **2011**, *257*, 5098–5103. [CrossRef]
24. Zhang, T.; Wang, Y.; Zhou, Y.; Song, G. Effect of temperature gradient in the disk during sintering on microstructure and mechanical properties of ZrC.sub.p/W composite. *Int. J. Refract. Met. Hard Mater.* **2009**, *27*, 126. [CrossRef]
25. Kim, J.-H.; Park, C.; Ha, D.; Kang, S. WC-toughened (Zr,W)C solid solution carbides. *J. Alloys Compd.* **2015**, *637*, 183–187. [CrossRef]
26. Yang, L.; Yu, T.; Li, M.; Zhao, Y.; Sun, J. Microstructure and wear resistance of in-situ synthesized Ti(C, N) ceramic reinforced Fe-based coating by laser cladding. *Ceram. Int.* **2018**, *44*, 22538–22548. [CrossRef]
27. Kharanzhevskiy, E.V.; Ipatov, A.G.; Makarov, A.V.; Gil'mutdinov, F.Z.; Soboleva, N.N.; Krivilyov, M.D. Effect of oxygen in surface layers formed during sliding wear of Ni–ZrO_2 coatings. *Surf. Coat. Technol.* **2022**, *434*, 128174. [CrossRef]
28. Deng, X.; Huang, L.; Wang, Q.; Fu, T.; Wang, Z. Three-body abrasion wear resistance of TiC-reinforced low-alloy abrasion-resistant martensitic steel under dry and wet sand conditions. *Wear* **2020**, *452–453*, 203310. [CrossRef]
29. Yazdani, S.; Vitry, V. Using Molecular Dynamic Simulation to Understand the Deformation Mechanism in Cu, Ni, and Equimolar Cu-Ni Polycrystalline Alloys. *Alloys* **2023**, *2*, 77–88. [CrossRef]

MDPI AG
Grosspeteranlage 5
4052 Basel
Switzerland
Tel.: +41 61 683 77 34

MDPI Books Editorial Office
E-mail: books@mdpi.com
www.mdpi.com/books

www.ingramcontent.com/pod-product-compliance
Lightning Source LLC
Chambersburg PA
CBHW041424210926
51850CB00013B/3351
* 9 7 8 3 7 2 5 8 3 8 0 3 5 *